Pharmacokinetics in Risk Assessment

Drinking Water and Health

Volume 8

Workshop Proceedings
Subcommittee on Pharmacokinetics in Risk Assessment
Safe Drinking Water Committee
Board on Environmental Studies and Toxicology
Commission on Life Sciences
National Research Council

NATIONAL ACADEMY PRESS
Washington, D.C. 1987

NATIONAL ACADEMY PRESS, 2101 Constitution Ave., NW, Washington, DC 20418

NOTICE: The project that is the subject of this report was approved by the Governing Board of the National Research Council, whose members are drawn from the councils of the National Academy of Sciences, the National Academy of Engineering, and the Institute of Medicine. The members of the committee responsible for the report were chosen for their special competences and with regard for appropriate balance.

This report has been reviewed by a group other than the authors according to procedures approved by a Report Review Committee consisting of members of the National Academy of Sciences, the National Academy of Engineering, and the Institute of Medicine.

The National Academy of Sciences is a private, nonprofit, self-perpetuating society of distinguished scholars engaged in scientific and engineering research, dedicated to the furtherance of science and technology and to their use for the general welfare. Upon the authority of the charter granted to it by the Congress in 1863, the Academy has a mandate that requires it to advise the federal government on scientific and technical matters. Dr. Frank Press is president of the National Academy of Sciences.

The National Academy of Engineering was established in 1964, under the charter of the National Academy of Sciences, as a parallel organization of outstanding engineers. It is autonomous in its administration and in the selection of its members, sharing with the National Academy of Sciences the responsibility for advising the federal government. The National Academy of Engineering also sponsors engineering programs aimed at meeting national needs, encourages education and research, and recognizes the superior achievements of engineers. Dr. Robert M. White is president of the National Academy of Engineering.

The Institute of Medicine was established in 1970 by the National Academy of Sciences to secure the services of eminent members of appropriate professions in the examination of policy matters pertaining to the health of the public. The Institute acts under the responsibility given to the National Academy of Sciences by its congressional charter to be an adviser to the federal government and, upon its own initiative, to identify issues of medical care, research, and education. Dr. Samuel O. Thier is president of the Institute of Medicine.

The National Research Council was organized by the National Academy of Sciences in 1916 to associate the broad community of science and technology with the Academy's purposes of furthering knowledge and advising the federal government. Functioning in accordance with general policies determined by the Academy, the Council has become the principal operating agency of both the National Academy of Sciences and the National Academy of Engineering in providing services to the government, the public, and the scientific and engineering communities. The Council is administered jointly by both Academies and the Institute of Medicine. Dr. Frank Press and Dr. Robert M. White are chairman and vice chairman, respectively, of the National Research Council.

This project has been funded by the U.S. Environmental Protection Agency under Contract No. 68-01-3169 with the National Academy of Sciences. The contents of this document do not necessarily reflect the views and policies of the Environmental Protection Agency, and an official endorsement should not be inferred.

Library of Congress Catalog Card Number 77-89284

International Standard Book Number 0-309-03775-1

Printed in the United States of America

List of Participants

SUBCOMMITTEE ON PHARMACOKINETICS

JAMES R. GILLETTE (*Co-Chairman*), Laboratory of Chemical Pharmacology, National Heart, Lung, and Blood Institute, National Institutes of Health, Bethesda, Maryland

DAVID J. JOLLOW (*Co-Chairman*), Medical University of South Carolina, Charleston, South Carolina

MELVIN E. ANDERSEN, Biochemical Toxicology Branch, Harry G. Armstrong Aerospace Medical Research Laboratory, Wright-Patterson Air Force Base, Dayton, Ohio

KENNETH B. BISCHOFF, Department of Chemical Engineering, University of Delaware, Newark, Delaware

MARY E. DAVIS, West Virginia University Medical Center, Morgantown, West Virginia

ROBERT L. DEDRICK, Biomedical Engineering and Instrumentation Branch, DRS, National Institutes of Health, Bethesda, Maryland

SEYMOUR L. FRIESS, Drill, Friess, Hays, Loomis & Shaffer, Inc., Arlington, Virginia

DANIEL KREWSKI, Environmental Health Directorate, Health Protection Branch, Health and Welfare Canada, Ottawa, Ontario, Canada

DANIEL B. MENZEL, Departments of Pharmacology and Medicine, Duke University Medical Center, Durham, North Carolina

RUSSELL A. PROUGH, University of Louisville, Louisville, Kentucky

RICHARD H. REITZ, Mammalian and Environmental Toxicology, Dow Chemical USA, Midland, Michigan

CURTIS C. TRAVIS, Health and Safety Research Division, Oak Ridge National Laboratory, Oak Ridge, Tennessee

GRANT R. WILKINSON, Department of Pharmacology, Vanderbilt University School of Medicine, Nashville, Tennessee

Advisers, Consultants, and Contributors

MARSHALL W. ANDERSON, Laboratory of Biochemical Risk Analysis, National Institute of Environmental Health Sciences, Research Triangle Park, North Carolina

MICHAEL J. ANGELO, Central Research Department, General Foods Technical Center, Tarrytown, New York

LESLIE BENET, University of California, San Francisco, California

JERRY N. BLANCATO, U.S. Environmental Protection Agency, Washington, D.C.

GARY E. BLAU, Dow Chemical USA, Midland, Michigan

J. R. BOGER III, Departments of Pharmacology and Medicine, Duke University Medical Center, Durham, North Carolina

ROBERT D. BRUCE, Miami Valley Laboratories, The Procter & Gamble Co., Cincinnati, Ohio

CHAO W. CHEN, U.S. Environmental Protection Agency, Research Triangle Park, North Carolina

HARVEY J. CLEWELL III, Environmental Toxicology Branch, Harry G. Armstrong Aerospace Medical Research Laboratory, Wright-Patterson Air Force Base, Dayton, Ohio

MURRAY S. COHN, U.S. Consumer Product Safety Commission, Washington, D.C.

JERRY M. COLLINS, Pharmacokinetics Section, Division of Cancer Treatment, National Cancer Institute, Bethesda, Maryland

RORY B. CONOLLY, Environmental Sciences, Northrop Services, Inc., Dayton, Ohio

KENNY CRUMP, K. S. Crump and Co., Ruston, Louisiana

RICHARD W. D'SOUZA, Miami Valley Laboratories, The Procter & Gamble Co., Cincinnati, Ohio

WILLIAM R. FRANCIS, Miami Valley Laboratories, The Procter & Gamble Co., Cincinnati, Ohio

MICHAEL L. GARGAS, Biochemiocal Toxicology Branch, Harry G. Armstrong Aerospace Medical Research Laboratory, Wright-Patterson Air Force Base, Dayton, Ohio

JAMES GIBSON, Chemical Industry Institute of Toxicology, Research Triangle Park, North Carolina

RICHARD C. GRAHAM, Environmental Sciences, Northrup Services, Inc., Research Triangle Park, North Carolina

DAVID HOEL, National Institute of Environmental Health Sciences, Research Triangle Park, North Carolina

FRED F. KADLUBAR, Division of Biochemical Toxicology, National Center for Toxicological Research, U.S. Food and Drug Administration, Jefferson, Arkansas

J. M. KOOTSEY, Department of Physiology, Duke University Medical Center, Durham, North Carolina

STAN L. LINDSTEDT, Department of Zoology and Physiology, University of Wyoming, Laramie, Wyoming

ROBERT J. LUTZ, Biomedical Engineering and Instrumentation Branch, National Institutes of Health, Bethesda, Maryland

FREDERICK J. MILLER, Health Effects Research Laboratory, U.S. Environmental Protection Agency, Research Triangle Park, North Carolina

PAUL F. MORRISON, Biomedical Engineering and Instrumentation Branch, National Institutes of Health, Bethesda, Maryland

DUNCAN J. MURDOCH, Environmental Health Directorate, Health Protection Branch, Health and Welfare Canada, Ottawa, Ontario, Canada

W. BROCK NEELY, EnviroSoft, Inc.

RICHARD J. NOLAN, Mammalian and Environmental Toxicology, Dow Chemical USA, Midland, Michigan

ELLEN J. O'FLAHERTY, Department of Environmental Health, Kettering Laboratory, University of Cincinnati College of Medicine, Cincinnati, Ohio

JOHN H. OVERTON, JR., Environmental Sciences, Northrup Services, Inc., Research Triangle Park, North Carolina

SANDY PANG, University of Toronto, Toronto, Ontario, Canada

DENNIS J. PAUSTENBACH, Environmental and Occupational Toxicology Environmental Health and Safety, Syntex, U.S.A., Palo Alto, California

CARL PECK, Uniformed Services College of Medicine, Bethesda, Maryland

ALAN B. PRITCHARD, Central Research Department, General Foods Technical Center, Tarrytown, New York

JOHN RAMSEY, Dow Chemical USA, Midland, Michigan

ALAN M. SCHUMANN, Mammalian and Environmental Toxicology, Dow Chemical USA, Midland, Michigan

ELAINE D. SMOLKO, Departments of Pharmacology and Medicine, Duke University Medical Center, Durham, North Carolina

HUGH SPITZER, Office of Research and Development, U.S. Environmental Protection Agency, Washington, D.C.

THOMAS B. STARR, Chemical Industry Institute of Toxicology, Research Triangle Park, North Carolina

JIM R. WITHEY, Environmental Health Directorate, Health Protection Branch, Health and Welfare Canada, Ottawa, Ontario, Canada

R. L. WOLPERT, Departments of Pharmacology and Medicine, Duke University Medical Center, Durham, North Carolina

JOHN F. YOUNG, Division of Reproductive and Developmental Toxicology, National Center for Toxicological Research, U.S. Food and Drug Administration, Jefferson, Arkansas

SAFE DRINKING WATER COMMITTEE

DAVID J. JOLLOW (*Chairman*), Medical University of South Carolina, Charleston, South Carolina

DAVID E. BICE, Lovelace Inhalation Toxicology Research Institute, Albuquerque, New Mexico

JOSEPH F. BORZELLECA, Virginia Commonwealth University, Richmond, Virginia

DAVID J. BRUSICK, Hazleton Laboratories America, Inc., Kensington, Maryland

EDWARD J. CALABRESE, North East Regional Environmental Public Health Center, University of Massachusetts, Amherst, Massachusetts

J. DONALD JOHNSON, School of Public Health, University of North Carolina, Chapel Hill, North Carolina

RONALD E. WYZGA, Electric Power Research Institute, Palo Alto, California

National Research Council Staff

RICHARD D. THOMAS, *Project Director*

BRUCE K. BERNARD, *Staff Officer*

EVELYN E. SIMEON, *Project Secretary*

Sponsoring Agency

JOSEPH COTRUVO and KRISHAN KHANNA, Office of Drinking Water, U.S. Environmental Protection Agency, Washington, D.C.

PETER PREUSS, Office of Research and Development, U.S. Environmental Protection Agency, Washington, D.C.

DONALD H. HUGHES, Scientific Committee, American Industrial Health Council, Washington, D.C.

H. B. MATTHEWS, Systemic Toxicology Branch, National Institute of Environmental Health Sciences, Research Triangle Park, North Carolina

E. SOMERS, Health Protection Branch, Health and Welfare Canada, Ottawa, Ontario, Canada

BARRY JOHNSON, Agency for Toxic Substances and Disease Registry, Centers for Disease Control, Atlanta, Georgia

BOARD ON ENVIRONMENTAL STUDIES AND TOXICOLOGY

DONALD F. HORNING (*Chairman*), School of Public Health, Harvard University, Boston, Massachusetts
ALVIN L. ALM, Thermal Analytical, Inc., Waltham, Massachusetts
RICHARD N. L. ANDREWS, Institute for Environmental Studies, University of North Carolina, Chapel Hill, North Carolina
RICHARD A. CONWAY, Union Carbide Corporation, South Charleston, West Virginia
WILLIAM E. COOPER, Michigan State University, East Lansing, Michigan
JOHN DOULL, University of Kansas Medical Center, Kansas City, Kansas
BENJAMIN G. FERRIS, School of Public Health, Harvard University, Boston, Massachusetts
SHELDON K. FRIEDLANDER, National Center Intermedia Transport Research, University of California, Los Angeles, California
BERNARD D. GOLDSTEIN, University of Medicine and Dentistry of New Jersey—Robert Wood Johnson Medical School, Piscataway, New Jersey
PHILIP J. LANDRIGAN, Mount Sinai Medical Center, New York, New York
PHILIP A. PALMER, E. I. Du Pont de Nemours & Co., Wilmington, Delaware
EMIL A. PFITZER, Hoffmann-La Roche, Inc., Nutley, New Jersey
PAUL PORTNEY, Resources for the Future, Washington, D.C.
PAUL RISSER, University of New Mexico, Albuquerque, New Mexico
WILLIAM H. RODGERS, School of Law, University of Washington, Seattle, Washington
F. SHERWOOD ROWLAND, University of California, Irvine, California
LIANE B. RUSSELL, Oak Ridge National Laboratory, Oak Ridge, Tennessee
ELLEN K. SILBERGELD, Environmental Defense Fund, Washington, D.C.
PETER S. SPENCER, Institute of Neurotoxicology, Albert Einstein College of Medicine, Bronx, New York

National Research Council Staff

DEVRA LEE DAVIS, *Executive Director*
JACQUELINE PRINCE, *Staff Assistant*

Preface

Human beings are exposed to many chemicals, natural and synthetic. To assess whether those chemicals are likely to produce adverse effects in human populations, toxicity tests are conducted in laboratory animals. In performing such tests, toxicologists assume that laboratory animals are reasonable surrogates for humans, that is, that most materials that have been found safe in laboratory animals will be safe in humans. Nevertheless, the possibility that reliance on data obtained solely with laboratory animals might lead to mistaken estimates of risk to humans has been of continuing concern, because interspecies and intraspecies differences in response clearly exist. Since some of these differences lie in how different species or individuals within a species handle xenobiotic chemicals, the new applications of pharmacokinetics discussed in these proceedings are exciting, because they give promise of major improvement in the prediction of human responses from animal data.

A wide variety of toxicity tests are performed routinely, but risk assessors usually focus attention on chronic tests, i.e., those in which a test material is administered to animals for most of their natural lifetimes. Such tests consume a large portion of the resources available for assessing risk, but are justified by the belief that they provide data that will be useful in predicting whether long-term exposure of humans to tested materials will increase the incidence of severely adverse conditions in humans, especially neoplastic diseases, pulmonary injuries, fetal abnormalities, and neurotoxicities.

Tests for carcinogenesis and other chronic conditions are typically started with homogeneous populations of young healthy animals. For practical reasons, only small numbers of doses in small numbers of animals can be evaluated. Consequently, doses are usually set as high as possible (e.g., the maximum tolerated dose), in the hope of providing maximal sensitivity, and the test agent is often administered by a route that is experimentally most convenient. The results of the completed tests are extrapolated to estimate the likelihood that similar responses will be produced in human populations exposed to lower concentrations of the same agent.

According to an NRC report prepared for the National Toxicology Program, adequate toxicologic data do not exist for most chemicals in commerce, and the available data seldom include noncancer effects, such as neurotoxicity or reproductive toxicity. Even when data are available, quantitative risk estimates still rely on assumptions concerning the mechanism of disease, dose-response relationships, effective dose in relation to exposure, etc., and therefore cannot be wholly accurate. Regulatory agencies are obliged to make assumptions that they believe to be conservative, so as to avoid underestimation of the risk to the human population. Unfortunately, overestimating a risk might unnecessarily eliminate jobs or commercially important materials, thereby decreasing our general standard of living. Hence, the more that we can replace empirical assumptions with experimentally validated procedures, the better off we will be in terms of both human health and economic well-being.

In developing the program of the workshop on pharmacokinetics in risk assessment, the Safe Drinking Water Subcommittee on Pharmacokinetics recognized that substances in the environment can lead to the expression of many kinds of toxicity through many mechanisms. It also recognized that humans can be exposed to substances at high concentrations for short periods (e.g., in industrial accidents, during transportation of chemicals, or in fires) or at low concentrations for long periods in various environmental settings. It acknowledged that risk assessors are required to make assessments on the basis of incomplete knowledge, particularly when the duration, route, and intensity of exposure differ from those tested in laboratory animals. In basing a risk assessment solely on long-term high-dose toxicity studies, the risk assessor is obliged to accept as working hypotheses that the incidence of response obtained in test animals receiving high doses can be used to provide valid estimates of the incidence of response in test animals exposed to much lower doses, that the response will also occur in humans, and that the consequences of exposure are independent of the route of exposure.

In light of these considerations, the subcommittee decided to start the workshop with a description of the background of quantitative risk assessment, the assumptions that must be made when knowledge is sparse, and the feasible contribution of pharmacokinetics. To put the role of pharmacokinetics into perspective, it was pointed out that biologic responses are governed by two general categories of factors: pharmacokinetic factors, which govern the concentration in target organs and the interaction of a biologically active substance with putative sites of action; and pharmacodynamic factors, which govern the sequence of events that result from the interaction and lead to the manifestation of the toxic response. The pharmacokinetic factors, in turn, may be subdivided into three groups: factors that determine the rate and extent of absorption of a substance from the site of administration; factors that govern the distribution and elimination of the substance and the formation, distribution, and elimination of biologically active metabolites, if any; and factors that affect the rate or extent of reaction of the substance with putative target substances, such as DNA, enzymes, and receptor sites.

The working hypotheses thus consist of many subhypotheses, each of which requires several assumptions. Pharmacodynamically, it is assumed that the toxicity seen in the test species will be seen in all species, that the mechanism of the toxicity will be the same in all species, and that other major kinds of toxicity will not occur. Pharmacokinetically, it is assumed that the bioavailability and time course of absorption, distribution, and elimination are generally the same in all species and hence that the exposure of the target to the ultimate toxicant will be similar. It is also inherently assumed that the concentration (or, when relevant, the area under the curve) of the toxicant in the target tissue is directly proportional to the applied dose.

During the last several decades, pharmacokineticists have developed several mathematical tools to evaluate factors that govern the time course of substances at putative sites of action. Each of the tools has its uses, but each has its limitations; both uses and limitations must be taken into account in selecting the tools that will be appropriate for specific risk assessments. The tools derive from two general approaches: compartmental analysis and physiologically based models. In practice, the pharmacokinetic models useful for risk assessment incorporate features from both general approaches, and the features that distinguish the two approaches have become blurred. The subcommittee therefore decided to include a session that traces the development of pharmacokinetic models and their application to risk assessment. Further, to illustrate the interrelationships between the two basic approaches to pharmacokinetics and the possibility that allometric methods of extrapolation are more valid with some mechanisms of elimination than with others, the subcommittee included sessions on the approaches used to generalize and extrapolate pharmacokinetic data.

Clearly, such extrapolations entail uncertainty—not only in theory, but also in the collection of data and in the sources of interspecies and intraspecies variation. Indeed, there are even uncertainties in what is assumed to be known of many of these uncertainties. Sessions were therefore designed to elucidate many of the sources of uncertainty and to illustrate the experimental use of physiologically based pharmacokinetics in exploring them.

The subcommittee recognized that scientists differ in their opinions of the validity of quantitative risk assessments and of the use of pharmacokinetic data in making them. To elicit those opinions and the concerns of scientists, the subcommittee included a poster session to broaden coverage of the subject; and the major issues raised by the posters were discussed in a plenary session.

Confidence in the validity of quantitative risk assessments will probably vary with the quantity and quality of toxicity data available and the nature of the toxicant under examination. Future development of the field of quantitative risk assessment depends on the growth of the capacity to decide which toxicants (and kinds of toxicity) yield reasonably accurate quantitative risk assessments and which do not. The final sessions of the workshop were designed to illustrate some of these problems and opportunities.

JAMES R. GILLETTE, *Co-chairman*
DAVID JOLLOW, *Co-chairman*
Subcommittee on Pharmacokinetics in Risk Assessment

Contents

Pharmacokinetics in Risk Assessment

Drinking Water and Health

PART I

Introduction: The Problem and an Approach

Risk Assessment: Historical Perspectives

Seymour L. Friess

Beginning in the 1930s in the United States and Europe, the protection of human health from chemicals in the workplace, the market place, and the environment became a commonly recognized international goal. The general approach toward that goal developed over time, but it was to be characterized roughly by processes that involved the development of some form of human dosage versus response relationships for undesirable health effects, the assessment of risk for those effects under specified modes of exposure to the chemical in question, and finally, the setting of permissible exposure limits for the chemical in various exposure situations based on some form of societal perception of acceptable risk.

To set this chain of procedures into action, it was first necessary to generate some form of trustworthy picture of dosage versus response for the most serious or most sensitive health effect that a chemical might produce in a human target. Indeed, since the early 1900s data had begun to accumulate on such health effects in human populations occupationally exposed to major industrial chemicals. In the 1930s and thereafter, data on dosage versus response were also generated in ever-increasing volume by dosing experiments in experimental animal systems under laboratory conditions. These toxicological experiments with rodents or larger mammals had the special merit of permitting exposure to much larger dosages and concentrations of chemical than adventitious occupational exposures of worker populations, and of providing precise measures of the total exposure of the biological target rather than the general guesses found in the early industrial epidemiological studies. The animal toxicology data, usually from subchronic or chronic exposure experiments, were always clearly indicated in the early studies as being pointed toward their ultimate use of predicting the risk of the corresponding health effects in human populations exposed to the chemical of concern. Toxicology was, and is, intended as a predictive science, with test data from animal experiments to be translated into assessments of risk for human and other populations. Generally, only the toxicological experiments furnished quantitative dose-response data for specific effects produced at specific target tissues or organs.

The risk assessment process, therefore, beginning roughly in the 1930s, took the form of an initial review of the epidemiological health effects data available for a given chemical in worker and user populations and of the dose versus response data generated in test animal systems. Somehow, usually by deliberations of a committee of specialists in the health sciences, the body of epidemiological and animal toxicological data for the chemical was assessed for scope and reliability, and then interpreted in terms of the most probable form for the dosage versus response relationship for each serious health effect in the human as a general target. For a given health effect, then, the relationship could be displayed either as a curve of dosage versus anticipated response or, in an attempt to linearize the relationship, as a curve of log dosage versus percentage response. The display could also take other forms, in parallel with modes of display developed in pharmacology.

Whatever the display mode, however, the predicted human dose-response curve was then used for two purposes. First, it could be used to predict human response amplitudes under a specified exposure scenario. Second, by accepting a 5 percent response amplitude as being essentially a no-effect level within the limits of biological variability in populations, the curve could be used to establish the human no-observable-effect level (NOEL). This procedure was, and is, a primitive quantitative risk assessment methodology.

A variant on this mode of generating human NOELs for specific health effects produced by a given chemical has also enjoyed wide international usage, beginning in the 1940s. Since the toxicological data base is usually far more extensive and more quantitative than that available from occupational epidemiological observations, the concept developed that human risk assessment (in the form of human NOELs) for health effects from a given chemical could be generated from the animal data base alone and later validated as human data accumulated. In this mode, if data exist for a spectrum of adverse health effects produced by a chemical in an assortment of test animal systems, either reversible or irreversible effects, the risk assessor selects the most serious health effect in the most sensitive animal species tested and uses the data to estimate the animal NOEL for that effect. In the choice among several health effects involving different tissues and organs, that effect may be selected which is manifested at the lowest dose range. Then, by application of a suitable safety factor (SF) or translation factor to the animal NOEL, a human NOEL is projected.

This translation procedure evolved. At least two key review papers in the 1950s serve as milestones in this evolution, that of Barnes and Denz (1954) and that produced by Lehman and colleagues (1959) at the U.S. Food and Drug Administration. In particular, the rationale for and size of the safety factor to be employed in the animal-human NOEL translation was developed largely at the hands of Lehman et al., although variants are still being discussed in today's literature. For a well-defined toxic action at a target tissue which appears to display a dose threshold and which is at least moderately reversible, a widely applicable SF of 100 was postulated, with a first factor of 10 for the NOEL translation from animal to human and a second factor of 10 to account for the variability in sensitivities in human populations. For more serious, irreversible types of effects, even including carcinogenesis in earlier considerations of risk assessment for this chronic effect, additional safety factors (range, 2–10 and higher) were factored into the fundamental SF of 100. For example, at times in the last two decades, total safety factors of 2,000 or 3,000 have been mentioned as being applicable to the apparent NOEL for tumorigenesis in a chronic animal bioassay of a chemical in converting to an estimate of a NOEL for tumorigenesis in human populations.

It should be noted that the simplistic form of risk assessment considered up to this point was always generated singly, chemical by chemical, and was also viewed as a prediction that should be validated or rejected as additional animal and human dose versus response data for a chemical were developed. When multiple chemical exposures were considered, the state of the art and knowledge only extended to the possibility of additivity for closely related structural analogs that produced simmilar effects on a given target tissue by similar interaction mechanisms. The possibilities of synergism, potentiation, or antagonism in multichemical exposures were discussed, but were rarely attacked in the form of a joint risk assessment.

An important point of departure from this relatively simplistic but practical methodology for risk assessment related to health effects from chemical exposures took place in the 1960s–1970s, with the evolution of the regulatory concept that there could be no such thing as a NOEL for chemical carcinogenesis in humans and no such observable as a practical threshold for the complex carcinogenic process in mammals. All exposures were conceived of as contributing finite increments of excess lifetime cancer risk from the chemical in question, regardless of whether repair processes were operable at some level after the initial attack on DNA. From this regulatory philosophy there evolved the process of modeling these excess lifetime risks for carcinogenesis in the humans based on postulated human exposure scenarios and the observed tumorigenic responses measured in chronic bioassay experiments with test animal systems (usually rodents). The modeling process has been labeled quantitative risk assessment for chemical carcinogenesis, and because it has moved in certain sectors of the public perception from being viewed as a predictive technique into the status of a supposedly factual presentation of real human risks, it deserves some explicit analysis, as follows.

1. The process starts with a chronic bioassay, usually in rodent systems, in which dosing with the test chemical extends over more than 0.5 lifetime of the species involved, with daily administered doses at the maximum tolerated level and one or more submultiples of the maximum tolerated dose (MTD).
2. The data, in the form of administered dosages versus tumorigenic responses, are then fitted to one or more modeling equations (for example, one-hit, multistage, Weibull) of a form in which the probability of excess cancer development is some function of the average daily lifetime dose.
3. Generally, additional restrictions are placed on this modeling step in the form of assumptions that the tumorigenesis process has no threshold (zero probability only at zero dose), and that the modeling equation is linear in the very low dose region, regardless of the curve's shape at the very high administered doses (MTD, 0.5 MTD, 0.33 MTD) under which the bioassay is performed.
4. The fitted equation of the modeler's choice is then used to extrapolate from the high-dose region of animal administered dose versus response down to theoretical response levels at a very low dose.
5. The assumption is then made (explicitly or implicitly) that the low-dose sector of the animal-fitted curve can be used to predict human risks for tumorigenesis by the test chemical at very low ambient exposure levels. In this translation process, it is then customary to make a gross correction for metabolic differences in handling of the chemical by the animal and the human in the form of a correction ratio applied to dose based on either relative body weights or body surface areas. The implicit assumption in this correction process is that the mechanisms of handling, target organ specificity, etc., in the two mammalian species are similar.
6. Finally, under a series of postulated human exposure scenarios, each of which leads to a calculated average daily lifetime dose (administered), the low-dose equation for the animal is used to predict a corresponding series of excess lifetime cancer risks in humans.

Regulatory communities worldwide now use these modeled risks for prioritizing their regulatory activities over a wide range of potentially carcinogenic chemicals in the environment or in the workplace, and for setting numerical triggers to be used in initiating rulemaking or restrictions on specific chemicals. These regulatory activities are important and are powerfully assisted by the quantitative risk assessment process. The process itself is being extended by investigators to cover simultaneous exposures to many potentially carcinogenic chemicals found at low levels in the environment, such as the assortment of chemicals found in drinking water supplies, by the use of risk combination techniques (e.g., Crouch et al., 1983).

It should be realized, however, that there are scientific problems inherent in the use of the modeling techniques based on animal bioassay data. To cite just a few: (1) There are no general experimental/theoretical justifications for the modeling assumptions, the validity of high dose-low dose extrapolations of the animal bioassay data, or animal to human translations; (2) the methodology provides no insight as to what delivered dose of what active material (original chemical or a metabolite) delivered to what target tissue should actually be modeled in a truly meaningful risk assessment; and (3) there is no assurance in any given risk assessment that modeling of the administered dosage data has any direct relationship to delivered dosage, for the test compound or an active metabolite. Indeed, recent examples abound to show a lack of administered dose/delivered dose congruence. All of these points have been well recognized and discussed in the 1970–1986 time frame by scientists concerned with making quantitative risk assessment more meaningful and sound.

Therefore, the need has been recognized for moving quantitative risk assessment to a more realistic dimension, particularly by the use of data from metabolic and comparative pharmacokinetic studies of a given chemical which, when combined with chronic bioassay data from animal experiments, can lead to knowledge about the active chemical species that reaches specific target tissues at measurable delivered concentrations (as a function of time) in the species of prime interest, humans.

The purpose of the workshop on which this volume is based was to review progress in this development of the risk assessment process and to probe the current strengths and weaknesses in this area.

REFERENCES

Barnes, J. M., and F. A. Denz. 1954. Experimental methods used in determining chronic toxicity. Pharmacol. Rev. 6:191–242.

Crouch E. A. C., R. Wilson, and L. Zeise. 1983. The risks of drinking water. Water Resources Res. 19:1359–1375.

Lehman, A. J., F. A. Vorhes, et al. 1959. Appraisal of the Safety of Chemicals in Foods, Drugs and Cosmetics. The Association of Food and Drug Officials of the United States. 107 pp.

Tissue Dosimetry in Risk Assessment, or What's the Problem Here Anyway?

Melvin E. Andersen

INTRODUCTION

The overall risk assessment process integrates hazard assessment data on chemical toxicity with exposure assessment information (Figure 1). Hazard assessment is the process by which the toxicity of a chemical is determined either by a series of bioassay experiments with intact test animals or by observing increased morbidity/mortality in exposed humans. Often there is no human epidemiology on particular chemicals, and risk managers have to rely solely on results of animal toxicity experiments for the hazard assessment. These animal experiments allow us to generate dose-response information on how much chemical is required to produce a specified degree of toxicity in test animals. The major challenges in the hazard assessment process are to generalize toxicity results in the test animal to (1) predict what will happen in test animals given much lower amounts of chemical; (2) predict what will happen in an entirely different species of animal, namely, humans; and (3) predict what will happen in a different species receiving a chemical by a route of administration different from that used in the animal studies. These are all problems of extrapolating beyond the conditions used in the actual toxicity studies to predict outcome under very different exposure conditions in a variety of species. What concepts tie these problems together and give us some confidence in the ultimate success of efforts to develop methods to conduct these extrapolations?

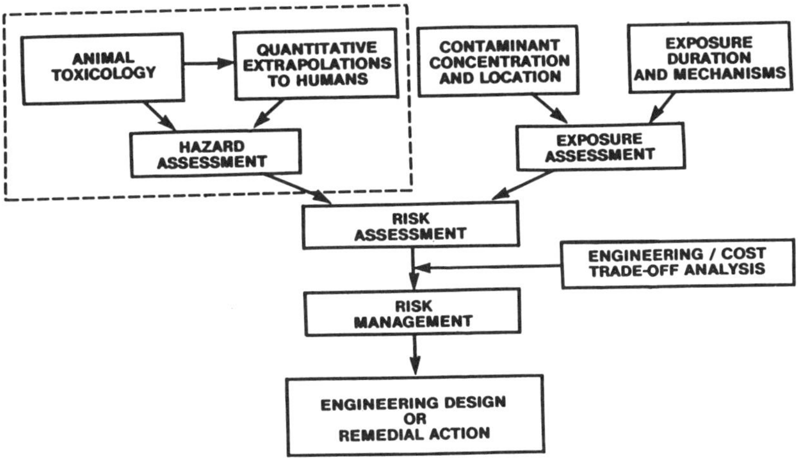

FIGURE 1 Elements of chemical risk assessment. The overall risk assessment process consists of both a hazard assessment and an exposure assessment. These provide information on which to make a risk analysis and give the risk manager detailed information on which to make decisions regarding acceptable environmental concentrations of a toxic chemical, cost-effective engineering design criteria for reducing effluent or emission concentrations, and the feasibility of replacing one chemical by another in a particular industrial application. Pharmacokinetic modeling is useful in hazard assessment where it can aid in estimating realistic measures of target tissue dose in exposed animals and be used to support extrapolations to estimate tissue dose in humans.

Basically, there seem to be two fundamental assumptions which toxicologists are forced to make in attempting quantitative extrapolations based on animal toxicity experiments. The first is that experimental animals are true surrogates for exposed humans. That is, chemicals would cause effects in the same tissues in humans as those tissues in which they cause effects in the exposed test animals, and the mechanisms of these effects would be qualitatively similar in the two different species. This assumption accepts that there is a qualitative similarity in effects in different species. There are instances where this assumption is suspect. For instance, vinyl chloride causes zymbal gland cancer in rats (Maltoni and Lefemine, 1974), but humans do not have this structure. However, vinyl chloride is obviously carcinogenic in a variety of animal species. In any case, the universal validity of this assumption of qualitiative similarity in toxicity is really not within the purview of the papers in this volume. It does seem to be valid in the great majority of instances. Instead, this volume focuses our attention on the other basic assumption that we are forced to make to conduct our risk assessment calculations.

This second tenet is that there is a quantitative equivalence in the tissue chemical exposure required to produce an equivalent intensity of biological effect in various species. This is the concept of tissue dose equivalence. More simply stated, all species are regarded to have equal sensitivity to the toxic chemical. Again, there are notable exceptions, such as the extreme interspecies differences in toxicity of 2,3,7,8-tetrachlorodibenzo-*p*-dioxin (Kociba and Schwetz, 1982). For more simple toxicities, related to reactive chemical moieties, this assumption seems entirely appropriate. In addition, the species and strain differences in dioxin toxicity might diminish substantially if the dose were expressed in relation to the concentration and affinity of the dioxin receptor(s) in these various animal species (Poland and Knutson, 1982). The catch, of course, is that tissue dose is not a simple concept and will be different for different chemicals. In fact, the real problem in hazard assessment is defining and measuring tissue dose under a variety of exposure conditions in several species.

A DOSE OF WHAT?

This hazard assessment process sounds deceptively simple. Determine the toxic tissue dose in the test species and calculate the exposure conditions under which this dose is likely to be achieved in humans. All we need to do is to define tissue dose. As a working definition, we can say that an appropriate measure of tissue dose is some measure of the intensity of chemical exposure which is directly linked to the biological processes leading to toxicity or tumor formation. With this definition, it is clear that some presumption of the mechanism of interaction between the chemical and the tissue is required before we can define tissue dose for any particular chemical.

What then are the primary processes by which chemicals interact with tissue constituents to cause biological changes in the tissue? The first process is by direct chemical reaction in which the toxic chemical reacts with and consumes cellular constituents (Figure 2). With this type of interaction the expected degree of damage, as loss of cellular constituents or accumulation of bound reactive intermediate, should be related to the time integral of tissue exposure to the reactive chemical. This time integral of tissue exposure is also called the area under the tissue concentration curve for the reactive chemical. The equations for reactivity in Figure 2 are true only for acute-exposure situations. In chronic administration, the equation should be expanded to include terms for the synthesis and normal catabolism of the macromolecules.

The second common process by which chemicals interact with tissue is by noncovalent binding to cellular receptor molecules. This is the mechanism by which dioxin is presumed to interact to initiate toxic changes in cells. This binding with concomitant changes in receptor occupancy causes

HOW DOES TISSUE DOSE ELICIT TOXIC RESPONSE?

● CHEMICAL REACTIVY:

$$T + MM \xrightarrow{k_1} T - MM \quad (DEPLETION)$$

$$\int_{MM_0}^{MM_t} \frac{d(MM)}{MM} = -\int_0^t k_1(T)dt$$

$$ln[(MM)_t \,/\, (MM)_0] = -k_1 \text{ (AUTC)}$$

● RECEPTOR BINDING:

$$T + R \overset{K_1}{\rightleftharpoons} TR \quad (Occupancy)$$

$$\frac{TR}{TR + R} = \frac{(T)}{(T) + K_1}$$

FIGURE 2 Tissue dose metrics and their relation to toxicity. Toxic chemicals interact with tissues by two general processes. In one case, chemical reactivity, the toxic chemical, T, reacts with cellular macromolecules, MM, to cause covalent binding of the toxic chemical and a depletion in the concentration of MM. Simultaneously, there are increases in the bound toxic chemical, $T\text{-}MM$, which for genotoxic carcinogens can be regarded as similar to a DNA adduct. Solution of the rate equation for loss of MM with time shows that the natural logarithm of the remaining MM is proportional to the second-order rate constant for the reaction of T with MM and the area under the tissue time course concentration of T (AUTC). In the second case, receptor binding, T binds to a receptor molecule, R, with a dissociation constant, K_1. Toxicity develops due to occupancy of the receptor with some attendant biological consequence. Occupancy is the ratio of bound receptor (TR) and total receptor ($TR + R$). As shown, this is equivalent to $T/(T + K_1)$, i.e., occupancy is determined by the free concentration of T and the binding constant.

some response on the part of the organism which is ultimately expressed as toxicity. The therapeutic action of most drugs is also related to specific receptor binding (Goldstein et al., 1974). With this type of interaction, the response of the cell is dependent on the occupancy of the receptor and occupancy is determined by the binding constant for the chemical and the

WHAT ARE SOME MEASURES OF TISSUE DOSE ?

— AUTC PARENT CHEMICAL
— AUTC STABLE METABOLITE
— TISSUE CONCENTRATION PARENT
— TISSUE CONCENTRATION METABOLITE
— AUTC REACTIVE METABOLITE

AUTC: AREA UNDER TISSUE CURVE

FIGURE 3 Some potential tissue dose metrics for toxic chemicals.

free concentration of chemical in the cell. Thus, there are a variety of quantitative measures of tissue dose which may be regarded as the appropriate measures of the intensity of tissue exposure (Figure 3). In this paper these parameters which are proportional to the relevant measure of tissue exposure are referred to as tissue dose metrics. These metrics include estimates of time integrals of tissue exposure to a chemical or its metabolite(s), concentrations of these materials in tissues, or receptor occupancy caused by the presence of these chemicals in tissues. The choice of which of these metrics to use as the appropriate measure of tissue dose now depends on some knowledge of the mechanism by which the toxic effects are induced.

In this usage mechanism does not mean an exhaustive, complete description of the entire set of events associated with toxicity. It relates instead to certain general aspects of the nature and causes of a particular toxic event. For instance, is the effect related to chemical reactivity or to occupancy of cellular receptor molecules? Is the effect associated with parent chemical or with a metabolite? If it is a metabolite, does the metabolite have a sufficiently long half-time in the body to circulate freely throughout the body or is it so reactive that it never leaves the organ(s) in which it is formed? In terms of the effects themselves, are they essentially reversible cytotoxic phenomena or irreversible carcinogenic transformations? If it is cancer induction, is the process one of direct genotoxicity or is it epigenetic in origin and associated with either induced hyperplasia as a result of cytotoxicity or with tumor-promoting effects of the chemical?

These are some of the more important aspects of mechanism that must be considered in establishing the correct metric for expressing tissue dose.

ISN'T THIS VOLUME ABOUT PHARMACOKINETICS?

The later portions of this chapter discuss tissue dose for several classes of chemical carcinogens. But right now, the reader may be wondering what tissue dose has to do with the theme of this volume—pharmacokinetics. Well, if determining and measuring the appropriate tissue dose is the problem in hazard assessment, then pharmacokinetic modeling is an indispensable tool in estimating tissue dose for a varity of exposure conditions. Pharmacokinetic modeling contributes to the process by which we translate from obvious measures of administered dose, such as amount of chemical instilled into the stomach or concentration in the inspired air, to estimate the more relevant measures of tissue dose which may not always be accessible to measurement by direct experimentation.

The development of a pharmacokinetic model for use in chemical risk assessment begins with identification of a toxic effect in a particular tissue. Based on some limited knowledge of the mechanism of toxicity and results from the literature, the appropriate measure of tissue dose is deduced, or alternatively, several potential measures of tissue dose might be proposed. Next, an analytical pharmacokinetic model should be developed to predict these relevant measures of tissue dose under a wide range of exposure conditions. What elements, if possible, need to be included in a useful pharmacokinetic model? It should contain structures to account for routes of administration, major storage tissues within the body, primary tissues involved with elimination, target tissues, and sufficient biochemical detail within target tissues to calculate the presumed measures of tissue dose. These models should be validated as much as possible with kinetic data or with ancillary experimentation to assess model parameters by experiments separate from the kinetic studies. The successful kinetic models can then be used to estimate tissue dose and correlate it with observed toxicity. Hazard assessment calculations for human exposures are subsequently conducted based on the expected human target tissue exposures under various exposure conditions.

In work in our laboratory in Dayton, Ohio, developing pharmacokinetic models for use in chemical risk assessments, we have relied heavily on use of so-called physiologically based pharmacokinetic models—PB-PK models (see H. J. Clewell III and M. E. Andersen, this volume). These models contain considerable physiological and biological information (Bischoff and Brown, 1966) and are amenable to interspecies extrapolation, a process which is essential for predicting human hazard based on results of animal toxicity studies (Dedrick, 1973). Gerlowski and Jain (1983) have provided a very good review of the status of PB-PK modeling of chemical disposition. These PB-PK models describe the body in terms of realistic tissue compartments with specified volumes, blood flows, partition coefficients, and tissue binding characteristics (Gargas et al., 1986; Ramsey and Andersen, 1984). Biochemical constants for metabolic pathways and for tissue binding can be included in the mass-balance equations for organs in which these interactions are important. For most of these metabolic pathways the important constants are the maximum velocity of the reaction (V_{max}; in milligrams per kilogram) and the binding affinity of the particular substrate for the metabolizing enzyme (in milligrams per liter). Complex metabolic pathways involving parallel or sequential reactions of the parent chemical or involving interactions between chemical metabolism and cofactor depletion can also be readily incorporated into these models, as necessary (H. J. Clewell III and M. E. Andersen, this volume).

The entire process of problem definition, tissue dose assignment, and pharmacokinetic model development can be captured in a simplistic flow diagram (Figure 4). In this representation the process of model formulation comes after evaluation of the nature of the problem and consideration of the impact of mechanism on the choice of tissue dose metric. It is followed by exercising the model, evaluating its success at predicting known kinetic and toxicity behavior, designing necessary experiments to collect crucial data for verifying or improving model performance, and refining the model when needed. A successful model can then be used as an integral part of the hazard assessment process. The take-home lesson here is that pharmacokinetic modeling is not some knee-jerk process where the investigator collects blood time course curves and draws limited inferences about the behavior of the chemical in the body by an abstract mathematical curve-fitting procedure. Instead, the pharmacokinetic modeling intended for risk assessment use is an integrating process which should be done early on before major data collection efforts. It should provide a comprehensive description of chemical disposition in target organs and be designed to predict human kinetic behavior when the biochemical metabolic constants and the tissue-binding characteristics of the chemical have been determined in human tissues.

The remainder of this chapter discusses tissue dose for various mechanisms of carcinogenesis, identifies essential elements required in PK models for tracking these particular forms of tissue dose, and emphasizes that pharmacokinetic model development will often suggest a need to collect critical metabolic or kinetic data that might not be available from the literature. In fact, it would be completely wrong to believe that PK models should be developed on existing toxicity data bases. The existing

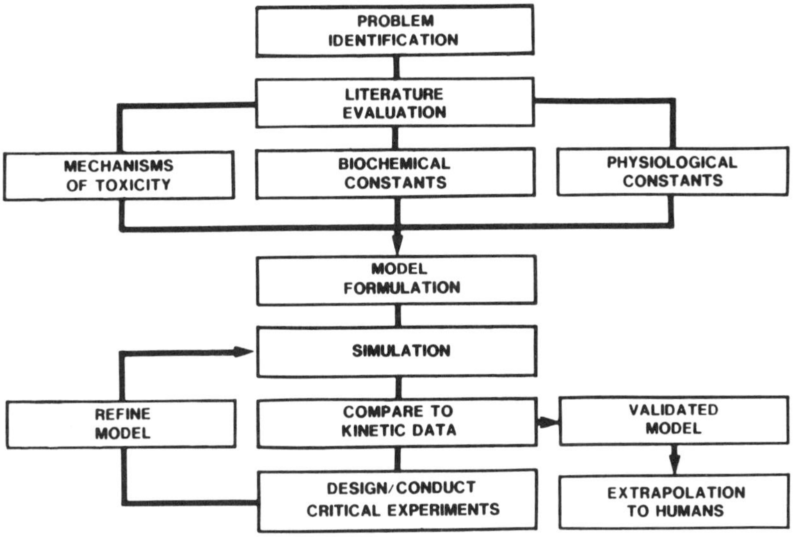

FIGURE 4 Simplified flow chart of the development of a pharmacokinetic model intended for use in risk assessment. Problem identification: the finding of a particular toxicity, in a particular organ(s), in a particular species. This effect is the benchmark on which the risk assessment will be conducted. Literature evaluation: the integration of available information about the mechanism of toxicity, the pathways of chemical metabolism, the nature of the toxic chemical species, the tissue-binding characteristics, and the physiological parameters of the target species. From these data a model is developed to estimate the appropriate measure of tissue exposure for a wide variety of exposure conditions. The essential elements to be included in such a model are outlined in the text. Before a PK model can be used in human risk assessment it has to be validated against kinetic, metabolic, and toxicity information and in many cases refined based on comparison with experimental results. The model development process can frequently be used to design critical experiments to collect data needed for final validation.

literature can be helpful for model definition, for drawing conclusions about the nature of appropriate measures of tissue dose, and for providing limited PK information, but it is also replete with experiments which are virtually useless for hazard assessment. If a new approach, such as PB-PK modeling, is proposed as an adjunct to existing hazard assessment techniques, it will have data requirements of its own and require some independent experimentation not previously conducted on a routine basis for each chemical for which a risk assessment was planned. In general, this does not mean that there has to be major new data acquisition needs for each PK model that might be developed for risk assessment use. For many cases, this will be only limited, critical experiments that are required to provide important constants for use in the PK model (Figure 4) or to fill data gaps identified in the literature survey.

GENOTOXIC CARCINOGENS

In terms of the chemical carcinogens themselves, there are two broad mechanisms by which chemicals cause cancer: by some direct chemical interaction with the DNA structures of the cell or by indirect effects on the cellular environment which increase tumor yield without direct chemical alteration of DNA. The former are called genotoxic carcinogens and the latter, epigenetic carcinogens (Weisburger and Williams, 1980). As might be expected, the distinction between these two categories of carcinogens is not always clear-cut. Many substances appear to possess properties characteristic of both categories of carcinogens. However, this division can be profitably examined in terms of the importance of a proper definition of tissue dose for both types of chemical carcinogens.

Genotoxic chemical carcinogens themselves can be further subdivided on the basis of whether parent chemical or a metabolite is the moiety that reacts with DNA. The possibilities include cases where parent chemicals, such as ethylene oxide or dimethylsulfate, are genotoxic; cases where stable metabolites, such as ethylene oxide formed from ethylene or butadiene epoxides formed from butadiene, are genotoxic; and cases where reactive, nonisolatable metabolites, such as the epoxide formed from vinyl chloride or the chloromethylgluthathione formed from methylene chloride, are presumed to be responsible for genotoxicity (Figure 5). These three possibilities for the nature of the DNA-reactive chemical need to be considered independently.

PARENT CHEMICAL

For the simplest case there is a chemical reaction between DNA and parent chemical leading to chemical alteration of the DNA which can cause mutation during cell replication. As discussed for cases of chemical reactivity, the tissue burden of altered DNA is expected to be associated with integrated tissue exposure to the DNA-reactive chemical. The modeling problem is to identify the chemical-specific tissue solubilities or tissue-binding characteristics and the distribution and activity of chemical-specific detoxifying enzymes in various tissues. The goal of the PK model is to identify and understand the metabolic and physiological processes that limit the action of the parent chemical in the cells.

Interspecies scaling is the determination of how target tissue exposure is affected by animal size for a particular administered dose. An attempt was made in volume 6 of *Drinking Water and Health* (National Research Council, 1986) to predict the interspecies scaling of tissue dose, depending on the nature of the toxic moiety—for example, parent or metabolite, etc. This analysis assumed that both metabolic and physiological clearances

GENOTOXIC INTERACTION

NATURE OF PROXIMATE CARCINOGEN

- **PARENT CHEMICAL:**
  - — ETHYLENE OXIDE
  - — DIMETHYLSULFATE
- **"STABLE" METABOLITE:**
  - — BUTADIENE
  - — ETHYLENE
- **"REACTIVE (NONISOLATABLE)" METABOLITE:**
  - — VINYL CHLORIDE
  - — METHYLENE CHLORIDE (?)

FIGURE 5 Classification of genotoxic carcinogens: Genotoxic chemicals are defined as those materials which are sufficiently reactive *in vivo* to react with DNA bases to form adducts. The presence of these adducts increases the probability of a mutational event during cell replication. The genotoxic entity can be parent chemical, a stable, freely circulating metabolite, or a reactive metabolite that is not sufficiently long-lived to isolate *in vivo*. Representative environmental chemicals from each of these classes of genotoxic carcinogens are listed here.

scale as a partial power of body weight and that both tissue volumes and genral volumes of distribution are directly proportional to body weight. For these idealized pharmacokinetic behaviors, a particular dose (in milligrams per kilogram) will give a larger area under the tissue curve for parent chemical in a larger species than in a smaller species. This occurs because the original dose is dispersed into a volume proportional to body weight, and animals, regardless of body weight, should attain the same initial internal concentration for a given dose (in milligrams per kilogram). Clearance, however, is expected to be proportionately larger in the smaller species because it increases as a fractional power of body weight, and therefore, the area under the blood and tissue curves will be smaller in the smaller animal species. This simplified PK analysis indicates that larger species are at greater risk (i.e., have larger area under the tissue curve) than are smaller animals for those instances where parent chemical is the DNA-reactive chemical. This behavior is consistent with the traditional surface area correction used for interspecies dose adjustments in the chemical risk assessments conducted by the Environmental Protection Agency (EPA). The use of this factor is usually justified by reference to studies on the interspecies differences in acute toxicity of a variety of chemicals used medically in cancer chemotherapy (Freireich et al., 1966).

STABLE METABOLITES

Ethylene oxide can also be produced *in vivo* by the oxidation of ethylene by microsomal metabolism. In developing a pharamacokinetic model for ethylene oxide as a DNA-reactive, stable metabolite, other PK modeling considerations become important. These include the rate of formation of the epoxide in various tissues, the stoichiometric yield of the epoxide from ethylene *in vivo*, and the distribution of the stable metabolite to target tissues. With butadiene, there are two epoxide metabolites that have genotoxic potential, and some provision for their differential DNA reactivities might have to be included in model development. A very elegant analysis of the relative risks of ethylene and ethylene oxide has been conducted by Bolt and Filser (in press). They developed pharmacokinetic models for both of these chemicals and attempted to predict ethylene exposure conditions that would yield carcinogenic tissue doses of the epoxide. They did not use a physiological model, and their results are limited to interpretation of the bioassay results in experimental animals. Nonetheless, their study is an excellent example of the use of sound pharmacokinetic principles in the analysis, interpretation, and design of toxicity experiments.

In instances in which the genotoxic chemical is a stable, freely circulating metabolite, the analysis of the effect of body size on tissue dose includes consideration of the metabolic formation of a DNA-reactive metabolite and its consumption by various clearance pathways. For the purposes of risk assessment, when there is no available information on the human population, it seems appropriate to assume that both the metabolic production and the clearance pathways are related to the same fractional power of body weight (National Research Council, 1986). Thus, equivalent doses on a body weight basis are expected to produce approximately equal tissue exposures expressed as area under the tissue curve of the genotoxic, stable metabolite. This suggests that larger animals should be at the same risk from equivalent doses of these chemicals as smaller animals. For this class of chemicals, the standard surface area correction used by EPA would overestimate the expected risk in humans based on extrapolation of toxicity results in small laboratory animals. The Food and Drug Administration approach which uses body weight for interspecies dose conversion is the more appropriate correction factor for this class of chemical carcinogens.

REACTIVE, NONISOLATABLE METABOLITES

In a recent paper we attempted to develop a strategy for conducting a pharmacokinetically based risk assessment for methylene chloride (Andersen et al., 1987). On the basis of a variety of kinetic and chemical arguments, we suggested that the carcinogenicity of methylene chloride was related to the metabolites produced by conjugation of parent chemical with glutathione. If this proposed mechanism of toxicity is correct, the appropriate measure of tissue dose should be the time integral of the tissue concentration of the glutathione conjugate. This material is too reactive to measure directly, and a surrogate measure of tissue concentration of this chemical must be utilized in place of its concentration. The surrogate dose metric that was developed based on kinetic principles was a ratio of the integral of the amount of chemical metabolized by this pathway in the target tissue divided by target tissue volume. This same approach could be used with other chemicals, like vinyl chloride, where the presumed genotoxic metabolite is also too short-lived to measure directly.

When reactive metabolites are associated with carcinogenicity, the simplified pharmacokinetic analysis of the effect of animal size on integrated tissue exposure suggests that larger species will be at proportionately less risk than smaller species (National Research Council, 1986). The reason for this dependence is that metabolic production (the numerator) is proportional to a fractional power of body weight, while tissue volume (the denominator) is directly proportional to body weight. The ratio of the two then decreases with increasing body weight. This approach to interspecies scaling for vinyl chloride was previously suggested by Gehring et al. (1978) based on somewhat different arguments.

The above examples point out that it is very difficult to depend on a single approach to interspecies scaling. When scaling strategies are developed based on generalized pharmacokinetic principles, there are several very different types of interspecies scaling behaviors depending on the nature of the DNA-reactive chemical—whether it is parent chemical, stable metabolite, or a highly reactive, nonisolatable metabolite. These differences should in some way be reflected in the process of hazard assessment when the mechanism of carcinogenicity of a chemical is fairly well-established. Universal reliance on the surface area correction, or any one particular adjustment factor, should be avoided; however, in the absence of information on the mechansim of toxicity, the surface area correction would at least err on the conservative side.

INTERCALATING AGENTS

Another group of direct-acting, DNA-interactive chemicals are the intercalating agents, represented by acridine-type dyes and related chemicals

(Rogers and Back, 1982). With these materials there is noncovalent bonding between the dye and DNA, and the interactions are probably best described by the mass action law with critical receptor site occupancy by the intercalated dye. For this type of tissue interaction we would need to know the dissociation constant(s) and binding capacity for the agent-DNA binding processes, and the time course of intercalator concentration in the target tissues. Tissue dose in this case is probably best represented as a time-weighted average receptor occupancy by the intercalating ligand. Thus, even for genotoxic chemicals there are possibilities that interactions can occur either by chemical reactivity or by mass action effects of particular chemical ligands. These two mechanisms lead to two very different expressions for tissue dose.

These estimates of tissue exposure with chemically reactive or intercalating agents can be used as the dose inputs to drive increased mutational frequency in biologically based cancer models such as that proposed by Moolgavkaar and Knudson (1981). Combining pharmacokinetic and pharmacodynamic modeling of the cancer process (Figure 6) promises to greatly improve our ability to conduct interspecies scaling and support risk assessment extrapolations. It is important to remember, however, that tissue dose will often be nonlinear with respect to administered dose, and it is clearly wrong to use administered dose uncritically in developing realistic cancer models. To a very great extent, it is only the availability of accurate pharmacokinetic descriptions of tissue exposure which permits validation of biologically motivated models of chemical carcinogenesis. In fact, it is essential to have an adequate understanding of the pharmacokinetic characteristics of target tissue exposure before pharmacodynamic models are developed for any kind of toxic response.

The mechanisms of carcinogenicity of directly acting genotoxic chemicals are still under active investigation in terms of fundamental questions about the nature of DNA adducts formed, rates of repair of damaged DNA, the presence of critical mutational sites on DNA, etc. Eventually, as more information is developed, it may even be possible to use the formation of particular adducts as the measure of tissue dose instead of the use of integrated tissue exposure. This would give metrics for tissue doses of carcinogens which were closer to the biological process of tumor induction.

EPIGENETIC CARCINOGENS

Despite the many outstanding questions with regard to the mode of action of genotoxic carcinogens and to the relative importance of particular DNA adducts, it is clear that a great deal more is known about the mechanisms of tumor initiation with these chemicals than about the detailed

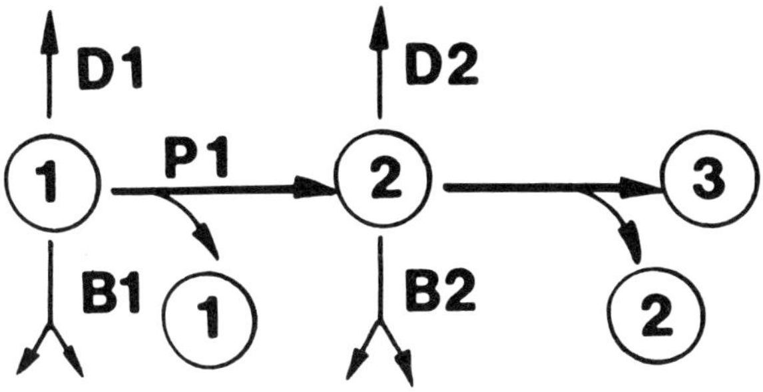

FIGURE 6 A schematic of a two-stage cancer model: Normal cells (1) have a basal birth rate (B1) and death rate (D1). There is also a background mutational rate (P1). The probability of undergoing a mutational event is the product of the birth rate and the mutational frequency (i.e., P1 × B1). The stage 2 cell (2) has undergone a single mutation but is not a tumor-forming cell. It has its own birth and death rate (B2 and D2) and some probability of undergoing a second mutational event (P2) to become a cancer cell capable of progressing to a tumor. Genotoxic carcinogens are expected to increase P1 and P2. The increase in P1 and P2 will be related to the integrated area under the tissue concentration curve of the DNA-reactive chemical. These tissue exposure estimates can be derived from physiologically realistic PK models. There are two major classes of epigenetic carcinogens discussed in this paper: those which cause cell toxicity with continuous cell division during chronic exposure, and those which act as promoters by increasing synthesis of new enzymes (or altered levels of existing enzymes). These mechanisms are more fully developed in the text. The former class of chemicals will primarily affect D1 and D2 with subsequent changes in B1 and B2. The latter group is believed to affect the balance of B2 and D2 to give the cells with single mutations a growth advantage. PK models will have to be developed to link tissue exposure, cell toxicity, enzyme induction, and changes in cell growth kinetics. These coupled PK and pharmacodynamic models of the cancer process should greatly improve cancer risk assessment for most chemical carcinogens.

mechanisms by which epigenetic carcinogens cause tumor development. In general, there seem to be two very different groups of epigenetic carcinogens. The first group consists of those chemicals that cause overt cytotoxicity and cancer appears secondary to chronic tissue damage. Chloroform and carbon tetrachloride are examples from this group (Reitz et al., 1982). The second group consists of the tumor promoters which interact with the cell in such a way to induce expression of new, characteristic sets of enzyme activities. The altered cellular environment caused by these promoters then leads to enhanced tumor yield under appropriate exposure conditions. Examples here include phenobarbital and dioxin. Tissue dosimetry for these epigenetic carcinogens will be more complex than it is for genotoxic carcinogens.

For epigenetic mechanisms, it will be necessary to include some kind of pharmacodynamic modeling along with the pharmacokinetic description. For cytotoxicity (Figure 6) it will be necessary to model the linkage between tissue reactivity of reactive chemical with depletion of critical cellular macromolecules, cell death, and attendant hyperplasia. This new birth rate function can then be used in a biologically motivated cancer model (see R. B. Conolly, R. H. Reitz, and M. E. Andersen, this volume). With promoters the proper tissue dose metric will be related to receptor occupancy and a resulting induction of new protein synthesis. The pharmacokinetic modeling strategy ultimately devised for these promoters will require physiologically accurate models for the processes involved in enzyme induction. In the two-stage cancer model, the action of these promoters is considered to be on the birth and death rates of the stage 2 cell, providing cells with a single mutation with a selective growth advantage. While dosimetry with these epigenetic carcinogens is more demanding than with genetic carcinogens, there still appears to be two dose metrics that emerge—integrated exposure to reactive chemical for cytotoxicants and time-weighted average receptor occupancy for chemicals such as dioxin and phenobarbital. These seem to be the primary expressions required for understanding our problem here: that is, just what is tissue dose? On the other hand, what can be said of pharmacokinetics in these cases. The modeling strategy is still the same—produce an integrated description of chemical disposition and tissue exposure which is readily amenable to interspecies extrapolation and in which all biochemical, physical chemical, biological, and physiological processes are as clearly defined as possible.

SUMMARY

Mechanistic information on the processes involved in cancer causation by a particular chemical is essential for defining the appropriate measure of target tissue dose. Tissue dose will usually either be a function of integrated tissue exposure or a function of the extent of receptor binding in a tissue. This latter metric of tissue dose is dependent on mass action principles and not simply on integrated tissue exposure. Measures of dose related simply to administered dose or total amount metabolized should be viewed with great caution unless there are compelling reasons for believing there is a direct correlation of these very coarse measures of dose with actual tissue exposure. Once the proper measure of dose is defined, a pharmacokinetic model should be developed to predict this dose metric for various exposure scenarios in a variety of species. The biological realism of physiologically based models confers on them certain advantages for use in the risk assessment arena. Finally, much better cancer risk assessments will be possible when validated pharmacokinetic models for tissue dose are used in conjunction with more biologically, realistic pharmacodynamic descriptions of the biological processes involved in chemical carcinogenesis.

REFERENCES

Andersen, M. E., H. J. Clewell III, M. L. Gargas, F. A. Smith, and R. H. Reitz. 1987. Physiologically-based pharmacokinetics and the risk assessment process for methylene chloride. Toxicol. Appl. Pharmacol. 87:185–205.

Bischoff, K. B., and R. G. Brown. 1966. Drug distribution in mammals. Chem. Eng. Prog. Symp. Ser. 62:33–45.

Bolt, H. M., and J. G. Filser. In press. Kinetics and disposition in toxicology. Example: Carcinogeinc risk assessment for ethylene. Proceedings of the International Conference on Contributions of Toxicology to Risk Assessment and Health Regulation. Arch. Toxicol.

Dedrick, R. L. 1973. Animal scale-up. J. Pharmacokinet. Biopharm. 1:435–461.

Freireich, E. J., E. A. Gehan, D. P. Rall, L. H. Schmidt, and H. E. Skipper. 1966. Quantitative comparison of toxicity of anticancer agents in mouse, rat, hamster, dog, monkey and man. Cancer Chemother. Res. 66:55–68.

Gargas, M. L., H. J. Clewell III, and M. E. Andersen. 1986. Metabolism of inhaled dihalomethanes in vivo: Differentiation of kinetic constants for two independent pathways. Toxicol. Appl. Pharmacol. 82:211–223.

Gehring, P. J., P. G. Watanabe, and C. N. Park. 1978. Resolution of dose-response toxicity data for chemicals requiring metabolic activation: Example—vinyl chloride. Toxicol. Appl. Pharmacol. 44:581–591.

Gerlowski, L. E., and R. K. Jain. 1983. Physiologically-based pharmacokinetic modeling: Principles and applications. J. Pharm. Sci. 72:1103–1126.

Goldstein, A., L. Aronow, and S. M. Kalman. 1974. Molecular mechanisms of drug action. Chapter 1 in Principles of Drug Action: The Basis of Pharmacology, 2nd ed. New York: John Wiley & Sons.

Kociba, R. J., and B. S. Schwetz.1982. Toxicity of 2,3,7,8-tetra-chlorodibenzo-p-dioxin (TCDD). Drug Metab. Rev. 13:387–406.

Maltoni, C., and G. Lefemine. 1974. Carcinogenicity bioassays of vinyl chloride. I. Research plan and early results. Environ. Res. 7:387–405.

Moolgavkaar, S. H., and A. G. Knudson, Jr. 1981. Mutation and cancer: A model for human carcinogenesis. J. Natl. Cancer Inst. 66:1037–1052.

National Research Council. 1986. Dose-route extrapolations: Using inhalation toxicity data to set drinking water limits. Pp. 168–225 in Drinking Water and Health, Vol. 6. Washington, D.C.: National Academy Press.

Poland, A., and J. C. Knutson. 1982. 2,3,7,8-Tetrachlorodibenzodioxin and related halogenated aromatic hydrocarbons: Examination of the mechanism of toxicity. Annu. Rev. Pharmacol. Toxicol. 22:517–554.

Ramsey, J. C., and M. E. Andersen. 1984. A physiologically based description of the inhalation pharmacokinetics of styrene in rats and humans. Toxicol. Appl. Pharmacol. 73:159–175.

Reitz, R. H., T. R. Fox, and J. F. Quast. 1982. Mechanistic considerations for carcinogenic risk estimations: Chloroform. Environ. Health Perspect. 46:163–168.

Weisburger, J. H., and G. M. Williams. 1980. Chemical carcinogens. Pp. 84–138 in J. Doull, C. D. Klaasen, and M. O. Amdur, Eds. Casarett and Doull's Toxicology: The Basic Science of Poisons, 2nd ed. New York: Macmillan.

PART II
Mathematical Modeling

Modeling: An Introduction

Ellen J. O'Flaherty

Fick's First Law states that the rate of diffusion of a solute down a concentration gradient is proportional to the magnitude of the gradient:

$$\frac{dM}{dt} = -DA\,\frac{dC}{dx}, \tag{1}$$

where M is mass, C is concentration, D is the diffusion constant with dimensions distance^2/time, A is the cross-sectional area of the diffusion volume, and dx is the distance over which the infinitesimally small concentration difference dC is measured. When Fick's First Law is restated for diffusion across a membrane barrier of thickness dx, the concentration gradient dC is approximated by the concentration difference across the membrane, $C_1 - C_2$, and DA/dx is the first-order transfer constant k_t for diffusion across the membrane, with dimensions distance^3/time, or volume/time:

$$\begin{aligned}\frac{dM}{dt} &= -\frac{DA}{dx}\,(C_1 - C_2)\\ &= -k_t\,(C_1 - C_2).\end{aligned} \tag{2}$$

Fick's Law states that transfer of freely diffusible molecules across a membrane should be first order. A large body of experimental observations supports this interpretation of the kinetic nature of diffusion. There are, of course, exceptions: Excretion from liver or kidney may not be first order, and gastrointestinal absorption may take place by active processes

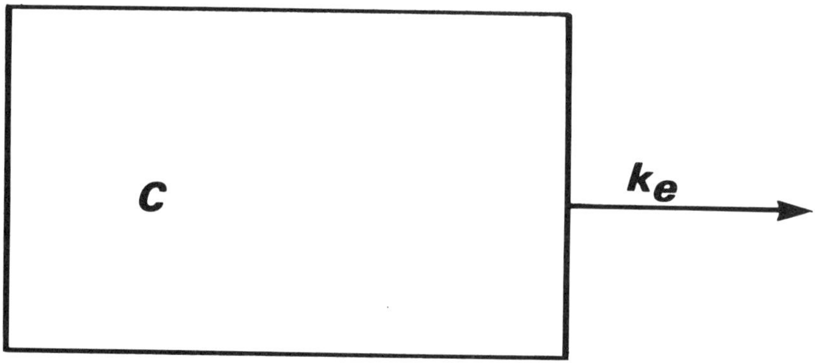

FIGURE 1

for a few chemicals. But, in general, it is reasonable to assume as a working hypothesis that absorption and distribution of exogenous chemicals are first order. The rate of diffusion is dependent on the partition coefficient and molecular size and configuration of the chemical, and on its degree of ionization.

Thus, diffusion out of a single compartment is a first-order process whose rate constant, k_t, depends both on the chemical and on the tissue and has the dimensions volume/time. If both sides of Equation 2 are divided by V so that it expresses the rate of change of concentration, not of mass—since concentration is what is measured *in vivo*—then the rate constant becomes the elimination rate constant k_e, with dimensions time^{-1} (Figure 1).

$$\frac{dC}{dt} = -k_e C. \tag{3}$$

Equation 3 is integrated to obtain the familiar first-order expression for C as a function of t,

$$C = C_0 e^{-k_e t}. \tag{4}$$

Equation 4 has a single exponential term so that if the natural logarithm of the concentration is plotted against time, the graph takes the form of a straight line whose slope is $-k_e$ and whose ordinate intercept is the logarithm of C_0. The half-life is estimated from the value of k_e, and the volume of distribution from the dose and the value of C_0. The model is, of course, the one-compartment body model with first-order elimination.

The body is not a single physiological compartment, however, and rarely behaves as if it were a single kinetic compartment. More sophisticated models of the body are created by the addition of peripheral compartments. The essence of different approaches to modeling lies in how these compartments are defined and in what kinds of variables—measurements of concentrations or amounts, or values of physiological parameters—are used to drive the development of a quantitative model.

In the 1940s and 1950s, it was recognized that concentration behavior in the central compartment of a multicompartment model could be represented by a sum of exponential terms like the single term describing the one-compartment model, one for each compartment in the model. Thus, for any model with more than one compartment, there will be more than one term, and the dependence of the logarithm of the concentration in the central compartment on time cannot be linear. Instead, it takes a curvilinear shape with a terminal straight-line portion. By a process variously known as feathering, peeling, or the method of residuals, curvilinear plots of ln C versus t were resolved into their component exponential terms. As many terms were included in this feathering process as were required to account fully for the curvature of the data. Such fits are now, of course, carried out by nonlinear regression computer programs, but there was a time when they were not.

In early modeling applications, model compartments were taken to have exact physiological correlates. Because of the correspondence between the number of compartments in a model and the number of terms required in the equation describing the model, the number of exponential terms necessary to account for the curvature in the data was taken to represent the number of distinguishable exchanges between the central compartment and peripheral tissues or organs, plus one term roughly equated with whole-body loss.

An example of this approach appears in a paper on the kinetics of the rapid phases of plasma free cholesterol turnover (Porte and Havel, 1961). Free cholesterol labeled with ^{14}C was incorporated into plasma lipoproteins *in vitro* and administered to dogs by intravenous injection. Resolution of the entire free cholesterol curve by successive subtraction of each component—that is, by feathering—gave five exponential terms with half-times of 4 min, 30 min, 65 min, 7 h, and 96 h. Porte and Havel compared these half-lives with turnover times reported for different pools of cholesterol, and concluded that the slowest, 96-h component represented metabolic turnover plus equilibration with very slowly exchanging compartments; the 7-h component represented formation of plasma ester cholesterol; the 65-min component represented exchange of free cholesterol between plasma and red blood cells; and the 30-min component represented exchange of free cholesterol between plasma and liver. The most rapid component, with a half-life of 4 min, could not be related to any known physiological compartment.

It quickly became apparent that forcing such a rigid correspondence between exponential terms and physiological compartments generated a number of problems, two of which are illustrated by the cholesterol ex-

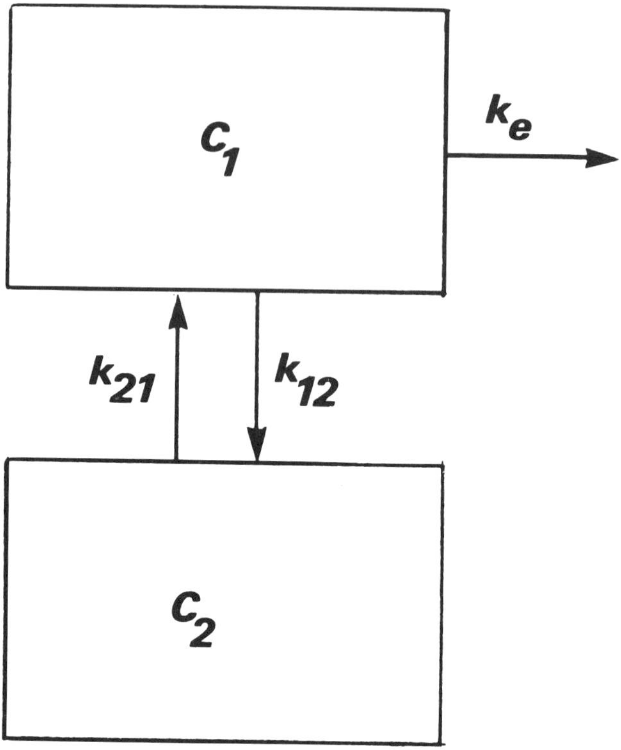

FIGURE 2

ample. It was not always possible to identify physiological correlates of exponential half-lives, particularly the shorter ones. And often more than one process—for example, metabolism and slow exchange—presented themselves as candidates for the source of an exponential term.

With general dissemination of explicit mathematical solutions of multicompartment models and recognition of the implications of these solutions, in the mid-1960s a reaction set in. If there is only one compartment, the half-life is the half-life of elimination, and dose/C_0 is the physiological volume of distribution. When there is more than one kinetically distinguishable compartment, the slopes and intercepts of the successive linear segments that are revealed by the curve-peeling process are expressed in appropriate units for calculation of half-lives and volumes of distribution. But these are not half-lives that are descriptive of a single process, nor are they physiological volumes of distribution.

The reason is apparent on consideration of the two-compartment model shown in Figure 2 and below.

$$\frac{dC_1}{dt} = k_{21}C_2 - (k_{12} + k_e)C_1, \text{ and}$$

$$\frac{dC_2}{dt} = k_{12}C_1 - k_{21}C_2. \tag{5}$$

Simultaneous integration of these two equations gives the explicit solution of the two-compartment model, in which the intercepts A_0 and B_0 and the kinetic rate constants α and β are expressed in terms of the rate constants k_{12} and k_{21} for transfer between compartment 1 and compartment 2, the elimination rate constant k_e, and the volumes of the compartments:

$$C_1 = A_0e^{-\alpha t} + B_0e^{-\beta t}$$

$$A_0 = \frac{D(\alpha - k_{21})}{V_1\,(\alpha - \beta)}$$

$$B_0 = \frac{D(k_{21} - \beta)}{V_1\,(\alpha - \beta)}$$

$$\alpha = 1/2\{(k_{12} + k_{21} + k_e) + [(k_{12} + k_{21} + k_e)^2 - 4k_{21}k_e]^{1/2}\}$$

$$\beta = 1/2\{(k_{12} + k_{21} + k_e) - [(k_{12} + k_{21} + k_e)^2 - 4k_{21}k_e]^{1/2}\}. \tag{6}$$

Alpha and β are functions of all of the rate constants, and A_0 and B_0 are functions of the volumes as well as of the rate constants. Being hybrid constants, they need have no direct physiological significance, although of course they reflect the biochemical and physiological basis of the chemical's disposition. Consequently, the volume of distribution (calculated as dose/B_0) and half-life (calculated as ln 2/β) need have no physiological correlates.

Other apparent volumes of distribution can also be calculated. All are constants that relate a concentration to an amount under a particular set of conditions. But because kinetically determined volumes of distribution usually do not correspond to real volumes of distribution, it became commonplace in the 1970s to consider them simply as proportionality constants.

The apparent volume of distribution is a useful pharmacokinetic parameter that relates the plasma or serum concentration of a drug to the total amount of drug in the body. Despite its name, this parameter usually has no direct physiologic meaning and does not refer to a real volume (Gibaldi and Perrier, 1975).

This philosophy of modeling was, of course, in some respects a reaction to what was correctly perceived as unproductive and in some cases misleading data interpretations as a result of insistence on too exact a correspondence between the terms of the equation describing loss of the chemical from the blood and the physiological nature of compartments—organs and tissues—exchanging with the blood. During this period also, it became commonplace to minimize the significance of half-lives derived from any but the terminal slope of the plasma or blood concentration curve. The terminal slope is used in the calculation of the biological half-life, which is generally accepted as an index of the persistence of the chemical in the body.

The impetus for physiological modeling arose independently of classical pharmacokinetics, and physiological modeling coexisted with classical pharmacokinetic approaches during the 1960s and 1970s. In the 1980s, it is beginning to emerge as the preeminent approach to pharmacokinetic modeling. Because physiologically based pharmacokinetic modeling has received so much recent attention, it is important to make the point that physiological pharmacokinetics and classical pharmacokinetics are not fundamentally incompatible. Although the philosophy behind the two approaches is different and dictates their application for different purposes, there is a direct link between the two approaches.

Let us return for a moment to Fick's First Law. Fick's First Law describes the change in *amount* of a chemical with time. Thus, in the closed two-compartment model, the rates of transfer across the membrane separating the two compartments are expressed in terms of mass and kinetic rate constants, or of concentrations and transfer constants or clearances:

$$\begin{aligned}\frac{dM_1}{dt} &= -k_{12}M_1 + k_{21}M_2 \\ &= -k_{12}V_1C_1 + k_{21}V_2C_2. \qquad (7)\end{aligned}$$

At steady state, when $dM_1/dt = 0$, $k_{12}V_1C_1 = k_{21}V_2C_2$; and since $C_1 = C_2$, $k_{12}V_1 = k_{21}V_2 = k_t$. The equality applies, of course, not only at equilibrium but at disequilibrium as well. Thus,

$$\frac{dM_1}{dt} = -k_t(C_1 - C_2). \qquad (8)$$

It is not coincidental that the transfer constant of classical pharmacokinetics has the dimensions of a flow rate. In the referent fluid volume for that flow rate lies the link between classical and physiological pharmacokinetics. If transfers are perfusion limited—that is, flow limited or first order—then the transfer constant is the rate of blood flow to the tissue. In the two-compartment closed model, let compartment 1 be the blood subcompartment and compartment 2 the fluid subcompartment of a tissue. Then, $k_{12}V_1$ is the rate of blood flow to the tissue, which in this model is equal to total blood flow since there is only one peripheral compartment.

Substituting blood flow rate for $k_{12}V_1$ and $k_{21}V_2$, we find that the rates of transfer out of the blood and into the tissue fluid are expressed as blood flow rate times the concentration difference:

$$\frac{dM_1}{dt} = V_1 \cdot \frac{dC_1}{dt} = -Q_B(C_1 - C_2). \tag{9}$$

With a single refinement, this is the fundamental equation of physiological pharmacokinetics. The refinement takes into account the fact that chemicals do not simply equilibrate between body fluids but, depending on their physicochemical properties, may be bound to tissue macromolecules or incorporated into tissue lipids. Thus, what is measured experimentally when a tissue is sampled is not C_2 but C_{II}, the concentration in the tissue including fluid subcompartment V_2 *and* bound or sequestered chemical. Because transfer is assumed to be flow limited, the concentration of the chemical in efferent blood from the tissue should be equal to its concentration in the fluid subcompartment of the tissue. Equilibration of the chemical between the fluid subcompartment of the tissue and its bound or sequestered forms is assumed to be very rapid, so that the partition coefficient $R = C_{II}/C_2 = C_{II}/$(concentration in efferent blood) can be measured and used to obtain C_2 at any time from a measurement of tissue concentration: $C_2 = C_{II}/R$. Often, the partition coefficient is determined in a separate *in vitro* vial equilibration experiment. Substituting C_{II}/R for C_2 in Equation 9, we obtain an expression for the rate of change in the amount of the chemical in blood or tissue as a function of blood flow rate, partition coefficient, and momentary blood and tissue concentrations:

$$\frac{dM_1}{dt} = V_1 \cdot \frac{dC_1}{dt} = -Q_B(C_1 - C_{II}/R).$$

The form of this equation suggests that it should be possible to substitute physiological values of flow rates and volumes, and values of partition coefficients, in order to obtain predictive, physiologically based pharmacokinetic models. In fact, this is the fundamental relationship on which such models are based. The expressions for all peripheral, nonelimination tissues will be of this form, with Q_B replaced by blood flow to the tissue in question. The equation for the blood will be more complex but is directly derivable from the same kinds of considerations of flow and partitioning. It will include contributions from major tissue groups characterized by different perfusion rates, and it may include input rate or elimination rate terms. The equation for the liver may also include terms for metabolism or for input by absorption from the gastrointestinal tract. Some of these terms, particularly those describing metabolism, may not be first order.

The difference between classical and physiological pharmacokinetic models, then, lies not in how they are constructed but in how they are driven. In classical pharmacokinetics, no effort is made to assign physiological correlates to model parameters. A compartment is simply defined as a volume (strictly speaking, as a fluid volume) that is kinetically homogeneous. It is generally recognized that only in a very few instances are more than three exponential terms required to describe satisfactorily, within the precision of the data, the behavior of a declining concentration curve; most often, two suffice. This understanding has given rise to a group of models in which the body is represented by a central compartment and one or two peripheral compartments which may be "shallow" or "deep"; i.e., they may exchange relatively rapidly or relatively slowly with blood plasma. Such important concepts as volume of distribution, biological half-life, clearance, integrated total exposure following a single dose, and achievement of steady state during chronic exposure arise naturally from these classical models. Their utility for characterization of the behavior of a chemical, and for comparison of its behavior with that of other chemicals, is firmly established.

Classical pharmacokinetic models support certain kinds of extrapolations in particular, extrapolation to different exposure conditions, with reasonable assurance. Capacity-limited or other nonlinear kinetic behavior can be incorporated into classical pharmacokinetic models. A specific advantage of the models is that because the kinetic characteristics of the compartments of which they are composed are not constrained, a best possible fit to a data set can be arrived at by varying the values of the parameters. Best estimates of parameter values can be compared across experimental conditions, treatments, or chemicals to establish whether apparent differences (effects) are statistically significant.

This strength of classical pharmacokinetic models is also their greatest weakness. Lacking a physiological or biochemical basis, the models cannot take into account intraspecies changes such as growth, sexual maturation, or aging, and cannot reliably be used in interspecies conversion of pharmacokinetic data. The need for interspecies conversion of laboratory animal data, in particular, has led to the development of physiological pharmacokinetic models, in which the unspecified compartments of the classical pharmacokinetic models are replaced by actual organs and tissues with their known blood flows. Because tissue volumes, blood flow rates, and enzyme activities can be varied only within physiological limits in these models, the models are not fit to experimental data in the classical sense. Instead, gross discrepancy between the predictions of a physiological model and experimental observation requires reformulation of the model in such a way as to account for the observed behavior.

In a sense, then, we have come full circle, from early insistence on correspondence between exponential terms and identifiable plasma-tissue interchanges to recognition that those interchanges do indeed give form to the plasma concentration curve, although not in the sense in which they were originally believed to do so. Physiological pharmacokinetic models have tremendous potential, particularly for species-to-species conversion of dose-effect data. But classical pharmacokinetic models still have their place. Specifically, they are amenable to statistical treatment and, thus, to hypothesis testing, whereas the purely physiological pharmacokinetic models are not as readily treated statistically.

Both physiological and classical pharmacokinetic models have valid applications today. Both are capable of predicting the dose delivered to a target organ, within somewhat different limits. The assumptions on which the physiological pharmacokinetic models are based make them uniquely suited to cross-species applications. On the other hand, the dependence of classical pharmacokinetic models on experimental measurement of concentration or amount makes them especially well suited to examination of questions about mechanisms of effects that involve changes in pharmacokinetic behavior.

REFERENCES

Gibaldi, M., and D. Perrier. 1975. Pharmacokinetics. P. 175. New York: Marcel Dekker.

Porte, D., Jr. and R. J. Havel. 1961. The use of cholesterol-4-C^{14}-labeled lipoproteins as a tracer for plasma cholesterol in the dog. J. Lipid Res. 2:357–362.

Physiologically Based Pharmacokinetic Modeling

Kenneth B. Bischoff

INTRODUCTION

Pharmacokinetic models are used to permit the rational prediction, as much as possible, of the events occurring during the processes of drug disposition throughout the body, thus yielding tissue levels. As will be seen, it appears that it is feasible to do this with certain parts of the overall problem, although other aspects are more elusive.

There are several specific reasons for pursuing this approach. One is the scientific intellectual satisfaction of having quantitative predictive models based on underlying knowledge, rather than the more empirical, curve-fitting approaches often used. The latter are always needed to some extent, of course, but should be minimized if possible. Another important purpose is to aid in the constant problems of interpreting animal experiments in drug screening, dosage regimen formulation, and similar matters. In quantitative terms this can be called *scaling* the results from one species to another, and ultimately to man (Dedrick, 1973a). It seems clear that it would be of benefit to have pharmacokinetic models that specifically incorporated known animal physiology and pharmacology in these endeavors. Yet another use of quantitative predictive models is in the development of optimal dosage regimens for clinical applications.

This chapter is primarily intended to document this approach which, as usual, is in many scattered publications. It is hoped that other investigators will find it useful to have these ideas, methods, formulations, and basic data gathered in one place.

The history and bases of physiological pharmacokinetics will be briefly reviewed, and some misconceptions will be pointed out, e.g., that membrane transport cannot be incorporated into these models and that only the flow-limited case can be handled. Several recent literature reviews will be given for those readers wanting further details on the modeling and/or specific drugs. This will be followed by a brief description of a few examples, and the chapter will conclude with my views of the most useful future research in the area.

The basic idea of physiological pharmacokinetics was to extend pharmacokinetic modeling so that quantitative aspects of other biological areas could be incorporated. For example, this includes what is known about physiological differences and similarities among species, membrane biophysics, biochemical kinetics, and others to be illustrated later. The approach will be to focus the models on anatomically real local tissue regions, including their blood flow, binding, and transport characteristics. Certain aspects are similar to the compartmental modeling methods of mathematical biology (see, e.g., Rescigno and Segre, 1966, or Riggs, 1970) or of what will be termed *classical pharmacokinetics*, which is primarily concerned with the prediction of blood levels of various dosage regimens (see Gibaldi and Perrier, 1982, for a comprehensive treatment).

Often, however, these compartments were rather abstract mathematical constructs, whose number and properties were only able to be ascertained by curve-fitting of experimental blood sample data. Useful insights into the quantitative operation of the body were obtained, although specific organ levels were usually not considered. Physiological pharmacokinetics, however, also attempts to predict the various organ and tissue levels, even extra- versus intracellular concentrations.

This concept of utilization of known anatomical and physiological functions as a basis for pharmacokinetic models was proposed early on by Teorell (1937). This remarkable work was not able to be fully utilized, however, because of the lack of reasonable computing capabilities. When computing capabilities became feasible, the number of differential equations that needed to be solved in comprehensive models was not of crucial importance, and multicompartment models based on known physiology were formulated by Bischoff and Brown (1966). The basis was to use a compartment as an actual local tissue region, as proposed by Bellman et al. (1960), although the term *physiological pharmacokinetics* was not used until about 1973 by Dedrick (1973b).

The philosophical basis of the present approach resides in chemical engineering modeling and design, in which several of the problems, such as combined flow, diffusion, and chemical reactions, are similar to the present problem (see Himmelblau and Bischoff, 1968).

BIOLOGICAL BASIS OF PHYSIOLOGICAL PHARMACOKINETICS

There are many similarities in the anatomy and physiology of mammalian species, and a general belief in this similarity has been the cornerstone of most biomedical research. Mammalian species share a remarkable geometric similarity. The same blood flow diagram could be used for all mammals, and most organs and tissues are similar fractions of the body weight. Major qualitative differences, such as the absence of a gallbladder in some species, are the exception.

In a classic article, Adolph (1949) summarized an orderly variation of numerous anatomic and physiologic properties with body weight. Many physiologic processes vary as the 0.7–0.8 power of body weight, and the anatomic variables show a more nearly first-degree dependence on body weight. The result of this is that the physiologic process per unit of body weight or per unit of organ weight tends to decrease as body size increases, although blood perfusion (in milliliters per minute gram of tissue) may only vary by a factor of 2 across a variety of species. It is well known, for example, that the mouse has a heart rate and a relative cardiac output about an order of magnitude higher than those of a human. This does not pose any theoretical limitation to the use of the mouse as a model for cardiovascular dynamics; however, it does require that appropriate time scaling be done if the results are to be generalizable.

Purely physiochemical interactions of exogenous chemicals with biological tissues and fluids might be expected not to show a great variation among species. We reported a pharmacokinetic model for thiopental (Bischoff and Dedrick, 1968) in which data from experiments as diverse as peanut oil-water distribution ratios and binding to bovine serum albumin and rabbit tissue homogenates were used to predict tissue and blood levels in the dog and the human.

The most significant species differences that can confound pharmacokinetic predictability are in the qualitative pathways and kinetic characteristics of metabolism. As discussed by Williams (1974), foreign organic compounds tend to be metabolized in two phases. Phase I reactions lead to oxidation, reduction, and hydrolysis products. Phase II reactions lead to synthetic or conjugation products that are relatively polar and thus more easily excreted by the kidney in the urine and, in some cases, the liver in the bile. Within this general framework, however, there are large species variations. Williams points out that species variations in phase I reactions are very common and often appear to be unpredictable. If a species difference is found for a particular compound, similar compounds may show similar variations. Phase II reactions are much more limited in number than phase I reactions, and it may be possible to identify patterns of these.

Krasovskii (1976) reviewed the hepatic microsomal enzyme activities of 15 enzymes in several mammalian species. He observed that all except phosphatase tended to decrease (in units of activity per kilogram of body weight) as body weight is increased. Despite these observations, it does not appear safe to make generalizations concerning rates of unknown metabolic reactions. Many exceptions could be found to the apparent tendency of the intrinsic rate to decrease with increasing body size; quite different qualitative pathways can dominate in different species, and toxicity sometimes can correlate with the concentration of an active intermediate that represents only a minor elimination pathway. Further aspects of these species similarities have been given by Dedrick and Bischoff (1980).

DEVELOPMENT OF MODELS

A comprehensive discussion of the details of constructing these pharmacokinetic models has been provided by Bischoff (1975). There are two parts to this: choice of body regions and compartments, and the formulation of the proper mathematical relations to represent the drug transport, clearance, etc., in these compartments. The second of these is the more straightforward, being based on the laws of physical chemistry and biology, for example; but, of course, we by no means have a satisfactory, complete quantitative description in hand. In fact, because of individual genetic differences, especially in the human population, we may never be able to precisely define all of the required parameters, as discussed by Gillette (1985). The first aspect of constructing these models, however, requires even more judgment as to what are the important features that must be considered for a given problem in pharmacology or toxicology.

CHOICE OF COMPARTMENTS

The natural basis for the choice of compartments is the anatomy and physiology of the body, from the cellular level to the whole body. It is clear that the main question is how much detail needs to be used to provide an adequate description of the events. Even though consideration of these events in individual cells throughout the whole body is not feasible, special collections of cells (e.g., tumors) may have to be modeled in this much detail. The main features of drug distribution, however, can often be described with models that have surprisingly little detail. Thus, certain parts of the body can be lumped together (e.g., an organ) and described by a single concentration level. We will often term these parts *body regions*, with a definite anatomic and physiologic basis, to avoid confusion

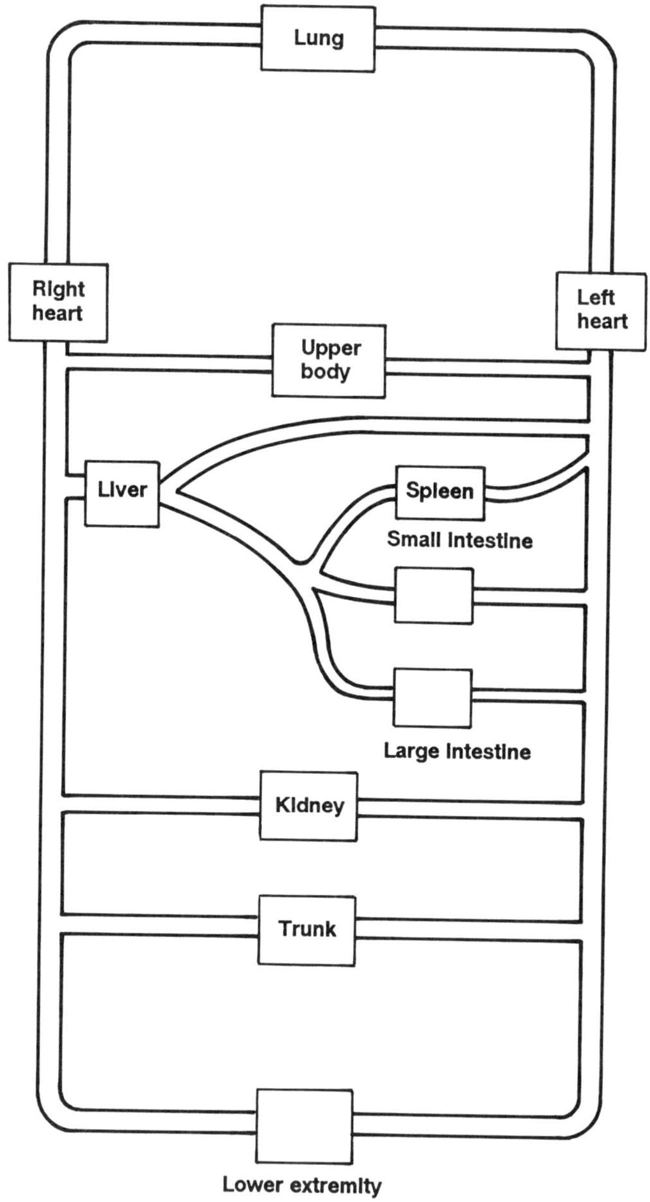

FIGURE 1 Flow diagram for mammals.

with the term *compartment*, which is often used in a more abstract sense. A general flow diagram for mammals is shown in Figure 1.

This still does not resolve the question of exactly how many body regions, or compartments, are needed. In fact, there is no simple way to decide this, because judgment is required as to the important aspects of the drug distribution events. We base the initial choice(s) on whatever is known about the physiochemical (binding, lipid solubility, ionization) and pharmacologic (mechanisms of transport, site[s] of action) properties of the drug. For example, if the drug is not lipid soluble, the details of the adipose tissues of the body are not particularly important.

One of the most difficult aspects concerns the importance of considering membrane resistances for adequate pharmacokinetic results (see Lutz et al., 1980). The main problem is the lack of knowledge of membrane transport for almost all drugs. The processes are usually agent, species, and tissue specific and often include saturable and/or active transport steps; and not much is known about generally useful models. Experimentation is difficult and tedious, leading to slow progress. Once again, however, there are some commonalities that can be used, and some workable models will be described later in this chapter. Because of these difficulties, it is often *expedient* to assume that the membrane resistances are not rate controlling, and the dominant variable is how much drug is presented to the tissue by the blood flow—the so-called flow-limited or perfusion models. It is most important, however, to state that this approximation is *not* fundamental to the physiological pharmacokinetic approach, and is commonly used merely for the lack of sufficient information.

Another basis for the initial determination of the number of required body regions is the speed or the time scale of events. Perhaps the simplest example of this is to consider the necessity of including the finite time of passage around the circulatory system (in man this is about 1 min). Thus, if one is attempting to describe tracer concentrations in an indicator-dilution experiment, the observations of interest are obviously occurring with a time scale the same as that of the mean circulation time, and the details of the transit times in the arteries and veins must be accounted for. (It should be mentioned here that if only an overall property [such as the total cardiac output] of the tracer concentration curves is of interest, then this detail may not be required.) Pharmacokinetic models containing this amount of anatomic geography are rather complicated, as is illustrated in Figure 2; the numbers in the compartments are estimates of the various volumes (in cubic centimeters) of the capillary, interstitial, and intracellular regions (from Bischoff, 1967).

The time scale of interesting events for most drugs is usually many minutes, hours, days, or even longer. After an hour, about 60 circulations have elapsed, and this is more than adequate to have ''mixed'' the drug such that blood in the circulatory system has an essentially overall uniform concentration of drug, even though there could be local arterial-venous

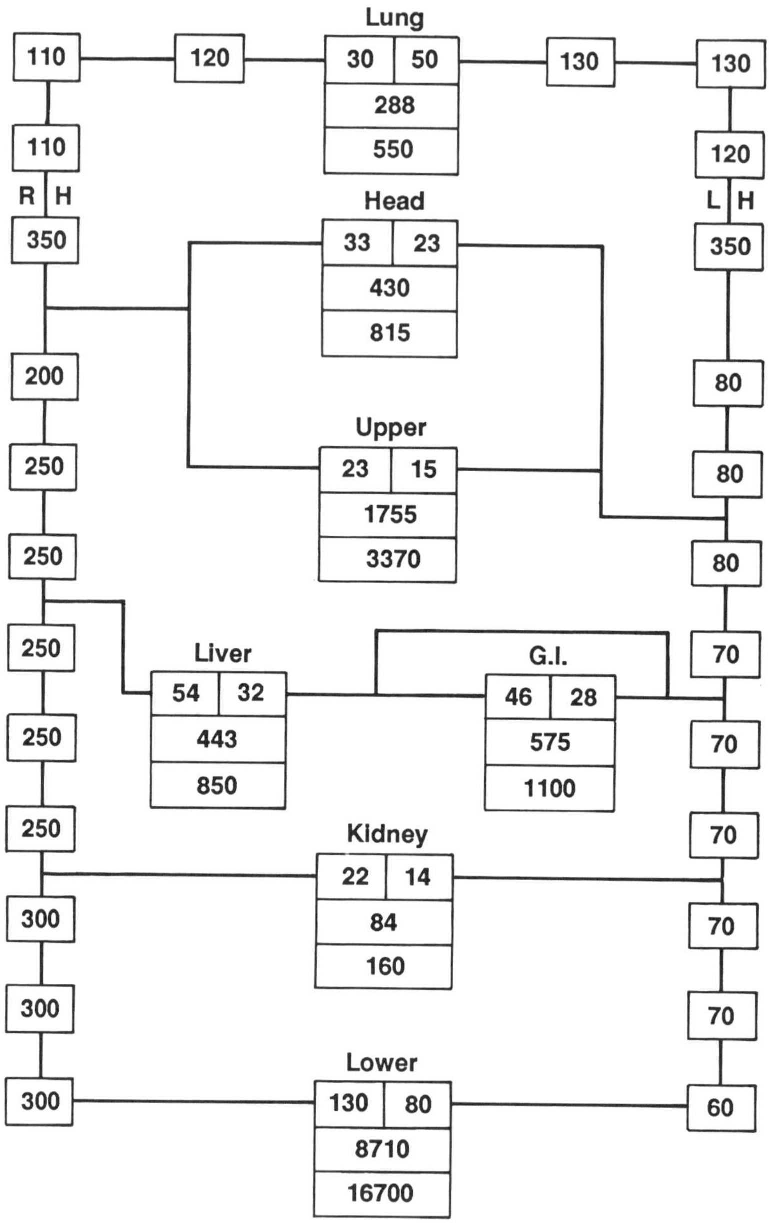

FIGURE 2 Flow diagram for disposition of rapidly acting substances—anesthetic agents or tracers (numbers are volume estimates, in cubic centimeters).

differences across certain organs. For a bolus injection, in fact, this is a good approximation after about three transient times. Thus, most pharmacokinetic models for drugs ignore the very early time events and lump the entire circulatory system into one blood pool. If the local (physiological) behavior is of interest, this approximation is probably not suitable. Also, to model very fast acting drugs, such as anesthetic agents, more detailed pharmacokinetic models than those normally used may be required. For most of the discussion here, however, only relatively long time scales will be considered.

If one carries this lumping to the extreme, the entire body can be assumed to have uniform (water) concentrations of the drug, with no detailed distinction between body regions. What is actually being assumed here will be discussed later, but pharmacologists know from experience that this approach often leads to reasonably accurate descriptions of the drug disposition. From the above discussion, one can reason that lumping is likely to be most useful for a drug that, for example, is rather slow acting, is primarily water soluble, and has no complicated transport mechanisms. Empirical evidence of this is that a plot of log concentration versus time gives a single straight line for a bolus injection. The mathematical equations resulting from this simple model are easy to manipulate and are the basis for most of the current detailed calculations of optimal dosage regimens by, for example, Krüger-Thiemer (1968, 1969).

However, it is probably more common to find that a semilog plot of blood (or plasma) concentration versus time has two rather pronounced phases with different slopes. The first, or earliest, time phase (sometimes denoted by α) is dominated by the distribution of drug between the central and peripheral parts of the body, and the second phase (or β) is dominated by the ultimate drug removal process(es). Riegelman and Rowland (1968a,b) have explored extensively the use of this two-compartment model for many specific drugs, and they have discussed the proper evaluation of the various parameters. A detailed analysis of this model was given by Bischoff and Dedrick (1970), and Wagner (1971) has provided a compendium of the detailed mathematical features of a collection of these and similar models.

Even though the biphasic model can often give a reasonable fit to pharmacokinetic data, the specific interpretation of the model parameters, such as that of the central compartment, is often ambiguous. Thus, one of the original goals of having a predictive model incorporating physiological and pharmacological data is not entirely met by this approach. It should be emphasized, though, that for semiempirical projections of clinical data, these models are often most convenient. For our purposes, however, we still wish to construct the pharmacokinetic models from (close to) first principles.

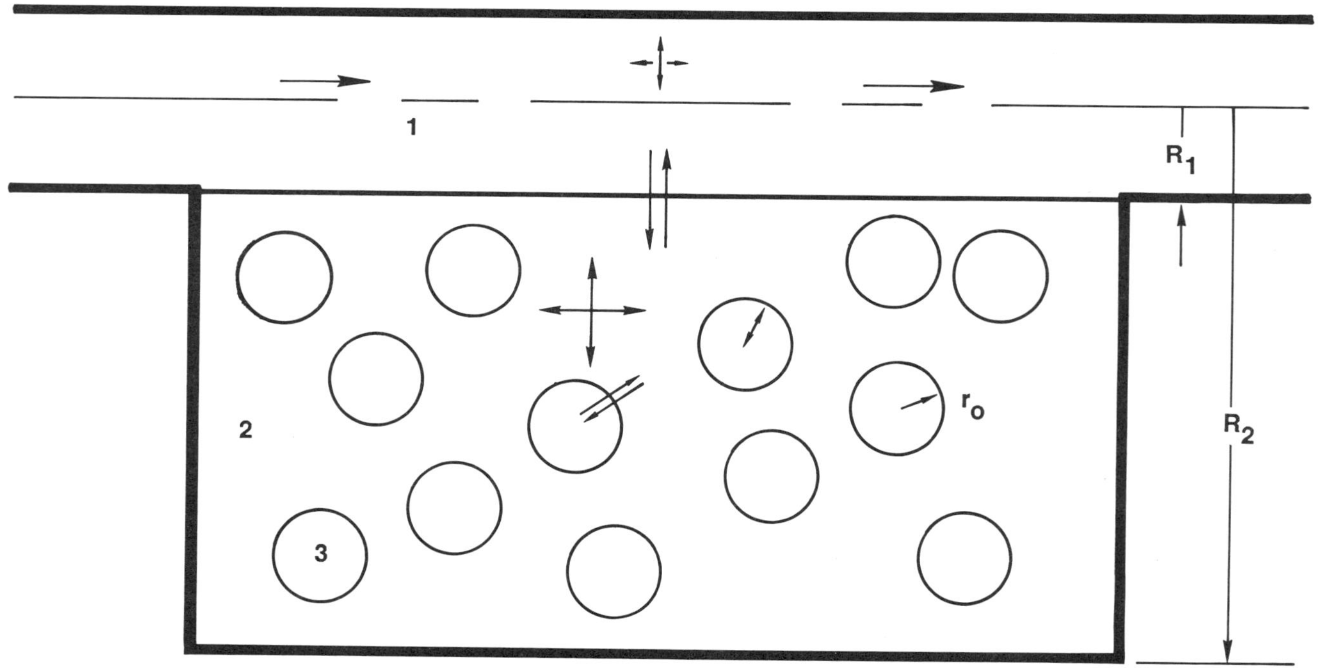

FIGURE 3 Schematic diagram of tissue regions, including intracellular and interstitial fluids.

The above matters are the most intellectually difficult on the theoretical side, because no one set of specific rules can enable us to automatically choose the appropriate compartments. The writing of the mathematical equations, and their use with experimental data, is more straightforward, although actually doing this may be equally difficult and certainly more time-consuming. I will now outline the modeling details to illustrate the concepts in a concrete manner. I will often refer to, or imply, the above general discussion in constructing the models, and thereby clarify the philosophy of the approach.

First, consider a body region consisting of cells randomly distributed in the interstitial fluid and supplied by a typical capillary, as shown in Figure 3. Even this may not be truly applicable to organs with a specific structure, such as the liver, but more appropriate models are only now being developed (see, e.g., Roberts and Rowland, 1986). Mass balances for the drug in each of the three components could be written, including three-dimensional diffusion, any flows of the fluids, membrane transport, etc., but we normally have nowhere near the amount of information to justify this. Therefore, some approximate form of these balances is used for a capillary, and these are then combined according to the architecture of the capillary bed, e.g., the Krogh tissue cylinder (see Leonard and Jorgensen, 1974, for a review of many of these approaches and a rather complete citation of the literature).

For drug pharmacokinetic models, this model of a body region is even further lumped, as shown in Figure 4, based on what is normally done in an experimental biopsy. Examples of the use of these conceptual diagrams will be given below.

BASIC MASS BALANCES

All of the above discussions and developments will now be incorporated into mass balance equations of various degrees of completeness that can be used in actual pharmacokinetic models. Only the fully lumped models for body regions will be considered, but the physiochemical complications will be retained. A complete set of anatomically and physiochemically complicated mass balances could be formulated, but these are probably much too involved for any practical use at this time.

The balances will explicitly account for free un-ionized, ionized, and bound forms of the drug, assuming that the ionized and bound forms are in instantaneous dynamic equilibrium with the free concentration. This seems to be a reasonable approximation right from the start, since the ionization and binding processes are normally very rapid.

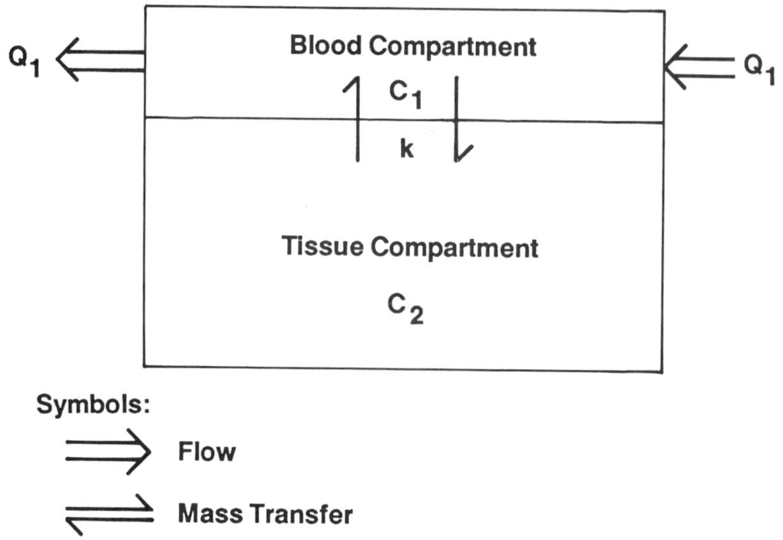

FIGURE 4 Typical lumped tissue region.

Mass Balance: Blood Pool

$$\underset{\text{(unbound drug)}}{w_B V_B \frac{dC_B}{dt}} + \underset{\text{(ionized drug)}}{w_B V_B \frac{dC_B{}^*}{dt}} + \underset{\text{(bound drug)}}{p_B V_B \frac{dx_B}{dt}}$$
$$= \underset{\text{(flow out to organs)}}{-Q_B(w_B C_B + w_B C_B{}^* + p_B x_B)}$$
$$+ \underset{\text{(flow in from organs)}}{\Sigma_i\, Q_i(w_{B_i} C_{B_i} + w_{B_i} C_{B_i}{}^* + p_{B_i} x_{B_i})} + \text{(injections)}. \tag{1}$$

Each of the terms in Equation 1 should be clear. To use the balance, the relationships between ionized and free drug concentrations, C^* (C, pH, . . .), must be known from the ionization equilibrium constant, and likewise the bound drug concentration, x (C, . . .), from Equation 2.

$$x = \sum_i \frac{S_i K_{A_i} C_A}{1 + K_{A_i} C_A}, \tag{2}$$

where S_i is the number of binding sites of type i. Equation 2 is valid for whole blood for a solute that instantaneously equilibrates with erythrocytes or for plasma when the solute is excluded from the erythrocytes. Cases with an intermediate degree of cell uptake are usually handled semiempirically.

The tissue regions will be described as above in the section "Choice of Compartments."

Mass Balance: Tissue Region *i*

For equilibrium blood:

$$V_{B_i}\frac{dq_{B_i}}{dt} = Q_i(q_B - q_{B_i}) - (jA)_{BI,i}, \tag{3}$$

where q_B is total concentration of drug and is equal to $w_B C_B + w_B C_B^* + p_B x_B$, $(jA)_{BI,i}$ is the membrane flux-area product and is equal to $PA(C_{B_i} - C_{T_i})$ for simple passive transport of free drug.

For interstitial fluid:

$$V_{I_i}\frac{dq_{I_i}}{dt} = (jA)_{BI,i} - (jA)_{IC,i}. \tag{4}$$

For intracellular fluid:

$$V_{C_i}\frac{dq_{C_i}}{dt} = (jA)_{IC,i} + \text{reaction, secretion, etc.} \tag{5}$$

In the above equations, the symbols are defined as follows: C, free (unbound) un-ionized concentration; C^*, free ionized concentration; x, bound concentration; V, volume; Q, flow rate; w, fraction of water; p, protein concentration; P, membrane permeability.

The combination of Equation 1 and Equations 3 to 5 for each of the appropriate body regions, together with Equation 2 and the ionization equilibria, would permit a solution for all of the various concentration levels. To actually do this, of course, numerical values of all the parameters must also be known. Rather good estimates are possible for the various organ volumes and flows, and plasma protein binding can be measured in *in vitro* experiments, as can ionization. The most difficult to obtain are tissue binding and the specific membrane permeabilities.

SIMPLIFICATIONS OF MASS BALANCES

Since the membrane permeabilities are not commonly available, Equations 3 to 5 given above are usually simplified into blood and tissue regions, also corresponding to usual biopsy methods. Then, Equations 4 and 5 are combined to give the following for tissue mass balance:

$$V_{T_i} \frac{dq_{T_i}}{dt} = (jA)_{BI,i} + \text{reaction, etc.} \tag{6}$$

where

$$V_{T_i} q_{T_i} = V_{I_i} q_{I_i} + V_{C_i} q_{C_i}. \tag{6a}$$

This, of course, only gives an approximation to the true intracellular concentration, unless the cell membrane permeability is very large (the mass transfer resistance is very low).

To completely get around this problem of membrane permeabilities, the concept of *flow-limited conditions* is most often introduced. Physically, this means that the membrane permeability is so large that any molecules that flow into the region have an easy time moving throughout the space. Thus, the part of the process that determines the drug concentration level is how much drug is able to flow in from the blood pool. A corollary is that the free concentration is essentially identical throughout the region, since this is the driving force for mass transfer across the membrane, for simple passive diffusion.

Mathematically, a very large permeability in Equations 3 and 6 is $PA \to \infty$, and $C_{B_i} = C_{T_i} \equiv C_i$ (equal free concentrations). Addition of Equations 3 and 6, then, eliminates the mass transfer term, and the result is:

$$(w_{B_i} V_{B_i} + w_{T_i} V_{T_i}) \frac{dC_i}{dt} + p_{B_i} V_{B_i} \frac{dx_{B_i}}{dt} + p_{T_i} V_{T_i} \frac{dx_{T_i}}{dt}$$
$$= Q_i (w_B C_B + p_B x_B - w_{B_i} C_i - p_{B_i} x_{B_i}) + r_i(C_i). \tag{7}$$

(The ionized moiety terms have been dropped for simplicity.) Thus, one equation results for each tissue region, but more importantly, the exact value of PA need not be known, just that it is very large.

Dedrick and Bischoff (1968) have shown that the criterion for flow limiting conditions is $(PA)_i/Q_i >> 1$. This relation is about what one would expect; that is, if the rate of mass transfer is much greater than the regional perfusion, one can assume that the former is relatively very rapid and that the only important resistance to transport is the flow limitation.

Some knowledge of the permeability is still required to use the criterion, and such data are scarce. Dedrick and Bischoff (1968) provided some information on how these can be estimated, and Lutz et al. (1980) have given specific values. Data from biological handbooks could also be recast into an appropriate form for the use of this criterion. Despite this there definitely is still a paucity of information and more studies are needed.

Flow-limited conditions will usually be assumed, but the true validity of this often cannot be assessed.

The opposite extreme of having the membrane permeability be the dominant mass transfer resistance has the implication that the blood concentration in the region is constant; there is essentially no concentration gradient across the region because more than enough drug is supplied by the blood flow. Mathematically, Equation 3 basically gives $C_{B_i} = C_B$, and then Equation 6 reduces to:

$$w_{T_i} V_{T_i} \frac{dC_{T_i}}{dt} + p_{T_i} V_{T_i} \frac{dx_{T_i}}{dt} = (PA)_i(C_B - C_{T_i}) + r_i(C_{T_i}). \qquad (8)$$

For no binding (or linear binding, see below), Equations 7 and 8 have essentially the same form as a function of C_B, and so the forms of the mathematical solutions would also be the same. With strong (nonlinear) binding, there could be some differences between the equations representing the two extreme limiting cases, but it is not clear how easily these differences might be observed. Thus, it is of interest that both limits have the same form of mathematical equation, except that the parameters have quite different physical meanings.

For flow-limited models, the most important property is the perfusion of a tissue. This then leads to four main compartments: the blood pool, highly perfused viscera, low-perfusion lean tissue, and low-perfusion adipose tissue. The last two would be kept separate for lipid-soluble drugs because of the greatly different physicochemical nature of the two types of regions. For primarily water-soluble drugs, the two low-perfusion regions could be combined.

Given the parameter values, sets of mass balances like the various combinations given above can be readily solved for the important interconnected body regions. Naturally, this question of parameter values is the crucial one, but it is also a key advantage of the physiological pharmacokinetics approach. Most of the parameters in the above equations are either known as average values for many animal species (e.g., volumes and flows) or can be measured in separate experiments (e.g., protein binding, although characterization of tissue binding is still difficult; see Jusko and Gretch, 1976, and Shen and Gibaldi, 1974). For a comprehensive discussion of scaling see Dedrick (1973), and for hemodynamic considerations see Wilkinson (1975).

Further predictions are possible for special cases. If the definition of tissue total concentration (Equation 6a) is combined with the equilibrium blood total concentration, the combined tissue level, as used in the flow-limited models, is obtained: $q_i = [(w_{B_i} V_{B_i} + w_{T_i} V_{T_i}) C_i + p_{B_i} V_{B_i} x_{B_i} + p_{T_i} V_{T_i} x_{T_i})]/(V_{B_i} + V_{T_i})$. If this is substituted into the complete mass balance for this situation, Equation 7, the following simple form is obtained (see Bischoff, 1975, for details):

$$V_B \frac{dq_B}{dt} = -Q_B q_B + \sum_i Q_i \frac{q_i}{R_i} + \text{injection, and} \qquad (9)$$

$$(V_{B_i} + V_{T_i}) \frac{dq_i}{dt} = Q_i\left(q_B - \frac{q_i}{R_i}\right) + r_i(\underset{\uparrow}{C_i}), \qquad (10)$$

where R_i = tissue/blood distribution ratio at equilibrium.

In these equations, R_i is equal to a constant only for linear binding, $x = KC$. It should be emphasized that Equations 9 and 10 involve no further assumptions beyond that of flow-limited conditions, which is really the only situation in which the total tissue concentration has a unique meaning. Equations 9 and 10 are valid for either a variable or constant R_i, although they are much more useful only in the latter case. The arrow under

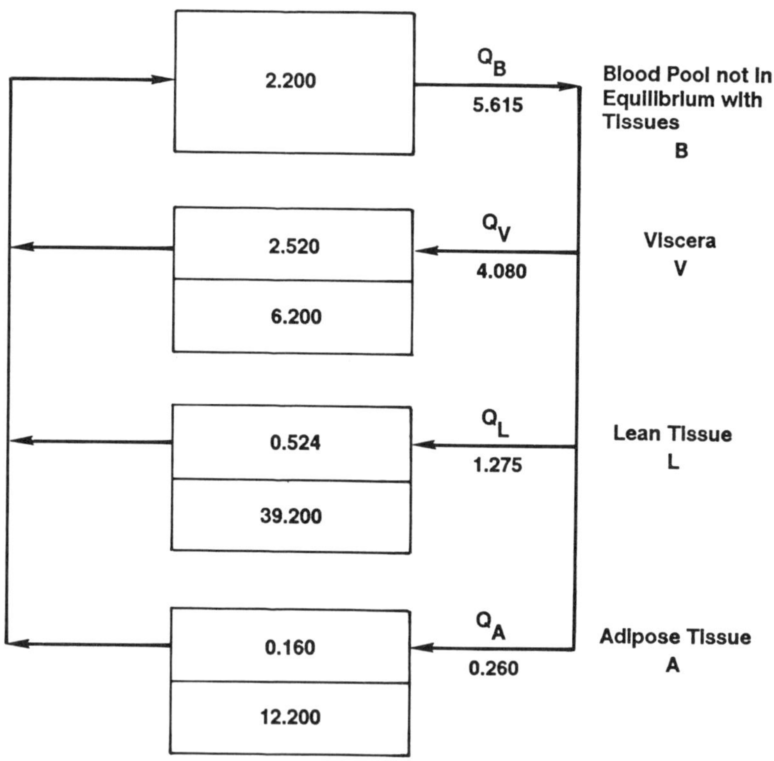

FIGURE 5 Model for thiopental pharmacokinetics.

the metabolic rate term indicates the question that the enzyme reactions presumably depend on the free concentration, and for nonlinear binding this has no simple relationship to the total concentrations in the rest of the mass balance. The practical importance of ignoring this and just redefining the rates in terms of q_i is not presently known.

EXAMPLES

Several sources of reviews and literature references now exist, and so only a brief set will be presented here. The second edition of the excellent book by Gibaldi and Perrier (1982) has a chapter, and other recent review articles have been presented by Himmelstein and Lutz (1979) and by Chen and Gross (1979); a comprehensive collection of drugs modeled by this approach is given by Gerlowski and Jain (1983). These provide a wealth of examples and sources of parameter values and other data, as do the usual physiological and pharmacological handbooks.

One of the first drugs simulated by us was thiopental (Bischoff and Dedrick, 1968). Figure 5 is the lumped, longer time scale model used, with an important compartment being the adipose tissue because the drug is highly lipid soluble. This drug is also strongly and nonlinearly bound, and so the detailed mass balances (Equations 1 and 7) were used, along with some approximate relations for the tissue binding. Figure 6 is one illustration of the success of the physiological model in simulation of the thiopental levels in the blood of humans and dogs. Similar comparisons for other tissues have been shown by Bischoff and Dedrick (1968). Bischoff et al. (1971) considered in some detail the pharmacokinetics of methotrexate. Because this drug appears to be essentially linearly bound for higher concentrations (an additive saturable binding term is needed for low concentrations), the simplified Equations 9 and 10 could be used in conjunction with the flow diagram shown in Figure 7. For low concentrations, the concentration-dependent term $R_i(C_i)$ was used. The mass balance equations are given below.

For plasma:

$$V_p \frac{dC_p}{dt} = (\text{injection}) + Q_L \frac{C_L}{R_L} + Q_K \frac{C_K}{R_K} + Q_M \frac{C_M}{R_M} - (Q_L + Q_K + Q_M)C_p. \tag{11a}$$

For muscle:

$$V_M \frac{dC_M}{dt} = Q_M\left(C_p - \frac{C_M}{R_M}\right). \tag{11b}$$

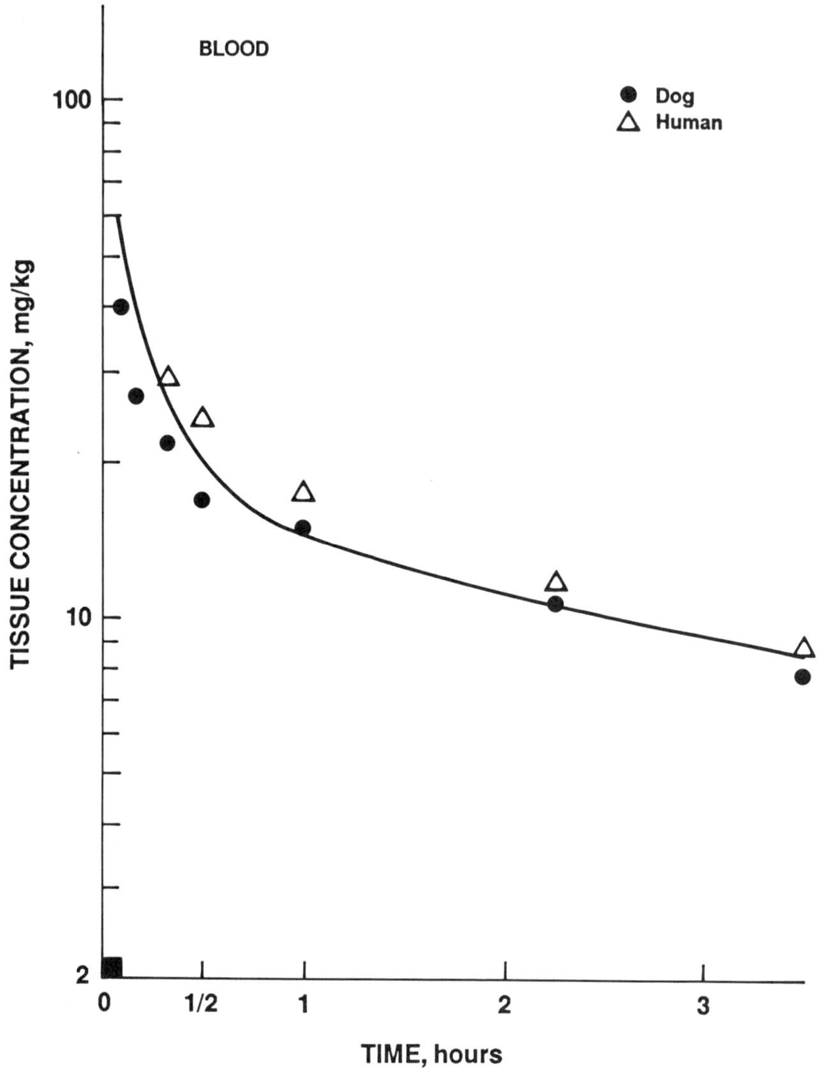

FIGURE 6 Comparison of data with model predictions for thiopental.

For kidney:

$$V_K \frac{dC_K}{dt} = Q_K\left(C_p - \frac{C_K}{R_K}\right) - k_K \frac{C_K}{R_K}. \tag{11c}$$

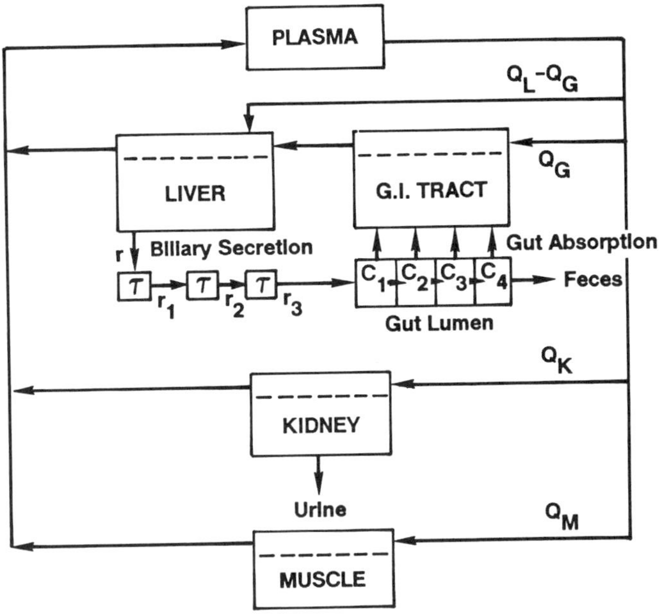

FIGURE 7 Model for methotrexate pharmacokinetics.

For liver:

$$V_L \frac{dC_L}{dt} = (Q_L - Q_G)\left(C_p - \frac{C_L}{R_L}\right) + Q_G\left(\frac{C_G}{R_G} - \frac{C_L}{R_L}\right) - r_0, \quad \text{(11d)}$$

where $r_0 = [K_L(C_L/R_L)]/[K_L + (C_L/R_L)]$.

For bile ducts:

$$\tau \frac{dr_i}{dt} = r_{i-1} - r_i, \ (i = 1, 2, 3). \quad \text{(11e)}$$

For gut tissue:

$$V_G \frac{dC_G}{dt} = Q_G\left(C_P - \frac{C_G}{R_G}\right) + \sum_{i=1}^{4} \frac{1}{4}\left(\frac{k_G C_i}{K_G + C_i} + bC_i\right) \cdot \quad \text{(11f)}$$

For gut lumen:

$$\frac{dC_{GL}}{dt} = \frac{1}{4} \sum_{i=1}^{4} \frac{dC_i}{dt}, \quad \text{(11g)}$$

$$\frac{V_{GL}}{4} \frac{dC_1}{dt} = r_3 - k_F V_{GL} C_1 - \frac{1}{4}\left(\frac{k_G C_1}{K_G + C_1} + bC_1\right), \text{ and} \quad \text{(11h)}$$

$$\frac{V_{GL}}{4}\frac{dC_i}{dt} = -k_F V_{GL}(C_{i-1} - C_i) - \frac{1}{4}\left(\frac{k_G C_i}{K_G + C_i} + bC_i\right), (i = 2, 3, 4). \tag{11i}$$

Most of the terms and their origins should now be clear. The new ones are as follow. In Equation 11c, the term $k_K C_K / R_K$ represents the renal excretion of methotrexate, which is close to the glomerular filtration rate; in Equation 11d–e, r_i is the bile flow of the drug, with the liver secretion being saturable at high doses (the three compartments in series represent the actual tubular or distributed nature of the real system); in Equation 11f–i, the motion down the gut lumen is modeled by four compartments, and the absorption term has both a saturable and a nonsaturable component that is important only for very high doses, presumably passive diffusion.

The various parameters in Equation 11a–i were either estimated from the values given earlier or were independently measured in the case of the complex secretion steps. Then, the pharmacokinetic behavior was able to be predicted in mice, rats, dogs, monkeys, and man over a dose range of 3,000, all with the same model and with a consistent set of parameters. Two examples are shown in Figures 8 and 9. Thus, the details of this model must be a reasonably faithful representation of the actual physiological and pharmacological events, and should be of aid in interpreting results of experiments. Also, valid predictions of local drug concentrations for various dosage regimens are possible.

The flow diagram in Figure 7 has a rather complicated configuration because of the importance of the enterohepatic cycle in the methotrexate pharmacokinetics. There was little direct metabolism of the drug, however, so metabolism is not a major route of elimination; the ultimate excretion was by the renal or the fecal route. Another example considers the opposite extreme, in which a straightforward anatomic flow diagram is the basis but the metabolism is dominant. Dedrick et al. (1972) have considered the drug cytosine arabinoside on the basis of the compartments seen in Figure 10. The sizes of the boxes in Figure 10 signify the relative importance of the various regions. Again, the same types of balances were used, but with metabolic terms in each based on known enzyme kinetics and levels. Figure 11 shows one prediction of the concentrations of cytosine arabinoside and its metabolite, uracil arabinoside.

A final reduction in the complexity of the models is possible when the excretion/metabolism processes are relatively slow compared with the intercompartment blood flows. In this case, the entire body has an essen-

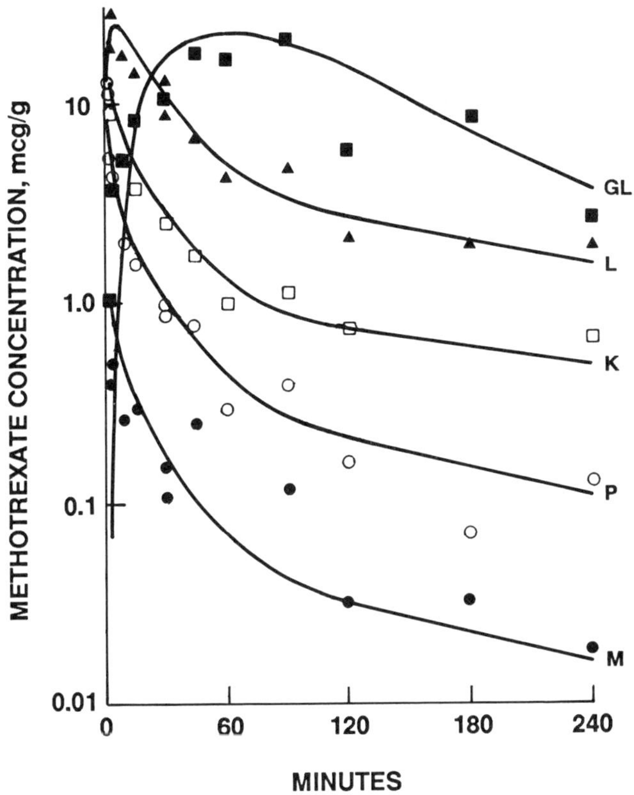

FIGURE 8 Comparisons of data with model predictions for methotrexate. Mice, 3 mg/kg. Abbreviations: GL, small intestine; L, liver; K, kidney; P, plasma; M, muscle.

tially identical time response, and a one-compartment whole-body model is useful. In terms of Equations 9 and 10 this implies that:

$$r_1 << Q_1 \rightarrow \infty, \; q_B \cong \frac{q_1}{R_1}, \tag{12}$$

and then the addition of all the tissue mass balances to that for blood gives:

$$[V_B + \sum_i (V_{B_i} + V_{T_i}) R_i] \frac{dq_B}{dt} = \Sigma \text{ (injections)} + \sum_i r_i(C_i). \tag{13}$$

The term in brackets would be the theoretical basis of the volume of distribution. An alternate form can be derived from Equations 1 and 7 in terms of free concentrations. These types of equations are very familiar

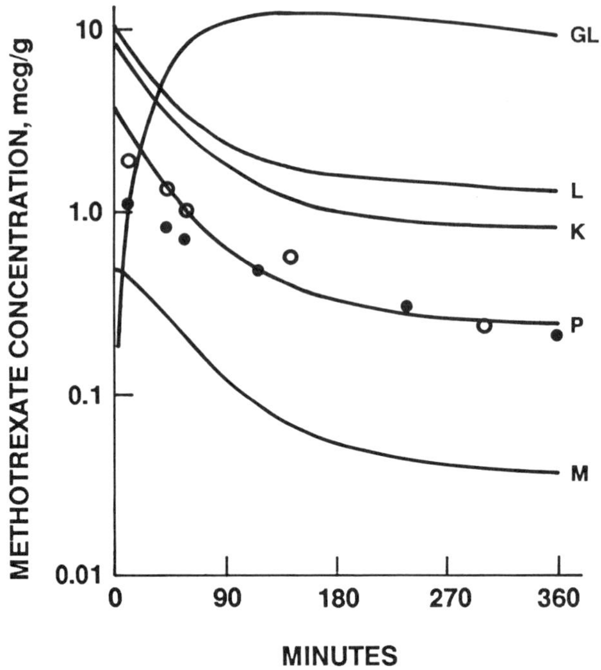

FIGURE 9 Comparison of data with model predictions for methotrexate. Man, 1 mg/kg.

in pharmacology, although they are usually used with empirical parameters, and therefore, no specific examples will be given.

DISCUSSION

Various special situations can require the use of combinations of all of the types of mass balance equations given above. The examples showed the type of reasoning used in several instances, although the flow-limited case was used in all of them. It appears that this quite often gives a good estimate of the overall drug concentrations throughout the body, even though the drug may be membrane limited in certain specific organs. Thus, a combination of the analyses illustrated in the examples, plus the use of Equation 6 for the specific region, might be a reasonable scheme for both the overall drug distribution and the details of, for example, tumor uptake. The work of Dedrick et al. (1975) is in some ways an illustration of this.

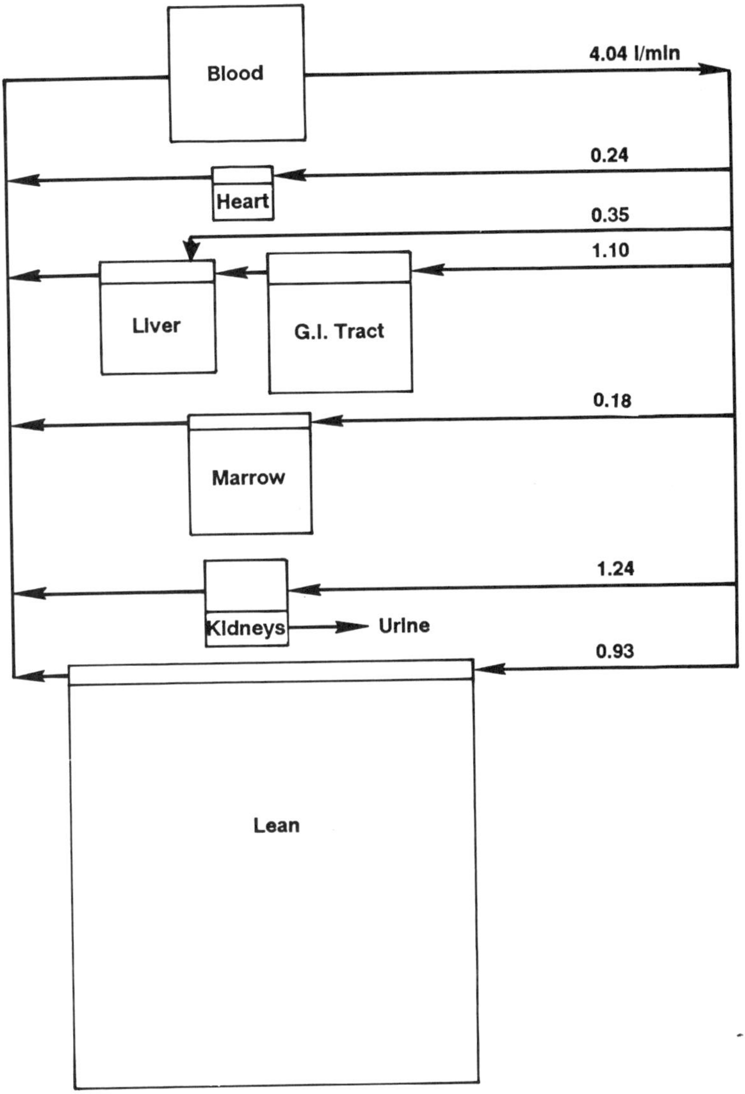

FIGURE 10 Model for cytosine arabinoside pharmacokinetics.

Finally, it should also be mentioned that the precise definition of the various anatomic regions is often somewhat flexible and can depend on the exact situation. For example, the major barrier to the transport of many drugs is not the capillary membrane but the cellular membrane; in this case, *tissue* can be defined as intracellular space, and *blood* can be defined as vascular plus interstitial space. The important point is that the physi-

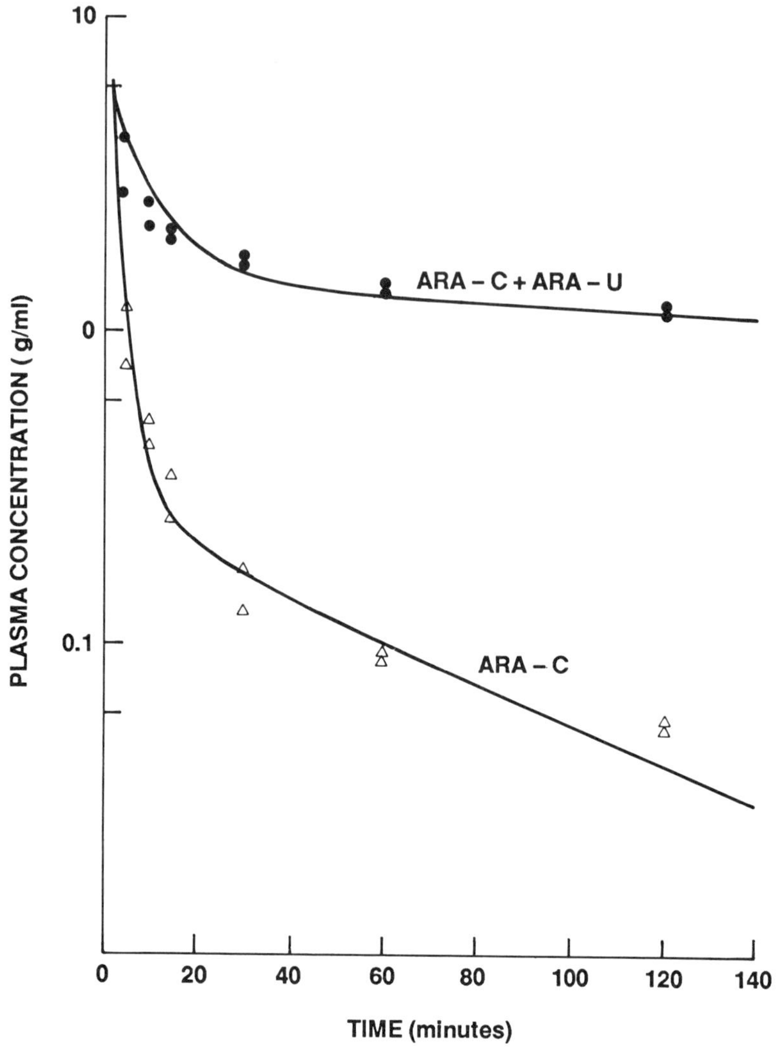

FIGURE 11 Comparison of data with model predictions for cytosine arabinoside and adenine arabinoside.

ological and pharmacological information should be used in formulating the model, so that the several goals mentioned in the introduction to this paper might be achieved. The reviews quoted above provide many examples of this; all of these are based on flow diagrams similar to those shown in Figure 7 or 10, with appropriate modifications for the specific drug.

FUTURE RESEARCH NEEDS

I would like to end with a brief discussion of my view of some of the important future research needs. The first is the use of nonlumped tissue region models, with spatial variation of concentrations, for certain critical tissues. This could be important in tubular-type regions or thick, quasi-homogeneous regions where diffusion must be accounted for other than across a thin membrane. Of course, this also includes more detailed descriptions of intracellular fluid, which are probably necessary to quantitate truly the biochemical drug effects. The above types of reaction-diffusion models lead to partial differential equations, which are more difficult mathematically, but with modern computing technology they should not cause severe problems in calculations. A related issue is the use of more realistic descriptions of the local blood flows in the microcirculation. This is not a totally new area, of course, because the Krogh tissue cylinder has been used for many years in physiology to model oxygen transport, and two papers in this volume are concerned with a very detailed distributed model of the lung to model transport and reaction of ozone (see J. H. Overton, R. C. Graham, and F. J. Miller and F. J. Miller, J. H. Overton, R. C. Graham, E. D. Smolko, and D. B. Menzel in this volume). A sort of middle ground model combining lumped compartments with distributed regions where needed was used by Flessner et al. (1984, 1985) to study peritoneal-plasma transport, and by Morrison and Dedrick (1986) to study the transport of cisplatin in the brain. The papers by Roberts and Rowland (1986) mentioned above also report comprehensive studies and refer to previous papers, one of which was an early approach by Pang and Rowland (1977) for hepatic metabolism. All this work is quite recent and has not yet been incorporated into very many pharmacokinetic studies.

A second broad area is the addition of more realistic biochemical rate expressions that involve known pathways and the like. This will virtually always lead to nonlinear terms, and so simple mathematical solutions will no longer be feasible. Finally, we must move forward with the modeling of drug effects with the same type of fundamental philosophy and combine these improved pharmacodynamic models with the pharmacokinetics to provide simulation of the actual problem: improved knowledge of tissue levels at the site of action for improved risk assessment. A few examples of this are discussed by Bischoff (1973).

REFERENCES

Adolph, E. F. 1949. Quantitative relations in the physiological constitutions of mammals. Science 109:579–585.

Bellman, R., R. Kalaba, and J. A. Jacquez. 1960. Some mathematical aspects of chemotherapy. Bull. Math. Biophys. 22:181–190.

Bischoff, K. B. 1967. Applications of a mathematical model for drug distribution in mammals. In Chemical Engineering in Medicine and Biology, D. Hershey, ed. New York: Plenum.

Bischoff, K. B. 1973. Pharmacokinetics and cancer chemotherapy. J. Pharmacokinet. Biopharm. 1:465–480.

Bischoff, K. B. 1975. Some fundamental considerations in the application of pharmacokinetics to cancer. Cancer Chemother. Rep. Part 1 59:777–793.

Bischoff, K. B., and R. G. Brown. 1966. Drug distribution in mammals. Chem. Eng. Prog. Symp. Ser. 66:33–45.

Bischoff, K. B., and R. L. Dedrick. 1968. Thiopental pharmacokinetics. J. Pharm. Sci. 57:1346–1351.

Bischoff, K. B., and R. L. Dedrick. 1970. Generalized solution to linear, two-compartment, open model for drug distribution. J. Theor. Biol. 29:63–83.

Bischoff, K. B., R. L. Dedrick, D. S. Zaharko, and J. A. Longstreth. 1971. Methotrexate pharmacokinetics. J. Pharm. Sci. 60:1128–1133.

Chen, H.-S. G., and J. F. Gross. 1979. Physiologically based pharmacokinetic models for anticancer drugs. Cancer Chemother. Pharm. 2:85–94.

Dedrick, R. L. 1973a. Animal scale-up. J. Pharmacokinet. Biopharm. 1:435–461.

Dedrick, R. L. 1973b. Physiological pharmacokinetics. J. Dyn. Syst. Meas. Cont. Trans. ASME Sept.: 255–257.

Dedrick, R. L., and K. B. Bischoff. 1968. Pharmacokinetics in applications of the artificial kidney. Chem. Eng. Prog. Symp. Ser. No. 84, 64:32–44.

Dedrick, R. L., and K. B. Bischoff. 1980. Species similarities on pharmacokinetics. Fed. Proc. 39:54–49.

Dedrick, R. L., D. D. Forrester, and D. H. W. Ho. 1972. *In vitro-in vivo* correlation of drug metabolism—deamination of 1-β-D-arabinofuranosylcytosin. Biochem. Pharmacol. 21:1–16.

Dedrick, R. L., R. L. Zaharko, R. A. Bender, W. A. Bleyer, and R. J. Lutz. 1975. Pharmacokinetic considerations on resistance to anticancer drugs. Cancer Chemother. Rep. 59:795–804.

Flessner, M. F., R. L. Dedrick, and J. S. Schultz. 1984. A distributed model of peritoneal-plasma transport: Theoretical considerations. Am. J. Physiol. 246:R597–R607.

Flessner, M. F., R. L. Dedrick, and J. S. Schultz. 1985. A distributed model of peritoneal-plasma transport; analysis of experimental data in the rat. Am. J. Physiol. 248:F413–F424.

Gerlowski, L. E., and R. K. Jain. 1983. Physiologically based pharmacokinetic modeling: Principles and applications. J. Pharm. Sci. 72:1103–1127.

Gibaldi, M., and D. Perrier. 1982. Pharmacokinetics, 2nd ed. New York: Marcel Dekker.

Gillette, J. R. 1985. Biological variation: The unsolvable problem in quantitation extrapolations from laboratory animals and other surrogate systems to human populations. Banbury Report 19: Risk Quantitation and Regulatory Policy. Cold Spring Harbor, N.Y.: Cold Spring Harbor Laboratory.

Himmelblau, D. M., and K. B. Bischoff. 1968. Process Analysis and Simulation. New York: John Wiley & Sons.

Himmelstein, K. J., and R. J. Lutz. 1979. A review of the applications of physiologically based pharmacokinetic modeling. J. Pharmacokinet. Biopharm 7:127–145.

Jusko, W. J., and M. Gretch. 1976. Plasma and tissue protein binding of drugs in pharmacokinetics. Drug Metab. Rev. 5:43–140.

Krasovskii, G. N. 1976. Extrapolation of experimental data from animals to man. Environ. Health Perspect. 13:51–58.

Krüger-Thiemer, E. 1968. Pharmacokinetics and dose-concentration relationships. Proceedings of the 3rd International Pharmacological Meeting, Sao Paulo, Brazil. Pp. 63–113 in Physico-Chemical Aspects of Drug Actions, Vol. 7. New York: Pergamon Press.

Krüger-Thiemer, E. 1969. Formal theory of drug dosage regimens. II. The exact plateau effect. J. Theor. Biol. 23:169–170.

Leonard, E. F., and S. B. Jorgensen. 1971. The analysis of convection and diffusion in capillary beds. Annu. Rev. Biophys. Bioeng. 3:293–339.

Lutz, R. J., R. L. Dedrick, and D. S. Zaharko. 1980. Physiological pharmacokinetics: An *in vivo* approach to membrane transport. Pharmacol. Ther. 11:559–592.

Morrison, P. F., and R. L. Dedrick. 1986. Transport of cisplatin in rat brain following microinfusion: An analysis. J. Pharm. Sci. 75:120–128.

Pang, K. S., and M. Rowland. 1977. Hepatic clearance of drugs. I. Theoretical considerations of a ''well-stirred'' model and a ''parallel tube'' model. J. Pharmacokinet. Biopharm. 5:625–653.

Rescigno, A., and G. Segre. 1966. Drug and Tracer Kinetics. Waltham, Mass.: Blaisdell.

Riegelman, S., and M. Rowland. 1968a. Shortcomings in pharmacokinetic analysis by conceiving the body to exhibit properties of a single compartment. J. Pharm. Sci. 57:117–123.

Riegelman, S., and M. Rowland. 1968b. Concept of a volume of distribution and possible errors in evaluation of this parameter. J. Pharm. Sci. 57:117–123.

Riggs, D. S. 1970. The Mathematical Approach to Physiological Problems. Cambridge, Mass.: MIT Press.

Roberts, M. S., and M. Rowland. 1986. A dispersion model of hepatic elimination, 1,2,3. J. Pharmacokinet. Biopharm. 14:227–260, 261–288, 289–308.

Shen, D., and M. Gibaldi. 1974. Critical evaluation of use of effective protein fractions in developing pharmacokinetic models for drug distribution. J. Pharm. Sci. 63:1698–1703.

Teorell, T. 1937. Kinetics distribution of substances administered to the body. Arch. Int. Pharmacodyn. Ther. 57:205–240.

Wagner, J. G. 1971. Biopharmaceutics and Relevant Pharmacokinetics. 1971. Hamilton, Ill.: Drug Intelligence Publications.

Wilkinson, G. R. 1975. Pharmacokinetics of drug disposition: Hemodynamic considerations. Annu. Rev. Pharm. 15:11–27.

Williams, R. T. 1974. Inter-species variations in the metabolism of xenobiotics. Biochem. Soc. Trans. 2:359–377.

PART III

Generalizations and Extrapolations

Allometry: Body Size Constraints in Animal Design

Stan L. Lindstedt

INTRODUCTION

In 1928 the English philosopher and physiologist J. B. S. Haldane wrote a captivating essay entitled, "On Being the Right Size." He began by discussing the dimensions of the giant Pope and Pagan from *Pilgrim's Progress*. If the giants were 10 times Christian's height, they would have been 1,000 (10 × 10 × 10) times his mass. Being of similar shape, Haldane concludes that the cross-sectional area of their bones would be but 100 times those of Christian. "As the human thigh bone breaks under about 10 times the human weight, Pope and Pagan would have broken their thighs every time they took a step. This is doubtless why they were sitting down in the picture I remember. But it lessens one's respect for Christian and Jack the Giant Killer" (Haldane, 1928). In the case of giants, as is a moose from the perspective of a mouse, bones are not built with geometric similarity, i.e., dimensional proportionality, rather bones of all species must be built with similarity of structural strength. This is why bones of larger animals are relatively more robust than those of smaller animals; the skeleton accounts for nearly 20% of an elephant's mass but less than 5% of a shrew's.

Perhaps no single factor is more dominant in constraining animal design than body size. Size-induced patterns have been identified for all aspects of animal design and function from structural dimensions, to life history characteristics, to pharmacokinetics. An animal's body size is certainly among it most prominent of all distinguishing features. Among the mammals, the 136,000-kg (150-ton) blue whale is 75 million times the mass of the 2-g Etruscan shrew, yet both share the same skeletal architecture, suite of organ systems, biochemical pathways, and even temperature of operation. What engineers have known for a long time, however, is only now gaining widespread consideration among biologists: there are trade-offs that must accompany size changes. The scaling up of a bridge, or a mammal, to a size 100,000 times larger than the original requires more than just the creation of bigger parts. Rather, those parts must be redesigned if they are to perform the same functions throughout a size range spanning several orders of magnitude in body mass. Further, changes in size result in shifts in optimal or preferred frequencies of use. It is becoming increasingly apparent that there are a suite of body size-dependent physical laws that dictate many features of animal design.

Size-dependent constraints of design are expressed in the form of allometric equations. *Allometry* (literally, ''of another measure'') describes the disproportionate changes in size (or function) that occur when separate isolated features in animals are compared across a range of body sizes. If all characteristics varied in direct proportion, that is, if the large animal were merely a scaled up exact replica of the small one, they would be built with *isometry*. Quantitatively, allometry takes the form of power law equations relating some variable of structure of function (Y) as a dependent function of body mass (M) in the form $Y = aM^b$, where a and b are derived empirically.

On logarithmic coordinates this equation describes a straight line with a slope of b. Therefore, the value of b describes the nature of the relationship. When b is near 1.0, Y scales as a fixed percentage of body mass (i.e., isometrically). Virtually all volumes or capacities, for instance, lung, gut, and heart volume, scale isometrically. Blood volume in all mammals is about 7% of body mass. If b is greater than 1, Y increases more rapidly than does mass in going from small to large animals. To maintain a constant safety factor of bone strength (discussed above), the mass of the skeleton varies with an exponent near 1.1 in both birds and mammals (Anderson et al., 1979). If b is between 0 and 1, a unit increase in mass is accompanied by a fractional increase in Y. The well-known Kleiber equation describing resting metabolism in homeotherms scales with an exponent of 3/4 (Kleiber, 1932); thus, weight-specific metabolism (rate of metabolism per unit mass) is much higher in small than in large animals. Finally, if b is negative, its absolute value is highest in the smallest animals. The exponent b is close to $-1/4$ for biological rates such as heart and respiratory rates.

Allometric equations are often limited in their use as descriptive tools to identify patterns of form and function. Recently, many of these equations have been compiled and presented in encyclopedic form (see, for example, Peters, 1983). Although these equations are most valuable for identifying interspecific patterns, there may be equally important information hidden within the variance around the equations. In looking for those species that do not conform to body size predictions, allometry can identify those animals that Calder (1984) refers to as "adaptive deviants," suggesting that the process of deviation from body size-expected patterns must be the result of selection rather than random variance. Finally, in addition to identification of patterns and those animals that deviate from the patterns, allometry may be most useful as a tool to identify interspecific constraints of design and function. Many of these are discussed in two recent books (Calder, 1984; Schmidt-Nielsen, 1984).

Allometric equations can furnish much more than empirical patterns that lack a conceptual foundation. Many structural relationships can best be explained allometrically. For instance, as the surface area of an animal, or any object, varies as the 2/3 power of its volume [i.e., suface area (Y) $= aM^{2/3}$] any increase in mass results in a decrease in relative surface area. Here we encounter an allometric "law" of sorts, namely, that size changes often require a concomitant mechanism to permit surface area to increase linearly with mass (i.e., as a fixed percentage of mass). A sufficient number of allometric equations have been generated to confirm quantitatively what Haldane (1928) proposed nearly 60 years ago: "The higher animals are not larger than the lower because they are more complicated. They are more complicated because they are larger. . . . Comparative anatomy is largely the story of the struggle to increase surface in proportion to volume." The design of the mammalian respiratory system provides a quantitative example of Haldane's declaration. Lungs with alveoli and pulmonary capillaries, and circulatory systems with capillaries and red cells, are prominent examples of structures that are necessary only as a price for large body size. Interestingly, in all cases the surface areas of the above-named structures vary with body mass exponents (b) near 1.0. Across 6 orders of magnitude in body size, metabolically important surfaces generally increase in direct proportion to volume. How can these apparent design patterns be of use in risk assessment?

SIZE, DESIGN, AND PHARMACOKINETICS

A hopeful result of the linking of pharmacokinetics and risk assessment will be to identify interspecific principles of design. Once identified, these may be extremely valuable in making species extrapolations with the eventual goal of risk assessment in humans. In suggesting a potential role for allometric analysis in that process, I will focus on three diverse examples from my own research which illustrate the value of allometry as a unique mechanism for identification of design constraints. In all cases, apparent principles of design surface only when viewed across a range of species.

Aerobic Energetics of Muscle *In Vivo*

The structural, contractile, and metabolic properties of skeletal muscle have been well characterized. The maximal cross-sectional force of all muscles is roughly constant, and mechanisms of cross-bridge cycling, adenosine triphosphate (ATP) synthesis, and use are well characterized (for reviews, see Peachey et al., 1983), yet *in vivo* muscle possesses some properties that are unpredictable from *in vitro* experimentation. In mammals, the external (weight-specific) work performed by locomotory muscles does not vary systematically with body size; however, the energy required to perform that work does. Small animals expend more energy for a given force production than do large animals. As a result, the energetic cost of locomotion (energy spent to move a unit mass a unit distance) and, therefore, the efficiency of locomotion are strongly body size dependent. There is an energetic cost associated with small body size, the source of which could be the obligate scaling of muscle contraction times. While all skeletal muscle can produce roughly the same maximal cross-sectional force, the power required to do so increases with increasing contraction velocity.

Because differences in energetics span more than an order of magnitude, by examining this question allometrically, we have the luxury of a favorable signal-to-noise ratio that is not present within any single species. It is possible to estimate the *in vivo* rate of muscle shortening by calculating a series of equations and combining these algebraically (after Stahl, 1962). Thus, ratios or products of allometric equations can be formed to predict functions that are either unmeasurable or impractical to measure. By doing so we found that variance in muscle biochemistry, structure, contractile properties, and whole animal maximum oxygen uptake all followed an identical pattern. Hence, there is a parallel (causal?) relationship among the rate of muscle shortening and the energy supplying oxygen consumption, volume density of ATP-synthesizing mitochondria, and the activity of myosin ATPase (Lindstedt et al., 1985) (Figure 1). When examined over a broad range of body sizes, constraints of design surface suggest a common scheme of muscle structure and function.

In this example, allometry is useful for making estimates for problems too demanding to be currently solved with direct measurements; it can thus be more than a descriptive tool.

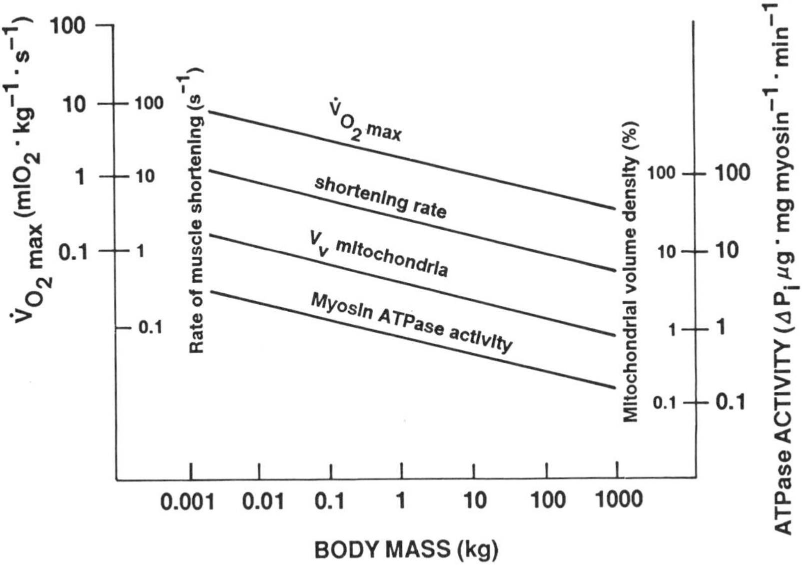

FIGURE 1 An allometic analysis of the design of the knee extensor muscles in mammals demonstrates remarkable structure-function consistency. Among terrestrial mammals, there is a constant relationship among oxygen consumption (V_{O_2max}), volume density (V_V) of mitochonodria, myosin ATPase activity, and the rate of muscle shortening. Although these parameters vary greatly as a function of body mass, within any given mammal their ratios remain nearly constant. (Used with permission of the *American Journal of Physiology*.)

Conflict of Physiological and Chronological Time

Four decades ago, Brody (1945) introduced the concept of "physiological time," in acknowledgment of a variable biological time scale that exists among organisms. Hill (1950) refined the concept by suggesting that within an organism physiological time may be as constant as is chronological time. Hill's speculation was that all physiological events are likely entrained to the same body size-dependant clock. Thus, critical biological times, such as gestation period or time for growth to maturity, as well as other temporally linked biological events, may *all* be constant if compared per unit of physiological time. Hill ended this most interesting paper by suggesting that 10 s to a large animal may be physiologically equivalent to 1 s in a small one, implying that they differ only in pace of life, not in absolute number of life's events.

A sufficient number of physiological rates and times have been measured to permit a quantitative examination of Hill's hypothesis. Among physiological events, all apparently occur with a body mass-dependent metric (Figure 2). In fact, we (Lindstedt and Calder, 1981) found that virtually all biological times do indeed vary with nearly the same body mass exponent (mean = 0.24) in both birds and mammals (Figure 3). It is assumed that this parallel scaling is the result of a common body size-dependent clock to which these events are all entrained (a *periodengeber*; see Lindstedt, 1985). Thus, whether the result of function is in a single tissue (e.g., muscle contraction times), single organ (e.g., cardiac cycle), organ systems (e.g., inulin or PAH clearance), the entire organism (e.g., growth times), or even populations of organisms (population doubling time), *all* biological times seem to vary as a consistent and predictable function of body mass.

There are at least two consequences of the regularity of physiological time that have direct bearing on pharmacokinetics and risk assessment.

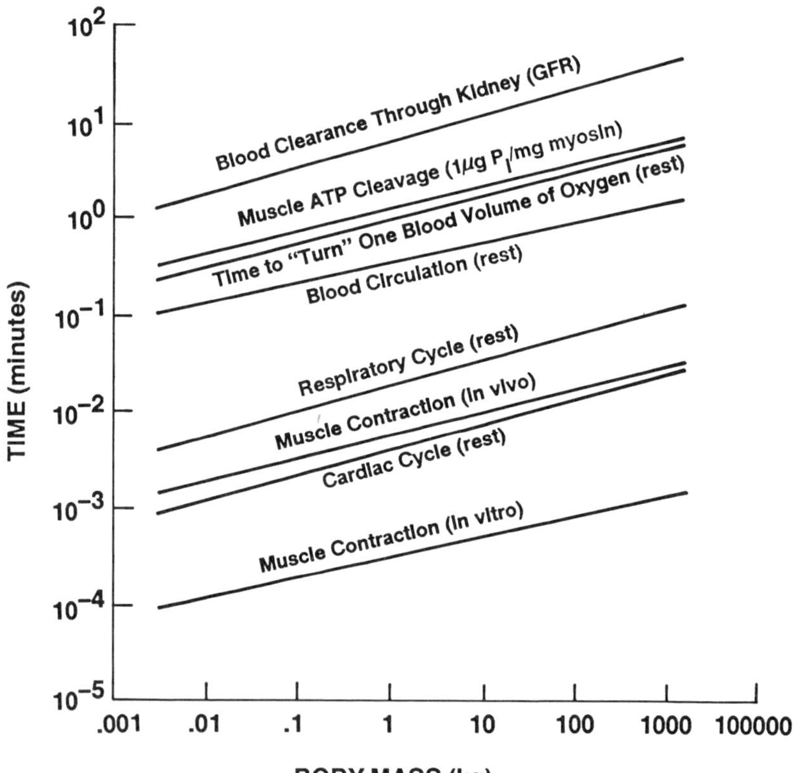

FIGURE 2 The duration of physiologically critical time periods are examined over a range of body masses. Equations are from many sources and have been uniformly extrapolated (when necessary) to encompass the range of body mass in terrestrial mammals.

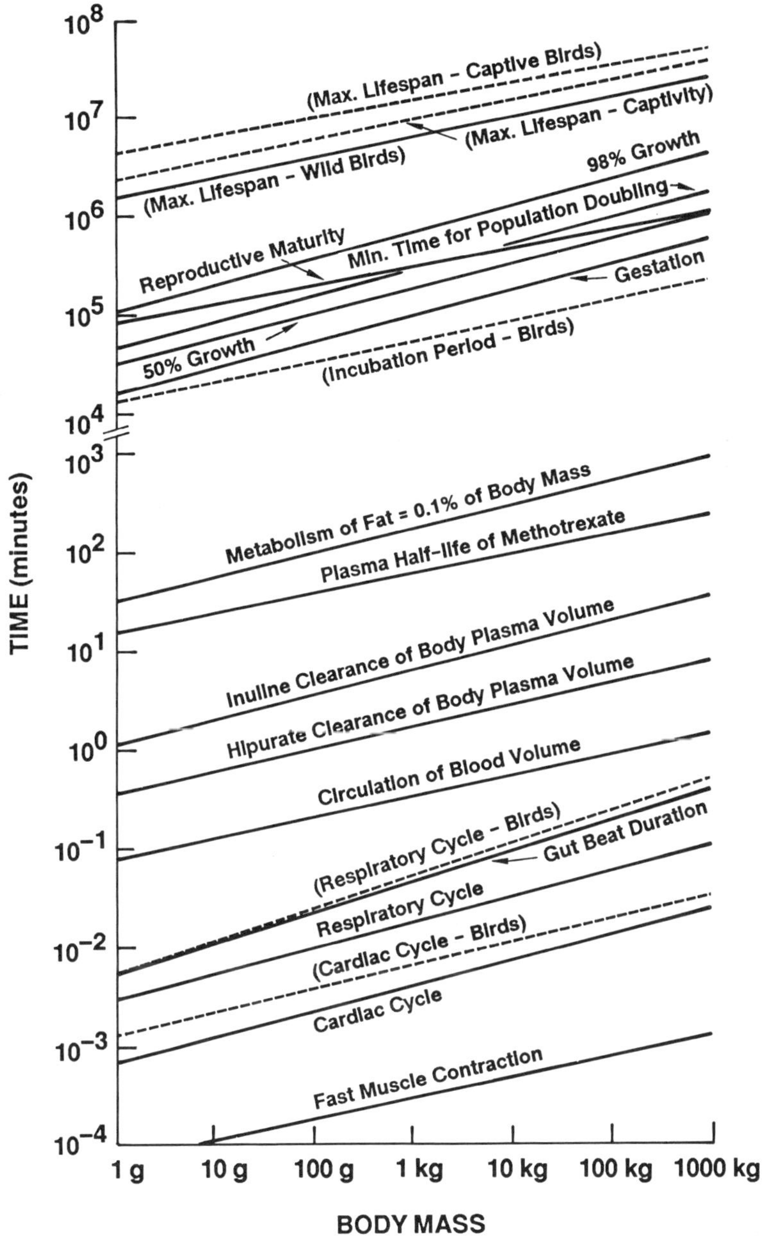

FIGURE 3 Biological times are shown for birds (dashed lines) as well as mammals (solid lines). These include strictly physiological times (muscle contraction) to strictly life history or ecological times (time to achieve reproductive maturity or population doubling). The slopes of these lines are nearly identical (mean = 0.24) in spite of the apparent unrelated nature of the times and methods of measurement. (Used with permission of the *Quarterly Review of Biology*.)

The first is that if volumes and capacities vary as linear functions of mass and times as mass raised to the 1/4 power, then volume rates (i.e., volume divided by time, such as clearance rates) must vary as $M^1/M^{1/4} = M^{3/4}$. Application of this physiological volume rate scaling will be discussed below. The second is that ratios of various temporal events are essentially constant, independent of body size, and independent of their absolute rate of occurrence (Figure 4). Mammals may indeed burn roughly the same

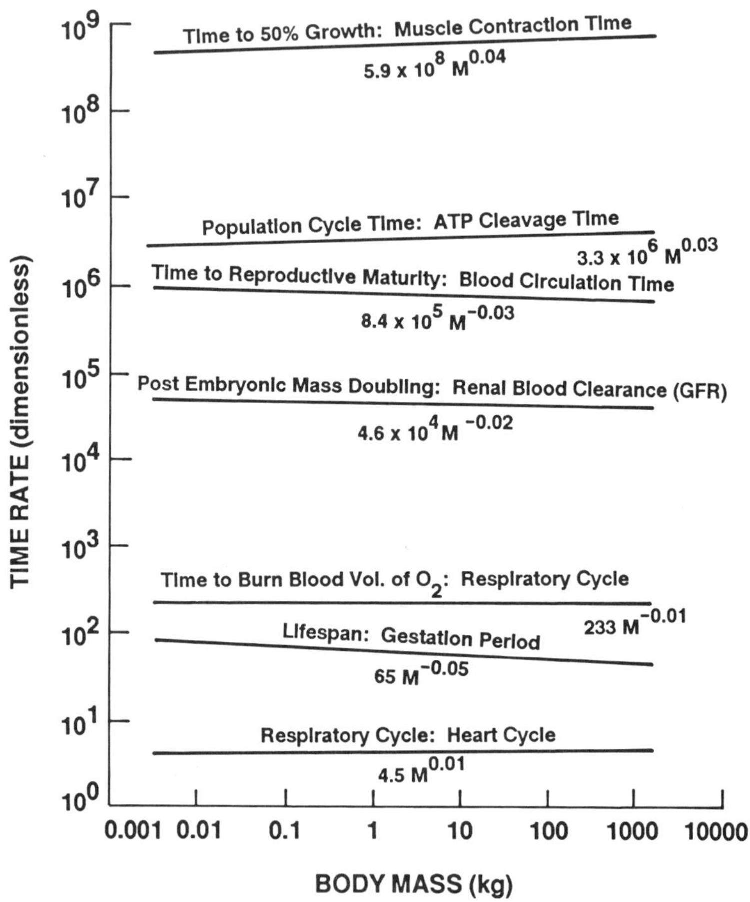

FIGURE 4 Ratios of any two physiological times are nearly constant, varying little from shrew to elephant. Thus, all mammals experience about four to five heart beats per breath, and all burn roughly the same number of calories per gram of tissue per lifetime. This constancy of times suggests that there is a body size-dependent clock to which all of life's events are entrained. Thus, physiological time is just as critical as chronological time and may be the best measure of time for cross-species comparisons.

number of calories per gram of tissue per lifetime, which itself is composed of a constant maximum number of cardiac, respiratory, and other cycles (Lindstedt, 1985; Lindstedt and Calder, 1981).

Thus, a second allometric law is identified; even though chronological time is body mass independent ($b = 0$), physiological time is not ($b \approx 1/4$). All animals must contend with identical lengths of days and seasons, but these span disproportionately longer physiological periods in small than in large animals (i.e., chronological/physiological time $= M^0/M^{1/4} = M^{-1/4}$). This simple principle may explain several interactions of time and size, from fasting survival times to the structure of the mammalian lung (Lindstedt, 1984; Lindstedt and Boyce, 1985). The ubiquitous presence and regularity of physiological time among animals argues strongly for its consideration as a constraint of design. That physiological time is just as critical as chronological time should have a profound effect on interpretation and extrapolation of experimental data.

In summary, through the use of allometric equations, a pattern surfaces, namely, that among the noise of various physiological rate functions there is a strong signal linking events as diverse as muscle contraction times occurring over periods of milliseconds with population doubling times requiring years. So salient is this pattern that physiological time must be recognized an identifiable characteristic of organisms.

Species Extrapolations, Physiological Time, and Pharmacokinetics

How does a consideration of allometry in general, and physiological time specifically, have an impact on pharmacokinetics? Like other rate functions, those involving the biochemistry of drug metabolism are inseparably bound to physiological but not chronological time. Uptake, processing, and excretion of drugs all transpire at rates that are generally directly proportional to one another and with body mass exponents characteristic of physiological time, $b \approx 1/4$ (Boxenbaum, 1982; Calder, 1984; Dedrick et al., 1970; Weiss et al., 1977). Virtually all pharmacokinetic variables from tissue dose (M. E. Anderson, this volume) to first-order kinetics (E. J. O'Flaherty, this volume) can best be interpreted through the perspective of physiological time. I naively present one caution and one possible application.

One danger exists in the way in which we perform and interpret our experiments. Doses (per kilogram) and incubation times are virtually always fixed, independent of both body mass and physiological time. Perhaps we are losing some information by comparing processes that transpire at mass-specific rates ($b = -1/4$)[1] using concentrations and times that

[1]Because rate is the reciprocal of time, the exponents describing biological rates are likewise the reciprocal of those describing times ($1/M^{1/4} = M^{-1/4}$).

are invariant ($b = 0$). This distinction becomes especially important when results are extrapolated over several orders of magnitude in body mass, for example, from rodents to humans. A simple example might best illustrate this point. Miners used canaries in caves as biomonitors because of their apparent increased sensitivity to toxicants; when the bird dropped dead the miners would make a hasty retreat from the mine. In fact, are canaries more sensitive or are they affected by the same absolute levels of toxicants over shorter (physiological) time periods? Pharmacokinetic effects are not simple dose effects, but are dose effects integrated over physiological time periods. Thus, when examined over physiological rather than chronological time, species variability in pharmacokinetic rates may disappear (see, e.g., Mordenti, 1985).

In a now classic paper, Dedrick et al. (1970) first made use of biological time scaling in their interspecific analysis of methotrexate pharmacokinetics. They found that plasma concentrations could be normalized if dose were divided by body mass raised to the 1/4 power ($b = 1/4$). Amid the noise of interspecific variation there emerged a strong signal; there is a hidden $M^{1/4}$ scaling linking diverse animals to a single pharmacokinetic pattern. In general, dose varies as a function of rate of clearance; hence, it would be expected that dose = quantity/time. If we correct for physiological rather than chronological time, then we must divide by $M^{1/4}$. A simple example best illustrates the point. Dedrick et al. (1973) carefully examined the clearance of 1-β-D-arabinofuranosylcytosine (Ara-C) in four mammalian species. They found that clearance rates varied nearly as mass raised to the 3/4 power, as would be predicted based on the volume and rate allometry discussed above. Thus, while absolute rates of clearance varied by nearly 1,000-fold, they did so in an identifiable pattern. If those data are normalized for size and physiological time scaling, clearance can be expressed per unit of body mass per unit of physiological time (Figure 5). By this manipulation, interspecific differences for this particular compound can be adequately explained solely as a function of size-dependent time scaling.

Several compounds have been examined through the perspective of what has since been called the Dedrick plot (Boxenbaum and Ronfeld, 1983). A key to understanding this type of manipulation is not that it makes all animals equal, rather that their differences actually become more pronounced when removed from inevitable, size-dependent variation resulting from physiological time scaling. For instance, if the resultant slope is not close to 0 (as it is in Figure 5), then we know immediately that there is a difference in the way small and large animals process a given drug. In the case of clearance data, a positive slope would imply that large animals are clearing the compound in faster physiological times; a slope below 0 would imply that smaller animals are doing so. In any case, the appropriate

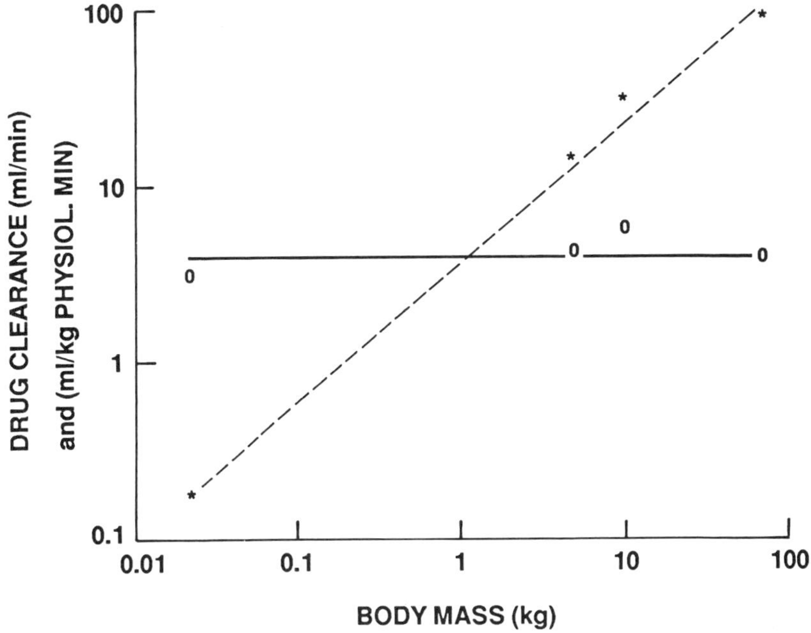

FIGURE 5 Clearance of 1-β-D-arabinofuranosylcytosine (Ara-C) in four species of mammals: mouse, monkey, dog, and man. Data are expressed in conventional units of milliliters per minute (asterisks and dashed line), as well as physiological time units of milliliters per kilogram physiological minute (open circles and solid line). Interspecific variation in the clearance rate of Ara-C virtually disappears when size-specific time scaling is considered (data are from Dedrick et al., 1973).

means of comparison is against a physiological rather than a chronological metric.

Perhaps physiological time can form the bases of comparison for risk assessment across species. The exponentially expanding pharmacokinetic data base may be outgrowing useful paradigms for interpretation. If comparisons across widely different taxa are possible, perhaps an equivalent means of comparison might be physiological time. In collaboration with several individuals at the Environmental Protection Agency (EPA) in Duluth, Minn., we have made a first attempt at such a comparison for the insecticide fenvalerate. Our preliminary results are somewhat promising because we have been able to directly equate results from different temperatures and modes of delivery in invertebrates (lobsters and flies) and vertebrates (rats and three species of fishes). We plotted the 50% lethal doses (LD_{50}s) as a function of physiological time, calculated as the time to metabolize a given quantity of oxygen proportional to body mass (roughly the reciprocal of weight-specific metabolism). The results suggest that the huge variability among animals may be largely accounted for ($r^2 = 0.82$) when physiological time is considered (Figure 6). The exception was the fly, which should have a much higher relative sensitivity (i.e., by insecticide design) and is about four orders of magnitude more sensitive than predicted by this analysis. While this approach is only a first rough attempt with a limited sample size, it may provide one potential means of cross-species pharmacokinetic comparison.

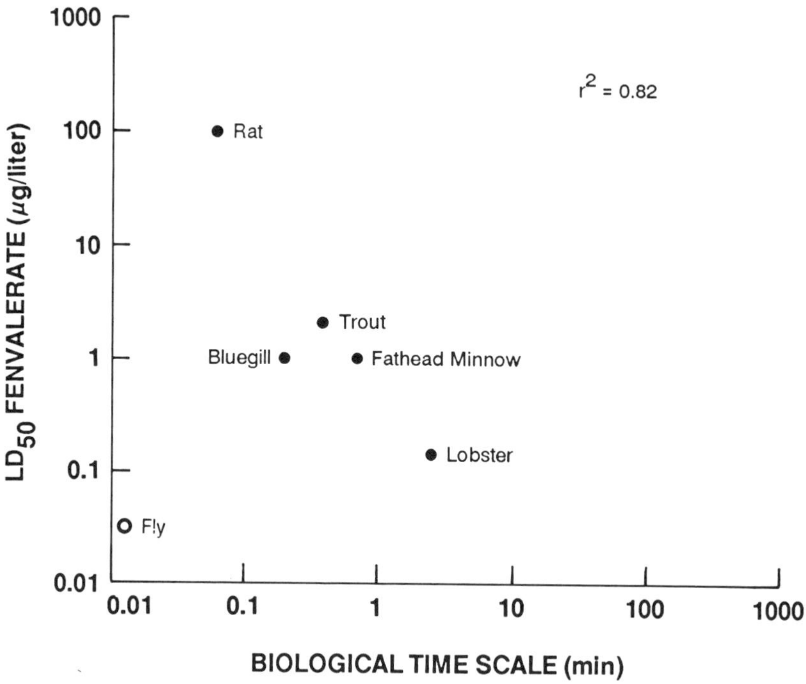

FIGURE 6 One method of comparing animals from diverse taxonomic groups may be on a single continuum of physiological time. Here LD_{50} values for fenvalerate are plotted as a function of one measure of time, namely, the time required to metabolize one body mass equivalent of O_2. These preliminary data suggest that sensitivity to fenvalerate decreases with decreasing body mass. They also suggest that the fly may be several orders of magnitude more sensitive to the drug than are the other species represented on this graph. For comparison purposes, man would be predicted to fall very near the point for trout (data are from the EPA laboratory in Duluth, Minn.).

CONCLUSIONS

In an essay entitled, "Comparative Physiology, Compared to What?," F. E. Yates (1979) summarized what should be the value of "comparative physiology." As a consequence of making comparisons (in design or function), physical laws and constraints often surface suggesting design principles that operate across species. Allometry is one of the most directly comparative of all tools available to biologists. MacArthur (1972) began his now classic book *Geographical Ecology* with the affirmation, "To do science is to search for repeated patterns in nature, not simply to accumulate facts." While the accumulation of facts will always be a necessary goal of science, the process of pattern identification continues to occur only infrequently.

Physiological time has emerged as one of those patterns in nature. It may prove to be of significant value in understanding interspecific pharmacokinetics. It is not proposed as a substitute for direct measurements; however, cross-species comparisons should be made by normalizing for physiological time first, assuming it is the "default value." If differences persist, those can be fairly attributed to species differences. In planning and executing experiments, researchers must be aware of the two important time scales that have an impact on animals and ask how each may affect interpretation and extrapolation of results. Lack of consideration of this pervasive principle is no less of an assumption. For instance, in designing experiments to assess lifetime exposure, perhaps the best technique for the extrapolation from risk to an experimental rat to a potential risk to humans is some measure of physiological time equivalence such as number of heartbeats (suggested by Blancato and S. L. Friess, personal communication) or a measure of drug or metabolite turnover time. If so, 2 years may be much shorter than researchers assume. Further, perhaps shrews or small birds, which have the most rapid physiological time scales, may be the ideal experimental animals for certain studies.

SUMMARY

One of the products of science is an exponentially increasing pool of knowledge. Hopefully one goal of science is to combine emergent facts in such a way to form patterns from which new conceptual hypotheses and predictions can be derived. One of the potentially valuable techniques for the identification of patterns is *allometry*.

Allometry literally means "of a different measure," in distinction to isometry. Functionally, allometry is an attempt to quantify the extreme importance body size has in dictating virtually all aspects of an animal's morphology, physiology, and ecology. Allometric equations can identify patterns of design and function that would be otherwise obscure. Further, they may be useful in identifying species that deviate from the pattern. Finally, allometric equations may be extremely useful in suggesting interspecific constraints of design and function. In this paper I present three examples.

1. Using a very simple model of a single mass and spring system, it is possible to couple the geometry of the mammalian knee with a description of muscle energetics to make an allometric estimation of locomotory energetics. A running mammal behaves like a tuned spring system with an optimal (and therefore most efficient) stride frequency, somewhat analogous to a resonating frequency (see Taylor, 1985). Rather than using allometry merely as a descriptive tool, we are able to make mechanical and bioenergetic predictions starting with a series of allometric equations.

2. If we examine a variety of rate functions across a wide range of species, another design constraint seems to surface, that of physiological time. While all animals experience the same chronological time of days or seasons, each animal apparently also possesses an internal body size-dependent clock to which physiological rates are entrained. From the rate of muscle contraction to life-span itself, physiological processes seem to be set to the same "periodengeber."

3. As an extension of this concept, we propose that species extrapolations may best be made as a function of physiological time (rather than dose or some other aspect of chronological time). Preliminary results for *Fenvalerate* suggest potential utility of this approach.

Physiological time is not a substitute for measurements, rather it is a strong enough design pattern that it should be considered in designing and interpreting any experiment.

REFERENCES

Anderson, J. F., H. Rahn, and H. D. Prange. 1979. Scaling of supportive tissue mass. Q. Rev. Biol. 54:139–148.

Boxenbaum, H. 1982. Interspecies scaling, allometry, physiological time, and the ground plan of pharmacokinetics. J. Pharmacokinet. Biopharm. 10:201–227.

Boxenbaum, H., and R. Ronfeld. 1983. Interspecies pharmacokinetic scaling and the Dedrick plots. Am. J. Physiol. 245:R768–R774.

Brody, S. 1945. Bioenergetics and Growth. New York: Reinhold.

Calder, W. A., III. 1984. Size, Function, and Life History. Cambridge, Mass.: Harvard University Press.

Dedrick, R., K. B. Bischoff, and D. S. Zaharko. 1970. Interspecies correlation of plasma concentration history of methotrexate. Cancer Chemother. Rep. 54:95–101.

Dedrick, R. L., D. D. Forrester, J. N. Cannon, S. M. Eldareer and B. Mellett. 1973. Pharmacokinetics of 1-β-D-arabinofuranosylcytosine (Arc-C) deamination in several species. Biochem. Pharmacol. 22:2405–2417.

Haldane, J. B. S. 1928. On being the right size. Pp. 952–957 in The World of Mathematics II, J. R. Newman, ed. (Reprinted 1956 by Simon and Schuster, New York.)

Hill, A. V. 1950. The dimensions of animals and their muscular dynamics. Sci. Prog. 38:209–230.

Kleiber, M. 1932. Body size and metabolism. Hilgardia 6:315–353.

Lindstedt, S. L. 1984. Pulmonary transit time and diffusing capacity in mammals. Am. J. Physiol. 246:R384–R388.

Lindstedt, S. L. 1985. Birds. Pp. 1–21 in Non-Mammalian Models for Research in Aging, F. A. Lints, ed. Basel: S Karger.

Lindstedt, S. L., and M. S. Boyce. 1985. Seasonality, body size and survival time in mammals. Am. Nat. 125:873–878.

Lindstedt, S. L., and W. A. Calder III. 1981. Body size, physiological time, and longevity of homeothermic animals. Q. Rev. Biol. 56:1–15.

Lindstedt, S. L., H. Hoppeler, K. M. Bard, and H. A. Thronson, Jr. 1985. Estimate of muscle shortening rate during locomotion. Am. J. Physiol. 249:R699–R703.

MacArthur, R. H. 1972. Geographical Ecology. New York: Harper and Row.

Mordenti, J. 1985. Forecasting cephalosporin and monobactam antibiotic half-lives from data collected in laboratory animals. Antimicrob. Agents Chemother. 27:887–891.

Peachy, L. D., R. H. Adrian, and S. R. Geiger. 1983. Handbook of Physiology, Skeletal Muscle. American Physiological Society. Baltimore: Willaims & Wilkins.

Peters, R. H. 1983. The ecological implications of body size. Cambridge: Cambridge University Press.

Schmidt-Nielsen, K. 1984. Scaling: Why is animal size so important? Cambridge: Cambridge University Press.

Stahl, W. R. 1962. Similarity and dimensional methods in biology. Science 137:205–212.

Taylor, C. R. 1985. Force development during sustained locomotion: A determinant of gait, speed, and metabolic power. J. Exp. Biol. 115:253–262.

Weiss, M. W., W. Sziegoleitt, and W. Forster. 1977. Dependence of pharmacokinetic parameters on the body weight. Int. J. Clin. Pharmacol. 15:572–575.

Yates, F. E. 1979. Comparative physiology: Compared to what? Am. J. Physiol. 237:R1–R2.

Prediction of *In Vivo* Parameters of Drug Metabolism and Distribution from *In Vitro* Studies

Grant R. Wilkinson

Physiologically based pharmacokinetic modeling is critically dependent on accurate estimates of the determining anatomical, physiological, and drug-specific parameters. Anatomical parameters, such as organ size and blood flow rates, are usually constant and well-established; if not, they can often be estimated directly or by allometry (see S. L. Linstedt, this volume). On the other hand, factors like drug binding to plasma and tissue constituents and elimination by various organs are not so readily available. In many instances these characteristics must be determined experimentally. In the past, this has largely involved *in vivo* measurements such as the determination of tissue/plasma partition coefficients or the apparent kinetic parameters of metabolism. Potentially, however, data obtained *in vitro* may, if appropriately interpreted, be applicable to the *in vivo* situation and can be rigorously incorporated into a physiologically based pharmacokinetic model to allow quantitative prediction. This approach was an implicit hope of early investigators (Bischoff and Dedrick, 1968; Dedrick et al., 1973), but unfortunately, it has not been explored as extensively as might be expected. The ever increasing number of drugs and other chemicals to which animals and humans are exposed and the need to assess their disposition, however, along with any toxicity, will necessitate that such *in vitro* studies become more widely applied. This is particularly the case in humans, in which direct exposure to a putative toxic agent may be limited, for example, to accidents or occupational activities for which the dose and other details of the exposure are not well known. Over the past several years, a limited number of *in vitro*/*in vivo* extrapolation studies have been undertaken that demonstrate the utility of the involved approaches, as well as their limitations. In this chapter I will review some of the findings and their potential for further application.

IN VITRO PREDICTION OF *IN VIVO* DRUG METABOLISM

Numerous *in vitro* systems are available to determine the metabolism of a drug, including simple aqueous solutions for measurement of nonenzymatically mediated reactions, purified enzyme preparations, crude tissue homogenates, microsomes, isolated intact and cultured cells, and tissue slices. Such variety indicates that no single system provides all the necessary information required for a complete understanding of the determining factors of drug biotransformation. Moreover, the purpose and experimental design of the large majority of studies with such systems is frequently not suitable for extrapolation of the findings to the *in vivo* situation. A critical requirement is that the kinetic parameters of metabolism, expressed in terms of a specific pathway(s) or the overall process, must be available. Generally, these are interpreted according to the well-known Michaelis-Menten mechanism so that the rate of metabolism $[(d[S]^{tot}/dt)V]$ can be described by Equation 1:

$$\frac{d[S]^{tot}}{dt} V = \frac{V_{max}\,[S]^{tot}}{K_m/f^u + [S]^{tot}}, \tag{1}$$

where V_{max} is the maximal rate of metabolism, K_m is the concentration of unbound substrate at half the maximal velocity, $[S]^{tot}$ is the total substrate concentration, and f^u is the fraction of the substrate that is unbound in the reaction medium. By normalizing the rate to the substrate concentration, metabolism can be expressed as a clearance value (Equation 2) which is frequently termed total intrinsic clearance (Gillette, 1971, 1984; Wilkinson and Shand, 1975):

$$\frac{d[S]^{tot}}{dt}\,\frac{V}{[S]^{tot}} = \frac{V_{max}}{K_m/f^u + [S]^{tot}} = \frac{f^u\,V_{max}}{K_m + f^u[S]^{tot}} = CL_{int}^{tot}. \tag{2}$$

Alternatively, the metabolic clearance rate can be related to the unbound drug concentration by the free intrinsic clearance rate ($CL_{int}^{tot} = f^u\,CL_{int}^{u}$). Account can also be taken that metabolism often involves more than one pathway, so that the overall intrinsic clearance in any system represents the sum of the individual processes (Equation 3):

$$CL_{int}^{u} = \sum_{i=1}^{i=n} \frac{V_{max_i}}{K_{m_i} + f^u[S]^{tot}}. \tag{3}$$

In many instances the driving concentration for metabolism, the unbound concentration, is far less than the K_m of the involved enzyme(s), and the kinetics become independent of substrate concentration. Under such linear or first-order conditions, the free intrinsic clearance reaches a constant maximum value equal to the ratio of the overall V_{max} to K_m (Equation 4):

$$CL^u_{int} = \sum_{i=1}^{i=n} \frac{V_{max_i}}{K_{m_i}} . \tag{4}$$

In principle, therefore, the means exists to relate readily determinable Michaelis-Menten enzyme kinetic constants to a well-understood concept and parameter of pharmacokinetics. Such V_{max} and K_m estimates can be directly incorporated into the differential mass-balance equation for total drug in a particular metabolizing organ(s) by using a perfusion-limited physiological model (Equation 5):

$$\frac{dC_T}{dt} V_T = Q\left[C_T/K_p - \frac{V_{max}(C_T/K_p)}{K_m/f^u + C_T/K_p} \right], \tag{5}$$

where C, K_p, and V represent the total substrate concentration, tissue/plasma partition coefficient, and volume of the eliminating tissue (T), respectively; and Q is the blood flow to the organ. This approach has been applied to studies with cytosine arabinoside, when the *in vitro* kinetics of deamination in a number of organs and in different species allowed good predictions of the *in vivo* situation (Dedrick et al., 1972, 1973). Similar findings were made with three barbiturates (Igari et al., 1982).

One possible reason for the paucity of this type of data is that Equation 5 provides no intuitive insights into drug elimination relative to the conventional quantitative estimates of this process, such as (organ) clearance (CL) and extraction ratio (E). Unless the equation is solved along with others in the model and the resulting blood concentration/time data are analyzed by conventional pharmacokinetic techniques, there is no easy way to assess the validity of the *in vitro* values relative to those determined from experimental data. Under steady-state conditions, Equation 5 can, however, be rearranged to provide an estimate of the extraction ratio (Equation 6) or clearance (QE) of an eliminating organ:

$$E = \frac{\dfrac{V_{max}}{K_m/f^u + C_T/K_p}}{Q + \dfrac{V_{max}}{K_m/f^u + C_T/K_p}} , \tag{6}$$

which under first-order conditions simplifies to:

$$E = \frac{CL_{int}^{tot}}{Q + CL_{int}^{tot}} = \frac{f_B^u\, CL_{int}^u}{Q + f_B^u\, CL_{int}^u}, \tag{7}$$

where f_B^u refers to the unbound fraction of drug in blood which is related to the more frequently measured value in the plasma (f_P^u) by the blood/plasma concentration ratio [$f_B^u = f_P^u/(B/P)$]. Equations 6 and 7 therefore provide the means to incorporate *in vitro* estimates of enzyme kinetics into a perfusion-limited model of an eliminating organ and to evaluate their appropriateness by comparing the predicted extraction ratio or clearance to that measured *in vivo*. This model has been termed the venous-equilibration or well-stirred model of organ elimination (Pang and Rowland, 1977; Rowland et al., 1973; Wilkinson and Shand, 1975). The feasibility of such an *in vitro*/*in vivo* predictive approach was first investigated with several drugs that had widely divergent extraction ratios (0.2 to >0.9) undergoing drug oxidation (Rane et al., 1977). The Michaelis-Menten kinetics of overall substrate disappearance were determined in a 9,000 × *g* supernate of rat liver; and after correction for dilution and protein recovery, the first-order, intrinsic clearance value (V_{max}/K_m), along with the unbound fraction, was incorporated into Equation 7, resulting in an *in vitro* predicted hepatic extraction ratio. Comparison of this with an experimentally determined steady-state extraction ratio in the isolated perfused rat liver showed excellent agreement ($r = 0.988$, $p < 0.001$). Subsequently, the same approach was applied to other drugs and metabolic pathways, as well as other hepatic preparations (Table 1). The prediction of metabolic elimination by other organs such as the lung and intestinal mucosa was also studied (Table 2).

Despite the relatively small number of these studies, some tentative conclusions can be drawn from their findings. First, and most important, the *in vitro* predictive approach using Equation 7 appears to provide a reasonable estimate of the *in vivo* situation for a number of drugs across a broad range of metabolic activity. This is particularly the case when the compounds are metabolized by the liver and involve the cytochrome P-450 system. Examples of other hepatic routes of metabolism are not sufficiently large to draw any conclusions, although the *in vitro* underestimation of glucuronidation (Rane et al., 1984) is consistent with the known lability of glucuronyltransferase activity. The situation in the lung is less clear, which may reflect the anatomical and biochemical heterogeneity of this organ.

The type of *in vitro* system on which the prediction is based may also be an important factor. Successful predictions require that the conditions of metabolism *in vitro* be identical to those *in vivo*. Unfortunately, these are never known, and this is the crux of the problem. Microsomal and purified enzyme preparations are usually investigated with a maximal

TABLE 1 Comparison Between the *In Vivo* Hepatic Extraction Ratio in the Rat and Other Species and That Predicted by the Venous Equilibration Model (Equation 7), Based on *In Vitro* Estimates of V_{max} and K_m

Drug	Preparation	*In Vitro* Prediction	*In Vivo* Measurement	Metabolic Pathway	Reference
Acetaminophen	Hepatocytes	0.7–0.72	0.67	Sulfation/gluronidation	Pang et al., 1985
Alprenolol	9,000 × *g* supernate	0.92	>0.90	Oxidation	Rane et al., 1977
	9,000 × *g* supernate	0.90	0.98		Skånberg, 1980
	Microsomes	0.6	0.98		
	Hepatocytes	0.75	0.98		
Antipyrine	9,000 × *g* supernate	0.04	0.01	Oxidation	Rane et al., 1977
Carbamazepine	9,000 × *g* supernate	0.05	0.04	Oxidation	Rane et al., 1977
Chlorpheniramine	600 × *g* supernate[a]	0.89	0.67–0.84	Oxidation	Huang et al, 1981
Diazepam	Microsomes	0.62	0.65	Oxidation	Igari et al., 1984
Felodipine	Microsomes	0.91	0.80	Oxidation	Bäärnhielm et al., 1986
	Microsomes[b]	0.80	0.83		
	Microsomes[c]	0.80	0.84		
Hexobarbital	9,000 × *g* supernate	0.44	0.33	Oxidation	Rane et al., 1977
5-Hydroxytryptamine	Homogenate	0.68	0.59	MAO	Wiersma and Roth, 1980
Lidocaine	9,000 × *g* supernate	0.80	>0.90	Oxidation	Rane et al., 1977
Morphine	Microsomes[d]	0.09–0.14	0.61–0.71	Glucuronidation	Rane et al., 1977
Phenacetin	Hepatocytes	0.54	0.87–0.95	Oxidation	Pang et al., 1985
Phenytoin	9,000 × *g* supernate	0.50	0.53	Oxidation	Rane et al., 1977
	Microsomes	0.03	0.05		Collins et al., 1978
Propranolol	9,000 × *g* supernate	0.83	>0.90	Oxidation	Rane et al., 1977

[a]Rabbit.
[b]Dog.
[c]Human.
[d]Monkey.

TABLE 2 Comparison Between *In Vivo* Extraction Ratio and That Predicted by the Venous Equilibration Model (Equation 7), Based on *In Vitro* Estimates of V_{max} and K_m

Drug	Preparation	*In Vitro* Predicticn	*In Vivo* Estimate	Metabolic Pathway	Reference
Rabbit Lung					
Benzopyrene 4,5-oxide	Microsomes	0.25 ± 0.05	0.20	Epoxide hydrolase	Smith and Bend, 1980
Benzopyrene 4,5-oxide	Cytosol	0.74 ± 0.06	0.44	Glutathione *S*-transferase	Smith and Bend, 1980
5-Hydroxytryptamine	600 × *g* supernate	0.03	0.43	MAO	Wiersma and Roth, 1980
Mescaline	600 × *g* supernate	0.14	0.14[a]	Oxidation	Hilliker and Roth, 1980
Rat Intestinal Mucosa					
Phenacetin	Intact mucusal cells	0.31–0 53	0.53	Oxidation	Klippert et al., 1982

[a]Actual value not reported but stated to be not significantly different from the predicted value.

concentration of substrate and under optimal biochemical conditions, in which the pH of the medium and the concentrations of cofactors may be much different from those in the *in vivo* situation. Purification steps may also affect the biochemical parameters. The latter may account, for example, for the almost 50% underestimate of phenytoin's hepatic extraction using microsomes (Collins et al., 1978), but not when a 9,000 × *g* supernate was used (Rane et al., 1977); preparation of microsomes reduces the V_{max} of phenytoin by about one-half without affecting the K_m (Kutt and Fouts, 1971). The use of intact cells such as hepatocytes may not necessarily overcome such problems because at least one additional factor is involved, namely, the ability of the drug and metabolites to diffuse into and out of the cells (deLannoy and Pang, 1986). If such diffusion is slow, then the apparent K_m value, but presumably not V_{max}, may differ from that observed in cell-free preparations. Other complicating factors may also occur so that the V_{max} or K_m values alone or together may be different from those in other *in vitro* systems (Gillette, 1984). Unfortunately, no general rules appear to exist that would indicate when hepatocytes, for example, reflect the *in vivo* situation more closely than do other preparations (Pang et al., 1985).

In addition to these factors, it must be recognized that the Michaelis-Menten mechanism is probably an oversimplification of actual intracellular events. The relationship between V_{max} and K_m may be quite complex and modified by changing the concentrations of, for example, cosubstrate, activators, and even products of the substrate and cosubstrate. Moreover, a multienzyme system is more frequently operative in a drug's overall metabolism than is a single enzyme. These enzymes may have widely different characteristics, including reaction mechanisms, so that any overall kinetic value like intrinsic clearance (Equation 4) may not accurately reflect in *in vivo* determinants (Gillette, 1984).

A further consideration in extrapolating *in vitro* biochemical data to the *in vivo* pharmacokinetic situation is the validity of the venous equilibration model (Equation 7) to describe organ elimination. In this model the organ is considered to be a single well-mixed compartment, and diffusion from the blood into the tissue is not considered to be rate-limiting. Accordingly, the concentration of unbound drug in the emergent blood is assumed to be in equilibrium with that in the tissue, and the concentration profile across the organ is constant. Two alternative perfusion-limited models have been developed, however, that have the merit that an arterio-venous concentration gradient exists across the organ. The sinusoidal perfusion or parallel tube model conceives of the organ as a large number of identical cylindrical tubes that are arranged in parallel with the cells (hepatocytes), each of which has the same eliminating activity, surrounding the cylinder. The relationship between intrinsic clearance and extraction is then given by Equation 8 (Bass et al., 1976; Pang and Rowland, 1977; Winkler et al., 1979):

$$E = 1 - e^{-\frac{f_B^u CL_{int}^u}{Q}}. \qquad (8)$$

By contrast, the dispersion model of elimination conceives of the organ as a packed-bed chemical reaction with nonideal flow (Roberts and Rowland, 1986). Two parameters characterize this model; the efficiency number (R_N) which describes the efficiency of drug removal and is equivalent under first-order conditions to $f_B^u CL_{int}^u/Q$, and an axial dispersion number D_N. The latter is a measure of dispersion or spread of residence times of drug molecules moving through the organ; the higher the value of D_N, the greater the degree of axial dispersion, reflective of functional heterogeneity. The mathematical solution for the extraction ratio is quite complex (Equation 9):

$$E = 1 - \frac{4a}{(1 + a)^2 \exp[(a - 1)/2D_N] - (1 - a)^2 \exp - [(a + 1)/2D_N]}, \qquad (9)$$

where $a = (1 + 4R_N D_N)^{1/2}$. Interestingly, when $D_N \rightarrow \infty$, i.e., when there is extensive axial dispersion, the model devolves to the venous-equilibration situation (Equation 7), whereas when $D_N \rightarrow 0$, i.e., axial dispersion is negligible, Equation 9 reduces to a form similar to that of the sinusoidal perfusion model (Equation 8).

In global terms, all three models predict similar relationships between the biological determinants of drug elimination, but they differ at the finer level. For example, an *in vitro* estimate of intrinsic clearance will predict a larger extraction ratio when incorporated into a sinusoidal perfusion model than if the same value is used in the venous equilibration model; the dispersion model provides an intermediate value. Moreover, the model discrepancies become larger as the intrinsic clearance increases, i.e., the greater the drug's extraction ratio. Discrimination studies to determine which model, if any, provides the best *in vitro*/*in vivo* predictions are, however, very limited. Recently, Roberts and Rowland (1985) concluded that the dispersion model was more consistent with published data for 10 drugs than either of the other two models, especially when the extraction ratio was greater than about 80%. Further studies of this aspect of prediction are clearly required.

Another modeling consideration requiring further investigation relates to the known functional heterogeneity of organs such as the liver. It is well established, for example, that hepatic perfusion is not uniform and intraorgan shunting is present; also, enzyme activity is not uniformly distributed in hepatocytes. The venous-equilibration model cannot take such factors into account, whereas they are implicit in the dispersion model; and the sinusoidal perfusion model can be modified to reflect such heterogeneity (Bass, 1980, 1983; Bass et al., 1978; Sawada et al., 1985). Studies have begun to address this problem (Pang, 1983; Pang et al., 1986), particularly with regard to the formation and elimination of different metabolites; however, it is not a factor that can be currently taken into account in *in vitro/in vivo* predictions.

Other unresolved problems of *in vitro* extrapolation of drug metabolism include consideration of metabolites. In the past, the primary focus has been on unchanged drug and its disposition, but the formation and subsequent fate of a metabolite, stable or reactive, may in certain cases be a more important consideration. Complications due to such factors as suicide-substrate inactivation of metabolizing enzymes and cofactor depletion need to be considered when certain types of nonlinear pharmacokinetics are likely. Finally, it is important to note that there appears to be only a single reported study (Bäärnhielm et al., 1986) where the metabolism of a drug in humans has been predicted from *in vitro* data. Human liver preparations, for example, are becoming increasingly available, and it would be a reasonable extension to apply modeling approaches to this critical area. This is particularly important with respect to interspecies extrapolation by using physiologically based pharmacokinetic models. The well-established variability in metabolism between species logically precludes the use of any allometric scaling factor. Accordingly, studies in human tissue(s) will be required.

IN VITRO PREDICTION OF *IN VIVO* DRUG BINDING AND DISTRIBUTION

Drugs and other xenobiotics invariably bind in a reversible fashion, described by the law of mass action, to a variety of blood and tissue constituents. Such binding determines the unbound fraction in the blood/plasma, and also distribution from the intravascular space into tissues. Accordingly, for physiologically based pharmacokinetic modeling to be successful, estimates of this binding and distribution are required.

The various *in vitro* techniques and interpretative approaches for characterizing the kinetics of drug binding to plasma proteins are well-established, as are various factors that modulate such binding (Reidenberg and Erill, 1986; Tillement and Lindenlaub, 1986). A limited number of examples exist in which *in vitro* parameters of linear or nonlinear plasma binding has been explicitly incorporated into physiologically based pharmacokinetic models of specific compounds (Bischoff and Dedrick, 1968; Engasser et al., 1981; Igari et al., 1983; Tsuji et al., 1983). But this factor is generally ignored, and only total drug is considered. Given the potential importance of plasma binding on the pharmacokinetics of drug elimination and distribution (Pang and Rowland, 1977; Wilkinson, 1983; Wilkinson and Shand, 1975), and the fact that only the unbound drug is considered to be biologically active, the omission of this factor in modeling is regrettable. But, more importantly from the extrapolation standpoint is the validity of the assumption that the binding parameters established *in vitro* reflect those *in vivo*.

The conventional concept of plasma binding and drug transport out of the vascular space is that only unbound drug is able to penetrate membranes, and therefore, the unbound fraction and concentration in the capillary blood are the important determinants. Moreover, binding equilibrium is maintained within the capillary so that the kinetics of binding is the same as that in systemic blood, and this can be estimated by an appropriate *in vitro* technique such as equilibrium dialysis. An increasing number of experimental studies with xenobiotics that are normally very extensively bound to plasma proteins (>99 percent), however, appear to be inconsistent with this dogma. Instead, it appears that bound drug is in some fashion intimately involved in uptake by such organs as the liver, heart, and brain, beyond its role as a passive store for replenishing the unbound moiety subsequent to its extraction. The mechanism involved in such transport is not well understood and may well differ, depending on the drug and the experimental situation (Jones et al., 1986; Pardridge, 1986; Weisiger, 1986). Nevertheless, it is clear that in this type of situation conventional modeling approaches will be of little value in predicting drug disposition.

For a noneliminating organ, the mass balance equation for drug transport is:

$$\frac{dC_T}{dt}V_T = Q(C_{in} - C_{out}). \tag{10}$$

Generally, it as assumed that distribution is perfusion limited, so that the emergent venous blood is in equilibrium with the average total concentration in the tissue (C_T), so that Equation 10 can be modified to the familiar form:

$$\frac{dC_T}{dt}V_T = Q\left(C_{in} - \frac{C_T}{K_p}\right), \tag{11}$$

where K_p is the tissue/blood (plasma) drug concentration ratio, also called the partition coefficient or solubility of drug in the tissue. Accordingly, knowledge of this parameter is critical in the physiological modeling procedure.

In practice, K_p is generally determined by direct measurement after drug administration and analysis of the arterial blood (plasma) and the tissues of interest. Sampling is usually performed after intravenous drug infusion to steady-state, although it is also possible to determine K_p after an intravenous bolus dose subsequent to the attainment of pseudo-equilibrium distribution. Theoretically, the partition coefficient during the terminal phase of elimination ($K_{p,app}$) is greater than the value obtained at steady state, the difference being smaller the more rapidly drug distributes into the tissue and the faster the rate of elimination (Equation 12):

$$K_{p,app} = \frac{K_p}{1 - \dfrac{\lambda_z}{k_T}}, \tag{12}$$

where λ_z and k_T are the first-order rate constant during the terminal elimination phase and for tissue uptake, respectively (Rowland, 1986). In most instances, it is likely that $k_T > \lambda_z$, so that the difference between K_p and $K_{p,app}$ is small, and experimental comparisons support this assumption (Lin et al., 1982; Schuhmann et al., 1987). It has also been demonstrated that determination of a partition coefficient is further complicated if elimination occurs in the tissue (Chen and Gross, 1979; Lam et al., 1982). In this case, the appropriate blood concentration used to estimate K_p should be that in the emergent venous blood of the organ rather than the arterial level (Rowland, 1986).

Under steady-state conditions, and recognizing that total drug concentration can be defined in terms of the unbound fraction and concentration, Equation 11 can be rearranged to show that K_p is equal to the ratio of the unbound fractions in the blood and tissue (f_B^u/f_T^u). Thus, in theory the *in vitro* estimation of the extent of a drug's plasma and tissue binding could be used to predict the *in vivo* partition coefficient. If valid, this approach would have considerable value in physiologically based pharmacokinetics.

The measurement of tissue binding and determining factors has not been as extensively explored as that involving plasma proteins. This may partly be explained by methodological problems, but mainly through the studies of Kurz and Fichtl (1983) and colleagues (Fichtl and Schuhmann, 1986; Schuhmann et al., 1987) these now appear to be mostly resolved. Tissue binding can be readily determined by ultrafiltration (Kurz and Fichtl, 1983) or equilibrium dialysis (Igari et al, 1982) of tissue homogenates, with appropriate correction for the effects of dilution (Kurz and Fichtl, 1983). Such binding demonstrates many of the characteristics of binding to plasma proteins, including saturability. Importantly, there does not, in general, appear to be any useful correlation between the two phenomena. Thus, plasma binding cannot be used to predict tissue binding. Also, binding to tissue constitutents appears to involve mechanisms besides simple hydrophobic interactions as related to lipid solubility measured by partitioning into organic solvents.

Good correlations (Figure 1) have been found between predicted K_p values based on *in vitro* measurement of f_P^u and f_T^u and those determined *in vivo* following intravenous infusion to steady-state of a number of drugs with widely different physicochemical and partitioning characteristics (Schuhmann et al., 1987). The best prediction was for muscle ($r = 0.93$), in which all of the data centered around a line with a slope of unity. A similar finding was obtained for the liver, except for drugs with a high extraction ratio in which the *in vivo* value was lower than predicted; this probably was due to the use of the arterial rather than the more appropriate

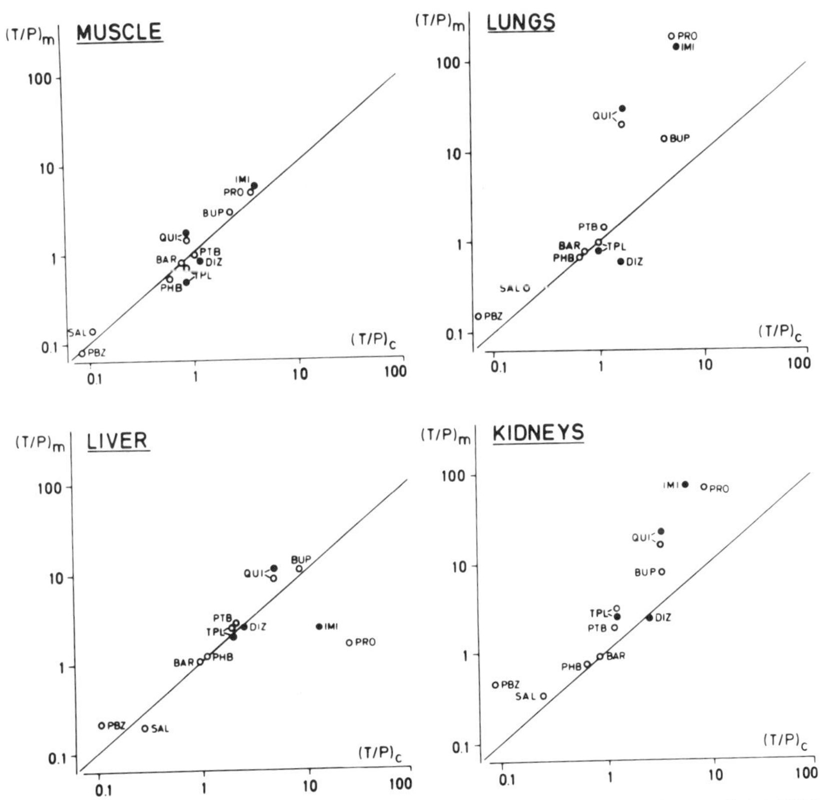

FIGURE 1 Correlation between experimentally measured $(T/P)_m$ and *in vitro*-predicted $(T/P)_c$ tissue partition coefficients for several drugs and tissues in the rabbit. Reproduced with permission from Schuhmann et al. (1987).

hepatic venous plasma concentration to estimate the *in vivo* partition coefficient. In the lung and kidney, the predicted K_p values of highly lipophilic, cationic drugs such as buphenine, quinidine, imipramine, and propranolol were lower than those measured *in vivo*. This probably reflects the loss of saturable uptake process in these homogenized tissues and the slow equilibration time for distribution *in vivo*.

These and other data (Harashima et al., 1983; Lin et al., 1982), although limited, indicate that it is possible to predict tissue partitioning *in vivo* for several tissues from their *in vitro* binding data, particularly if the drug is anionic or neutral. With cationic drugs, distribution into muscle, and possibly liver, is also well predicted; but discrepancies are possible with organs such as lung and kidney, in which active uptake processes may be present *in vivo*. A similar situation would be expected if tissue distribution is not perfusion limited; i.e., diffusion across the capillary epithelium is slow. Interestingly, preliminary findings indicate that binding of drugs to various different tissues are well-correlated and similar (Fichtl and Schuhmann, 1986). This raises the possibility that drug distribution to a variety of organs could be predicted by using a single tissue such as muscle. As an extension of this hypothesis, Schuhmann et al. (1987) have recently found that the unbound volume of distribution of drugs *in vivo* could be predicted by their binding to rabbit muscle tissue *in vitro*. Moreover, interspecies differences in tissue binding do not appear to be as pronounced as those in plasma binding. Partitioning of a large number of drugs into muscle tissue of rats, rabbits, and humans is very similar (Fitchtl and Schuhmann, 1986). If further substantiated, this would simplify considerably the interspecies extrapolation problem.

CONCLUSION

In summary, the concepts and experimental techniques for predicting the *in vivo* distribution of a drug into tissues and its elimination by metabolism by using *in vitro* data have been established. Experimental data to substantiate and validate these approaches are still relatively limited, but they are sufficiently encouraging to suggest practical feasibility. Clearly, much more investigation is required, particularly with respect to delineation of the approaches' limitations. In this regard it is important to define the goals and recognize the purposes of any *in vitro*/*in vivo* prediction. As the precision and accuracy requirements become more rigorous, the more difficult it will be to make a prediction. On the other hand, much useful information and insight can be gained from appropriately interpreted *in vitro* data, even though the prediction is not necessarily exact. Finally, it must be appreciated that interindividual variability, particularly in drug metabolism, may be quite extensive, and predictions based on a single or average set of data may be of little relevance to the individual.

REFERENCES

Bäärnhielm, C., H. Dählback, and Skånberg, I. 1986. *In vivo* pharmacokinetics of felodipine predicted from *in vitro* studies in rat, dog and man. Acta Pharm. Toxicol. 59: 113–122.

Bass, L. 1980. Flow dependence of a first-order uptake of substances by heterogeneous perfused organs. J. Theor. Biol. 86:363–376.

Bass, L. 1983. Saturation kinetics in hepatic removal: A statistical approach to functional heterogeneity. Am. J. Physiol. 244:G583–G589.

Bass, L., S. Keiding, K. Winkler, and N. Tygstrup. 1976. Enzymatic elimination of substrates flowing through the intact liver. J. Theor. Biol. 61:393–409.

Bass, L., P. Robinson, and A. J. Bracken. 1978. Hepatic elimination of flowing substrates: The distributed model. J. Theor. Biol. 72:161–184.

Bischoff, K., and R. L. Dedrick. 1968. Thiopental pharmacokinetics. J. Pharm. Sci. 57:1346–1351.

Chen, H.-S. G., and J. F. Gross. 1979. Estimation of tissue-to-plasma partition coefficients used in physiological pharmacokinetic models. J. Pharmacokinet. Biopharm. 7:117–125.

Collins, J. M., D. A. Blake, and P. G. Egner. 1978. Phenytoin metabolism in the rat. Pharmacokinetic correlations between *in vitro* hepatic microsomal enzyme activity and *in vivo* elimination kinetics. Drug Metab. Dispos. 6:251–257.

Dedrick, R. L., D. D. Forrester, and D. H. W. Ho. 1972. *In vitro-in vivo* correlation of drug metabolism—deamination of 1-β-D-arabinofuranosylcytosine. Biochem. Paramacol. 21:1–16.

Dedrick, R. L., D. D. Forrester, J. N. Cannon, S. M. El Dareer, and L. B. Mellett. 1973. Pharmacokinetics of 1-β-D-arabinofuranosylcytosine (Ara-C) deamination in several species. Biochem. Pharmacol. 22:2405–2417.

deLannoy, I. A. M., and K. S. Pang. 1986. Presence of a diffusional barrier on metabolite kinetics: Enalaprilat as a generated versus preformed metabolite. Drug Metab. Dispos. 14:513–520.

Engasser, J. M., F. Sarhan, C. Falcoz, M. Minier, P. Letourneur, and G. Siest. 1981. Distribution, metabolism, and elimination of phenobarbital in rats: Physiologically based pharmacokinetic model. J. Pharm. Sci. 70:1233–1238.

Fichtl, B., and G. Schuhmann. 1986. Relationships between plasma and tissue binding of drugs. Pp. 255–271 in Symposia Medica Hoeschst, Vol. 20, Protein Binding and Drug Transport, J.-P. Tillement and E. Lindenlaub, eds. Stuttgart: F. K. Schattauer Verlag.

Gillette, J. R. 1971. Factors affecting drug metabolism. Ann. N.Y. Acad. Sci. 179: 43–66.

Gillette, J. R. 1984. Problems in correlating *in vitro* and *in vivo* studies of drug metabolism. Pp. 235–252 in Pharmacokinetics: A Modern View, L. Z. Benet, G. Levy, and B. L. Ferraiola, eds. New York: Plenum.

Harashima, H., Y. Sugiyama, Y. Sawada, T. Iga, and M. Hanano. 1983. Comparison between *in vivo* and *in vitro* tissue-to-plasma unbound concentration ratios ($K_{p,f}$) of quinidine in rats. J. Pharm. Pharmacol. 36:340–342.

Hilliker, K. S., and R. A. Roth. 1980. Prediction of mescaline clearance by rabbit lung and liver from enzyme kinetic data. Biochem. Pharmacol. 29:253–255.

Huang, S.-M., Y. C. Huang, and W. L. Chiou. 1981. Oral absorption and presystemic first-pass effect of chlorpheniramine in rabbits. J. Pharmacokinet. Biopharm. 9: 725–738.

Igari, Y., Y. Sugiyama, S. Awazu, and M. Hanano. 1982. Comparative physiologically based pharmacokinetics of hexobarbital, phenobarbital and thiopental in the rat. J. Pharmacokinet. Biopharm. 10:53–75.

Igari, Y., Y. Sugiyama, Y. Sawada, T. Iga, and M. Hanano. 1983. Prediction of diazepam disposition in the rat and man by a physiologically based pharmacokinetic model. J. Pharmacokinet. Biopharm. 11:577–593.

Igari, Y., Y., Sugiyama, Y. Sawada, T. Iga, and M. Hanano. 1984. *In vitro* and *in vivo* assessment of hepatic and extrahepatic metabolism of diazepam in the rat. J. Pharm. Sci. 73:826–828.

Jones, D. R., S. D. Hall, R. A. Branch, E. K. Jackson, and G. R. Wilkinson. 1986. Plasma binding and brain uptake of benzodiazepines. Pp. 311–324 in Symposium Medica Hoeschst, Vol. 20, Protein Binding and Drug Transport, J.-P. Tillement and E. Lindenlaub, eds. Stuttgart: F. K. Schattauer Verlag.

Klippert, P., P. Borm, and J. Noordhoek. 1982. Prediction of intestinal first-pass effect of phenacetin in the rat from enzyme kinetic data—correlation with *in vivo* data using mucosal blood flow. Biochem. Pharmacol. 31:2545–2548.

Kurz, H., and B. Fichtl. 1983. Binding of drugs to tissues. Drug. Metab. Rev. 14:467–510.

Kutt, H., and J. R. Fouts. 1971. Diphenylhydantoin metabolism by rat liver microsomes and some effects of drug or chemical pretreatment on diphenylhydantoin metabolism by rat liver microsomal preparations. J. Pharmacol. Exp. Ther. 176:11–26.

Lam, G., M.-L. Chen, and W. L. Chiou. 1982. Determination of tissue to blood partition coefficients in physiologically-based pharmacokinetic studies. J. Pharm. Sci. 71: 454–456.

Lin, J. H., Y. Sugiyama, S. Awazu, and M. Hanano. 1982. *In vitro* and *in vivo* evaluation of the tissue-to-blood partition coefficient for physiological pharmacokinetic models. J. Pharmacokinet. Biopharm. 10:637–647.

Pang, K. S. 1983. The effect of intercellular distribution of drug-metabolizing enzymes on the kinetics of stable metabolite formation and elimination by the liver: First-pass effects. Drug. Metab. Rev. 14:661–76.

Pang, K. S., and M. Rowland. 1977. Hepatic clearance of drugs. I. Theoretical considerations of a "well-stirred" model and "parallel tube" model. Influence of hepatic blood flow, plasma and blood cell binding, and the hepatocellular enzymatic activity on hepatic drug clearance. J. Pharmacokinet. Biopharm. 5:625–653.

Pang, K. S., P. Kong, J. A. Terrell, and R. E. Billings. 1985. Metabolism of acetaminophen and phenacetin by isolated rat hepatocytes: A system in which spatial organization inherent in the liver is disrupted. Drug Metab. Dispos. 13:42–50.

Pang, K. S., J. A. Terrell, S. D. Nelson, K. F. Feuer, M.-J. Clements, and L. Endrenyi. 1986. An enzyme-distributed system for lidocaine metabolism in the perfused rat liver preparation. J. Pharmacokinet. Biopharm. 14:107–130.

Pardridge, W. M. 1986. Transport of plasma protein-bound drugs into tissues *in vivo*. Pp. 277–292 in Symposia Medica Hoeschst, Vol. 20, Protein Binding and Drug Transport, J.-P. Tillement and E. Lindenlaub, eds. Stuttgart: F. K. Schattauer Verlag.

Rane, A., G. R. Wilkinson, and D. G. Shand. 1977. Prediction of hepatic extraction ratio from *in vitro* measurement of intrinsic clearance. J. Pharmacol. Exp. Ther. 200: 420–424.

Rane, A., J. Säwe, B. Lindberg, J.-O. Svensson, M. Garle, R. Erwald, and H. Jorulf. 1984. Morphine glucuronidation in the rhesus monkey: A comparative *in vivo* and *in vitro* study. J. Pharmacol. Exp. Ther. 229:571–576.

Reidenberg, M. M., and S. Erill. 1986. Drug-Protein Binding. New York: Praeger.

Rowland, M. 1986. Physiologic pharmacokinetic models and interanimal species scaling. Pharmacol. Ther. 29:49–68.

Roberts, M. S., and M. Rowland. 1985. Correlation between *in vitro* microsomal enzyme activity and whole organ hepatic elimination kinetics: Analysis with a dispersion model. J. Pharm. Pharmacol. 38:177–181.

Roberts, M. S., and M. Rowland. 1986. A dispersion model of hepatic elimination. 1-3. J. Pharmacokinet. Biopharm. 14:227–308.

Rowland, M., L. Z. Benet, and G. G. Graham. 1973. Clearance concepts in pharmacokinetics. J. Pharmacokinet. Biopharm. 1:123–136.

Sawada, Y., Y. Sugiyama, Y. Miyamoto, T. Iga, and M. Hanano. 1985. Hepatic drug clearance model: Comparison among the distributed, parallel-tube and well-stirred models. Chem. Pharm. Bull. 33:319–326.

Schuhmann, G., B. Fichtl, and H. Kurz. 1987. Prediction of drug distribution *in vivo* on the basis of *in vitro* binding data. Biopharm. Drug. Dispos. 8:73–76.

Skånberg, I. 1980. Metabolism of two beta-adrenoceptor antagonists, alprenolol and metoprolol, in different species. *In vitro* and *in vivo* correlations. Acta Universitatis Upsaliensis (Abstracts of Upsalla Dissertations from the Faculty of Pharmacy), Vol. 50.

Smith, B. R., and J. R. Bend. 1980. Prediction of pulmonary benzo(a)pyrene 4,5-oxide clearance: A pharmacokinetic analysis of epoxide-metabolizing enzymes in rabbit lung. J. Pharmacol. Exp. Ther. 214:478–482.

Tillement, J.-P., and E. Lindenlaub, ed. 1986. Symposia Medica Hoeschst, Vol. 20, Protein Binding and Drug Transport. Stuttgart: F. K. Schattauer Verlag.

Tsuji, A., T. Yoshikawa, K. Nishide, H. Minami, M. Kimura, E. Nakashima, T. Terasaki, E. Miyamoto, C. H. Nightingale, and T. Yamana. 1983. Physiologically based pharmacokinetic model for β-lactam antibiotics. I. Tissue distribution and elimination in rats. J. Pharm. Sci. 72:1239–1252.

Weisiger, R. A. 1986. Non-equilibrium drug binding and hepatic drug removal. Pp. 293–310 in Symposia Medica Hoeschst, Vol. 20, Protein Binding and Drug Transport, J.-P. Tillement and E. Lindenlaub, eds. Stuttgart: F. K. Schauttauer Verlag.

Wiersma, D. A., and R. A. Roth. 1980. Clearance of 5-hydroxytryptamine by rat lung and liver: The importance of relative perfusion and intrinsic clearance. J. Pharmacol. Exp. Ther. 212:97–102.

Wilkinson, G. R. 1983. Plasma and tissue binding considerations in drug disposition. Drug Metab. Rev. 14:427–465.

Wilkinson, G. R., and D. G. Shand. 1975. A physiological approach to hepatic drug clearance. Clin. Pharmacol. Ther. 18:377–390.

Winkler, K., L. Bass, S. Keiding, and N. Tygstrup. 1979. The physiologic basis for clearance measurements in hepatology. Scand. J. Gastroentrol. 14:439–448.

Dose, Species, and Route Extrapolation: General Aspects

James R. Gillette

The objective of any pharmacokinetic study is to describe as simply as possible the factors that govern the time course of the concentrations of biologically active form(s) of a substance at putative action sites and to relate these concentrations to the incidence and magnitude of the biological responses. The biologically active form(s) may be the parent substance and/or one or more of its metabolites. In developing pharmacokinetic models to achieve this objective, investigators must consider a host of interrelated complex factors and events that could conceivably affect the time course of the parent substance and its biologically active metabolites at the action sites and to decide whether such factors are sufficiently important to include in the model.

DIFFERENT PROBLEMS AND OBJECTIVES, DIFFERENT MODELS

Clearly, the complexity of suitable models differs with the substance. A model that is suitable for describing the pharmacokinetics of substances, like ozone, which might be expected to act almost solely at the site of administration (such as lung) would clearly differ from a model designed to describe the pharmacokinetics of a substance (like halogenated hydrocarbons and heavy metals) that persists for several years in the body or in the food chain and causes toxicity in remote organs of the body. A model designed to describe the pharmacokinetics of a parent substance would be inadequate to describe the pharmacokinetics of a toxic metabolite. A model designed to describe the pharmacokinetics of a stable metabolite would differ from one designed to describe the pharmacokinetics of a suicide inhibitor. A model designed to describe the kinetics of a single dose of a substance may or may not be appropriate for the description of the kinetics of a substance entering the body by a series of repeated doses or at a continuous rate.

Different Mechanisms

When a substance enters the body by a series of repeated doses until a steady state is achieved, the investigator must also consider whether the incidence or magnitude of the biological response is most closely related to the maximum concentration, the average concentration, the minimum concentration, or the total dose of the biologically active form of the toxicant. The response will be most closely related to the maximum concentration of the unbound biologically active forms (1) when the response is manifested virtually instantaneously after the active forms interact reversibly with receptor sites, (2) when the response is due to irreversible interaction of a substance (or enzyme) that turns over rapidly in the body, or (3) when the biologically active form(s) causes physiological or biochemical changes that exceed the capacity of homeostatic mechanisms to adjust to them and result in irreversible damage. The response will be most closely related to the average concentration of the unbound biologically active form(s) when the response is due to certain irreversible mechanisms when the rates of replacement or repair of the action sites are slow. The response will be most closely related to the minimum concentration of the unbound active form(s) (1) when the response is caused by noncompetitive, but reversible, inhibition of an enzyme that alters the concentration of a vitally important endogenous substance, or (2) with certain kinds of irreversible mechanisms. The response will be more closely related to the total dose of the biologically active forms when the response is due to an irreversible accumulation of toxic products in the body. In this context, the accumulation of transformed cells that serve as clones for tumors can be viewed as the accumulation of toxic products.

GENERAL PHYSIOLOGICALLY BASED PHARMACOKINETIC MODELS

Unquestionably, the approach used in developing physiologically based pharmacokinetic models is the most versatile of all the approaches used in solving pharmacokinetic problems. As pointed out by K. B. Bischoff (this volume), one simply writes an appropriate set of differential equations, based on the conservation of mass law, which describe the rates of changes in the amounts of the substance and its metabolites in various compartments in blood, extracellular fluid, and cells within an organ and allows the computer programs to integrate the equations by iterative procedures. The investigator thus is not limited to first-order reactions that have simple analytic solutions. A complete description of all of the events that can occur within an organ, however, can be very complex. Sets of equations could be written to describe the rates of change in the amounts of various metabolites within cells and other compartments of the organ. Other sets of equations could be written to include changes in concentration of the substance and metabolites as they pass from the proximal to the distal portion of capillaries in a given organ.

Thus, the number of differential equations used by the investigator to describe a physiologically based pharmacokinetic model can be virtually infinite. The solution of such a model, however, would also require knowledge of the values of all of the rate constants that relate the amount of the substance and its metabolites in the various compartments to the rates of transfer into and out of each of the various compartments. Most of these rate constants, however, are not only unknown but also virtually unattainable.

Simplification of Models

Clearly, the investigator wishes to simplify the system to a minimum number of differential equations that still would provide all of the relevant information. How the model can be simplified and still be valid, however, will differ with the substance and the organ and the time frame within which the investigator wishes to focus attention.

Most simplifications are based on the realization that when a substance is constantly infused through an organ most reactions that occur within the intraorgan compartment will approach "virtual equilibria" and "virtual steady states" within the time required for the blood to pass through the capillary bed of the organ. For example, theoretical calculations of the rates of formation and dissociation of complexes of various small molecular weight substances with various proteins, including enzymes, will usually be considerably faster than the transit time of blood through organs. Thus, these complexes are usually represented by steady-state conditions.

Rates of Formation of Complexes

The validity of this conclusion can be illustrated by the events that occur immediately after a substrate is added to a solution containing an enzyme. Under these conditions the following differential equation can be written:

$$d[ES]/dt = \{[Et] - [ES]\}[S]k_1 - [ES](k_2 + k_3), \quad (1)$$

$$d[ES]/dt = [Et][S]k_1 - [ES](k_1[S] + k_2 + k_3), \text{ or} \quad (1a)$$

$$d[ES]/dt = [Et][S]k_1 - [ES]k_1\{[S] + [(k_2 + k_3)/k_1]\}. \quad (1b)$$

But $(k_2 + k_3)/k_1$ may be set to K_m and thus,

$$d[ES]/dt = [Et][S]k_1 - [ES]k_1([S] + K_m). \quad (1c)$$

Integration of Equation 1c gives:

$$[ES] = \frac{[Et][S]}{[S] + K_m}[1 - e^{-k_1([S]+K_m)t}]. \quad (1d)$$

Inspection of the exponent reveals that the half-time required to approach the maximum concentration of $[ES]$ is dependent on the value of k_1, as well as on $[S]$ and K_m. Although the value of k_1 may vary markedly, most values of k_1 between proteins and small molecular weight substances are between $10^5\ M^{-1}\ s^{-1}$ to $10^8\ M^{-1}\ s^{-1}$ (Taylor, 1972). Thus, for an enzyme in which the K_m is greater than 10^{-5}, the half-time for the approach to a steady state usually should be less than a second; thus, a virtual steady state usually would be achieved within 5 s. After a virtual steady state is achieved, the rate of metabolism of the substrate can be frequently expressed by the usual form of the Michaelis-Menten equation.

Equations analogous to Equation 1a can be written to describe the rate at which a substance may become reversibly bound to each of the components in blood, extracellular fluid, and cells in an organ. The time required to approach a virtual steady state depends largely on the second-order rate constant for the formation of the complex and the equilibrium constant of the complex. Since the average residence time of blood within capillaries is probably in the range of 1 to 10 s in most organs, one can be reasonably assured that virtual steady-state values can be assumed when either the unbound concentration or the $1/K_a$ values for reversible binding are greater than $10^{-5}\ M$. But one would worry when they are less than $10^{-8}\ M$. Equilibrium constants rather than rate constants of association and dissociation are thus generally used in the models, but the rate constant of dissociation for ultrahigh-affinity binding sites may still be important during the terminal phases of elimination, when $[S]$ decreases below the $1/K_a$ values or during constant exposure to very small doses.

Diffusional Barriers and Modified Fick's Law

Whether an investigator wishes to incorporate diffusional barriers in the model depends on the time required to achieve the virtual steady state in "filling" the extracellular and intracellular spaces. According to Fick's

Law of Diffusion, the rate of passive transfer of a substance across a thin membrane is usually written as,

$$\frac{dS}{dt} = \frac{DA}{X}(C_1^u - C_2^u), \qquad (2)$$

where D is the diffusivity index, A is the area of the membrane, and X is the thickness of the membrane. Since a membrane is usually considered to be lipoidal in character, the value of D is usually considered to be proportional to the partition of the substance between the cell membrane and water, which is frequently estimated from the partitioning of the substance between oil and water. When the substance is a weak acid or a weak base, it is usually assumed that only the neutral form of the substance is able to pass through semipermeable membranes and that the concentration of the neutral form can be estimated from the total concentration of the unbound substance by means of the Henderson-Hasselbalch equation. The ionized forms of many substances, however, can pass through semipermeable membranes, even though the diffusivity index for the ionized form may be orders of magnitude smaller than that for the neutral form. Thus, for weak acids with pK_a values that are several units smaller than 7, and weak bases with pK_a values that are several units larger than 7, the rate of diffusion can be governed predominantly by the concentration of the ionized form, despite its low diffusivity index. Moreover, membranes can differ markedly in their structure and function. For example, capillaries usually have intercellular spaces, which allow passage of both polar and nonpolar substances. Moreover, the plasma membranes of cells frequently permit the passage of very small substances through the membrane. Thus, the value of the effective D depends on a number of the diffusion indices representing different parts of the membrane (Pang and Gillette, 1979). For these reasons a modified version of Fick's Law of Diffusion is more appropriate (Figure 1). Thus, the passage of the substance across a membrane can be described by:

$$dS/dt = (D_1A/X)C_1^u - (D_2A/X)C_2^u, \qquad (2a)$$

in which both D_1 and D_2 incorporate not only the heterogenous nature of biological membranes but also the effects of different pH values and other factors that affect electrochemical gradients at the two interfaces of the membrane. In Equation 2a, both C_1^u and C_2^u are the concentrations of the unbound substance at the two interfaces of the membrane; they do not include the amount reversibly bound to proteins and other components in the compartments.

The rate at which the concentration of unbound substance changes in compartment 2 can be expressed by:

MODIFIED FICKS LAW OF DIFFUSION

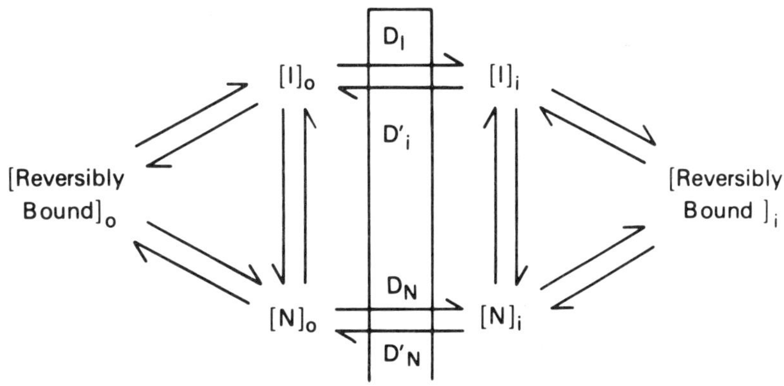

$$\frac{dSi}{dt} = [I]_o\left(\frac{D_I A}{x}\right) + [N]_o\left(\frac{D_N A}{x}\right) - [I]_i\left(\frac{D_I A}{x}\right) - [N]_i\left(\frac{D_{N'} A}{x}\right)$$

At Equilibrium

$$[I]_o\left(\frac{D_I A}{x}\right) + [N]_o\left(\frac{D_N A}{x}\right) = [I]_i\left(\frac{D_{I'} A}{x}\right) + [N]_i\left(\frac{D_{N'} A}{x}\right)$$

or

$$[I]_o\left(\frac{D_I}{D_N}\right) + [N]_o = [I]_{i'}\left(\frac{D_{I'}}{D_{N'}}\right) + [N]_i$$

N = Neutral : I = Ionized

FIGURE 1 Modified version of Fick's Law of Diffusion.

$$V_2^u\, dC_2^u/dt = (D_1 A/X)\, C_1^u - (D_2 A/X) C_2^u, \tag{2b}$$

where V_2^u is the apparent volume of distribution of the unbound substance in compartment 2. Since C_2^u is expressed as the concentration of unbound substance within the compartment, V_2^u would be defined as the total amount of substance in compartment 2 at any given time divided by C_2^u at that time. Under conditions in which C_1^u is constant, integration of Equation 2b would give the equation:

$$C_2^u = \frac{D_1 C_1^u}{D_2}[1 - e^{-(D_2 A/XV_2^u)t}]. \tag{2c}$$

Thus, the approach to a virtual equilibrium in the system depends on the apparent volume of compartment 2, but the ratio of the concentrations of unbound substance in the two compartments at equilibrium does not.

When a substance leaves compartment 2 by first-order processes, such as metabolism, Equation 2c would be modified to:

$$C_2^u = \frac{(D_1\, A/X)C_1^u}{(D_2\, A/X) + CL_{20}^u}\,[1 - e^{-\{[(D_2\, A/X) + CL_{20}^u]/V_2^u\}t}], \qquad (2d)$$

where CL_{20}^u represents the metabolic clearance. When a substance leaves the compartment by passage into another fluid, such as lymph or the cerebral spinal fluid, the CL_{20}^u is replaced by CL_{20}^u/f^u. Thus, inclusion of the clearance term actually decreases the time required to come to a virtual steady state. But because of the ratio of the concentrations of unbound substance in the two compartments, a virtual steady state would be dependent on the relative values of $(D_1^u\, A/X)$ and CL_{20}^u.

Whether the approach to a virtual steady state for the diffusion of substance is less than a few seconds, and therefore can be excluded from consideration by the investigator, depends not only on the substance but also on the organ and even perhaps on the types of cells within an organ. Obviously, a lipid-soluble substance in an organ in which the apparent volume of distribution is small is unlikely to be diffusion limited. Indeed, the attempts to demonstrate diffusion-limited metabolism with such substances have invariably failed. With polar substances, however, the conclusion is less certain. Obviously, diffusion limitations do take place with substances that are apparently excluded from the brain; in such cases, the low brain concentrations are maintained at low levels predominantly by the cerebral spinal flow. Moreover, the rate of metabolism of polar substances in liver can be diffusion limited (deLannoy and Pang, 1986).

Simple PB-PK Models

Frequently, pharmacokineticists that develop physiologically based pharmacokinetic models (PB-PK models) simply assume that the substance in the capillaries of an organ distribute instantaneously to all compartments within the organ, that metabolism within the cell is diffusion independent, and that the approach to virtual steady states in all organs is limited solely by blood flow rates. They describe the approach to a virtual steady state of the entire organ by a very simple differential equation. For a well-stirred model, in which the total concentration of a substance in capillary blood is assumed to equal its total concentration in venous blood leaving the organ,

$$\frac{V_i\, dC_i}{dt} = Q_1 C_a - \frac{Q_i C_i}{R_i} - \frac{V_{max(i)} C_i}{R_i K_{m(i)} + C_i} - \frac{CL_{int(i)} C_i}{R_i}, \quad (3)$$

where V_i is the actual volume of the organ, C_i is the amount of a substance in the organ at any given time divided by the volume of the organ, Q_i is the blood flow rate through the organ, $V_{max(i)}$ and $K_{m(i)}$ are Michaelis-Menten parameters, and $CL_{int(i)}$ is the intrinsic clearance of other processes of elimination from the organ. R_i is the ratio of C_i to the total concentration in the blood within the organ, $C_{v(i)}$.

Basic Parameters f^u and R

Although within the limitations of the assumptions Equation 3 is correct as written, it nevertheless is deceptive in that certain relationships may not necessarily be obvious. Equation 3 could just as easily be written as:

$$\frac{R_i V_i\, dC_{v(i)}}{dt} = Q_i\, [C_a - C_{v(i)}] - \frac{V_{max(i)}\, C_{v(i)}}{K_{m(i)} + C_{v(i)}} - CL_{int(i)}\, C_{v(i)}. \quad (3a)$$

In this form of the equation it becomes obvious that the K_m value and the $CL_{int(i)}$ value are expressed in terms of the total concentration of the substance. But it is the unbound concentration in capillary blood, which in turn is assumed to be its unbound concentration within the cells of the organ, that governs the activity of the enzyme within the cells. Thus, there would be no way of relating a K_m value or the CL_{int} value obtained from *in vitro* experiments with the K_m and the CL_{int} values used in Equations 3 and 3a. To emphasize this point, Equation 3 more properly should be written as:

$$\frac{V_i\, dC}{dt} = Q_i C_a - \frac{Q_i C_i}{R_i} - \frac{V_{max(i)} C_i}{[R_i\, K_{m(i)}/f^u] + C_i} - \frac{CL_{int(i)} f^u C_i}{R_i}, \text{ or} \quad (3b)$$

$$\frac{R_i V_i dC_{v(i)}}{dt} = Q_i\, [C_a - C_{v(i)}] - \frac{V_{max(i)}\, f^u C_{v(i)}}{K_m + f^u C_{v(i)}} - CL_{int(i)}\, f^u C_{v(i)}, \quad (3c)$$

where f^u is the unbound fraction of the substance in blood.

The term C_i represents the total concentration of the substance in the organ. It thus does not necessarily represent the total concentration of the substance in any compartment within the organ. Moreover, R_i is assumed to be a constant, but because reversible binding of substances to proteins and other components in blood plasma, erythrocytes, and tissue cells may be nonlinear, it should be regarded as a mathematical function, which under certain conditions may be virtually a constant. Furthermore, the rate of metabolism is assumed to be independent of the rates at which the substance diffuses through capillary and cellular membranes. When the rate of metabolism of polar substances in liver is thought to be diffusion limited (deLannoy and Pang, 1986), the equations would have to be modified.

Nonlinear Kinetics and Lost Concepts

When the concentration of substances approach or exceed the K_m values of enzymes or transport systems or the $1/K_a$ values of binding sites in blood or tissues, the meanings of such terms as organ clearances, organ availabilities, organ extraction ratios, volumes of distribution, total body clearances, and biological half-life become obscure. Moreover, the usefulness of measurements of area under the curves of the substance in blood and tissues in evaluating many of these parameters, as well as bioavailability, is largely lost. When the concentrations of substances and their metabolites are small, however, physiologically based, linear compartment models can be developed, which provide information that cannot be obtained from either general pharmacokinetic approach alone.

INTERFACE BETWEEN PB-PK MODELS AND CLEARANCES

When $C_{v(i)}$ never exceeds a value equal to about 10% of the smallest K_m of any drug-metabolizing enzyme in an organ, the rate of metabolism by the enzyme can be written as a linear equation:

$$f^u C_{c(i)} V_{max}/K_m = f^u C_{v(i)} CL_{int}, \qquad (4)$$

and we can rewrite Equation 3c as:

$$\frac{R_i V_i \, dC_{v(i)}}{dt} = Q_i[C_a - C_{v(i)}] - f^u C_{v(i)}(\text{Sum } CL_{int}). \qquad (4a)$$

Organ Availabilities (*F*), Extraction Ratios (*E*), and Clearances (*CL*)

After a virtual steady state is achieved in the organ, $R_i V_i \, dC_{v(i)}$ can be set equal to 0, and the resulting equation can be written as:

$$Q_i\{1 - [C_{v(i)}/C_a]\} = [f^u C_{v(i)}/C_a] \text{ Sum } CL_{int} = CL_i. \qquad (4b)$$

The term $C_{v(i)}/C_a$ is frequently called the organ availability, F_i, and can be obtained by another rearrangement of the equation:

$$F_i = C_{v(i)}/C_a = Q_i/(Q_i + f^u \text{ Sum } CL_{int}). \quad (4c)$$

Substitution of Equation 4c into Equation 4b provides the following equation:

$$CL_i = Q_i[1 - F] = Q\,E_i = \frac{Q_i f^u \text{ Sum } CL_{int}}{Q_i + f^u \text{ Sum } CL_{int}}, \quad (4d)$$

where E_i is referred to as the extraction ratio of the substance. Equation 4d is the same equation for a well-stirred model discussed by G. R. Wilkinson (this volume) and is the link between the differential equations used in general physiologically based pharmacokinetic models and the equations used in the physiologically based linear compartment pharmacokinetic models.

Physiologically Based Linear Compartmental Pharmacokinetic Models

In these models investigators simply assume that all elimination processes that occur in organs approach virtual steady states rapidly. They visualize how the individual organ clearances can be incorporated into a total body clearance term and obtain a first-order elimination constant by dividing the total body clearance by an apparent volume of an empirical central compartment that includes not only the blood and the organs of elimination but also the nonelimination organs that would be included in this apparent volume of distribution.

Validity of the Assumption of Virtual Steady State

In order to gain an insight into the time required to attain a virtual steady state in an organ, let us assume that QC_a remains constant during the approach to virtual steady state and integrate Equation 4a. We would obtain:

$$C_v = \frac{Q_i C_a}{Q_i + f^u \text{ Sum } CL_{int}}(1 - e^{-kt}), \quad (5)$$

where,

$$k = \frac{Q_i + f^u \text{ Sum } CL_{int(i)}}{R_i V_i}. \quad (5a)$$

Thus, the value of k, which governs the time required to approach a virtual steady state, is dependent on R_i, even though the steady-state value of C_v is not. The longest time required to approach a virtual steady state can be estimated by setting $k = Q_i/R_i V_i$. The effect of R_i on the maximal $t_{0.5}$

for the approach to virtual steady state in the organ can be estimated from the relationship, $t_{0.5} = R_i \ln 2/(Q_i/V_i)$. For example, most estimates of the blood rate through the liver of different animals, including humans, are usually between 1 to 3 ml min^{-1} g^{-1}. On setting $Q_i/V_i = 1.0$ min^{-1}, we can calculate that $\ln 2/(Q_i/V_i) = 0.693$ min^{-1}. Thus, we can calculate that the longest $t_{0.5}$ for a substance with a $R_i = 10$ in liver would be about 7 min. Similar estimates can be made for other organs in the body. The assumption that the major organs of elimination (namely liver, lung, and kidney) attain virtual steady states by the time that the first blood sample is taken thus will be reasonably valid for most substances.

Calculation of Other Compartmental Model Parameters

The rate constants for the other compartments in the model can also be derived with the same parameters that are required for the solution of the simple physiologically based pharmacokinetic models. Each of the organs (or group of organs) would represent a separate compartment. The respective rate constants for the substance entering the organ from the central compartment would be (Q_i/R_cV_c), and the rate constants for the substance leaving the organ and entering the central compartment would be Q_iF/R_iV_i. After substitution of these rate constants into the appropriate La Place transform equation, the hybrid rate constants used in compartmental models can be obtained by finding the roots of the resulting equation. These in turn can be used to calculate the coefficients of the various phases. When there are many compartments in the model, however, the process can be quite laborious.

Approximations of Terminal Half-Lives

As long as the rate constant of absorption is not the smallest rate constant, an approximation of the half-life of the terminal phase can be obtained rather easily from the relationship:

$$A \ln 2 < t_{0.5\ (\text{terminal})} < (A + B) \ln 2, \tag{4e}$$

where,

$$A = (R_cV_c + \text{Sum } R_iV_i)/CL, \text{ and} \tag{4f}$$

$$B = \text{Sum } (R_iV_i/Q_i). \tag{4g}$$

In Equation 4f the value of A is the transit time of a linear model. During the workshop on which the volume is based, Leslie Benet suggested this value as an estimate of the minimum half-life of a substance in the body.

Approximate Time Required to Approach Steady State

During repeated administration of a substance, the amount of substance eliminated from the body during the dosage interval approaches the amount of substance administered during the dosage interval. The time required to approach within 95% of the steady-state value can be estimated by multiplying the sum of A and B in Equation 4e by 3.

LINEAR PHARMACOKINETIC SYSTEMS

Even though all processes that govern the pharmacokinetics of substances in organs do not always follow first-order kinetics, physiologically based pharmacokinetic models of first-order systems provide insights of relationships that are not always obvious. From these relationships, the investigator can identify parameters that are useful in comparing data for consistency and in making inferences and extrapolations that would be difficult, if not impossible, to obtain from direct experimentation.

Among the most useful are the relationships:

$$CL = \underset{\text{(single dose)}}{F\,Dose/(AUC)_a} = \underset{\text{(repeated doses)}}{\{F(Dose/\tau)/[(AUC)_a/\tau]\}_{ss}}$$

$$= \underset{\text{(constant infusion)}}{F\,k_0/C_{a(ss)}}, \qquad (6)$$

where F is the fraction of the dose that reaches arterial blood, $(AUC)_a$ is the area under the concentration-time curve in arterial blood, τ is the dosage interval between doses when the substance is administered repetitively, k_0 is the constant rate of infusion, and $C_{a(ss)}$ is the steady-state concentration ultimately achieved during constant infusion; $C_{a(ss)}$ implies that all organs and compartments within the organs are in steady state. These relationships should be valid as long as all processes that govern the parameters follow first-order kinetics and remain constant during the time frame under consideration.

The term steady state, as used in $\{F(Dose/\tau)/[(AUC)_a/\tau]\}_{ss}$, simply means that the administration of a substance is repeated until the amount of substance eliminated from the body during the dosage interval equals the amount of substance administered at regular intervals. The relationship may express a variety of different dosage schedules. For example, an individual may be exposed to a substance for 8 h/day for 5 days a week. In this situation the dosage interval would be 1 week, and the dose would be the total dose received during the week.

Total Body Clearance

In these equations, the total body clearance that governs the value of $C_{a(ss)}$ includes all of the processes by which the substance is eliminated from the body and not just the metabolic clearances by various organs in the body. Almost invariably investigators would consider the contributions of excretion into urine of the unchanged substance and perhaps the contribution of biliary excretion. But they frequently ignore elimination of volatile substances into exhaled air and through the skin, and of lipid-soluble substances directly into the gastrointestinal tract across the intestinal wall. Indeed, Fick's Law of Diffusion suggests that any site of administration that requires the passage of a substance across a biological membrane should also be considered as an organ of elimination. In fact, even saliva can be visualized as a fluid of elimination; indeed, assays of substances in saliva have been used as indirect measures of the concentration of unbound drug in blood (although with mixed success) (Horning et al., 1977).

One conclusion that results from these considerations is that the unbound concentration of an absorbed substance in arterial blood can never be greater than the maximum concentration of the substance at the site of administration, unless there are active transport systems or ion trapping effects due to pH differences. The maximum concentrations of metabolites, however, may exceed the concentration of the parent substance.

Importance of the Unbound Concentration of Substances

Because substances are usually bound to many components within tissues of an organ, in addition to possible reversible binding to action sites, measurements of the total concentration of the substance within an organ is proportional to the magnitude of the response only to the extent that the total concentration of the substance is proportional to the concentration of the substance in water in the immediate environment of the action site. *Investigators interested in biologically relevant concentrations thus should focus attention on the factors that govern the concentration of the unbound substances at action sites rather than on factors that govern the total concentration of the substance in organs containing the action sites.*

Classification of Organs; Routes of Administration

Although the concept of *AUC* values have usually been restricted to the concentration of substances in arterial blood, the concept can be used to estimate *AUC* values of substances in blood within capillaries of various organs and within the total organ. In this approach it is useful to classify the organs according to different categories (Gillette, 1985) (Figure 2). After a substance is absorbed from the site of administration, it enters a series of organs before it reaches the arterial circulation. All of the organs through which the substance has passed have been called *first-pass organs*. In turn, the first-pass organs can be classified into two groups: elimination organs and nonelimination organs. The elimination organs contain either mechanisms that remove the substance from the organ, such as metabolism or excretion into bile or expired air, or nonelimination organs, which do not contain any of these processes. After the substance reaches the arterial circulation, it is distributed to other organs in the body, which have been called non-first-pass organs. In turn, the non-first-pass organs can be divided into rapidly equilibrated and slowly equilibrated organs according to their *Q*/*VR* values. Rapidly and slowly equilibrated organs are further subdivided as nonelimination and elimination organs. Some organs, such as the brain, are visualized as elimination organs, because substances can be removed from the organ in fluids other than blood, even though the clearances by these processes do not contribute to the total body clearance of the substance. According to this classification, some organs such as the kidney are always non-first-pass organs. The lungs are always first-pass organs, except in certain experimental procedures. Some organs, such as the intestinal mucosa, the liver, and the skin, can be either first-

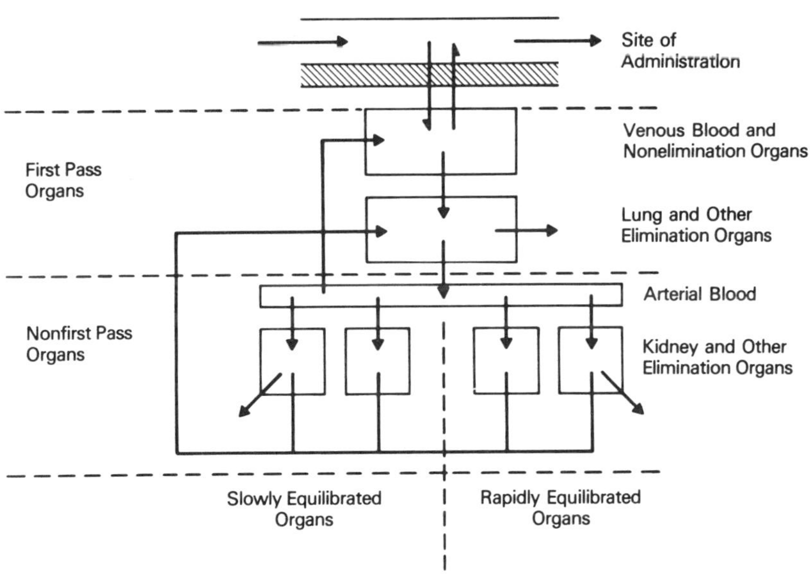

FIGURE 2 Relationships among sites of administration, first-pass organs, and non-first-pass organs in physiologically based models.

pass organs or non-first-pass organs, depending on the route of administration.

Non-First-Pass, Nonelimination Organs

Under steady-state conditions, there is no net flux of a substance between the blood and the tissues within a nonelimination organ. It follows, therefore, that the steady-state concentration of the substance within the capillaries of nonelimination organs that are also non-first-pass organs is identical to that in arterial blood. It also follows from the identities shown in Equation 6 that this is also true for the *AUC* values after a single dose and the steady-state *AUC* values during repetitive administration. This principle is the basis for the rationale that measurements of the *AUC* in venous blood passing through nonelimination, non-first-pass organs provide valid estimates of the *AUC* value in arterial blood, even though the shape of the time course of the concentrations in the arterial and the venous blood may differ. It also follows from this principle that the value of *R* for a given organ in linear reversible binding models can be estimated from the ratio of the area under the concentration curve of the substance in the organ to the area under the concentration curve in the blood. The value of *R* obtained from this relationship should be constant, provided that the concentration of reversibly bound substance to components in both blood and the tissues in the organ are directly proportional to the concentration of unbound substance (Gillette, 1984, 1985). Nonlinear, dose-dependent changes in the elimination of the substance in other organs, however, would not change the value of *R* in a nonelimination organ.

As long as changes in reversible binding to various components in different organs and blood do not result in changes in the steady-state concentration of unbound substance in arterial blood, they do not affect the steady-state concentration of unbound substance at action sites in non-first-pass organs that are not elimination organs. The steady-state concentration of unbound substance in arterial blood, however, is governed only by the available dose and the clearances of the elimination organs. Inspection of Equation 4d reveals that under steady-state conditions, the equation for the concentration of unbound substance in blood or in the organ does not contain any term for reversible binding of the substance in the organ cells. Thus, alterations of reversible binding to tissues other than the blood usually do not affect the average concentration of unbound substance in blood. Indeed, they do so only when the organ clearance is nonlinear and the changes in tissue binding result in oscillations in the concentration that affect any of the organ clearances.

The principle also predicts that the steady-state concentration of substance in both extracellular and cellular water (i.e., unbound concentration) in the organ is identical to that in blood, provided that the classical equation of Fick's Law of Diffusion is valid. It would not be valid, however, if there were ion trapping of weak acids and weak bases due to differences in pH between cellular and extracellular water of tissues in the organ or if there were active transport systems in cellular membranes (Figure 1).

Range of Maximum and Minimum Unbound Concentrations in Nonelimination Organs and Repetitive Administration

Linear compartmental models, based on the assumption that the classical form of Fick's Law of Diffusion is valid (i.e., $C_1^u = C_2^u$ at equilibrium), reveal that after a substance is administered repetitively until a steady state is achieved, the maximum concentration of unbound substance in any nonelimination compartment cannot exceed the maximum concentration of unbound substance in the central compartment during the dosage interval. In fact, when the maximum concentration of unbound substance is reached in any nonelimination organ, the concentrations of unbound substance in blood and the organ should be equal, a principle that provides another way of measuring R (i.e., $R = C_{i(max)}/C_{b(\text{at } t_{max})}$). Moreover, the minimum concentration of unbound substance in any nonelimination compartment cannot be less than the minimum concentration of unbound substance in the central compartment. Thus, under steady-state conditions the maximum and minimum concentrations of the unbound substance in the central compartment serve as the boundaries between which the concentration of unbound substance must be in any nonelimination compartment in the body (Gillette, 1984, 1985) (Figure 3).

When these principles are applied to physiologically based pharmacokinetic models, it follows that during a dosage interval under steady-state conditions, the maximum and minimum concentrations of unbound substance in arterial blood will be between the maximum and minimum concentrations of unbound substance in any non-first-pass, nonelimination organ. The maximum concentration of unbound substance in such organs will be between the maximum concentration and the AUC/τ value of the unbound substance in arterial blood, and the minimum concentration of unbound substance in such organs will be between the minimum concentration and the AUC/τ value of unbound substance in arterial blood. In organs with small $R \ln 2/(Q/V)$ values, the changes will tend to mimick the arterial concentrations rather closely. In organs with large $R \ln 2/(Q/V)$ values, the concentration of unbound substance will remain rather constant during the dosage interval and will mimick the AUC/τ value. From these considerations, it becomes evident that much information can be gained concerning the concentration of unbound substance in non-first-pass, nonelimination organs from studies of the concentration of unbound

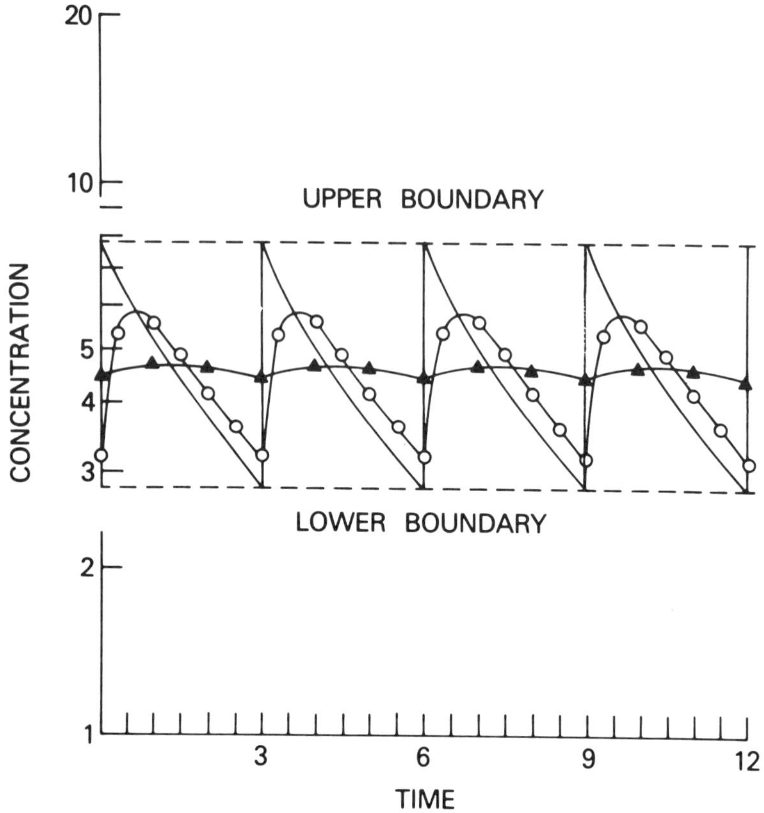

FIGURE 3 Steady-state maximum and minimum concentrations of the diffusible form of substances administered intravenously. The concentrations represent the boundaries between which the concentrations must lie in non-first-pass, nonelimination organs when the substance enters and leaves cells solely by passive diffusion. Blood (------); shallow compartment (○----○); deep compartment (▲-------▲).

substance in arterial blood, even when the *R* values for various organs are not known.

Non-First-Pass, Elimination Organs

According to the well-stirred model, the concentration of a substance in blood within the capillaries of an elimination organ is assumed to be the same as the concentration of the substance in venous blood leaving the organ. Thus, under steady-state conditions, the concentration of the substance in blood within the capillaries can be estimated from the arterial concentration and the ratio $C_{out}/C_{in} = F$. It follows, therefore, that the AUC_{org} within the capillaries can be estimated from the AUC of the substance in arterial blood by the relationship:

$$AUC_{org} = (AUC)_a F_{org}. \tag{7}$$

As pointed out by G. R. Wilkinson (this volume), many investigators have questioned the validity of the well-stirred model as a universally valid concept. Some have suggested the parallel tube model, in which the concentration of the substance in blood decreases gradually as the substance passes from the proximal to the distal ends of the capillaries. Others have suggested models that provide estimates between the well-stirred model and the parallel tube model. In the parallel tube model the average concentration of the substance in blood within the capillaries [$C_{(av)}$] can be estimated from the relationship:

$$C_{(av)} = (C_{in} - C_{out})/\ln(C_{in}/C_{out}), \text{ or} \tag{7a}$$

$$C_{(av)}/C_{in} = (1 - F)/\ln(1/F). \tag{7b}$$

As F approaches 1.0, however, both $C_{(av)}$ and C_{out} approach C_{in}, and thus, the distinctions among the models become less important when the value of $1 - F$ is small.

Portions of the blood pass through several organs from the time it leaves the arterial circulation until it reenters the arterial circulation. For example, blood passes through the intestinal mucosa into the portal vein and then through the liver and lungs before it reenters the arterial circulation. When several of these organs serve as elimination organs, adjustments must be made to relate the arterial concentration to the concentration entering the various sequential organs (Gillette, 1982; Rowland and Tozer, 1980).

FIRST-PASS, NONELIMINATION ORGANS

A substance can pass through several organs from the site of administration to the arterial circulation. If a substance is not eliminated in an organ in this pathway, that organ would be considered a first-pass, nonelimination organ.

The steady-state rate at which a substance enters a first-pass organ depends on two rates: the rate at which the substance enters the organ directly from the site of administration and the rate at which the substance enters the organ from the arterial circulation. If none of the first-pass organs were elimination organs, to determine the concentration of the substance entering a first-pass, nonelimination organ, under steady-state conditions, we can write:

$$C_{in} = [Rate_{(abs)} + Rate_{(art)}]_{ss}/Q_{org} = [Rate_{(abs)ss}/Q_{org}; + C_{a(ss)}. \quad (8)$$

The blood passing through the site of administration, however, may have passed through an elimination organ before it reached the first-pass, nonelimination organ, and some of the blood coming from the arterial circulation may have passed through organs of elimination before it reached the first-pass, nonelimination organ. Thus, adjustments may have to be made for both elements governing the concentration of substance entering the organ:

$$C_{in} = [F_1\ Rate_{(abs)(ss)}]/Q_{org} + F_2 C_{a(ss)}, \quad (8a)$$

where F_1 is the fraction of the absorbed dose that enters the organ, and F_2 reflects a rather complex function that represents an effective availability, which has been called a poly-input availability (Gillette, 1982).

As with non-first-pass, nonelimination organs, the steady-state concentration of the substance in blood within the capillaries of the organ and in venous blood leaving the organ is the same as that entering the organ. Thus, all of the conclusions that were valid for non-first-pass, nonelimination organs are also valid for first-pass, nonelimination organs, after the input concentration has been modified. In addition, however, the term $F_1\ Rate_{(abs)(ss)}/Q_{org}$ represents the minimum AUC value that the substance can attain even when the clearance by organs not contributing to F_1 are so large that C_a is virtually negligible. Indeed, the relative contributions of the two terms to the total steady-state concentration represents the difference by which different routes of administration may affect the steady-state concentration of the substance in first-pass organs.

First-Pass, Elimination Organs

According to the well-stirred model, the steady-state concentration of the substance in blood within the capillaries of a first-pass, elimination organ can be described by:

$$C_{out} = F_{org} C_{in}, \quad (9)$$

where C_{in} is described by Equation 8a.

According to the parallel tube model, the average concentration of the substance in blood within the capillaries of the organ, however, should be described by:

$$C_{(av)} = C_{in}\ [(1 - F)/\ln (1/F)]. \quad (9a)$$

ROUTE-TO-ROUTE EXTRAPOLATION

When the various pathways along which blood must pass from the various sites of administration to the arterial circulation are considered, it becomes evident that the lungs are a first-pass organ by almost any route of administration. Thus, the general equation for the steady-state concentration of a substance in arterial blood takes the form of:

$$C_{a(ss)} = \frac{k_0 F_{pre} F_L}{CL} . \tag{10}$$

CL is the total body clearance, which includes the clearance by the lung which in turn depends on the intrinsic clearances of exhalation and metabolism, F_{pre} is the fraction of the dose entering the body that enters the lung, and F_L is the availability of the substance in lung. For a well-stirred model:

$$F_L = \frac{k_0 Q_{co}}{Q_{co} + CL_{(exhalation)} + CL_{(met)}} , \tag{10a}$$

where Q_{co} is the cardiac output, and the other clearances are intrinsic clearances of exhalation and metabolism.

The effects of various routes of administration on the steady-state concentration of the substance in arterial blood thus depends on whether the route of administration changes F_{pre}.

Lungs and Skin Administration

When a volatile substance is inhaled and absorbed into blood in the lungs, the value of F_{pre} would be 1.0. When the same substance is absorbed through the skin, the value of F_{pre} depends on the fraction of the absorbed dose that escapes metabolism in skin and enters the lungs. Usually the fraction of the dose that is eliminated from the body by the skin by diffusion and metabolism is assumed to be negligible, and thus, the value of F_{pre} in this situation is assumed to be 1.0. Thus, the steady-state concentration obtained by two routes of administration (skin and lung) should be identical as long as all processes follow first-order kinetics and equal amounts of the substance are absorbed by the two routes of administration.

Whether an inhalation experiment provides sufficient information to predict the steady-state concentrations in blood depends on how the experiment is performed. Placing animals in a closed container and measuring the area under the concentration curve of the substance in the container provides an estimate only of the total metabolic clearance of the substance. It does not estimate the total body clearance, because the unchanged substance passes from the animals back into the container and thus does not measure the clearance of exhalation. Estimation of the total body clearance requires either measurements of the AUC in the blood of the animals or a combination of estimations of the amount of unchanged substance that would be exhaled as well as the amount metabolized.

Oral Administration

When a substance is administered orally, the steady-state concentration in arterial blood depends on where the substance is absorbed. If it is absorbed only in the mouth, the effect on F_{pre} would depend only on the fraction of the dose that escaped metabolism in the membrane lining of the mouth, but if the substance is absorbed in the stomach or the gastrointestinal tract (excluding certain parts of the rectum), the value of F_{pre} depends not only on the fraction of the absorbed dose that escaped metabolism by these tissues but also by the liver, because the venous blood draining these organs enters the portal circulation.

Biliary Excretion

Whether biliary excretion contributes to F_{pre} depends on whether the substance excreted into bile is reabsorbed. If it were completely reabsorbed, biliary excretion would not contribute to F_{pre}, although biliary excretion would affect the shape of the concentration-time curve during single or repeated administration of the substance.

Location of Organs of Elimination

The effects of different routes of administration on the steady-state concentration of the substance in first-pass organs depends on which organs contribute to the total body clearance of the substance. If the substance were eliminated from the body by only one organ in the body, the steady-state concentration of the substance in that organ would be independent of the route of administration, even though the organ may be a first-pass organ after one route of administration but a non-first-pass organ by another. The concentration of the substance in first-pass, nonelimination organs, however, would still depend on the route of administration.

The ratio of the steady-state concentrations of the substance in blood within first-pass elimination organs becomes much more complicated when the substance is eliminated by several organs of the body. For example, the well-stirred model predicts that when the same amount of a substance is absorbed and the substance is eliminated by the lungs, kidney, and liver, the ratio of the steady-state concentrations of the substance in blood within the sinusoids of the liver would be:

$$\frac{C_{H,ss(p.o.)}}{C_{H,ss(inh)}} = \frac{1}{F_L}\left[1 + \frac{Q_R}{Q_H}(E_L F_R + E_R) + \frac{(Q_{co} - Q_H - Q_R)E_L}{Q_H}\right]. \quad (11)$$

If the substance were eliminated from the body solely by lung and liver, the equation would become:

$$\frac{C_{H,ss(p.o.)}}{C_{H,ss(inh)}} = \frac{1}{F_L}\left[1 + \frac{(Q_{co} - Q_H)E_L}{Q_H}\right], \tag{11a}$$

where Q_{co} is the cardiac output, the E terms are extraction ratios, and the F terms are organ availabilities.

Notice that Equation 10 does not contain any terms representing the elimination of the substance by non-first-pass organs and Equation 11 does not contain any terms representing the extraction of the substance by the liver. Thus, the effects of different routes of administration on the steady-state concentrations of a substance in an organ depend solely on the elimination of the substance in organs other than the one under consideration.

Inspection of Equations 10 and 11 also reveals that the effects of different routes of administration depend on the organ availabilities (i.e., the F values). It follows, therefore, that when the apparent total body clearance, as measured by the dose/AUC value, is considerably smaller than the slowest blood flow rates through any organ of elimination, the value of all elimination organs is virtually 1.0 and the elimination of the substance is flow independent. Under these conditions any difference between the AUC values observed after different routes of administration can be attributed to differences in the fraction of the dose that is absorbed from the site of administration or to the effects of different rates of absorption on saturable processes of elimination.

Linear Steady-State Models for Metabolite Formation and Elimination

The rate of formation of a metabolite in cells within an organ obviously depends on the concentration of its precursor in the cells, and thus, the discussion of the factors that govern the concentration of the parent substance in the previous sections are relevant to the formation of its metabolites. The pharmacokinetics of metabolites, however, are far more subtle and frequently not intuitively obvious. If a metabolite were formed directly from the parent substance in a homogenous group of cells within a single non-first-pass organ, a linear, well-stirred, steady-state concentration of the metabolite in arterial blood can be described by Equation 12 (Gillette, 1985; Pang and Gillette, 1979):

$$C^M_{a(ss)} = \frac{k_0 F^S_a F^S_{orgJ} f^u CL^{S\to M}_{int} F^M}{CL^S CL^M}, \tag{12}$$

where F^S_a represents the fraction of the dose of the substance that enters the arterial circulation and reaches the organ in which the metabolite is formed. F^S_{org} is the organ availability for the substance, and f^u is the unbound fraction of the substance. $CL^{S \to M}_{int}$ is the intrinsic clearance of the enzymes that catalyze the formation of the metabolite; F^M is an availability term that describes the fraction of the metabolite formed in the cells that leaves the cells and eventually enters the arterial circulation; and CL^S and CL^M are the total body clearances of the substance and its metabolite, respectively. In Equation 12, the terms $k_0\, F^S_a\, F^S_{org}\, f^u / CL^S$ are those that govern the steady-state concentration of unbound parent substance within the organ. The steady-state rate of formation of the metabolite can be expressed by the steady-state concentration of the unbound substance and the intrinsic clearances of the enzymes that catalyze the formation of the metabolite. These terms also can be expressed as the rate of infusion times the fraction of the dose that is converted to the metabolite in the organ, $F^{S \to M}$. Thus, Equation 12 can be written as:

$$C^M_{a(ss)} = k_0\, F^{S \to M}_{org}\, F^M. \tag{12a}$$

This form of the equation points out the invalidity of the assumption of many investigators that only the activity of the enzyme(s) that catalyzes the formation of a toxic metabolite governs the steady-state concentration of the toxic metabolite. Instead, Equation 12a demonstrates that as long as F^M and CL^M remain constant, changes in the value of $CL^{S \to M}_{int}$ affect the steady-state concentration of the toxic metabolite in arterial blood only to the extent that such changes result in changes in $F^{S \to M}_{org}$. Indeed, if the formation of the metabolite were the sole mechanism of elimination of the parent substance, $F^{S \to M}_{org}$ would be 1.0, and the steady-state concentration of the metabolite in arterial blood would be completely independent of the activity of the enzyme that catalyzed the formation of the metabolite, as long as the rate of infusion did not exceed the V_{max} of the enzyme (see Figure 7).

TABLE 1 Functional Classification of Metabolites

1. Ultrashort-lived metabolites
   Metabolites never leave the enzyme (suicide enzyme inhibitors).
2. Short-lived metabolites
   Metabolites never leave the cell.
3. Intermediately lived metabolites
   Metabolites never enter aortic blood.
4. Long-lived metabolites
   Negligible amounts of metabolites excreted into air, bile, and urine.
5. Ultralong-lived metabolites
   Metabolites extensively excreted into air, bile, or urine.

Stable Metabolites

When the value of F^M approaches 1.0, the general conclusions of the relationships between the steady-state concentration of unbound substance in blood to the concentration of unbound substance in nonelimination organs are also applicable to the metabolite. For example, under steady-state conditions the maximum concentration of unbound metabolite in an organ that does not participate in either the formation or the elimination of the metabolite will never exceed the highest concentration of the unbound metabolite in blood, and the minimum concentration of unbound metabolite under these conditions will always exceed the minimum concentration of metabolite in blood during a dosage interval, unless there are ion-trapping effects or active transport systems in the organ.

Categories of Unstable Metabolites

With many toxic metabolites, however, the value of F^M may approach 0, in which case the metabolite would be undetectable in blood. It is, therefore, useful to devise functional categories of metabolites according to the extent to which the metabolites can leave the enzyme that catalyzes their formation, leave the cells in which they are generated, and leave the organ in which they are generated (Table 1). The approaches used to study the kinetics of metabolites that may be appropriate to one of these categories may not be appropriate for studying another category. For example, when it can be demonstrated that a metabolite never leaves the cells in which it was generated, the total body clearance of the metabolite would be restricted to the cells in which the metabolite was generated, and the investigator would focus attention on such intracellular mechanisms. Several equations describing the kinetics of metabolites in different categories have been devised (Gillette, 1985). Such considerations are especially relevant in focusing attention on potential target organs of toxic, chemically reactive metabolites. If toxic, reactive metabolites never leave the cells in which they were formed, then the direct toxic effects would be restricted to those cells containing the enzymes that catalyzed their formation. *Alterations in the activities of the enzymes that catalyze the formation of the metabolite in cells of different organs thus may alter the relative toxicities in different target organs and would be unpredictable from pharmacokinetic studies solely of the parent substance.* By contrast, the interorgan differences in the toxicity of long-lived metabolites depend largely on the biochemistry and physiology of the target organ.

DOSE EXTRAPOLATIONS

Increasing the dose of a substance can have profound effects on virtually every aspect of pharmacokinetics, including the rates and extents of absorption from various sites of administration, the extent of reversible binding of the substances to binding components in blood and other organs, the rates of formation of different metabolites, and the rates of excretion of the substance and its metabolites into bile and urine.

Unknown Biologically Active Forms

It is important to realize that when the biologically active form(s) is not known, only the dose-dependent effects on the rate and extent of absorption can be unequivocably related to the incidence and intensity of toxicities. It is frequently impossible to decide whether the response will be directly proportional to dose, greater than proportional to dose, or less than proportional to dose when increasing doses result in the approach to saturation of various enzymes and active transport systems. With nonlinear models the concentration of unbound parent substance becomes proportionately greater as the dose is increased. But the increase in the concentration of various metabolites may be either greater or less than proportional to the dose. It is also possible that under certain conditions the concentration of a metabolite may remain directly proportional to the dose, even when the concentration of its precursor is not. Under other conditions, the concentration of a metabolite may be larger or smaller than proportional to the dose, even though the concentration of its precursor is virtually proportional to the dose.

When pharmacokinetic studies are made a part of a toxicity study, such discrepancies frequently aid the investigator in elucidating the biologically active forms that cause toxicity. But without such prior knowledge, the regulator is faced with a host of possible interpretations that are frequently contradictory.

When considering the effects of increasing doses and concentrations of compounds, such as enzymes, on saturable processes, many investigators focus attention on a single enzyme, apparently without realizing that knowledge of the Michaelis-Menten parameters of a single enzyme alone seldom, if ever, provides sufficient information to predict either the concentration of the parent substance or any of its metabolites.

Dose-Dependent Absorption of Insoluble Substances

According to Fick's Law of Diffusion, the rate of absorption by passive diffusion should always be directly proportional to the concentration of diffusible forms of a substance, provided that the substance does not alter the characteristics of the membrane. Most dose-dependent kinetics of absorption occur with insoluble compounds, in which a steady state is established between the rate of dissolution of the crystals of the substance and the rate of diffusion across the pulmonary membranes, gastrointestinal membranes, or skin (Pang and Gillette, 1979; Rowland and Tozer, 1980). The question then arises as to which of these processes becomes the predominant rate-limiting step in governing the rate of absorption. Two extreme cases are represented in Figure 4. In the first case, the substance has a high oil/water partition ratio and therefore would be rapidly absorbed, but the rate of absorption may be limited by the rate of diffusion from the bulk liquid phase to the membrane interface. Thus, the rate of absorption of a substance in the gastrointestinal tract frequently may be limited by the motility of the gastrointestinal tract. In a well-mixed absorption compartment, however, the rate of absorption may be limited by the rate of dissolution of the substance from its crystalline form. In this case, however, the rate of absorption would be dependent more on the size of crystals and other physicochemical characteristics of the dosage preparation than on the factors that govern Fick's Law of Diffusion.

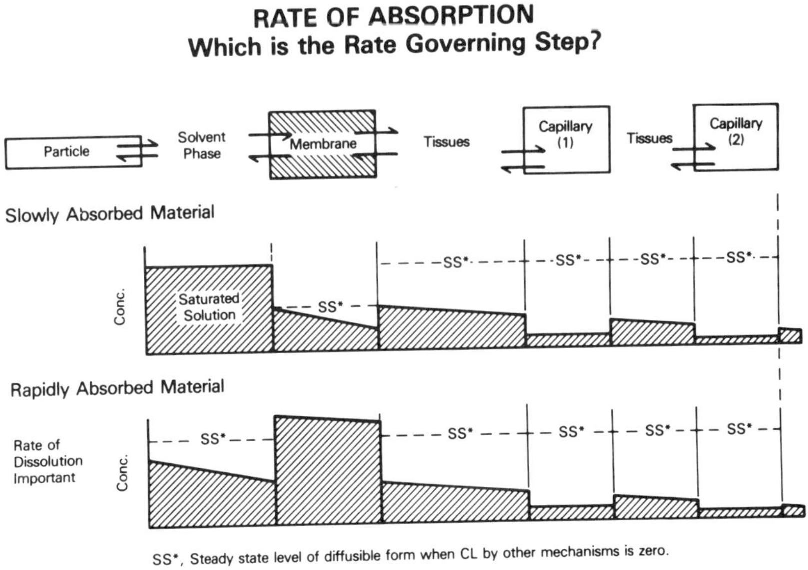

FIGURE 4 Rates of dissolution and rates of diffusion as rate-limiting steps in absorption.

Fraction of the Dose Absorbed and Bioavailability

The fraction of the dose that is absorbed, however, depends not only on the rate of absorption diffusion and facilitative and active transport systems, but also the rate at which the substance is removed from the site of administration removed by other processes.

Bioavailability and Gastrointestinal Absorption

In the gastrointestinal tract, the substance may be eliminated with the feces. In addition, however, the substance may be unstable at pH values occurring in the stomach, or it may be metabolized by intestinal flora and/or intestinal mucosa. These latter processes confuse the interpretation of pharmacokinetic studies of metabolites, because it is frequently difficult to determine whether a metabolite is formed in the gastrointestinal tract and then absorbed or is absorbed and then converted to a metabolite within other organs of the body (Rowland and Tozer, 1980).

Biliary excretion and passive diffusion from the body through the gastrointestinal wall further complicate the interpretation of bioavailability studies, particularly when these processes represent major contributions to the total body clearance.

Bioavailability and Absorption Through Skin

Studies on the absorption of insoluble substances through the skin are especially difficult to interpret and apply to risk assessment studies. Although it might be assumed that a substance must dissolve into a liquid vehicle before it can be absorbed, the solubility in that vehicle seldom is known. Because it is the concentration of the substance in the vehicle as well as the partition ratio of the substance between the vehicle and the lipoidal membrane that presumably governs the rate of diffusion, extrapolations to different vehicles in which the substance may dissolve to different extents would be difficult. Moreover, the fraction of the dose that is absorbed would depend on the time at which the substance is washed off or otherwise removed from the skin. Whatever the actual kinetics may be, the assumption that the rate of absorption would be directly proportional to the total amount of material suspended in the dosage form will not always be valid.

Because the rate of absorption of substances across thin membranes is usually directly proportional to the oil/water partition ratios, it might be assumed that this would also be true for thick membranes, such as the skin. But when a membrane contains several layers of cells, the substance must pass through not only lipoidal membranes but also aqueous phases before it reaches the capillary beds within the membrane. When the membrane consists of several layers, the rate of diffusion depends not only on the diffusivity through the lipoidal layer but also the diffusivity through aqueous layers. The rate of diffusion of substances with very high oil/water partition ratios, therefore, can be limited by its concentration and diffusivity through the aqueous phases (Figure 5). Thus, there frequently is an optimum oil/water partition ratio for absorption through thick membranes (Houston et al., 1974).

The rate of absorption through thick membranes, however, is seldom limited by the blood flow rate through the membrane. Under conditions in which this would likely occur, the substance simply diffuses more deeply into the membrane and enters capillaries in the deeper regions of the membrane.

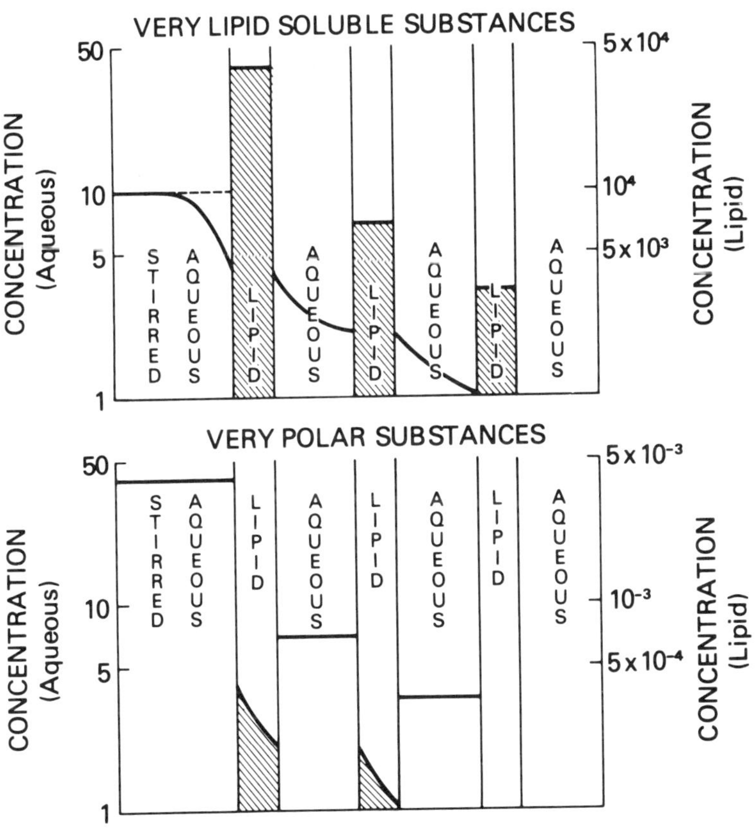

FIGURE 5 Diffusion through thick membranes.

Dose Dependencies in Reversible Binding of Substances to Proteins and Other Macromolecules in Blood and Organs

Increases in the dose of a substance may result in concentrations of unbound substances in blood and organs that approach or exceed the $1/K_a$ values of the various binding components in blood and the organs. Because the effects are dependent on concentration, however, the effect is dependent on the rate and the extent of absorption of the substance from the site of administration, as well as the rates of distribution and elimination of the substance.

The effects of increases in concentration on the value of the organ partition ratio R depend on the K_a values of the various binding components, the number of binding sites with a given K_a value, and the location of the binding sites. Suppose, for example, that the sites with the highest K_a value were located in the blood, the sites with the second highest K_a value were located in the tissues of the organ, the sites with the third highest K_a value were in the blood, and the sites with the fourth highest K_a value were in the tissues of the organ. As the concentration of unbound substance is increased from a very low level, the value of R would first increase, then decrease, then increase, and then decrease. Because the relative distribution of binding sites within an organ may differ with the organ, the pattern of changes may differ with the organ. Thus, as long as the relative as well as the total number of the binding sites and their respective K_a values remain unknown, the effects of an increase in the concentration of the substance on the value of R in various organs are unpredictable. The situation becomes even more complex if we include competitive binding with the metabolites of the substance. It follows, therefore, that the effects of increasing doses on the shape of the time course of the total concentration of the substance in various organs and blood are unpredictable. Moreover, the relationship between the total and the unbound concentrations in various organs varies not only as the concentration approaches the maxima in various organs but also as the concentrations in the organs decline after the administration of bolus doses of the substance. Thus, the time course of the unbound concentration should be simulated separately in order to relate them to the response.

The effects of concentration-dependent changes in the reversible binding to blood components on the maximum, minimum, and average concentrations of unbound substance after a steady state is achieved during repeated administration of the substance largely depend on the influence that reversible binding of the substance to blood components has on the rate of absorption of the substance and the extraction ratios and location of the elimination organs. If saturation of the binding sites decreases the rate of absorption without having much effect on the extraction ratio of a first-pass organ, the maximum concentration may be less than proportional to the dose. But if saturation of the binding sites has little influence on the rate of absorption of the substance and markedly decreases the extraction ratio of a first-pass organ, the maximum concentration may be greater than proportional to the dose. Thus, the effects of saturation of binding sites to blood components is unpredictable.

Saturation of binding sites in tissues within the organs, however, causes greater increases than expected in the magnitude of the oscillations between the maximum and minimum concentrations of unbound substance in the blood and the organs during the dosage interval, which may be relevant to the magnitude of rapid responses initiated by reversible binding to receptor sites.

DOSE-DEPENDENT CHANGES IN METABOLISM OF SUBSTANCES AND THEIR METABOLITES

Michaelis-Menten Kinetics

When pharmacokineticists discuss the implications of dose-dependent metabolism of substances, they usually focus attention on the classical Michaelis-Menten equation and note that when the concentration of the substance approaches infinity, the rate of metabolism approaches a constant, namely, V_{max}. But they frequently fail to emphasize the importance of other mechanisms of elimination of the parent compound in governing the concentration of the substance under steady-state conditions. Moreover, most substances can be metabolized by several enzymes in the body, each one of which has its own substrate specificity and its own set of Michaelis-Menten parameters for any given substrate.

It may be useful for the purpose of illustration to visualize a model in which a substance is constantly infused intravenously until a steady state is achieved and the substance is eliminated from the body by being excreted unchanged into urine and into exhaled air and by metabolism in the liver. In this model, the value of f^u is assumed to be 1.0 in order to preclude any complicating effects on the clearances caused by dose-dependent, reversible binding. The steady-state equation for such a system would be:

$$C_{a(ss)} = \frac{k_0 F_L}{CL_H + CL_R + (Q_{co} - CL_H - CL_R)\, E_L}, \qquad (13)$$

where Q_{co} is the cardiac output.

If the substance were metabolized either by a single enzyme or by a group of enzymes with the same K_m value, we can write the equation for hepatic clearances as:

$$CL_H = \frac{Q_H \text{ Sum } V_{max}}{Q_H K_m + (\text{Sum } V_{max}) + Q_H C_{out(ss)}}, \text{ or} \tag{14}$$

$$CL_H = \frac{CL_{int(H)}}{1 + (CL_{int}/Q)_H + [C_{out(ss)}/K_m)]H}, \tag{14a}$$

where $CL_{int} = \text{Sum}(V_{max}/K_m)$.

The term $C_{out(ss)}$ requires the solution of the quadratic equation:

$$C_{out(ss)} = 0.5\Bigg[C_{in(ss)} - K_m [1 + (CL_{int}/Q)] + \left(\{C_{in(ss)} - K_m [1 + (CL_{int}/Q)]\}^2 - 4\, K_m C_{in(ss)}\right)^{0.5}\Bigg]. \tag{14b}$$

Simulations based on these equations would indicate that when $CL_{int} \gg 0$, the hepatic clearance approaches hepatic blood flow at very low values of $C_{\mathrm{out(ss)}}$. But when $C_{\mathrm{out(ss)}}$ approaches virtual infinity, the hepatic clearance approaches 0. Substitution of these values into Equation 13 would thus provide an estimate of the maximum range of effects that an increase in the rate of infusion would have on the steady-state concentration of the substance.

Studies designed to estimate only the Michaelis-Menten parameters V_{max} *and* K_m *or the clearance of a single organ, however, do not provide sufficient information to predict steady-state concentrations of a given substance in arterial blood and other organs at different doses*. The values of the clearances of other organs that contribute to the total body clearance and dose-dependent effects on these clearances must also be known and reported to accomplish this objective.

Intraorgan Localization of Enzymes

Although the clearance of a substance by an organ in most cases can be adequately described by well-stirred models, the kinetics describing the formation and elimination of metabolites requires more sophisticated models when the organ availabilities of either the parent substance or the metabolite are small. Under such conditions the intraorgan localization of the enzymes (Baron and Kawabata, 1983) may be important. As pointed out by G. R. Wilkinson (this volume), the changes in liver metabolites of highly extracted substances caused by changes in either the blood flow rate or the fraction of unbound drug are not accurately predicted by either the well-stirred model or the parallel tube model. Instead, the experimental availabilities are usually between the values predicted by the two models. The parallel tube model, however, provides an easy way of illustrating the effects of an increase in the dose (and thus the steady-state concentration of the parent substance) on the exit concentrations of metabolites formed by different enzymes preferentially distributed in different zones of an organ (Figure 6). For example, imagine that an enzyme (Figure 6, enzyme 1) that formed one metabolite (M_1) was localized preferentially in the periportal zone of the liver (the proximal end of the tube) and that another enzyme (Figure 6, enzyme 3) that formed another metabolite (M_3) was localized in the pericentral zone (the distal end of the tube). Also imagine that the intrinsic clearance of enzyme 1 was considerably greater than the hepatic blood flow rate. When the rate of infusion is slow, virtually all of the substance entering the liver would be converted to M_1, because very little of the substance would reach enzyme 3. When the infusion rate is increased, however, the concentration may approach or exceed the K_m value of enzyme 1, and the available fraction reaching enzyme 3 would increase. Thus, the pattern of metabolites exiting the liver would change as the infusion rate increased, even when the K_m values of the two enzymes are virtually identical.

If the K_m values of the two enzymes differed, the effects of an increase in the infusion rate would depend on which enzyme had the lower K_m value. The increase in the hepatic availability of the substance caused by increasing the infusion rate would become evident at lower infusion rates when enzyme 1 has the lower K_m value than when enzyme 3 has the lower K_m value. Moreover, the shape of the changes in the relative proportions

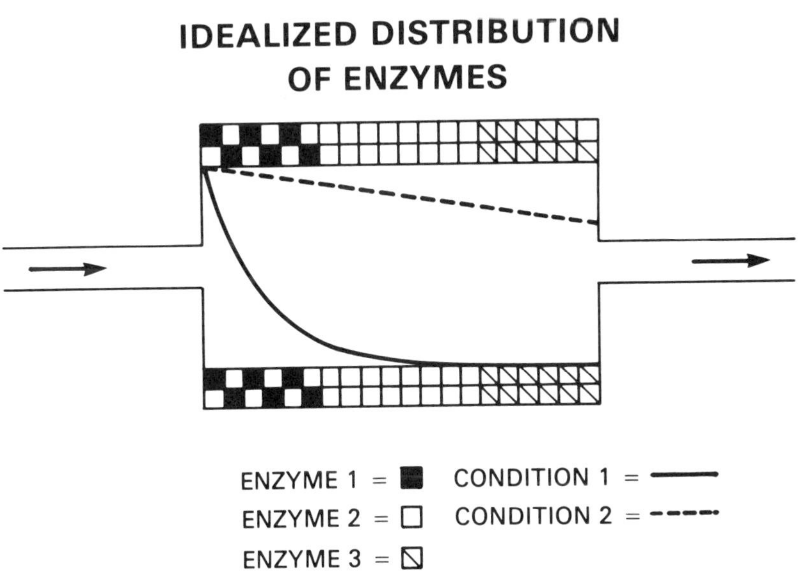

FIGURE 6 Idealized intraorgan distribution of enzymes.

of the metabolite depends on which of the enzymes has the lower K_m value as both enzymes approach saturation.

The parallel tube model also predicts that the location of enzymes that catalyze the metabolism of a metabolite affects the apparent organ availability of the metabolite, when the intrinsic clearance of the enzyme that catalyzes the metabolism of the metabolite greatly exceeds the hepatic blood flow. If enzyme 1 generates the metabolite and enzyme 3 metabolizes it (model 1), very little of the metabolite would escape the liver. But if enzyme 3 generates the metabolite and enzyme 1 metabolizes it (model 2), virtually all of the metabolite would escape the liver. If the K_m value of the enzyme that generates the formation of the metabolite were always larger than the K_m value of the enzyme that catalyzes the metabolism of the metabolite, then an increase in the rate of infusion would result in increases in the availability of the metabolite in model 1, but not in model 2. An increase in the rate of infusion would decrease the hepatic clearance of the metabolite in both models.

It is unlikely that enzymes would be segregated as markedly as has been assumed in models 1 and 2, and thus, any equation that described these events would not be universally valid. Nevertheless, the distribution of enzymes within organs is known to influence the availability of metabolites, although the extent of the influence is unpredictable (Pang, 1983).

Fraction of the Dose Principle

Although many studies of toxic metabolites focus attention on the enzymes that catalyze the formation of the toxic metabolites, it is important to realize that any alteration in the activity of the various enzymes that catalyze the metabolism of any of the precursors of the toxic metabolite will affect the steady-state concentration of the toxic metabolite only to the extent that the alterations cause a change in the fraction of the dose that is converted to the metabolite. According to this general principle, approaches to the saturation of enzymes that catalyze the metabolism of a precursor to innocuous metabolites are just as important as approaches to the saturation of enzymes that catalyze the reactions that lead to the toxic metabolites. This principle can be illustrated by simulations of models based on Equations 12 and 13, in which a substance is constantly infused intravenously into a rat until a steady state is achieved and the substance is eliminated from the body by a combination of excretion unchanged into urine and metabolism by two enzymes in the liver (Figures 7 and 8). Enzyme 1 is assumed to form a toxic metabolite, whereas enzyme 2 is assumed to form a nontoxic metabolite. The renal clearance is assumed to be independent of the concentration of the parent substance, and the

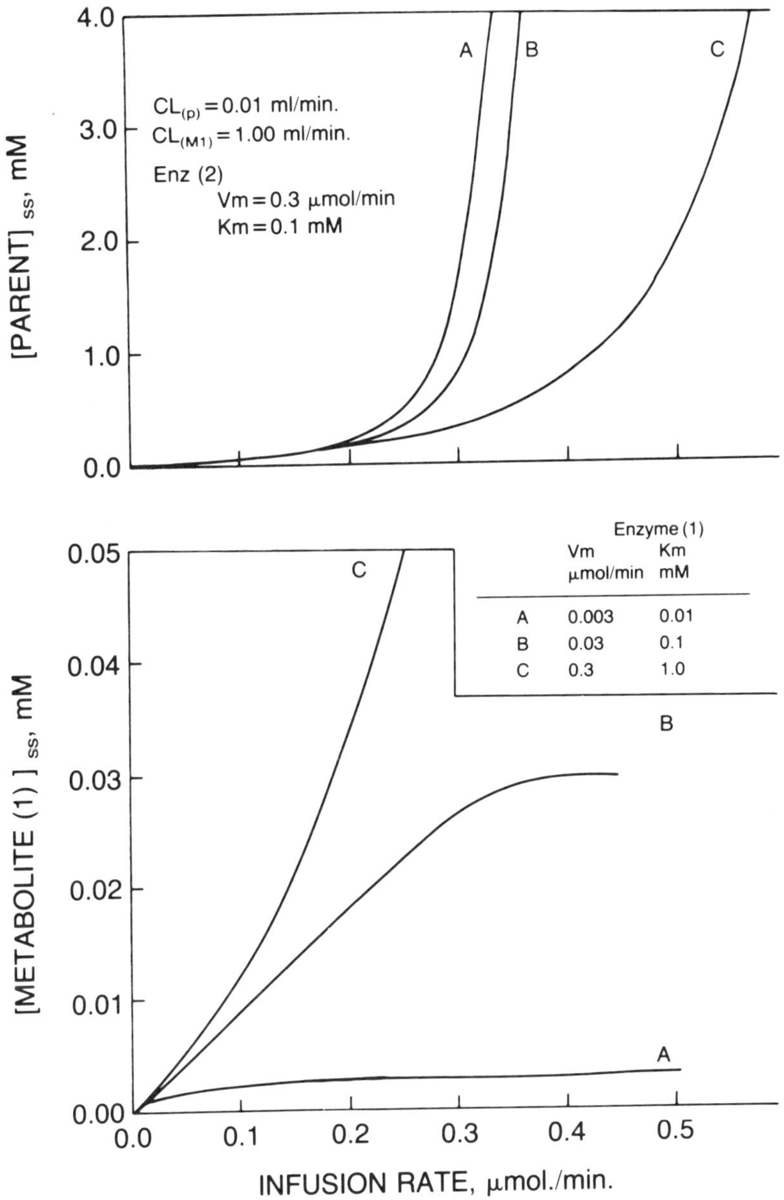

FIGURE 7 Importance of alternative mechanisms of elimination: low renal clearance. Simulations of the steady-state concentrations of parent compound and toxic metabolites when the toxic metabolites are formed by different enzymes (1A, 1B, and 1C) with different V_{max} and K_m values. CL_P is the renal clearance of the parent compound, and enzyme 2 catalyzes the formation of a nontoxic metabolite. CL_{M_1} is the total body clearance of the toxic metabolite (M_1).

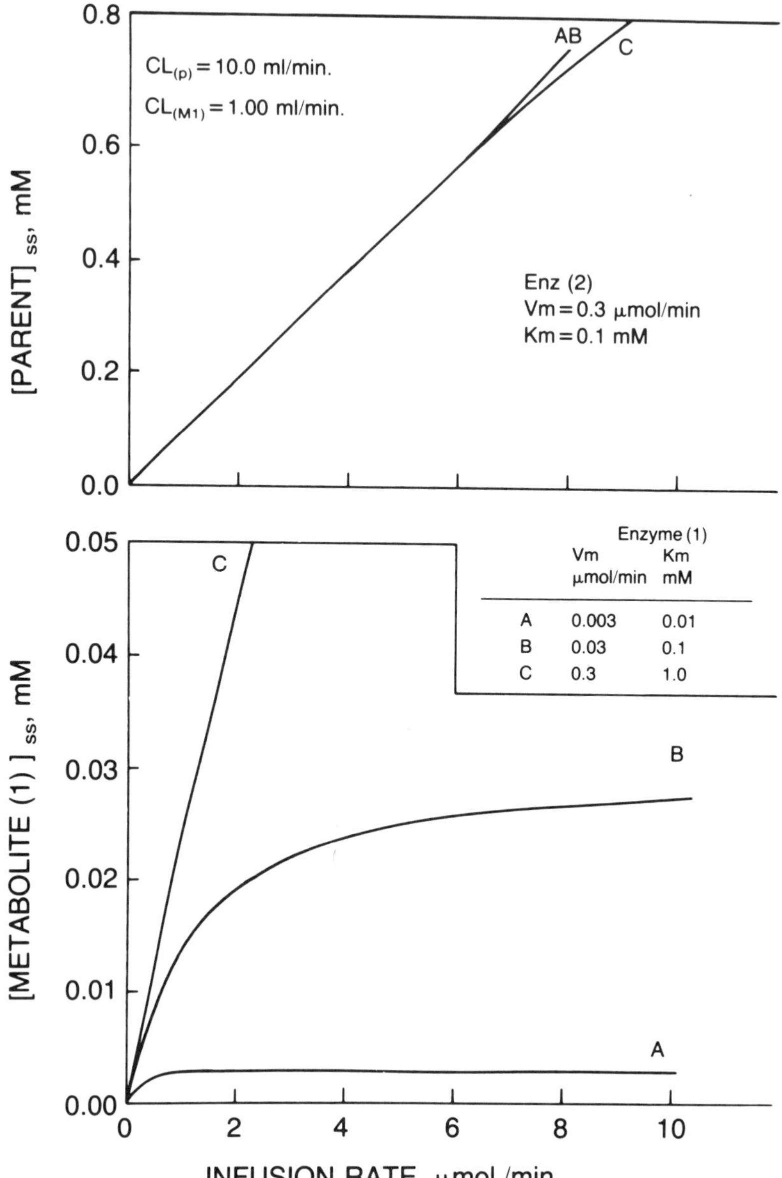

FIGURE 8 Importance of alternative mechanisms of elimination: high renal clearance. Simulations of the steady-state concentrations of parent compound and toxic metabolites when the toxic metabolites are formed by different enzymes (1A, 1B, and 1C) with different V_{max} and K_m values. CL_P is the renal clearance of the parent compound, and Enzyme 2 catalyzes the formation of a nontoxic metabolite. CL_{M1} is the total body clearance of the toxic metabolite (M_1).

V_{max} and K_m values of enzyme 2 are the same in all simulations; the intrinsic clearance (V_{max}/K_m) value of enzyme 2 was set at 3.0 ml/min, which is well below the hepatic blood flow in rats (about 25 ml per 300-g rat). Enzyme 1 is assumed to have any one of three different K_m values, but the intrinsic clearances were set at 0.3 ml/min in the three situations. The total body clearance of the toxic metabolite is assumed to be the same in all simulations and independent of the concentration of the metabolite. Thus, in these simulations low infusion rates would result in the same steady-state concentrations of the toxic metabolite, regardless of the K_m value of enzyme 1.

First Set of Simulations

In this set of simulations, the renal clearance was set at 0.01 ml/min, which is about the rate of urine formation in rats (Figure 7). Thus, the value is about the expected renal clearance of a lipid-soluble substance that entered the glomerular filtrate, but was reabsorbed with water during the concentration of the filtrate. Because the sum of the intrinsic clearances of enzymes 1 and 2 was set at 3.3 ml/min, at low infusion rates the substance is eliminated from the body predominantly by metabolism in the liver.

Enzyme 1 K_m Equals Enzyme 2 K_m

In simulation B (Figure 7), the K_m of enzyme 1 is set at the same K_m as enzyme 2. In this situation the effects of an increase in the infusion rates on the rates of formation of the metabolites formed by the two enzymes would be identical. The increase in the steady-state concentration of the metabolite remains virtually directly proportional to the rate of infusion until the rate of infusion approaches the sum of the V_{max} values of the two enzymes (see Gillette, 1986; O'Flaherty, 1986). This follows from the definition of a steady state; i.e., the rate of elimination of the substance must equal the rate of infusion. Because in this particular model the rate of elimination is predominantly due to metabolism of the substance up to an infusion rate that equals the sum of the V_{max} values, the substance simply accumulates in the animal until it reaches a concentration at which the rate of elimination by the combination of metabolism and excretion equals the rate of infusion. The fraction of the dose that is converted to the toxic metabolite over this range of infusion rates, however, would remain virtually constant; and therefore, the steady-state concentrations of metabolites 1 and 2 would be directly proportional to the rate of infusion.

Most of the increase in the steady-state concentration of the precursor occurs within a relatively small range of infusion rates. The equation $V_{max}/$

$(K_m + f^u S)$, which defines the effective CL_{int}, predicts that as the concentration of unbound substance increases from very low values up to a value equal to the K_m, the value of the effective CL_{int} decreases by 50%. If the total body clearance were governed solely by the enzyme, we can calculate that an infusion rate equal to 50% of the V_{max} value of the enzyme would result in a steady-state concentration of the substance that would be only double that predicted by the relationship $k_0/(V_{max}/K_m)$. When the clearances by other routes of elimination are very small compared with V_{max}/K_m values of the enzymes, however, the steady-state concentration of the substance would approach values predicted by k_0/CL when the infusion rates exceed the V_{max} value of the enzyme. Thus, only a doubling of the infusion rate would be required to go from $0.5V_{max}$ to V_{max}. Such large increases in the concentration of the parent substance probably would result in toxic effects of the parent substance rather than a delayed toxicity of the metabolite. The narrowness of the dosage range, however, implies the existence of a virtual dose threshold.

This discussion should not be construed as indicating that the concentration of a substance administered repetitively may never be severalfold larger than the K_m value of the dominant enzyme during the dosage interval under steady-state conditions. For example, if a substance were injected into a one-compartment model repeatedly until a steady state were achieved, the following equation based on the integrated form of the Michaelis-Menten equation can be written,

$$\left(\frac{V_{max}\,\tau}{Dose/\tau}\right)_{ss} = 1 + \frac{K_m}{f^u C_{ss}}, \tag{15}$$

in which:

$$C_{ss} = [(C_0 - C_\tau)/\ln (C_0/C_\tau)]_{ss}, \tag{15a}$$

where C_0 is the concentration of the substance immediately after the injection of the substance at the beginning of the dosage interval, and C_τ is the concentration of the substance at the end of the dosage interval under steady-state conditions.

Thus, as long as $(Dose/\tau) < V_{max}\,\tau$, the system theoretically can attain a steady state regardless of the dosage interval and the volumes of distribution that relate the amount of the substance in the compartment to the concentrations of the substance. The equation is not valid for multi-compartment systems, and thus, iterative approaches would be needed to simulate not only the approach to steady states but also the concentration changes within a dosage interval. Nevertheless, Equation 15 may be useful for estimating a range of permissible dosage schedules in toxicity studies from *in vitro* estimates of V_{max} and K_m values. Caution should be used, however, because the estimates of V_{max} and K_m obtained *in vitro* do not always reflect the V_{max} and K_m values *in vivo* and because the *in vivo* V_{max} and K_m values may change during the course of the study. When the pharmacokinetic study is not an integral part of the toxicity study, even greater caution should be used in interpreting the results. Differences in the V_{max} values between the groups of animals in either enzymes 1 or 2 that changed the Sum V_{max} value would affect the dose-response curve whether the toxicity was caused by the parent substance or a metabolite of enzyme 1, because such differences could result in a change in the proportion of the dose that was metabolized by the enzymes and would affect the dose at which the steady-state concentration of the substance would no longer be linearly related to the dose. Differences in the V_{max} value in enzyme 2 without proportional increases in enzyme 2 would also affect the dose-response curve of a toxicity caused by a metabolite of enzyme 1.

In simulation A (Figure 7) the K_m value of enzyme 1 was set at 0.01 m*M*, an order of magnitude below that of enzyme 2. Thus, the steady-state rate of formation would approach V_{max} (enzyme 1) at rather low rates of infusion. With further increases in the rate of infusion, the fraction of the dose converted to the toxic metabolite would be decreased, but the value of $F^{S \rightarrow M}$ would tend to decline linearly until V_{max} (enzyme 2) was reached and then would decline rapidly with further increases in the infusion rate. In such situations, it would obviously be better to focus attention on the value of V_{max} (enzyme 1) and to ignore the variations in the relationships between the rate of synthesis of the toxic metabolite and the dose.

When the pharmacokinetic studies are an integral part of the toxicity, it may be possible to distinguish between toxicities caused by the metabolites of enzyme 1, enzyme 2, and the parent substance. The maximum concentration and thus, possibly, the maximum intensity of the toxicity caused by a metabolite of enzyme 1 would occur well below the K_m value of enzyme 2. By contrast, the intensity of a toxicity caused by a metabolite of enzyme 2 may be proportional to the dose, and the dose-intensity relationship of a toxicity caused by the parent substance may be very sharp and appear to have a dose threshold. When the pharmacokinetic studies are not a part of the toxicity studies and the toxic form is unknown, such issues become clouded and risk estimators would not know which pharmacokinetic parameters should be used in their calculations. Moreover, they could be led to the wrong conclusion if the pharmacokinetic studies were focused only on the parent substance.

In simulation C, the V_{max} of enzyme 1 equals that of enzyme 2, but the K_m of enzyme 1 is an order of magnitude larger than that of enzyme 2. With increasing rates of infusion, the steady-state concentration of the metabolite of enzyme 1 would deviate from a linear relationship, but would approach an asymptote up to a rate of infusion that equalled the sum of the V_{max} values of the two enzymes. By contrast, the steady-state concentration of the parent substance would approach concentrations governed by k_0/CL more gradually than in simulations A and B. Thus, as the perfusion rate is increased, the fraction of the dose converted to a metabolite by enzyme 1 increases until the V_{max} of enzyme 1 is reached and then decreases. Thus, attempts to relate the toxicity to any of the forms of the toxicant would be difficult.

Second Set of Simulations

For the simulations shown in Figure 8, the clearance for the excretion of unchanged substance into urine (or air) was increased to 10 ml/min. This is about what would be expected for the clearance of a polar substance that is excreted by an active transport system in kidneys, but for the purpose of illustration it has been assumed that an increase in the infusion rate would not affect the clearance of the unchanged substance. The V_{max} and K_m values for enzyme 1 and enzyme 2 were identical to those used in the simulations shown in Figure 7. In this set of simulations, the maximal fraction of the dose that undergoes metabolism by the two enzymes at low infusion rates would be about 25% and would decrease as the infusion rates were increased in all three simulations. As a result the steady-state concentrations of the parent substance are virtually proportional to the infusion rates regardless of the degree of saturation of the enzymes. Although an increase in the renal clearance does not significantly change the shape of curve A, the steady-state rate of infusion required to achieve the V_{max} of enzyme 1 is significantly larger in Figure 8. Moreover, the shapes of curves B and C as well as the rates of infusions required to achieve V_{max} values of enzymes 1 and 2 are changed.

From comparisons of the simulations in Figures 7 and 8, it should be obvious that studies of the kinetic parameters of all processes of elimination of the substance and its biologically active metabolites are needed to predict their steady-state concentrations. Knowledge of the kinetic parameters of enzymes without knowledge of the clearance by other pathways of elimination of the parent substance is also necessary. Pharmacokinetic studies that do not provide such information in a readily understandable form are thus virtually uninterpretable and largely useless to the risk assessor.

Concentration-Dependent Metabolite Elimination

An increase in the rate of infusion of a substance may also result in dose-dependent elimination of a metabolite and thereby affect the steady-state concentration of the metabolite. For example, if the metabolite were eliminated from the body predominately by an enzyme with a smaller K_m value than that of the enzyme that catalyzed its formation, the steady-state concentration of the metabolite would increase. When the metabolite is formed and eliminated by different enzymes, the maximum extent to which the steady-state concentration of the metabolite can increase depends on the clearance values of other pathways of elimination of the metabolite. Indeed, relationships can be written as:

$$[M]_{ss}^{max} = V_{max}^{S \to M} / [(V_{max}^{M \to 0} / [M]_{ss}^{max}) + CL_{other}^{M}], \qquad (16)$$

which can be solved by a quadratic equation. Without an alternative pathway of elimination, a steady state would not be achievable when the rate of formation of the metabolite exceeds $V_{max}^{M \to 0}$, and the concentration of the metabolite would increase to infinity.

If the metabolite were formed and eliminated by the same enzyme, however, a steady state could be achievable, because the parent substance and the metabolite would serve as mutual competitive inhibitors as the steady-state concentration of the substance is increased. If the metabolite were eliminated solely by the enzyme, the steady-state concentration of the metabolite can be predicted by the equation:

$$[M]_{ss} / [S]_{ss} = (V_{max} / K_m)^{S \to M} / (V_{max} / K_m)^{M \to 0}. \qquad (17)$$

Thus, the steady-state concentration of the metabolite would be directly proportional to the steady-state concentration of the substance regardless of the infusion rate of the substance. But because the metabolite serves as a competitive inhibitor of the metabolism of the parent substance, it affects the relationship between the steady-state concentration of the parent substance and the rate of infusion.

Dose-Dependent Cofactor Depletion

Although most simulations of dose-dependent kinetics utilize a simplistic form of the Michaelis-Menten equation, in which both V_{max} and K_m are viewed as constants, it is important to realize that there are many different enzyme mechanisms. Equations that relate the rate of metabolism to the substrate concentration for some of these mechanisms may contain terms in which the substrate concentration is raised to the second or even greater powers, but fortunately, these mechanisms are rarely encountered in the metabolism of foreign substances. It is important, however, to realize that the concentrations of cofactors within cells may limit the rates of metabolism of foreign compounds. In such instances, the kinetics of the processes that govern the concentration of the cofactors within cells can play a dominant role.

In many enzyme mechanisms, the relationship between the rate of metabolism of a substrate and the concentration of the cofactor in the immediate environment of the enzyme can be expressed by the following Michaelis-Menten equation:

$$Rate^{S \to M} = \frac{\left(\dfrac{E_t K_1 k[C]}{K_4 + K_5[C]}\right)[S]}{\left(\dfrac{K_2 + K_3[C]}{K_4 + K_5[C]}\right) + [S]} = \frac{V_{max}[S]}{K_m + [S]}, \qquad (18)$$

where E_t is the total amount of enzyme; K_1, K_2, K_3, K_4, and K_5 are ratios of sets of rate constants, the meanings of which vary with the mechanism of the enzyme; and k is a first-order rate constant.

Inspection of Equation 18 reveals that the simple Michaelis-Menten equation would be valid only when either of two situations occur: (1) The rate of metabolism of substrate may change $[C]$ but the value of $[C]K_3$ is always much greater than that of K_2, and the value of $[C]K_5$ is always much greater than that of K_4. In this situation the rate of metabolism would be virtually independent of $[C]$. (2) The rate of metabolism of the substrate does not perceptibly change $[C]$.

It is evident, therefore, that the validity of the simple form of the Michaelis-Menten equation depends on the kinetics of the mechanisms that govern the concentration of the cofactor within cells.

In the absence of the foreign substance, the amount of a cofactor is presumably governed by its rate of synthesis from endogenous precursors ($Pre\ k^{pre \to C}$), the rate at which the cofactor is converted back to its precursors ($V_c[C]_{ss}\ k^{C \to pre}$), and the rate at which it is consumed by other endogenous reactions in the cells ($V_c[C]_{ss}\ k^{C \to 0}$). If the amount of the cofactor can be assumed to remain in a virtual steady state, the equation can be written as:

$$V_c[C]_{ss} = \frac{Pre\ k^{pre \to C}}{k^{C \to pre} + k^{C \to 0}}, \qquad (19)$$

where V_c is the volume of distribution of the cofactor.

Immediately after the injection of the foreign compound, however, the rate of change in the concentration of the cofactor can be written as:

$$\frac{V_c\ C}{dt} = Pre\ k^{pre \to C} - V_c[C]\ (k^{C \to pre} + k^{C \to 0}) - Rate^{S \to M}, \qquad (20)$$

where $Rate^{S \to M}$ is defined by equation 18.

Integration of Equation 20 can be performed only by an iterative procedure, but it describes the approach to a virtual steady state, after which time:

$$V\,[C]_{ss} = \frac{Pre\ k^{pre \to C}}{k^{C \to pre} + k^{C \to 0} + (Rate^{S \to M}/V_c[C]_{ss})}. \qquad (20a)$$

Inspection of this equation reveals that when $k^{C \to pre} + k^{C \to 0} \gg Rate^{S \to M}/V_c[C]_{ss}$, simulations based on the simple Michaelis-Menten equation are valid. But when $k^{C \to pre} + k^{C \to 0} \ll Rate^{S \to M}/V_c[C]_{ss}$, the rate of metabolism is governed by the rate of synthesis of the precursor.

Studies of acetaminophen metabolism performed in various laboratories have demonstrated that the rates of formation of all three major metabolites of the drug may be governed by dose-dependent cofactor depletion. The formation of acetaminophen sulfate may be limited by the cofactor phospho-adenosyl-phospho sulfate (PAPS), and ultimately on the total body pool of sulfate (Galinsky and Levy, 1981; Hjelle et al., 1985; Levy et al., 1982). The formation of acetaminophen glucuronide may be limited by uridine diphosphate glucuronate (UDPGA) (Hjelle et al., 1985; Jollow et al., 1982; Price and Jollow, 1982, 1984), and ultimately on the processes that govern intracellular uridine triphosphate (UTP) and glucose concentrations. Acetaminophen is also converted to *N*-acetyl imidoquinone, which reacts both enzymatically and nonenzymatically with glutathione (Mitchell et al., 1973). The rate of formation of the glutathione conjugate thus may be limited by glutathione. Since the hepatotoxicity caused by acetaminophen is thought to be mediated by *N*-acetyl imidoquinone (Mitchell et al., 1973), it is noteworthy that dose-dependent depletion of cofactors for all three reactions would tend to increase the virtual steady-state concentration of the toxic form of the drug.

Suicide Inhibitors

Occasionally a substance is converted to a metabolite that never leaves the enzyme that catalyzes its formation and thereby causes virtual irreversible inhibition of the enzyme. Among the substrates that inactivate enzyme by this mechanism are many cholinesterase and monoamine oxidase inhibitors. A general equation that describes this kind of inhibitor is:

$$\frac{dE}{dt} = k_0^E - E\left(k_E + \frac{k_c F f^u[S]}{K_m + f^u[S]}\right), \qquad (21)$$

where k_0^E is the steady-state rate of synthesis of the enzyme, E is amount of active enzyme at any given time, k_E is the first-order rate constant that governs the normal catabolism of the enzyme, k_c is the catalytic constant of the enzyme, and F is the fraction of the metabolite that results in the inactivation of the enzyme. The kinetics of inactivation depend on the processes that govern the concentration of the precursor substance. Thus, the factors that govern the amount of active enzyme at any given time frequently are complex and are best estimated by iterative procedures rather than integral calculus. In certain situations, however, solutions may be obtained by integral calculus. For example, if $f^u[S]$ were held constant, the following equation can be written:

$$E_t = [k_0^E/k] + [E_0 - (k_0^E/k)]\, e^{-kt}, \tag{21a}$$

where E_0 is the initial amount of active enzyme, and

$$k = k_E + \frac{k_c F f^u[S]_{ss}}{K_m + f^u[S]_{ss}}. \tag{21b}$$

Inspection of Equation 21a reveals that the approach to the new steady state should be first order and that at the new steady state the amount of active enzyme should equal k_0^E/k. Moreover, provided that the rate of synthesis of the enzyme does not change, the effectiveness of the suicide inhibitor can be estimated from the relationship $E_{ss}/E_0 = k_E/k$.

In cell-free systems, there presumably would be neither synthesis nor catabolism of the enzymes. Moreover, the value of $[S]$ can be held constant. Thus, under these conditions Equation 21 can be modified to:

$$\frac{dE}{dt} = \frac{E k_c F[S]_{ss}}{K_m + [S]_{ss}}, \tag{21c}$$

which may be integrated to:

$$E = E_0 e^{-k(inact)t},$$

where,

$$k_{(inact)} = \frac{k_c f^u[S]_{ss}}{K_m + f^u[S]_{ss}}. \tag{21d}$$

Thus, the amount of active enzyme decreases exponentially, but the rate constant depends on $[S]_{ss}$. Plots of $1/k_{(inact)}$ versus $1/[S]_{ss}$, however, provide estimates of the relationships between k_c, F, and K_m, i.e.:

$$1/k_{(inact)} = \frac{1}{k_c F}\left(1 + \frac{K_m}{[S]_{ss}}\right). \tag{21e}$$

Under another set of conditions in which (1) most of the precursor of the suicide inhibitor is eliminated from the body predominantly by a mechanism other than metabolism by the inhibitable enzyme, (2) the elimination of the precursor can be described by a linear one-compartment system, (3) both k_0^E and k_E are negligible during the elimination of the precursor from the body, the approximate amount of active enzyme may be estimated by:

$$\ln (E/E_0) = \left(\frac{k_cF}{k_{el}}\right) \ln\left(\frac{f^u [S]_0 e^{-k_{el}t} + K_m}{f^u [S]_0 + K_m}\right), \tag{21f}$$

where k_{el} is the first-order elimination constant and $[S_0]$ is the initial concentration of the precursor.

If the value of $S_0 \ll K_m$, however, the approximate amount of active enzyme may be estimated by:

$$\ln (E_0/E) = \frac{k_cFf^u}{K_m} \mid AUC_{ss} |_0^t. \tag{21g}$$

In Equation 21g, the precursor need not be restricted to a single compartment.

Equation 21g permits direct comparisons between *in vivo* and *in vitro* values, because k_cF/K_m can be estimated *in vitro* from Equation 21e.

SPECIES-TO-SPECIES EXTRAPOLATIONS

It may be useful to describe categories of the various factors that govern the pharmacokinetics of substances and their biologically active metabolites according to the extent to which we believe they would contribute to differences between individuals in the human population and to differences between experimental animals and subpopulations of human beings.

Extrapolations in the Absorption of Substances

Some of the factors that govern the pharmacokinetics of a substance and its metabolites are dependent predominantly on the physical chemical characteristics of the substances and the physiological characteristics of animals. For example, the rates of absorption of a substance may be dependent on its solubility, the rates of dissolution of the dosage form of the substance, and the lipid/aqueous buffer partition ratio. In the gastrointestinal tract, the extent of absorption may be limited by the residence time of the substance in the tract. Although the residence time may be affected by species differences in the length of the intestines, it is frequently governed to a greater extent by either diarrhea or constipation and to differences in binding to constituents in food and the gastrointestinal motility. Indeed, the extent of absorption of a substance in a given individual may vary markedly from day to day, depending on the times at which the substance is swallowed and the meal is eaten, as well as on the composition of the meal.

Some substances are unstable under the acidic conditions that exist in the stomach or, as in the case of nitrite, react with other substances to form precursors of mutagens. Other substances are metabolized by bacterial flora that either inactivate the parent substance or convert the substance to biologically active metabolites. Daily changes in the relative residence times of such substances in the stomach, the small intestine, and the large intestine can result in marked daily differences in the extent to which these reactions can occur in an individual. Moreover, there are species and perhaps individual differences in the intestinal flora (Drasar et al., 1970). The extent to which various reactions occur within the intestines thus can vary with the species. But the extent to which metabolism by intestinal flora would contribute to differences in formation of toxic metabolites also depends on the consumption of antibiotics.

Extrapolations of Interorgan Distribution of Substances

As pointed out above, most phamacokineticists who use physiologically based pharmacokinetic models assume that a substance entering an organ is distributed virtually instantaneously and that equilibration of the substance between that in blood and that within nonelimination organs is governed solely by the blood flow rate and the partition ratio between the total concentration of the substance and the total concentration in blood, i.e., R. Although models in which this assumption has been used have been surprisingly successful in predicting the time course of substances in various organs, various tests of the assumption suggest that the proportionality with blood flow rates through slowly equilibrating organs is really an indirect measure of other phenomena that govern the distribution of the substance. Indeed, if the assumption were entirely valid, then venous blood taken from the arms of human subjects should reflect the concentration of the substance in the muscles of the arms and hands rather than the concentration in arterial blood, assuming, of course, that the arteriovenous shunts in muscles are virtually nonfunctional. The findings that the concentrations of substances in venous blood are reasonably good estimators of the concentrations of substances in central compartments thus raise doubts concerning the validity of the assumption. Moreover, the assumption predicts that vigorous exercise, which results not only in an increase in cardiac output but also in a shift in the relative rates of blood flow to muscles and vital organs, should markedly hasten the entrance of substances into muscles. But whether exercise does change the pharmacokinetics of drugs to the extent that would be expected is questionable. By contrast, it has been suggested that changes in the hepatic blood flow caused by drugs or meals can affect the bioavailability of highly extracted drugs. Regardless of what governs the distribution of substances into slowly equilibrating nonelimination organs, the fact that the kinetics appear to be related to baseline blood flow rates through such organs suggests that some allometric extrapolations between animal species may be valid.

Valid and Invalid Extrapolations of Allometric Methods

In most instances, there are little interspecies differences in the gross composition of individual organs. Therefore, as long as the tissue to blood partition ratios in any given organ depend predominantly on the partition of substances between cell water and neutral fat in adipocytes and on reversible binding to various components in an organ and blood, it seems likely that little difference in the partition ratios of different organs will be found between different animal species. When the partition ratio is markedly affected by the presence of active transport systems and/or the presence of metabolic enzymes, however, the validity of the assumption that allometric methods provide accurate extrapolations is less certain. In most instances, the organs in which such active transport systems and metabolic enzymes exist are usually small and therefore usually make only minor contributions to the total body volume of distribution of the substance. But clearly, the presence of active transport systems in certain cells within an organ can govern which cells are at risk. Indeed, much of the tissue specificity in the toxicity of paraquat is thought to be due in part to the transport systems in certain lung cells (Rose et al., 1976).

Relevance and Irrelevance of Distributional Rate Constants

Because tissue to blood partition ratios are components of the rate constants that describe the entrance and exit of substances in nonelimination organs, they can have a profound influence on the time course of the substance in different organs after a single administration of the substance. Indeed, some substances may have remarkably long half-lives in the body, even when the hepatic clearance approaches hepatic blood flow, because the partition ratios in many of the organs are very high. Moreover, the combination of blood flow and partition ratios can have a profound influence on the approach to steady state during repeated administration of the substance. After a steady state is achieved and the substance is withdrawn, the time required for the substance to be virtually eliminated from the body also depends on the half-life of the terminal phase of elimination. When the investigator is primarily concerned with these phases of the time course of the concentration of a substance and its metabolites, then knowledge of the values of the organ to blood partition ratios are important. In these instances it is important to establish the validity of allometric approaches of extrapolating the value of the blood flow rates and the values of $R_{organ/blood}$ ratios from species to species.

When investigators are primarily concerned with studies in which the substance and its biologically active metabolites are in steady states during most of the time of the study, however, knowledge of the blood flow rates and the $R_{organ/blood}$ values may diminish in importance. Under steady-state conditions, species differences in the $R_{organ/blood}$ ratios govern the magnitude of the oscillations between the maximum and minimum concentration of unbound substance in organs; the larger the values of $R_{organ/blood}$ the smaller the magnitude of the oscillation. Moreover, species differences in the $VR_{organ/blood}$ ratios, particularly in rapidly equilibrating organs, affect the magnitude of the oscillations in the maximum and minimum concentrations of unbound substance in blood. But the magnitude of the oscillations affects the average concentration of unbound substance in nonelimination organs only to the extent that they affect the clearance (*CL*) and the availability values of the organs of elimination.

SPECIES DIFFERENCES IN THE ELIMINATION OF FOREIGN COMPOUNDS

To the extent that species differences in the magnitude of toxicities are due to pharmacokinetic factors, it is likely that most differences are due to factors that govern the total body clearances of substances and the formation and elimination of biologically active metabolites.

Elimination by Excretion into Urine, Air, and Bile

To the extent that substances are excreted unchanged from the body by processes that depend predominantly on physiological processes of animals and the physical chemical characteristics of the substance, it seems likely that allometric methods for extrapolating from one species to another provide reasonably valid results. For example, substances eliminated into exhaled air are governed predominantly by the air/water partition ratio of the substance, the total volume of distribution, the rate of respiration, and the cardiac output (McDougal et al., 1986). Excretion of unbound substances into urine is frequently governed predominantly by the glomerular filtration rate and the lipid solubility of the substance.

Substances that are actively transported into bile (Smith, 1973) and urine, however, are subject to marked species differences; but the rates of excretion of many of these substances may be limited by the renal and hepatic blood flow rates. When this occurs species differences may appear to be related when allometric methods are used to estimate blood flow rates. When the rate of elimination of substances by the liver is limited by the blood flow rate, however, care should be used in the interpretation of extrapolations based on allometric methods. In such situations, the terminal half-life of the substance may appear to be predicted by the allometric methods, but there still may be marked species differences in the bioavailability of substances administered orally that may not be predicted by the allometric methods.

Elimination by Metabolism

It is doubtful that allometric methods are universally valid for predicting species differences in the metabolism of foreign compounds, because there can be marked differences in the metabolism of foreign compounds even between individuals and strains of the same species. Before discussing species differences in metabolism, it is useful to discuss factors that contribute to these intraspecies differences. Despite the emphasis that pharmacokineticists place on flow-limited metabolism, it is important to realize that the metabolism of most foreign compounds is slow compared with the blood flow rates. Thus, $R_{organ/blood}$, f^u in blood, or blood flow rates may be largely irrelevant to discussions of the species differences in pharmacokinetic factors that contribute to species differences in the toxicities of substances that are administered repetitively to test animals during toxicity studies lasting several weeks or months.

INDIVIDUALS AND STRAIN DIFFERENCES IN METABOLISM OF FOREIGN COMPOUNDS

During the past three decades, it has become obvious that foreign compounds are seldom metabolized in the body by a single enzyme. Many compounds are metabolized by combinations of reactions, such as oxidation, dehydrogenation, hydrolysis, sulfation, glucuronidation, and glutathionyl conjugation reactions. Each of these general reactions can be catalyzed by several different enzymes. Indeed, several isozymes catalyze the transfer of sulfate from PAPS to phenols and alcohols (Jakoby et al., 1980); several catalyze the transfer of glucuronic acid from UDPGA to phenols and alcohols (Burchell et al., 1985); several catalyze the transfer of glutathione to various kinds of foreign compounds (Jakoby et al., 1980); and several catalyze the hydrolysis of esters and amides (Heymann, 1980). Even the term cytochrome P-450 is now known to represent at least nine different isozymes in rat liver (Levin et al., 1985). Each of the isozymes of the different categories of enzymes has its own substrate specificity. Some of the isozymes are highly specific and catalyze the metabolism of only a few substances, but others catalyze the metabolism of many different substrates. For each substrate, each isozyme has its own set of rate constants that govern the apparent V_{max} and K_m values. A substrate can be metabolized by several different enzymes and isozymes that may or may not have similar apparent K_m values. Each isozyme can form several metabolites from a given substrate, but the relative rates of formation of the metabolites may differ with the isozyme. In some instances, the pattern of metabolites formed by different isozymes is virtually identical; in other instances, it is markedly different.

ISOZYMES WITH DIFFERENT K_m VALUES

The total rate of metabolism at any given concentration of a substrate within cells is governed not only by the amounts of the various isozymes that are able to metabolize the substrate but also by their apparent K_m values. For example, the 7-hydroxylation of 2-acetylaminofluorene in rabbit liver microsomes is catalyzed by at least two isozymes with K_m values of 0.5 and 147 μM (Thorgeirsson, 1985). Thus, the relative contributions of the various isozymes to the total metabolism of the substrate may vary with the substrate concentration. At high substrate concentrations, the rate may be governed predominantly by isozymes with large V_{max} and K_m values, whereas at low substrate concentrations the rate may be governed predominantly by isozymes with low V_{max} and K_m values. The presence of such enzymes is frequently difficult to detect from *in vivo* pharmacokinetic studies, especially when the studies are interrupted before the concentrations of unbound substance have not declined to levels below the K_m values of the high-affinity enzymes.

SEX DIFFERENCES

The amounts of many of the isozymes in various organs are under hormonal control. For example, the amounts of some isozymes of cytochrome P-450 may be altered by the administration of growth hormone (Gustafsson et al., 1985), testosterone, or estradiol. Thus, the relative distribution of the cytochrome P-450 isozymes (at least in rat liver) changes as the animals mature and results in marked sex differences in the metabolism of some compounds but not in others.

INDUCERS

The amounts of some isozymes may also be changed by foreign compounds entering the body from the environment. Many foreign compounds increase the amounts of some isozymes (Conney, 1982) but decrease the amounts of others, presumably by interacting either directly or indirectly with regulatory genes. Different inducers affect different regulatory genes and thus may cause increases and decreases in several isozymes, but the changes in the amounts of different isozymes vary with the inducer (Thomas et al., 1986). There are strain differences not only in structural genes that govern the amino acid sequence of the various isozymes but also in regulatory genes. Thus, at a given dose, an inducer may cause marked alterations in the pattern of isozymes in one strain of animals but not in another.

INTERORGAN DIFFERENCES IN METABOLISM

Although liver is rightly considered the major organ involved in the metabolism of most foreign compounds, drug-metabolizing enzymes, including isozymes of cytochrome P-450, are also present in other organs. However, the relative distribution of the isozymes varies with the organ. For example, 4-ipomeanol, a substance produced in moldy sweet potatoes that causes lethal pulmonary injury in cattle, rats, mice, and hamsters, is converted to a toxic metabolite in the Clara cells in rat lungs (Boyd, 1980) by an isozyme(s) that has a K_m value about one-tenth that of the predominant isozyme(s) in rat liver that catalyzes the formation of the metabolite in rat liver (Boyd et al., 1978). Moreover, inducers generally alter the isozymes in different organs to different extents. For example, pretreatment of rats with 3-methylcholanthrene increases the rate of formation of the toxic metabolite of ipomeanol in liver microsomes, but not that in lung microsomes. As a result, liver becomes a target organ, but the toxicity in lung is diminished (Boyd, 1980). Nevertheless, 3-methylcholanthrene does increase the amount of some of the isozymes in extrahepatic organs. For example, pretreatment of rats with 3-methylcholanthrene increases the rate of metabolism of 3,4-benzo(a)pyrene by microsomes of many organs, including intestinal mucosa and lung. By contrast, pretreatment of rats with phenobarbital also increases the activity of the isozymes that catalyze the formation of the toxic metabolite of ipomeanol in liver microsomes, but not the isozymes that catalyze its formation in lung. But, because pretreatment with phenobarbital causes an even greater increase in the glucuronidation of ipomeanol *in vivo*, it decreases the amount of metabolite formed in both liver and lung and thus decreases the toxicity in lung, while the liver remains unaffected.

INTERSTRAIN DIFFERENCES

Interstrain differences in the relative amounts of various isozymes may result in either small or large differences in rates of metabolism, depending on the substrate and the animal species. In rats and mice interstrain differences in the metabolism of several substrates, such as hexobarbital, aminopyrine, and acetanilide, by liver microsomes appear to be between two- and threefold (Testa and Jenner, 1976). Whether this would be true for all substrates, however, is open to debate. Within a group of eight strains of rabbits, the metabolism of hexobarbital, aminopyrine, and aniline by liver microsomes also varied between two- and threefold, but there was a 20-fold interstrain difference in the metabolism of amphetamine.

Polymorphisms in Animals

A given kind of metabolite of some substances may be formed predominantly by a single enzyme in the body. When this occurs a change in a structural gene would cause the synthesis of a different allozyme that may or may not have marked differences in the metabolism of the substrate. On one hand, purification of certain allozymes of cytochrome P-450 have virtually identical kinetic characteristics. But on the other hand, allozymes of *N*-acetyltransferase, which catalyzes the acetylation of many drugs and other foreign compounds, have been isolated from phenotypically different rabbits and have been found to have markedly different kinetic characteristics (Andres and Weber, 1986). The V_{max} values of the allozyme isolated from slow acetylators (*rr*) for both *p*-aminobenzoic acid and procainamide were an order of magnitude below the V_{max} values of the allozyme isolated from fast acetylators (*RR*). But the K_m value of the *rr* allozyme for *p*-aminobenzoic acid was also about an order of magnitude smaller than that of the *RR* allozyme, whereas the K_m value of the *rr* allozyme for procaine amide was double that of the *RR* allozyme. Thus, the intrinsic clearances (V_{max}/K_m) of the two allozymes were only slightly different for *p*-aminobenzoic acid but markedly different for procainamide. There is therefore little difference between the phenotypes in the acetylation of *p*-aminobenzoic acid *in vivo*, but more than an order of magnitude difference in the acetylation of procainamide.

Polymorphisms and Environmental Differences in Humans

A combination of environmental and genetic polymorphisms are known to contribute to the individual variability in the metabolism of foreign compounds in humans. For example, smoking and consumption of certain vegetables and open-fire-broiled meat are known to alter the metabolism of many foreign compounds (Conney et al., 1980; Kappas et al., 1977). Moreover, polymorphisms are also known to exist (Kalow, 1962); those associated with the acetylation of arylamines, the hydrolysis of succinyl choline and paroxon, and the hydroxylation of debrisoquine (Smith, 1985) have been the most extensively studied. There is also suggestive evidence of other polymorphisms in the human population in the metabolism of foreign compounds, but these have been less well documented. The combination of both environmental and genetic differences undoubtedly accounts for the observation by clinicians that individual differences in drug metabolism can range from 4- to 40-fold, depending on the drug (Sjoqvist et al., 1976; Vesell and Penno, 1983). In fact, the debrisoquine polymorphism has been reported to cause as much as a 300-fold difference in the hydroxylation of debrisoquine and quanoxan between two scientists in the same laboratory (Idle, 1980). The variance in metabolism can be even greater in extrahepatic tissues; placentas from a group of women, composed of both smokers and nonsmokers, had as much as a 400-fold difference in the rates of metabolism of benzo(*a*)pyrene (Conney, 1982).

Interspecies Differences in the Metabolism of Foreign Compounds

Over the past several decades there has been a plethora of studies of the metabolism of foreign compounds in various animal species. But there does not appear to be any consistent relationship that would justify the assumption that species differences can always be related by allometric methods. The half-lives of some drugs are longer in humans than in rats, which would tend to support the concept that allometric methods might be useful in making such extrapolations. But the half-lives of other drugs are virtually identical in rats and humans in other studies, and the half-lives of still other drugs are longer in rats than in humans (Smith and Caldwell, 1977).

Species differences in the pattern of metabolites excreted into urine are even more extensive. For example, in a literature survey in which the patterns of the urinary metabolites of 23 compounds excreted by rats and humans were compared, only four patterns were considered sufficiently similar to rate a good correlation, whereas eight patterns were considered completely different (Smith and Caldwell, 1977).

When the toxicity of a substance is caused by a minor pathway of metabolism, studies of pharmacokinetic parameters and the pattern of urinary metabolism may not always provide sufficient information to predict toxicities. For example, at doses of acetaminophen that cause hepatotoxicity in mice, but not in Sprague-Dawley rats, there are such small differences in both the half-life of the drug and the pattern of its urinary metabolites that it would have been difficult to conclude from those data alone that the species difference in toxicity was due to differences in the metabolism (Gillette, 1977). Other studies, including the measurements of declines in glutathione concentrations and covalent binding to hepatic proteins, were needed to reveal the relevant species differences in metabolism that caused the species differences in toxicity (Mitchell et al., 1973; Potter et al., 1974).

There also may be species differences in the effects of inducers. For example, pretreatment of mice with phenobarbital increases the hepatotoxicity of acetaminophen by causing a preferential increase in the formation of the chemically reactive metabolite, but pretreatment of hamsters with phenobarbital decreases the hepatotoxicity of acetaminophen by preferentially accelerating the glucuronidation of the drug (Gillette, 1977). Thus, it is not always possible to extrapolate the effects of inducers from one species to another.

Molecular biologists have begun to classify the isozymes of cytochrome P-450 according to the degree of similarity in their amino acid sequences. Isozymes having less than 36% homology are considered to represent different families. According to this classification there are at least eight families of cytochrome P-450 isozymes in mammals (Nebert et al., in press). Within each family there may be several subfamilies. Whether a given inducer increases the synthesis of the same family of isozymes, or whether it even increases all of the members of the same family, has not yet been established. But it is clear that allozymes in different animal species and the isozymes induced in the different species do not necessarily have the same substrate specificity. Moreover, when the isozymes do metabolize the same substrate, they do not always form the same relative amounts of metabolites or possess the same V_{max} and K_m values. For example, isozymes analogous to cytochrome P-450c and cytochrome P-450d induced by 3-methylcholanthrene in rats convert propranolol almost exclusively to desisopropyl propranolol in rabbit liver microsomes, almost exclusively to 4-hydroxypropranolol and 5-hydroxylpropranolol in guinea pig microsomes, and to all three metabolites in liver microsomes of rats and mice (H. A. Sasame, unpublished results).

GENERAL COMMENTS

The utopian objective of a quantitative risk assessment is to be able to predict the incidence rates of toxicities in a human population that is exposed to low doses of toxicants solely from the results of a single toxicity study in a single strain of test animal subjected to relatively large doses of toxicants. Various mathematical models have been developed by statisticians as aids in making extrapolations from high doses to low doses and from populations of small laboratory animals to human populations. That there are uncertainties in the validity of the values obtained with such mathematical models is recognized by nearly everyone. But there is considerable disagreement among scientists concerning the magnitude of the uncertainty. At one extreme, some scientists apparently believe that the values should be accurate within a very narrow range (say, about twofold), and thus focus attention on factors that would affect the calculated value very little. At the other extreme, some scientists apparently believe that the uncertainties are so great that any estimate obtained with the mathematical models is virtually useless.

To many scientists, it seems evident that the degree of inaccuracy of the values calculated by quantitative risk assessment methods differs with the toxicant, the mechanism of the toxicity evoked by the toxicant, and the situation. It therefore is important to delineate those kinds of toxicants and mechanisms of toxicity in which we can be reasonably confident that the degree of inaccuracy is small from those in which the degree of uncertainty is likely to be large. To this end, however, it is necessary to understand precisely what the problems are and whether the various mathematical models adequately address the problems.

One of the problems is that the shape of a dose-response curve may be a composite of at least three different kinds of dose-response relationships. Indeed, the meaning of a dose-response curve is largely dependent on how the response is recorded. The meaning of each type can be illustrated by an idealized experiment in which a drug acts by binding reversibly to a set of receptor sites in a group of identical animals.

1. In this idealized situation, the magnitude of the response is presumably directly proportional to the number of receptor sites occupied by the drug. Thus, if the action sites act independently of one another, the fraction of the total number of receptor sites that are occupied by the substance at any given concentration of unbound substance will usually follow the Law of Mass Action, i.e.:

$$(\text{Effect/Maximum Effect}) = [D]/(K_d + [D]), \qquad (22)$$

where $[D]$ is the concentration of the unbound drug, and K_d is the dissociation constant of the drug-receptor site complex. Notice that at low concentrations of the drug the magnitude of the effect should be directly proportional to the concentration of the unbound drug. Thus, the claim that there is an inherent dose threshold for the action of a reversibly acting drug is fallacious. What is meant by a no-observed-effect level (NOEL) is either that the magnitude of the effect at very low concentrations can become so small that it cannot be detectable or that the response is not always directly proportional to the number of receptor sites occupied. Consider, however, that in a group of completely identical animals, the magnitude of the response should be the same in all animals.

2. A dose-duration of response curve depends on the rates at which the drug enters the blood and rises to its maximum concentration in the immediate environment of the action sites and then declines to a concentration that exerts a magnitude of response below which the effect is no longer detectable. Consider, however, that a given dose administered identically

to a group of identical animals should result in identical durations of response in all of the animals in the group.

3. In a dose-incidence of response curve, the proportion of the group of animals receiving a given dose of the drug is evaluated. Inherent in this type of dose-response curve is the assumption that there is some dose that results in a magnitude of response below which the response either is not detectable or is not biologically significant. In the idealized experiment, however, there would be no observable effect in any of the animals until a certain critical dose is administered, but at slightly larger doses the response would be observable in all of the animals. Hence the slope of the dose-incidence of response curve in the idealized experiment would be infinite. In the real world, the slope of the dose-incidence of response curve thus is a measure of the precision with which the substance is administered and the response recorded, and the homogeneity of the population of animals in response to the substance.

Although the mathematics required to describe other mechanisms of response would be different, the basic meanings of the dose-magnitude of response and the dose-incidence of response curves are applicable to all mechanisms of response.

Many risk assessments for carcinogens are based on the multistage model, which is described by the equation:

$$P = 1 - e^{-X}, \tag{23}$$

where

$$X = q_0 + q_1 d + q_2 d^2 \ldots . \tag{23a}$$

In Equation 23 P is defined as the probability of occurrence of a tumor in an individual animal exposed to a daily dose (d) of a chemical for a lifetime, and the q values are parameters estimated from the experiment. At first glance Equation 23 may appear to be reasonable, but in light of the above discussion, Equation 23 is based on inconsistent logic, because a dose-incidence of response curve implies the existence of a NOEL below which tumors would be undetectable. Indeed, it seems more logical to treat data obtained in a study of carcinogenesis as a dose-magnitude of response curve in which the number of tumors observed during the course of the study is related to the numbers of cells at risk (cells). Accordingly, we could write an equation:

$$\text{Tumors/Cells} = X = q_0 + q_1 d + q_2 d^2 + \ldots . \tag{24}$$

According to this view, however, the development of tumors would be considered a rare event, and thus a group of 100 virtually identical animals receiving the same dose in a bioassay for carcinogenesis would be visualized as a single animal, which is 100 times as large and the number of animals actually employed.

The two approaches thus are conceptually very different and lead to different interpretations of the slope of a dose-response curve; in the dose-incidence of response curve, the slope is a measure of the homogeneity of the animals used in the bioassay, whereas in the dose-magnitude of response curve the slope assumes no heterogeneity within the population. Nevertheless, the two approaches can be interrelated by normalizing the number of cells to the number of cells at risk per animal, that is:

$$\text{(Tumors/Cells)/(Cells/Animal)} = \text{(Tumors/Animal)} = X. \qquad (25)$$

Whereas, at low values of X:

$$P = (1 - e^{-X}) \simeq X. \qquad (26)$$

Thus, in practice the two concepts usually provide virtually the same estimate. But the practice of counting an animal with multiple tumors only once in a bioassay, though defensible as a practical matter, is not defensible from a theoretical point of view.

Another problem that must be considered is that Equations 23a and 24 probably are not completely valid in describing all of the possible ways in which the value of X is curvilinear with increasing doses. An increase in dose can alter not only pharmacokinetic parameters but also pharmacodynamic factors. For example, perhaps high doses of a substance may cause so much damage to DNA that the rates of repair approach maximal values. Some promoters can act by releasing growth factors, but the rate or extent of release of these factors may not be directly proportional to the dose.

On the other hand, the bioassay coupled with pharmacokinetic studies may not always reveal the presence of high-affinity, low-capacity enzymes that can be of predominant importance at low doses but only of trivial significance at high doses. Nor will the effects of dose-dependent pharmacokinetic parameters be accurately predicted by a function of X represented by a polynomial.

In calculating the P (probablity) values, statisticians do incorporate interanimal variability within error functions. Moreover, they can differentiate between models through the use of maximum likelihood methods, but marked interanimal variability decreases their ability to distinguish between plausible models in the bioassay. Whatever the problems that occur in the interpretation of experiments may be, however, it is important to remember that the calculated values for P represent the values for the *animals used in the bioassay under the conditions of the experiment.*

In bioassays lasting several days, weeks, months, or years, either the processes that govern the pharmacodynamic factors or the processes that govern the pharmacokinetic factors may be altered during repeated administration of high doses of the substance under investigation. Indeed, studies of alterations in the concentrations of several hormones that are thought to influence the manifestation of tumors are sometimes incorporated into the protocol of the bioassay. But it is also true that repeated administration of substances can alter processes that govern relevant pharmacokinetic parameters, such as kidney and pulmonary function or the activity of enzymes that govern the total body clearances of a parent compound and its biologically active metabolites and the pattern of metabolism of various tissues. Moreover, if short-lived metabolites are thought to cause the toxicity, it may be necessary to assess the activity of enzymes that catalyze the formation of the metabolite, inactivation of the enzymes in potential target cells and organs, as well as the activity of enzymes that account for most of the total body clearances of the precursors of the ultimate carcinogen. *Thus, pharmacokinetic studies performed solely with untreated animals may be largely irrelevant to the quantitative risk assessment process. Instead, the pharmacokinetic studies should be repeated at intervals throughout the course of the bioassay.* If the pharmacokinetic factors do change during repeated administration, then the investigator is faced with the problem of deciding whether the changes would markedly influence the magnitude of toxic response, and if they do how they would affect the shape of the dose-magnitude of response curve.

In the extrapolation of the calculated values of tumors/cells in Equation 24 from high doses in animals to low doses in humans, it is important to remember that the sequence is an extrapolation from high doses to low doses for the animals in the bioassay and then an extrapolation from low doses in the experimental animals to low doses in the human population. Thus, to the extent that pharmacokinetic parameters contribute to interspecies differences in the magnitude of the response, the differences usually can be expressed by differences in the parameters of linear models. Hence, most of the complexities in pharmacokinetics due to dose-dependent nonlinearities that are important in understanding the shape of the dose-magnitude of response curve in the animal assay are usually not relevant in extrapolating from low doses in animals to low doses in the human population.

Nevertheless, in making an extrapolation from the animals used in the bioassay to the human population, investigators are faced with two problems: (1) Are the pharmacodynamic and pharmacokinetic factors that govern the magnitude of response at low doses of the toxicant in the test animals markedly different from those in the mean of the human population? (2) What is the range of the differences in these factors in the human population relative to the range of the differences in the animal population used in the bioassay?

Within this context, it is important to stress that any given pharmacokinetic and toxicity study in research animals is performed in animals that are phenotypically virtually identical and maintained under environmental conditions that are kept as homogeneous as possible. By contrast, the human population is composed of individuals that are genetically heterogeneous, are of different ages and sizes, are either female or male, may suffer from various diseases, have different personal habits including smoking or exercise, eat different diets, and live in different environments that are heterogeneously distributed throughout the world. All of these factors have been shown or suspected to contribute to differences in the pharmacodynamics or the pharmacokinetics of foreign compounds. It is also important to realize that the way that pharmacokinetic data are reported in the literature stresses the mean value within given populations. Outliers are usually ignored, on the assumption that they result from analytical or sampling errors. The possibility that they represent polymorphisms in either the pharmacodynamic or pharmacokinetic factors of toxicants is frequently ignored, even though such subpopulations may include virtually all of the individuals suffering from the toxicity of a substance in the human population. Elucidation of such polymorphisms, however, can only occur by studying the human population.

Unfortunately, science has not yet been able to develop a universally valid approach for predicting the range of variability within the human population in the disposition of all foreign compounds. Indeed, the range of variability within the human population is known to vary with the foreign compound.

In attempting to extrapolate estimates from animals to the mean of the human population, various investigators have pointed out that interspecies differences in many physiological processes, including cardiac output, organ sizes, blood flow rates, and basal metabolism rates, may be related to the surface area of the animals and physiological time. In recent years, a few pharmacokinetic studies have provided data that would tend to support the idea that such extrapolations may be valid for those compounds. *But the finding that interspecies differences in the pharmacokinetics of some compounds can be related by allometric methods should not be construed as meaning that allometric methods are valid for predicting interspecies differences in the pharmacokinetics of all foreign compounds.* Indeed, the predominance of evidence in the field of drug metabolism that has accumulated in the past indicates that allometric methods would be virtually useless for the prediction of metabolism of many substances as a mean of the human population, much less the range of values within the human population. It therefore seems important to delineate situations in which allometric methods are likely to provide reasonably valid extrapolated values from those in which the validity of extrapolations obtained by such methods would be highly questionable.

It seems likely that interspecies differences in those processes that depend predominantly on the physical chemical properties (such as solubility in water and lipid/aqueous partition ratios) and basic physiological processes (such as rates of respiration, blood flow rates through various organs, and renal clearances, particularly those dependent predominantly on glomerular filtration rates) may be predictable by allometric methods with a reasonable degree of precision (e.g., within a fivefold range of confidence). Thus, allometric methods should provide reasonably valid predictions when interspecies differences in the toxicity are predominantly due to pharmacokinetic factors (as opposed to pharmacodynamic factors) and are caused solely by a parent compound that is eliminated predominantly unchanged into exhaled air, urine, or possibly, feces. Moreover, allometric methods can also provide reasonably valid predictions when the toxicity is caused solely by a parent compound that is eliminated by flow-limited metabolism in non-first-pass organs; under these conditions marked interspecies differences in the intrinsic clearances of the enzymes that metabolize the compound can occur without having marked interspecies differences in the total body clearance or the biological half-life of the compound.

I am less hopeful that allometric methods will always provide valid interspecies extrapolations when toxic parent substances are eliminated from the body predominantly by enzymes having intrinsic clearances much less than the blood flow rates through the organs of elimination or when the toxicity is caused by metabolites of the toxicant. In such cases allometric methods may provide reasonably valid interspecies extrapolations for some toxicants, but not for others.

To differentiate between these broad categories, however, it would be necessary to perform well-integrated pharmacokinetic and toxicity studies that would elucidate whether the toxicity is caused by the parent substance, one or more of its metabolites, or a combination of the parent substance and its metabolite.

When the mechanism of toxicity is not known and the sources of interspecies and intraspecies differences in the response to the toxicant are not clearly understood, extrapolations based on allometric methods represent only first guesses of the incidence rates of the toxicity in the human population that we hope will be reasonably valid for most compounds. Perhaps some of the extrapolations may be validated by studies of the toxicants in humans that are accidentally exposed. In such cases Bayesian approaches to evaluate the mean values and ranges of pharmacodynamic and pharmacokinetic parameters in humans may be useful.

At the present time, however, the hope of the risk assessor is not whether the extrapolated incidence rate is completely accurate. Instead, the hope is that the assessment is not so far wrong that the incidence rate exceeds a de minimus value. Although it would be nice to think that the de minimus value could be set at an arbitrary level (such as 1:1,000,000), as a practical matter it is really limited to the ability of epidemiological methods to detect toxicities over background values. Even though allometric methods will not always provide accurate assessments, there does not appear to be any reasonable alternative to using them. Thus, in the absence of complete information concerning the mechanisms of toxicity and the pharmacokinetic and pharmacodynamic factors that govern the manifestations of the toxicities, it is perhaps advisable to try to establish a consensus among scientists of arbitrary broad ranges of uncertainty, even though such ranges cannot be rigorously defended by science. When the mechanism of toxicity becomes known and if it can be established that intraspecies and interspecies differences in the pharmacodynamic and pharmacokinetic factors can be predictable within narrower ranges of uncertainty for given toxicities caused by given toxicants, the arbitrary range of uncertainty can be narrowed for that toxicity and toxicant. Unless such a system is established, I believe that the field of quantitative risk assessment will remain highly controversial. Whatever system is ultimately established, I believe it inevitable that some mistakes will be made and therefore that quantitative risk assessments will never replace the need for epidemiological studies or toxicity reporting systems.

REFERENCES

Andres, H. H., and W. W. Weber. 1986. N-Acetylation pharmacogenetics: Michaelis-Menten constants for arylamine drugs as predictors of their N-acetylation rates *in vivo*. Drug Metab. Dispos. 14:382–385.

Baron, J., and T. T. Kawabata. 1983. Intratissue distribution of activating and detoxicating enzymes. Pp. 105–135 in Biological Basis of Detoxication, J. Caldwell and W. B. Jakoby, eds. New York: Academic Press.

Boyd, M. R. 1980. Biochemical mechanisms in chemical-induced lung injury: Roles of metabolic activation. CRC Crit. Rev. Toxicol. 103–176.

Boyd, M. R., L. T. Burka, B. J. Wilson, and H. A. Sasame. 1978. *In vitro* studies on the metabolic activation of the pulmonary toxin, 4-isomeanol by rat lung and liver microsomes. J. Pharmacol. Exp. Ther. 207:677–686.

Burchell, B., M. R. Jackson, S. M. E. Kennedy, L. McCarthy, and G. C. Barr. 1985. Characterization and regulation of hepatic UDP-glucuronyltransferase. Pp. 212–220 in Microsomes and Drug Oxidations, A. R. Boobis, J. Caldwell, F. DeMatteis, and C. R. Elcombe, eds.

Conney, A. H. 1982. Induction of microsomal enzymes by foreign chemicals and carcinogenesis by polycyclic aromatic chemicals. G. H. Clowes Memorial Lecture. Cancer Res. 82:4875–4917.

Conney, A. H., M. K. Buening, E. J. Pantuck, C. B. Pantuck, J. A. Fortner, K. E. Anderson, and A. Kappas. 1980. Regulation of human drug metabolism by dietary factors. Pp. 147–167 in Environmental Chemicals, Enzyme Function and Human Disease, Ciba Foundation Symposium, Vol. 76.

deLannoy, I. A. M., and K. S. Pang. 1986. Commentary: Presence of a diffusional barrier on metabolite kinetics: Enalaprilat as a generated versus performed metabolite. Drug Metab. Dispos. 14:513–520.

Drasar, B. S., M. J. Hill, and R. E. O. Williams. 1970. The significance of the glutflora in safety testing of food additives. In Metabolic Aspects of Food Safety, J. C. Rose, ed. Ixford and Edinburgh: Blackwell.

Galinsky, R. E., and G. Levy. 1981. Dose and time-dependent elimination of acetaminophen in rats: Pharmacokinetic implications of cosubstrate depletion. J. Pharmacol. Exp. Ther. 219:14–20.

Gillette, J. R. 1977. The phenomenon of species variations; problems and opportunities. Pp. 147–168 in Drug Metabolism from Microbe to Man, D. V. Parke and R. S. Smith, eds. London: Taylor & Francis.

Gillette, J. R. 1982. Sequential organ first-pass effects: Simple methods for constructing compartmental pharmacokinetic models from physiological models of drug disposition by several organs. J. Pharm. Sci. 71:673–677.

Gillette, J. R. 1984. Solvable and unsolvable problems in extrapolating toxicological data between animal species and strains. Pp. 237–260 in Drug Metabolism and Drug Toxicity, J. R. Mitchell and M. G. Horning, eds. New York: Raven.

Gillette, J. R. 1985. Pharmacokinetics of biological activation and inactivation of foreign compounds. Pp. 30–70 in Bioactivation of Foreign Compounds, M. W. Anders, ed. New York: Academic Press.

Gillette, J. R. 1986. Significance of covalent binding of chemically reactive metabolites of foreign compounds to proteins and lipids. Pp. 63–82 in Biological Reactive Intermediates III, J. R. Kocsis, D. J. Jollow, C. M. Witmer, J. O. Nelson, and R. Snyder, eds. New York: Plenum.

Gustafsson, J.-A., C. MacGeoch, and E. T. Morgan. 1985. Isolation characterization and regulation of sex specific isozymes of cytochrome P-450 catalyzing 15 beta-hydroxylation of steroid sulfates and 16 alpha-hydroxylation of 4-androstene-3,17-dione. In Microsomes and Drug Oxidations, A. R. Boobis, J. Caldwell, F. DeMatteis, and C. R. Elcombe, eds. London: Taylor & Francis.

Heymann, E. 1980. Carboxylesterases and amidase. Pp. 291–323 in Enzymatic Basis of Detoxication, Vol. II. New York: Academic Press.

Hjelle, J. J., G. A. Hazelton, and C. D. Klaassen. 1985. Acetaminophen decreases adenosine 3′-phosphate 5-phosphosulfate and uridine diphosphoglucuronic acid in rat liver. Drug Metab. Dispos. 13:35–41.

Horning, M. G., L. Brown, J. Nowlin, K. Lertratanagkoon, P. Kellaway, and T. E. Zion. 1977. Use of saliva in therapeutic drug monitoring. Clin. Chem. 23:157–164.

Houston, J. B., D. G. Upshall, and J. W. Bridges. 1974. A re-evaluation of the importance of partition coefficients in the gastrointestinal absorption of nutrients. J. Pharmacol. Exp. Ther. 189:244–254.

Idle, J. R. 1980. Discussion. P. 284 in Environmental Chemicals, Enzyme Function and Human Disease. Ciba Foundation Symposium, Vol. 76.

Jakoby, W. B., R. D. Sekura, E. S. Lyon, C. J. Marcus, and J.-L. Wang. 1980. Sulfotransferases. Pp. 199–228 in Enzymatic Basis of Detoxication, Vol. II, W. B. Jakoby, ed. New York: Academic Press.

Jollow, D. J., S. Roberts, V. Price, S. Longacere, and C. Smith. 1982. Pharmacokinetic considerations in toxicity testing. Drug. Metab. Rev. 13:983–1007.

Kalow, W. 1962. Pharmacogenetics: Heredity and the Responses to Drugs. Philadelphia: W. B. Saunders.

Kappas, A., A. P. Alvarez, K. E. Anderson, W. A. Garland, E. J. Pantuck, and A. H. Conney. 1977. The regulation of human drug metabolism by nutritional factors. Pp. 703–708 in V. Ullrich, A. Hildebrandt, I. Roots, R. W. Estabrook, and A. H. Conney, eds. Drug Microsomes and Drug Oxidations. New York: Pergamon.

Levin, W., P. E. Thomas, L. M. Reik, D. E. Ryan, S. Bandiera, M. Haniu, and J. E. Shively. 1985. Immunochemical and structural characterization of rat hepatic cytochrome P-450. Pp. 983–1007 in Microsomes and Drug Oxidations, A. R. Boobis, J. Caldwell, F. DeMatteis, and C. R. Elcombe, eds. London: Taylor & Francis.

Levy, G., R. E. Galinsky, and J. H. Lin. 1982. Pharmacokinetic consequences and toxicologic implications of endogenous cosubstrate depletion. Drug Metab. Rev. 13:1009–1020.

McDougal, J. M., G. W. Jepson, H. J. Clewell III, M. G. MacNaughton, and M. E. Andersen. 1986. A physiological pharmacokinetic model for dermal absorption. Toxicol. Appl. Pharmacol. 85:286–294.

Mitchell, J. R., D. J. Jollow, W. Z. Potter, J. R. Gillette, and B. B. Brodie. 1973. Acetaminophen-induced hepatic necrosis. IV. Protective role of glutathione. J. Pharmacol. Exp. Ther. 187:211–217.

Nebert, D. W., M. Adesnick, M. J. Coon, R. W. Estabrook, F. J. Gonzalez, F. P. Guengerich, I. C. Gunsalus, E. F. Johnson, B. Kemper, W. Levin, I. R. Phillips, R. Sato, and M. R. Waterman. In press. The P-450 gene superfamily. Recommended nomenclature. DNA.

O'Flaherty, E. J. 1986. Dose dependent toxicity. Comments Toxicol. 1:23–34.

Pang, K. S. 1983. Fate of xenobiotics: Physiological and kinetic considerations. Pp. 213–250 in Biological Basis of Detoxication, J. Caldwell and W. B. Jakoby, eds. New York: Academic Press.

Pang, K. S., and J. R. Gillette. 1979. Sequential first-pass elimination of a metabolite derived from a precursor. J. Pharmacok. Biopharm. 7:275–290.

Potter, W. Z., S. S. Thorgeirsson, D. J. Jollow, and J. R. Mitchell. 1974. Acetaminophen-induced hepatic necrosis. V. Correlation of hepatic necrosis, covalent binding and glutathione depletion in hamsters. Pharmacology 12:129–143.

Price, V. F., and D. J. Jollow. 1982. Increased resistance of diabetic rats to acetaminophen-induced hepatotoxicity. J. Pharmacol. Exp. Ther. 220:504–513.

Price, V. F., and D. J. Jollow. 1984. Role of UDPGA flux in acetaminophen clearance and hepatotoxicity. Xenobiotica 14:553–559.

Rose, M. S., E. A. Lock, L. L. Smith, and I. Wyatt. 1976. Paraquat accumulation: Tissues and species specificity. Biochem. Pharmacol. 25:419–423.

Rowland, M., and T. N. Tozer. 1980. Clinical Pharmacokinetics: Concepts and Applications. Philadelphia: Lea & Febieger.

Sjoqvist, F., O. Borga, and M. L. Orme. 1976. Fundamentals of clinical pharmacology. In Drug Treatment: Principles and Practice of Clinical Pharmacology and Therapeutics, G. S. Avery, ed. Seaforth, Australia: AIDS.

Smith, R. L. 1973. Pp. 76–93 in The Excretory Function of Bile: The Elimination of Drugs and Toxic Substances in Bile. New York: John Wiley & Sons.

Smith, R. L. 1985. Genetic polymorphisms of drug oxidation in man. Pp. 349–360 in Microsomes and Drug Oxidations, A. R. Boobis, J. Caldwell, F. DeMatteis, and C. R. Elcombe, eds. London: Taylor & Francis.

Smith, R. L., and J. Caldwell. 1977. Drug metabolism in non-human primates. Pp. 331–356 in Drug Metabolism from Microbe to Man, D. V. Parke and R. L. Smith, eds. London: Taylor & Francis.

Taylor, P. W. 1972. Fast reactions—flow and relaxation methods. Pp. 351–380 in Methods in Pharmacology, Vol. 2, Physical Methods, A. Schwartz, series ed., C. F. Chignell, ed. New York: Appleton-Century-Crofts.

Testa, B., and P. Jenner. 1976. Pp. 361–384 in Drug Metabolism: Chemical and Biological Aspects. New York: Marcel Dekker.

Thomas, P. E., L. M. Reik, S. L. Maines, S. Bandiera, D. E. Ryan, and W. Levin. 1986. Antibodies as probes of cytochrome P-450 isozymes. Pp. 95–106 in Biological Reactive Intermediates III, J. J. Kocsis, D. J. Jollow, C. M. Witmer, J. O. Nelson, and R. Snyder, eds. New York: Plenum.

Thorgeirsson, S. S. 1985. Kinetics of acetylaminofluorene hydroxylation reactions in microsomes and drug oxidations. Pp. 320–329 in A. R. Boobis, J. Caldwell, F. DeMatteis, and C. R. Elcombe, eds. Microsomes and Drug Oxidations. London: Taylor & Francis.

Vesell, E. S., and M. B. Penno. 1983. Intraindividual and interindividual variations. Pp. 369–410 in Biological Basis of Detoxication, J. Caldwell and W. B. Jakoby, eds. New York: Academic Press.

Dose, Species, and Route Extrapolation Using Physiologically Based Pharmacokinetic Models

Harvey J. Clewell III and *Melvin E. Andersen*

INTRODUCTION

We can distinguish four types of extrapolations involved in assessing the expected human risk associated with exposure to environmental chemicals. They consist of (1) predicting the low-dose response in experimental animals based on observed responses at very much higher doses, (2) predicting the response in the human population based on the results in the test species, (3) predicting the risks associated with the anticipated human exposure route based on toxicity observed when a different route of exposure is used in the animal toxicity studies, and (4) predicting human response at realistic, discontinuous environmental exposures based on animal results in well-controlled, much more easily characterized exposures. In this paper, these are called, respectively, the dose, species, route, and exposure scenario extrapolations. Each of these extrapolative steps is important and must be conducted on the basis of the best available, scientifically justifiable procedures if the final exposure limits are to have validity and enjoy consensus support from government, industry, and the concerned public. Before considering the application of physiologically based pharmacokinetic modeling for performing these extrapolations, we will briefly review current practices in each of the four areas.

Much of the introductory material for this paper is from H. J. Clewell and M. E. Andersen. 1985. Risk assessment extrapolations and physiological modeling. Toxicol. Ind. Health. 1:111–131.

Dose

Many risk assessments are based on the results of animal experimentation conducted at very high daily doses compared with those likely to be encountered in human exposures. Typically, extrapolations are based on administered dose, and a linear extrapolation through zero is utilized to predict the incidence of a particular effect in the low-dose region. A variety of statistical models is used for this extrapolation, and the estimated risk can differ by several orders of magnitude, depending on the model. Often the response data are extrapolated by using a parametric fit (e.g., the multistage model), and the linear, nonthreshold model is assumed to apply in the low-dose region. It is hoped that if nonlinearities exist, the linear model will tend to err conservatively. The potential problems are twofold. On the one hand, saturation of metabolism could lead to situations in which a linear extrapolation was not conservative, particularly if a reactive metabolic intermediate was the active moiety (Ramsey and Reitz, 1981). On the other hand, in cases in which an effective threshold for toxicity appears to exist, the use of an overly conservative estimate could result in an unnecessarily restrictive regulatory decision.

Species

Lack of controlled human exposure data on most toxic chemicals makes it necessary to infer human susceptibility from animal results. The dose administered to the test species is generally converted to an equivalent human dose on the basis of either body weight (in milligrams per kilogram) or surface area, where surface area is taken to be proportional to body weight raised to the 2/3 power. The latter scaling factor is generally justified on the basis of the studies by Freireich et al. (1966), who examined the interspecies differences in toxicity of a variety of antineoplastic drugs. This "surface area adjustment" is often appropriate for a toxic chemical detoxified by metabolism. It has come to be applied routinely, however, even in the case of chemicals with toxic metabolites for which such a surface area dependence would not be expected (M. E. Andersen, this volume). For a chemical that demonstrates significant interspecies variation in toxicity in animal experiments, the most susceptible species is generally used as the reference for this extrapolation. Because of the recognized uncertainty involved, a safety factor of 10 to 1,000 or more may often be applied.

Route

In some cases, there are no animal data corresponding to the expected human exposure route. For example, in developing a surface water standard for a compound that has only been studied via inhalation, the relationship between the inhalation and oral routes of administration must be estimated. The surface water concentration is then calculated from the acceptable oral dose by using some assumed level of human consumption. Methods for relating inhalation and oral doses for systemic toxicants usually assume some sort of direct correspondence based on total administered dose, calculated uptake, or achieved blood levels (EPA, 1984). Again, a safety factor is often applied, reflecting the increased uncertainty.

Exposure Scenario

There are many other ways in which animal studies may differ from expected human exposure scenarios, chiefly relating to the frequency and duration of exposure. Examples include estimating lifetime carcinogenic risk from studies of less than lifetime duration, correlating 50% letal doses (LD_{50}s) for acute toxicity to no-observed-adverse-effect levels for chronic exposure, and adjusting workplace exposure limits to reflect changes in work shifts. The chosen relationships may be statistical, semiempirical, or just plain intuitive.

Pharmacokinetically Based Extrapolations

In contrast to these rule-of-thumb approaches, the fundamental assumption in conducting pharmacokinetic extrapolations is that a particular effective tissue dose in one species or by one route of administration is equally effective in another species or if obtained by a different route of administration. For all of these questions of how to extrapolate from animal experiments to estimate human risk, we must know how to calculate the effective dose of a chemical reaching appropriate target tissues under any exposure condition. The basic issue, then, is the relationship between the administered dose and some delivered or effective dose. In the past it has been tacitly assumed that the effective dose was the same as the administered dose. It is now clear that the relationship between these two expressions of dose is complex (Andersen, 1981). There are many factors involved in determining this relationship, and they are functions of both the chemical and the organism, leading to a complexity which defies any general description. For example, saturation of metabolism can lead to apparently nonlinear dose-response behavior. For some chemicals, this relationship can be further complicated by induction or inhibition of the relevant enzyme systems as well as by depletion of necessary cofactors. Metabolic first-pass effects can lead to variations in bioavailability both within and between routes. The direction of the effect of these factors can be to either increase or decrease relative toxicity, depending on whether the toxicity results from the parent chemical, a stable metabolite, or a transient intermediate. What is needed is a framework with which to describe the actions of all these important factors in a quantitative fashion. The development of such a framework is one of the purposes of physiologically based pharmacokinetic modeling.

PHYSIOLOGICALLY BASED PHARMACOKINETIC MODELING

Pharmacokinetics is the study of the time course for the absorption, distribution, metabolism, and elimination of a chemical substance in a biological system. Implicit in any pharmacokinetic description is the assumption that the response of some target tissue can be related to the concentration profile of the active form of the substance in that tissue. Pharmacokinetic models generally are divided into two categories: compartmental and physiological. A typical compartmental model attempts to relate the blood or tissue concentration profile of the parent or the metabolite to the administered dose of the parent chemical by using a set of mathematical equations. The parameters for these equations are determined from experiments following the time course of the chemical in body fluids and occasionally in specific tissues. A simple model might consist of just two compartments: a central compartment in equilibrium with the blood, and a peripheral compartment whose concentration can be related to the central compartment by rate constants in each direction. The volumes of the compartments and the values of the rate constants are adjusted to fit the experimental data, after which the model can be used for interpolation and limited extrapolation.

Physiologically based pharmacokinetic models differ from the conventional compartmental models in that they are based to a large extent on the actual physiology of the organism (Figure 1). Instead of compartments defined by the experimental data itself, actual organ and tissue groups are used with weights and blood flows from the literature (Bischoff and Brown, 1966; Himmelstein and Lutz, 1979). Instead of composite rate constants determined by fitting the data, actual physical-chemical and biochemical constants of the compound are used. The result is a model that predicts the qualitative behavior of the experimental time course without being based on it. Refinement of the model to incorporate additional insights gained from comparison with experimental data yields a model that can be used for quantitative extrapolation well beyond the range of experimental conditions.

The chief advantage of a physiologically based model is its greater predictive power. Because fundamental metabolic parameters are used, dose extrapolation over ranges in which saturation of metabolism occurs is possible. Because known physiological parameters are used, a different species can be modeled by simply replacing the appropriate constants

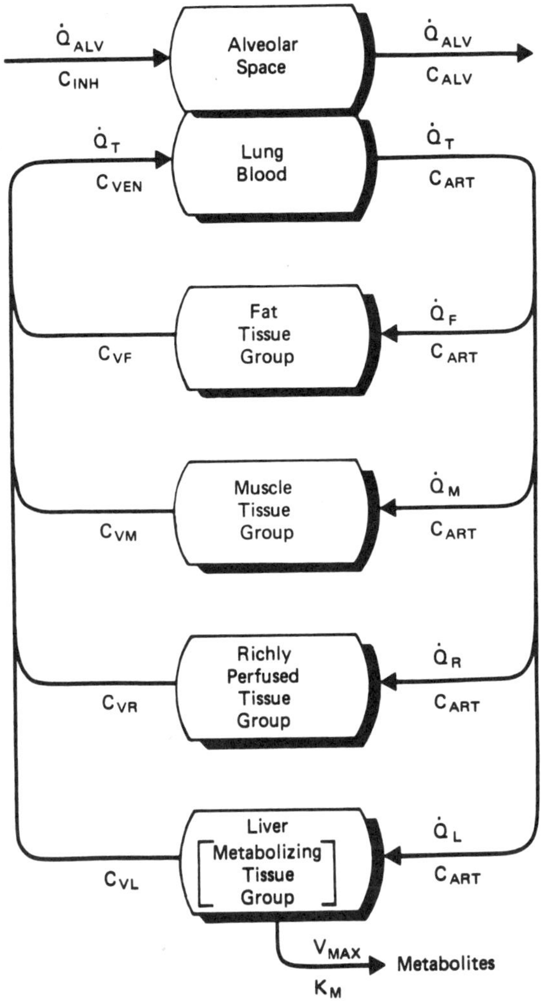

FIGURE 1 Diagram of a generic physiologically based pharmacokinetic model for volatile organic chemicals. All of the models described in this paper are adaptations of this simple model, which was used by Ramsey and Andersen (1984) to investigate the pharmacokinetics of styrene. In this description, groups of tissues are defined with respect to their volumes, blood flows (Q), and partition coefficients for the chemical. The uptake of vapor is determined by the alveolar ventilation (Q_{alv}), cardiac output (Q_t), blood: air partition coefficient, and the concentration gradient between arterial and venous pulmonary blood (C_{art} and C_{ven}). Metabolism is described in the liver, with a saturable pathway defined by a maximum velocity (V_{max}) and affinity (K_m) and, when necessary, with a first-order pathway (data not shown). The mathematical description assumes equilibration between arterial blood and alveolar air, as well as between each of the tissues and the venous blood exiting from that tissue. Reproduced with permission from Ramsey and Andersen (1984).

(Dedrick, 1973). Similarly, the behavior for a different route of administration can be determined by adding equations which describe the nature of the new input function. The extrapolation from one exposure scenario (say, a single 6-h exposure) to another (e.g., a repetitive 6-h exposure, 5 days a week for the life of the animal) is relatively easy and only requires a little ingenuity in writing the equations for the dosing regimen in the kinetic model (D. J. Paustenbach, H. J. Clewell, M. L. Gargas, and M. E. Andersen, this volume).

Because measured physical-chemical and biochemical parameters are used, the behavior for a different chemical can quickly be estimated by determining the appropriate constants. An important result is the ability to reduce the need for extensive range-finding experiments with new chemicals. The process of selecting the most informative experimental design is also facilitated by the availability of a predictive pharmacokinetic model. Perhaps the most desirable feature of a physiologically based model is that it provides a conceptual framework for employing the scientific method in which hypotheses can be described in terms of biological processes, predictions can be made on the basis of the description, and the hypotheses can be revised on the basis of comparison with experimental data.

The trade-off against the greater predictive capability of physiologically based models is an increased number of parameters and equations. Values for many of the parameters, however, particularly the physiological ones, are already available in the literature; and several techniques have been developed for rapidly determining the compound-specific parameters. There is even a prospect that predictive models can be developed based entirely on data obtained from *in vitro* studies. For volatile liquids, the type of chemicals with which we are most familiar, tissue partition coefficients can be determined by a simple *in vitro* technique called vial equilibration (Sato and Nakajima, 1979b), and tissue metabolic constants can be determined by a modification of the same technique (Sato and Nakajima, 1979a). Alternatively, other rapid *in vivo* approaches for determining metabolic constants can be used either based on steady-state (Andersen et al., 1984b) or gas uptake (Andersen et al., 1980; Filser and Bolt, 1979; Gargas et al., 1986a,b) experiments. The resulting set of constants together with the general physiological parameters provide a model of parent chemical behavior and rate of metabolism which can be predictive of kinetic behavior at various concentrations, for various dose routes, in a variety of species, with any number of exposure scenarios. In essence, the same approach can be used with nonvolatile xenobiotics; but at present experiments to determine solubility, tissue binding, and metabolic constants are not as easily conducted with these chemicals as they are with the gases and volatile liquids. Nonetheless, there are now several very good examples of physiologically based models which describe the kinetics of important nonvolatile environmental contaminants, including kepone, polybrominated biphenyls, and polychlorinated dibenzofurans (Bungay et al., 1981; King et al., 1983; Tuey and Matthews, 1980).

DOSE-ROUTE EXTRAPOLATION

We have described a generic physiologically based pharmacokinetic model for volatile organic chemicals (Clewell and Andersen, 1986). This generic model predicts blood and tissue concentrations of parent chemical and the rate of metabolism in the liver. Nonlinear kinetic behavior, the high-dose to low-dose extrapolation problem, is accounted for by including terms for two metabolic pathways in the liver tissue, one of which is saturable and the other of which is strictly first order.

Dose-route extrapolations can be conducted by using the inhalation description and adding appropriate equations representative of other routes of administration. Intravenous (i.v.) injection is easily described by a constant rate of infusion into mixed venous blood. Oral administration in a water vehicle can be modeled by first-order uptake from a bolus gastrointestinal dose, with the incoming chemical presumed to appear in the liver. This is done because portal blood flow goes to the liver before it is recirculated in the systemic circulation. Predictions of the inhalation, i.v., and oral kinetics of dibromomethane (Figure 2) from the model agree very well with data collected in our laboratory (Clewell and Andersen, 1986). In this case, oral kinetics were not independently predicted. The inhalation model with first-order input from the gut was manipulated by adjusting the uptake rate constant until the best description of the oral uptake data was obtained. It is worth noting that the similar data for oral administration in an oil vehicle could not be successfully simulated by assuming either first-order or zero-order input (see also Ramsey and Andersen, 1984). Additional work is needed to understand the effect of administering a chemical dissolved in oil on its uptake kinetics.

This simple, generic description provides a prototypical predictive model for a wide variety of very important gases and vapors—materials such as trichloroethylene, perchloroethylene, benzene, chloroform, methylene chloride—all of which have been identified as water-borne environmental contaminants in various groundwater samples. We have now collected the partition coefficients and metabolic constants for all these chemicals and used these constants to construct simple, four-compartment, physiologically based models. In our laboratory, it now takes about 2 weeks to develop the metabolic and solubility constants needed to develop a physiological model for these volatile chemicals. Of course, for some chemicals metabolism may not be adequately represented by the two pathways de-

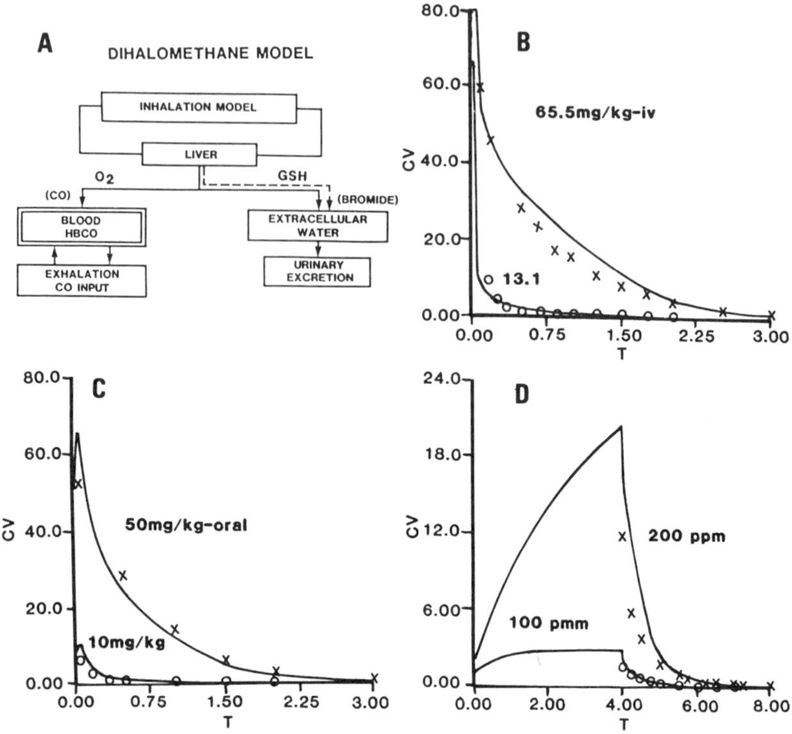

FIGURE 2 Dose-route extrapolation. (A) For the dihalomethanes, metabolism procedes by two pathways. A saturable oxidative pathway yields carbon monoxide (CO) and halide ion (bromide in the case of dibromomethane) and a first-order glutathione (GSH) conjugation pathway produces halide but not CO. By varying the input function in a physiologically based pharmacokinetic model that includes both of these metabolic pathways, the time course of dibromomethane can be predicted for a variety of exposure routes. (B) Dibromomethane concentration (in milligrams per liter) in mixed venous blood of Fischer 344 rats given intravenous injections of 65.5 and 13.1 mg/kg. Solid lines are the predictions of the model, and points represent the average of two (top curve) or three (bottom curve) animals. (C) Dibromomethane concentration (in milligrams per liter) in mixed venous blood of Fischer 344 rats given oral doses of 50 and 10 mg/kg administered as a saline solution. Solid lines are the predictions of the model assuming a first-order uptake from the stomach (10/h), and points represent the average of four (top curve) or two (bottom curve) animals. (D) Dibromomethane concentration (in milligrams per liter) in mixed venous blood of Fischer 344 rats during and after 4-h inhalation exposures at 200 ppm and 100 ppm. Solid lines are the predictions of the model, and points represent the average of three animals. Reproduced with permission from Clewell and Andersen (1986).

scribed above. In this case additional experiments must be performed to characterize the metabolic conversion of the particular chemical under study.

This basic model has several tissue groups which are lumped together according to their perfusion and solubility characteristics. Each of these several compartments is described by a single mass-balance differential equation. It would be possible to describe individual tissues in each of the lumped compartments. For instance, fat could be broken down to perirenal, epidydimal, brown fat, etc. This detail is usually unnecessary unless some particular tissue in a lumped compartment is the target tissue. One might want to separate brain from other well-perfused tissues if the model were for a chemical that had a toxic effect on the central nervous system. More biochemical and physiological detail could then be incorporated for the target tissue. With brain there might be reason to explicitly define a blood:brain barrier or to include terms which describe saturable binding of the chemical to specific receptors in the tissue. Increasing the number of subcompartments does increase the number of differential equations required to define the model. However, within reason, the number of equations does not pose any problem. It is relatively straightforward to solve the small groups of equations that describe most physiological models of xenobiotic disposition, requiring only a personal microcomputer and readily accessible software.

By adding another compartment, the basic inhalation model can easily be extended to predict the kinetics for uptake of the chemical through the skin. In this case a diffusion-limited compartment is used to represent the skin as a portal of entry. With this minimal change the model has been used to describe the absorption of chemical vapors through the skin of rats fitted with masks to prevent inhalation exposure (McDougal et al., 1986). The dermal absorption of dibromomethane vapors (Figure 3) was very accurately predicted by this model with a skin permeation coefficient of 1.32 cm/h.

EXPOSURE SCENARIO EXTRAPOLATION

The basic inhalation model can also be enlarged to focus on kinetics of metabolites or on the amount of chemical metabolized in a given exposure scenario. For example, our more complete model of the brominated dihalomethanes (Andersen et al., 1984a; Gargas et al., 1986b) not only tracks the parent chemical (e.g., dibromomethane or bromochloromethane) but also two metabolites: carbon monoxide (CO) and bromide ion (Br). Two metabolic pathways are still described: a saturable oxidative pathway that produces carbon dioxide (CO_2), CO, and Br and a first-order conjugative pathway which produces only CO_2 and Br (Figure 2). The metabolite model includes a fairly complete description of the fate of the CO produced and is able to predict the fraction of hemoglobin tied up as carboxyhemoglobin at any time, as determined by the current rate of CO production (and inhalation of CO, if appropriate), the competition of oxygen and CO for hemoglobin, and the rate of exhalation of unbound

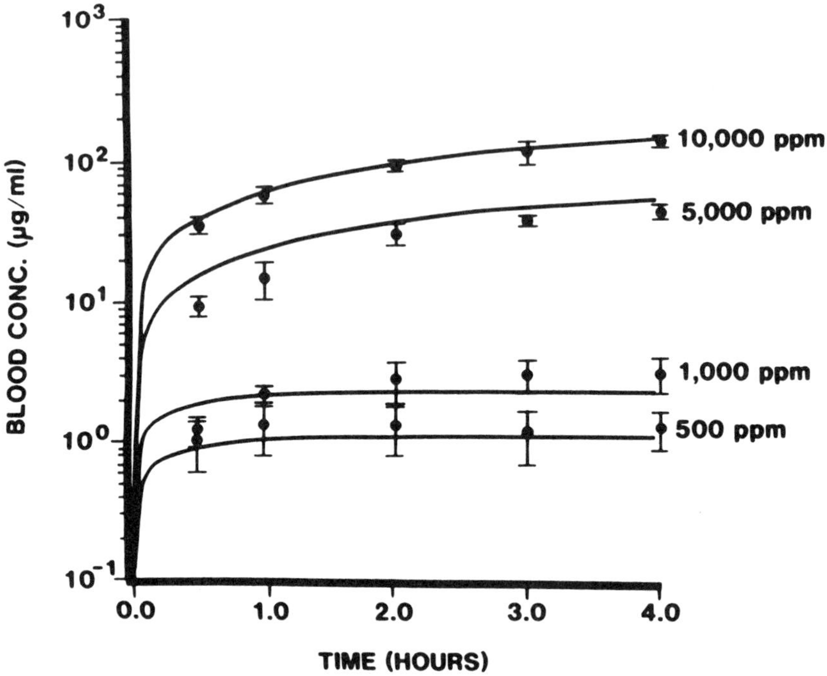

FIGURE 3 Dose-route extrapolation (continued). Dibromomethane concentration (in milligrams per liter) in mixed venous blood of Fischer 344 rats whose skin only was exposed to dibromomethane vapor at 10,000, 5,000, 1,000, and 500 ppm. Rats were protected against inhalation of the vapors by masks. Solid lines show the predictions of the model using a skin permeation coefficient of 1.32 cm/h, and symbols represent the mean and standard deviation for five or six animals. To describe uptake of vapors through the skin, a diffusion-limited skin compartment was added to the basic dihalomethane model. Reproduced with permission from McDougal et al. (1986).

CO. This model has also been successfully applied to a variety of exposure routes: inhalation, intravenous, oral, and dermal. As a challenge of the ability of this model to accurately predict kinetics for very different exposure scenarios, we performed short-duration, high-concentration exposures of rats to methylene chloride and bromochloromethane (Andersen et al., 1984a). The model correctly predicted the appearance of an increased fraction of carboxyhemoglobin in the blood which was maintained for several hours after the exposure was terminated (Figure 4).

SPECIES EXTRAPOLATION

To demonstrate interspecies extrapolation with the dihalomethane model, predictions of the model were compared with data available from studies

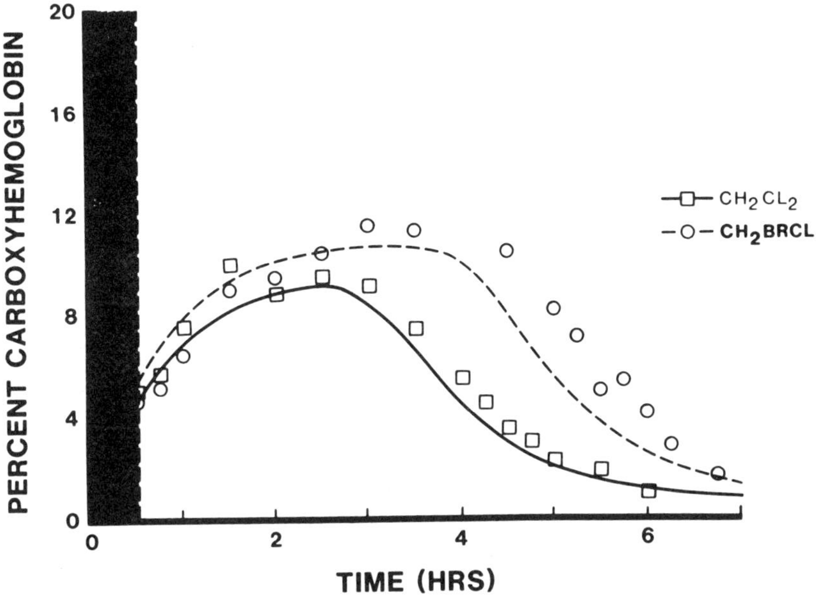

FIGURE 4 Exposure scenario extrapolation. Percent carboxyhemoglobin in the blood of Fischer 344 rats following a very short (0.5 h), high-concentration (5,000 ppm) exposure to bromochloromethane (CH_2BrCl) or methylene chloride (CH_2Cl_2). The shaded area indicates the exposure period, lines are the predictions of the model, and points represent the average of two (CH_2BrCl) or three ($CHCl_2$) animals. To predict the time course of blood carboxyhemoglobin, the basic dihalomethane model was expanded to include a description of the fate of the CO produced by the oxidative pathway, including competition with oxygen for binding to hemoglobin and exhalation of unbound CO. The longer postexposure metabolism of bromochloromethane as compared with methylene chloride reflects its greater tissue solubility. Adapted from Andersen et al. (1984a).

in which human volunteers were exposed to methylene chloride (Andersen et al., 1986b). For the purpose of this extrapolation, tissue solubilities in humans were made the same as in the rat, tissue volumes were considered proportional to body weight, and flows were considered proportional to body weight raised to the 0.7 power (see Adolph, 1949). The human blood:air partition coefficient of 9.7 was determined in our laboratory (Andersen et al., 1986b). Metabolism of methylene chloride to carbon monoxide was scaled by assuming that affinity (K_m) did not change from species to species, while V_{max} in humans could be estimated from human exposure data and was found to be about 119 mg/h (Andersen et al., 1986b). Usually, it is more difficult to determine how to scale V_{max} and K_m because there is less reason to believe that biochemical constants for xenobiotic metabolism vary coherently from species to species (Dedrick and Bischoff, 1980). This is especially true for molecules with structures that are much more complex than that of methylene chloride. It may be possible to estimate metabolic parameters on small biopsied human tissue samples or to evaluate them in many different mammalian species to determine if there is extensive interspecies variation. The approach utilized with methylene chloride provided a very good prediction of human pharmacokinetics (Figure 5), both for the parent chemical and for exhaled carbon monoxide.

DOSE EXTRAPOLATION

The versatility of the physiological approach for examining the kinetics of volatile compounds can also be appreciated by considering another experimental design for conducting an inhalation exposure. In the usual exposure design, animals are exposed to a constant concentration of chemical for a specified period of time. A less frequently used exposure design is to expose animals in a small chamber into which is injected a known amount of test chemical. The chamber concentration in this experiment declines from the initial value as the chemical is absorbed into the tissues of the animals, and there is further loss from the chamber if the chemical is metabolized. These types of experiments are concerned with the uptake of chemical into the rats from the chamber atmosphere. In view of this, the technique has been referred to as gas uptake. From a mathematical point of view, this differs from the constant concentration exposure model by the addition of one more mass-balance equation describing the chamber atmosphere itself. While this exercise looks like a form of dose-route extrapolation in which the input function is slightly altered, this exposure system actually provides a very convenient method to estimate the kinetic constants for metabolism of a variety of volatile organic chemicals (Gargas et al., 1986a). In addition, because metabolism of many of these chemicals is readily saturated, this procedure also enables us to illustrate the extrapolation from high-concentration behavior in which metabolism is saturated to lower concentrations and in which metabolism is first order with respect to the chemical substrate.

Gargas et al. (1986b) described the closed-chamber kinetics of methylene chloride (Figure 6). There is a marked concentration dependence on the observed rate of loss of this chemical from the chamber. At high concentrations, oxidative metabolism is saturated, and uptake is primarily determined by the solubilities of the test chemical in the blood and tissues of the rats and by the rate of the first-order metabolic pathway. At the intermediate concentrations, at which there is extensive curvature, we observe the transition of the oxidative pathway from saturation to first-order behavior; and the shape of the curve is critically dependent on the capacity of the oxidative pathway. Finally, at the lowest initial chamber

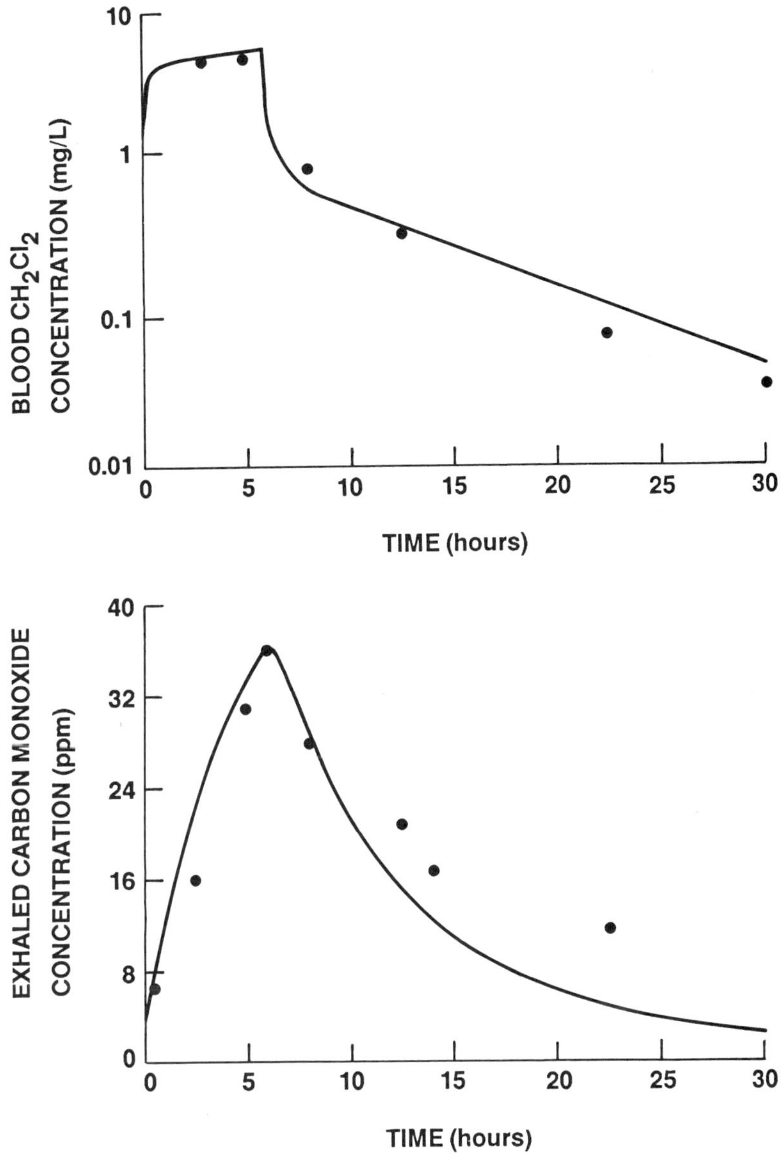

FIGURE 5 Species extrapolation. Methylene chloride concentration (in milligrams per liter) in mixed venous blood (A) and carbon monoxide concentration (in parts per million) in exhaled air (B) from human volunteers during and after a 6-h inhalation exposure to 350 ppm methylene chloride. Solid lines are the predictions of the dihalomethane model, and points represent the average of four individuals (Andersen et al., 1986b). The model was scaled to humans on the basis of both direct human data and established allometric relationships, as described in the text.

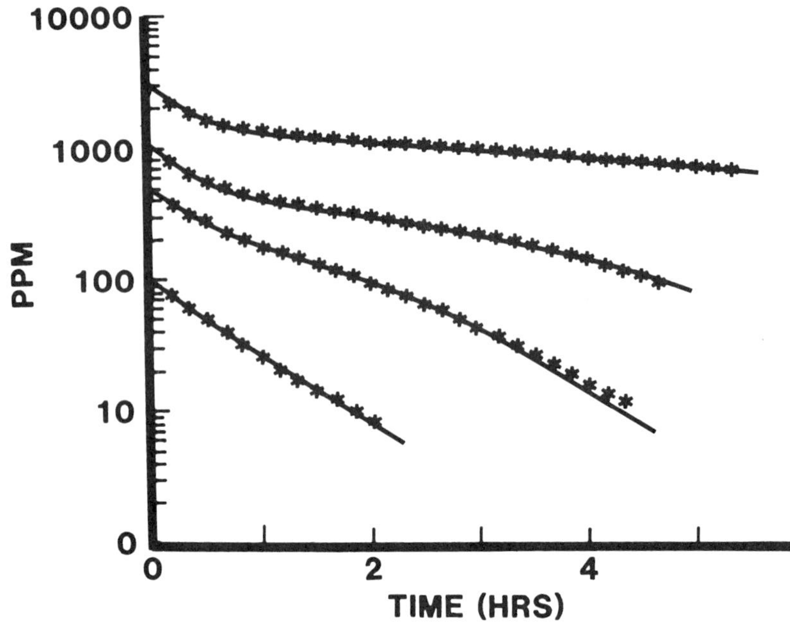

FIGURE 6 Dose extrapolation. Concentration (in parts per million) of methylene chloride in a closed, recirculated chamber containing three Fischer 344 rats. Initial chamber concentrations were (from top to bottom) 3,000, 1,000, 500, and 100 ppm. Solid lines show the predictions of the model for a V_{max} of 4.0 mg/kg/h, a K_m of 0.3 mg/liter, and a first-order rate constant of 2.0/kg/h, while symbols represent the measured chamber atmosphere concentrations. The closed-chamber experiment was described mathematically by adding a compartment to the basic model that followed the chamber atmosphere. The initial decrease in chamber concentration results from the uptake of chemical into the animal tissues. Subsequent uptake is a function of the metabolic clearance in the animals, and the complex behavior reflects the transition from partially saturated metabolism at higher concentrations to linearity in the low-concentration regime. Reproduced with permission from Gargas et al. (1986b).

concentration, the entire system is linear and the uptake curves are predominantly determined by the animal's ventilation rate and the binding affinity of the metabolizing enzyme for the substrate. These curves can be quantitatively analyzed with a physiological model to estimate the values of the kinetic constants for metabolism. The tissue partition coefficients are determined experimentally, and physiological parameters are estimated based on a combination of literature values and laboratory experience derived from the modeling process. The physiological model is then exercised for various values of the kinetic constants until a single choice of constants provides good agreement with the entire set of uptake data. With methylene chloride the uptake curves were best described by contributions from both a saturable and a strictly first-order component for metabolism. The kinetic constants for the capacity-limited oxidative pathway were 4.0 mg/kg/h and 0.3 mg/liter for V_{max} and K_m, respectively. The first-order constant had a value of 2/kg/h (Gargas et al., 1986b). In our laboratory in Dayton, Ohio, we have now estimated kinetic constants for about 30 chemicals by means of these gas-uptake techniques (Gargas et al., 1986a). The technique is very rapid and straightforward for chemicals with appropriate tissue partition coefficients and vapor pressure characteristics.

SUICIDE ENZYME INHIBITION

Saturable metabolism is by no means the only mechanism by which nonlinearities arise in the kinetics of disposition of various volatile chemicals. In fact, we have studied two other capacity-limited metabolic systems with these same gas-uptake methods. The first is suicide enzyme inhibition caused by metabolism of *cis*- and *trans*-1,2-dichloroethylene. Both of these materials are initially metabolized by microsomal oxidation, but reactive metabolites produced during their metabolism react with and destroy the active site of the metabolizing enzyme (Andersen et al., 1986c). This behavior was only uncovered because the uptake curves were analyzed with a physiological pharmacokinetic model instead of a compartmental model. These chloroethylenes are metabolized by a single, high-affinity, saturable pathway. With the appropriate values of tissue partition coefficients and physiological parameters and assuming a single saturable pathway, it was impossible to generate a good fit to the experimental data. When an attempt was made to fit the high-concentration data preferentially (Figure 7), the actual data points fell off more rapidly at the beginning of the experiment, and then the decline of chemical in the chamber slowed down more than could be accounted for by the standard physiological model. This indicated that the rate constant of metabolism was decreasing with time. At the same time, the model's consistent underprediction of metabolic clearance for the two lower concentration experiments indicated that this time-dependent decrease was less severe for lower concentrations of chemical. Together these observations suggested that enzyme destruction was occurring. Other experiments with mixed atmospheres in the chamber confirmed the loss of chloroethylene-metabolizing capacity (Andersen et al., 1986c).

The use of the physiological model allowed us to investigate the nature of the inhibition and the relative rates at which enzyme inactivation proceeded. We tested four possibile mechanisms for the suicide inactivation, and only one was able to reproduce the time course behavior observed by gas-uptake analysis. The successful description assumed that reactive me-

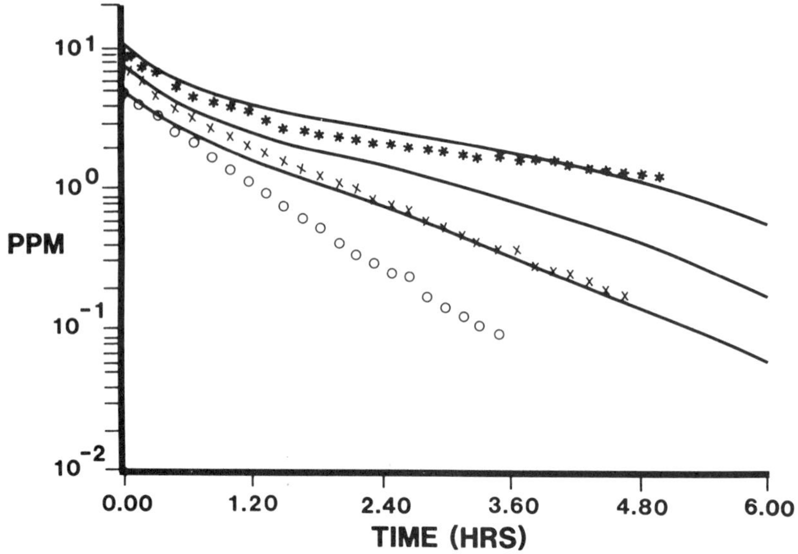

FIGURE 7 Enzyme inhibition. Concentration (in parts per million) of *trans*-1,2-dichloroethylene (*trans*-DCE) in a closed, recirculated chamber containing three Fischer 344 rats (Andersen et al., 1986c). Initial chamber concentrations were (top to bottom) 12, 8, and 5 ppm. Symbols represent the measured chamber atmosphere concentrations, and the solid lines are the best result that could be obtained from an attempt to fit all of the data with a single set of metabolic constants by using the closed-chamber model described in the legend to Figure 6. As described in the text, the systematic discrepancy between the model and the data provided an indication that the simple description of metabolism in the model was inadequate for this chemical.

tabolites produced by 1,2-dichloroethylene metabolism reacted with the enzyme-substrate complex to inactivate metabolizing enzyme. In addition, it was also necessary to include enzyme resynthesis in the model to obtain an accurate representation of the experimental data (Figure 8). These interactions were included simply by enlarging the mass-balance equation for the liver to include the biochemical events involved with enzyme inactivation. For these chemicals that are very efficient suicide inhibitors, one of the important kinetic constants for risk assessment is the resynthesis rate for new enzyme since resynthesis becomes the limiting step for metabolism at high concentrations of inhibitor. In the case of these studies with *trans*-1,2-dichloroethylene, the estimated zero-order resynthesis rate in male Fischer 344 rats was 2.5% of the steady-state activity per hour. These experiments also showed the *trans* isomer to be a much more active suicide inhibitor than the *cis* isomer of dichloroethylene.

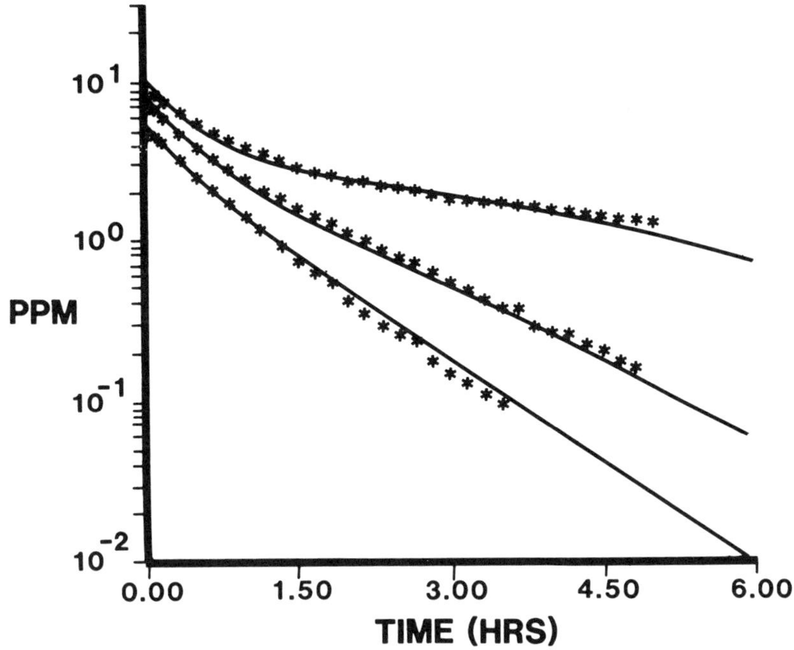

FIGURE 8 Enzyme inhibition (continued). Symbols represent the same experimental data as described in the legend to Figure 7. In this case, however, the lines show the predictions of a closed-chamber model in which the mathematical description included inactivation of the metabolizing enzyme by reactive metabolites assumed to be produced during the metabolism of *trans*-DCE. Four alternative mechanisms for the suicide inactivation were actually tested, and only one was able to coherently describe the entire data set. In the successful description, the rate of enzyme inactivation was proportional to a second-order rate constant (k_d) times the square of the instantaneous rate of metabolism, representing the reaction of free metabolite with the enzyme-substrate complex. The model also included a zero-order rate of enzyme resynthesis (k_s) during the exposure. The curves shown were obtained with a V_{max} of 3.0 mg/kg/h, a K_m of 0.1 mg/liter, a k_d of 400, and a k_s of 0.025/h. The suggestion of enzyme inactivation was also demonstrated experimentally by the inhibition of trichloroethylene metabolism after preexposure to 10–20 ppm *trans*-DCE (Andersen et al., 1986c).

GLUTATHIONE DEPLETION

Another example of capacity-limited metabolism was observed during studies of the gas-uptake behavior of several chemicals that are known to produce depletion of hepatic glutathione. The conjugation pathway for the reaction of methylene chloride and glutathione regenerates glutathione, but with other volatile chemicals, such as 1,2-dichloroethane and allyl chloride, the conjugation reaction consumes glutathione. In gas-uptake experiments with allyl chloride, both the oxidative and the conjugative (glutathione) pathways appear to be dose dependent (Figure 9). The data points were obtained in the uptake studies, and the smooth curves in Figure 9 are the best-fit curves, assuming a saturable pathway and a first-order pathway with a rate constant that is independent of concentration. With this model there were systematic errors about the predicted curves. The prediction at high concentrations was lower than the data points, and at intermediate concentrations the prediction was uniformly higher than the

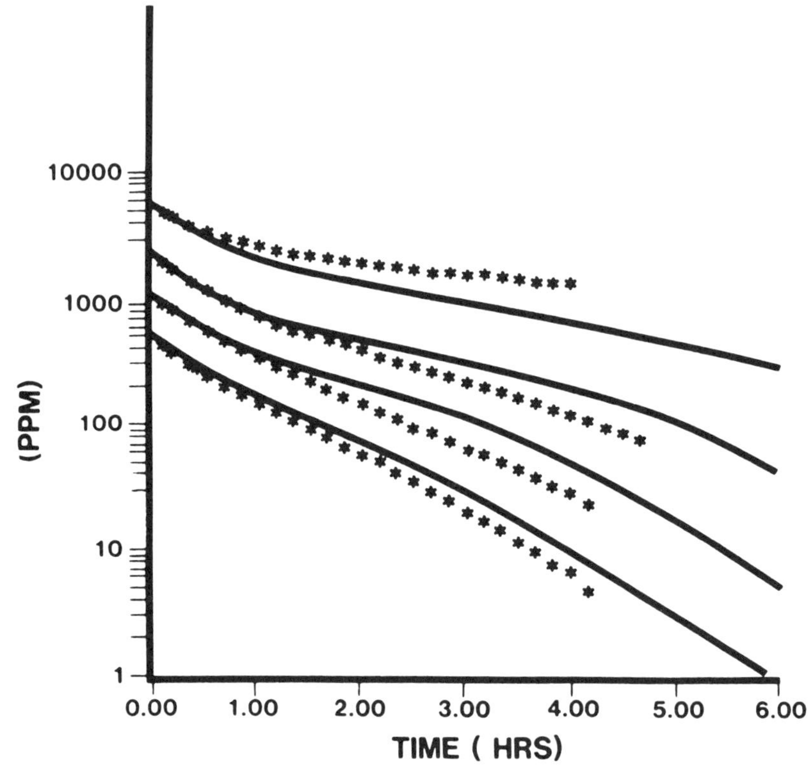

FIGURE 9 Cofactor depletion. Concentration (in parts per million) of allyl chloride in a closed, recirculated chamber containing three Fischer 344 rats (Andersen et al., 1986a). Initial chamber concentrations were (top to bottom) 5,000, 2,000, 1,000, and 500 ppm. Symbols represent the measured chamber atmosphere concentrations. The curves represent the best result that could be obtained from an attempt to fit all of the data with the same set of metabolic constants by using the closed-chamber model described in the legend to Figure 6. In this case, the apparent dose-dependent nature of the discrepancy between the model and data suggested the presence of a second capacity limitation on metabolism not included in the original description. Because this indication was consistent with other experimental evidence that the metabolism of allyl chloride consumes glutathione, the mathematical model of the closed-chamber experiment was expanded to include a more complete description of the glutathione-dependent pathway.

data. A much better fit could be obtained by setting the first-order rate constant to a lower value at the higher concentration. This approach provided a better correspondence between data and the smooth curves from the model, but did not provide biological information about the mechanism(s) by which the rate of this second pathway diminishes with concentration.

To generate a model for examining the biological basis of the kinetic behavior, it was necessary to model the time dependence of hepatic glutathione. The basic model used for this description (Andersen et al., 1986a) had a zero-order production of glutathione and a first-order consumption rate that was increased by reaction of the glutathione with allyl chloride. In the final model, glutathione resynthesis was regulated by controlling the concentration of the rate-limiting enzyme for glutathione biosynthesis. The production of this enzyme was inversely related to the instantaneous glutathione concentration. This description, coupling the loss of allyl chloride from the chamber and depletion of the glutathione concentration in the liver, provided a much improved correspondence between the data and the predicted behavior (Figure 10). Of course, the improvement in fit was obtained at the expense of adding several new glutathione-related constants. While this does add more freedom to the model for fitting the uptake data, it also suggests that we can generalize the behavior and predict both allyl chloride and hepatic glutathione concentrations during constant concentration inhalation exposures. Model output for expected end-exposure hepatic glutathione concentrations compared very favorably with actual data that was obtained by J. Waechter at Dow Chemical Co., Midland, Mich. (Table 1). Once again, the experiments with allyl chloride can be considered as they relate to estimations of risk. At high concentrations, the ability of a tissue to produce the glutathione conjugate becomes a function of the maximum resynthesis rate of the glutathione cofactor in that particular tissue. If the glutathione conjugate is the toxic moiety, tissue toxicity may well be dependent on the tissue capacity for glutathione resynthesis at high substrate concentrations.

CONCLUSION

The mathematical structure of physiological pharmacokinetic models is somewhat more complex than that of simpler one-, two-, or three-compartment models that have closed-form solutions. The solution of these physiological models requires numerical integration of a series of nonlinear, simultaneous differential equations. One advantage of these more computationally demanding models is that they can be explicitly designed to allow for the processes of extrapolation that are so necessary for rational risk assessements. These extrapolations are high dose to low dose, dose

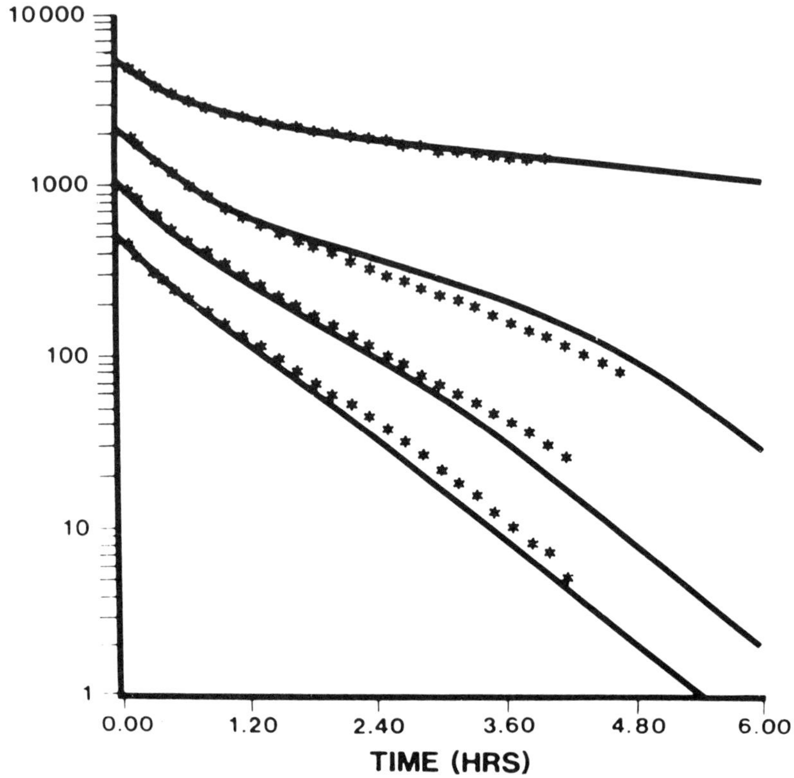

FIGURE 10 Cofactor depletion (continued). Symbols represent the same experimental data as described in the legend to Figure 9. The curves show the predictions of the expanded model (Andersen et al., 1986a), which not only included depletion of glutathione by reaction with allyl chloride but also provided for the regulation of glutathione biosynthesis on the basis of the instantaneous glutathione concentration, as described in the text.

route, interspecies, and dose rate. A second advantage is that these models can easily be expanded to include more detailed information on the chemistry and biochemistry of the test chemical and the test animal. This progressive expansion of a simple model to include more detail was seen with both the suicide inhibitors and glutathione depletion. In both cases, the crucial role of the model in the conduct of the scientific method was apparent. The mathematical model gives quantitative form to the investigator's conception of the biological system, allowing him or her to develop testable, quantitative hypotheses, to design informative experiments, and to recognize inconsistencies between theory (model) and data. As the models become more complex, they necessarily contain larger

TABLE 1 Predicted Glutathione Depletion Caused by Inhalation Exposures to Allyl Chloride

Concentration (ppm)	Depletion (μM) Observed	Predicted
0	7,080 ± 120	7,088[a]
10	7,290 ± 130	6,998
0	7,230 ± 80	7,238[a]
100	5,660 ± 90	5,939
0	7,340 ± 180	7,341[a]
1,000	970 ± 10	839
0	6,890 ± 170	6,890[a]
2,000	464 ± 60	399

Note: Glutathione depletion data were graciously supplied by John Waechter, Dow Chemical Co., Midland, Mich.

[a]For the purpose of this comparison, the basal glutathione consumption rate in the model was adjusted to obtain rough agreement with the controls in each experiment. This basal consumption rate was then used to simulate the associated exposure.

numbers of physiological, biochemical, and biological constants. The task with model development is to keep the description as simple as possible and to keep an eye out to ensure identifiability of new parameters that are added to the models. Every attempt should be made to obtain or verify model constants from experimental studies separate from the modeling exercises themselves. In the final analysis, complex models may be the price we have to bear for living in a complex world. This complexity can be handled by good experimental design; competent, critical accumulation of necessary model input; and honest attempts to develop generalizable descriptions that support extrapolation to exposure conditions relevant to exposed human populations.

Toxicological studies are the cornerstone of any risk analysis and provide dose-response curves on which risk analyses must be based. In contrast, pharmacokinetic models are interpretive tools to be used in conjunction with toxicity and mechanistic studies. The use of predictive physiological kinetic models in risk assessment is predicated on a very simple premise: An effective dose in one species or under a particular condition is expected to be equally effective in another species or under some altered exposure condition. While this is obviously not true for some chemicals, like dioxin with its great species specificity (Kociba and Schwetz, 1982), the premise does appear valid for many chemicals and especially for those solvent chemicals whose toxicity is due to formation of reactive metabolites. Physiological models can be used to redefine the dose-effect relationship based on effective dose. It is important to remember that kinetic studies alone cannot determine which parameter(s) one should regard as the appropriate measure of effective dose. Mechanistic studies and biological plausibility are required to develop the argument that one measure of dose is more correlated with toxicity than another. The kinetic analysis can then be used to clarify the relationship between effect and the more meaningful measure of effective dose.

Clearly, physiologically based pharmacokinetic models will not remove all of the uncertainty from the risk assessment process. In fact, in a way they introduce new uncertainties: the adequacy of the model, the accuracy of the parameters in the model, and the appropriateness of the chosen measure of effective dose. The rationale for using physiologically based pharmacokinetic models in risk assessment is that they provide a documentable, scientifically defensible means of bridging part of the gap between animal bioassays and human risk estimates. In particular, they move the risk assessment from the administered dose to a dose more closely associated with the toxic effect by explicitly describing their relationships as a function of dose, species, route, and exposure scenario. The next step, from the dose at the target tissue to the actual toxic event, is the subject of pharmacodynamic modeling; and the nature of this relationship is an area of considerable uncertainty. Nevertheless, risk estimates must continue to be made as the need arises, on the basis of what is known at that time, and in the most scientifically defensible manner available. Every effort must be made to apply scientific principles throughout the risk assessment process; to document the assumptions, the decisions, and the uncertainties at each step; and to provide this information to the risk manager in a form which allows him or her to weigh the predicted risks along with the uncertainties in the assessment to arrive at a final decision concerning acceptable exposure levels. Substituting conservatism for science throughout the risk assessment process severely restricts the utility of the results. Many important risk management decisions—such as prioritizing hazardous waste sites or deciding which solvent to use for an industrial process—require accurate comparative risk estimates of the relevant chemicals, not just individually conservative estimates.

REFERENCES

Adolph, E. F. 1949. Quantitative relations in the physiological constitutions of mammals. Science 109:579–585.

Andersen, M. E. 1981. Saturable metabolism and its relationship to toxicity. CRC Crit. Rev. Toxicol. 9:105–150.

Andersen, M. E., M. L. Gargas, R. A. Jones, and L. J. Jenkins, Jr. 1980. Determination of the kinetic constants of metabolism of inhaled toxicant *in vivo* based on gas uptake measurements. Toxicol. Appl. Pharmacol. 54:100–116.

Andersen, M. E., R. L. Archer, H. J. Clewell, and M. G. MacNaughton. 1984a. A physiological model of the intravenous and inhalation pharmacokinetics of three dihalomethanes—CH_2Cl_2, $CH_{2r}Cl$, and CH_2Br_2—in the rat. Toxicologist 4(1):111. (Abstract 443.)

Andersen, M. E., M. L. Gargas, and J. C. Ramsey. 1984b. Inhalation pharmacokinetics: Evaluating systemic extraction, total *in vivo* metabolism and the time course of enzyme induction for inhaled styrene in rats based on arterial blood:inhaled air concentration ratios. Toxicol. Appl. Pharmacol. 73:176–187.

Andersen, M. E., H. J. Clewell, M. L. Gargas, and R. B. Conolly. 1986a. A physiological pharmacokinetic model for hepatic glutathione (GSH) depletion by inhaled halogenated hydrocarbons. Toxicologist 6(1):148. (Abstract 598.)

Andersen, M. E., H. J. Clewell, M. L. Gargas, F. A. Smith, and R. H. Reitz. 1986b. Physiologically-based pharmacokinetics and the risk assessment for methylene chloride. Toxicol. Appl. Pharmacol. 87:185–205.

Andersen, M. E., M. L. Gargas, and H. J. Clewell. 1986c. Suicide inactivation of microsomal oxidation by cis- and trans-dichloroethylene (C-DCE and T-DCE) in male Fischer rats *in vivo*. Toxicologist 6(1):12. (Abstract 47.)

Bischoff, K. B., and R. G. Brown. 1966. Drug distribution in mammals. Chem. Eng. Prog. Symp. Ser. 62(66):33–45.

Bungay, P. M., R. L. Dedrick, and H. B. Matthews. 1981. Enteric transport of chlordecone (Kepone) in the rat. J. Pharmacokinet. Biopharm. 9:309–341.

Clewell, H. J., and M. E. Andersen. 1986. A multiple dose-route physiological pharmacokinetic model for volatile chemicals using ACSL/PC. Pp. 95–101 in Languages for Continuous System Simulation, F. D. Cellier, ed. San Diego: Society for Computer Simulation.

Dedrick, R. L. 1973. Animal scale-up. J. Pharmacokinet. Biopharm. 1:435–461.

Dedrick, R. L., and K. B. Bischoff. 1980. Species similarities in pharmacokinetics. Fed. Proc. 39:54–59.

EPA (Environmental Protection Agency). 1984. National primary drinking water regulations; volatile synthetic organic chemicals. Fed. Regist. 49:24330–24355. (40 CFR Part 141.)

Filser, J. G., and H. M. Bolt. 1979. Pharmacokinetics of halogenated ethylenes in rats. Arch. Toxicol. 42:123–136.

Freireich, E. J., E. A. Gehan, D. P. Rall, L. H. Schmidt, and H. E. Skipper. 1966. Quantitative comparison of toxicity of anticancer agents in mouse, rat, hamster, dog, monkey and man. Cancer Chemother. Rep. 50:219–244.

Gargas, M. L., M. E. Andersen, and H. J. Clewell. 1986a. A physiologically based simulation approach for determining metabolic constants from gas uptake data. Toxicol. Appl. Pharmacol. 86:341–352.

Gargas, M. L., H. J. Clewell, and M. E. Andersen. 1986b. Metabolism of inhaled dihalomethanes *in vivo*: Differentiation of kinetic constants for two independent pathways. Toxicol. Appl. Pharmacol. 82:211–223.

Himmelstein, K. J., and R. J. Lutz. 1979. A review of the application of physiologically based pharmacokinetic modeling. J. Pharmacokinet. Biopharm. 7:127–145.

King, F. G., R. L. Dedrick, J. M. Collins, H. B. Matthews, and L. S. Birnbaum. 1983. Physiological model for the pharmacokinetics of 2,3,7,8-tetrachlorodibenzofuran in several species. Toxicol. Appl. Pharmacol. 67:390–400.

Kociba, R. J., and B. S. Schwetz. 1982. Toxicity of 2,3,7,8-tetrachlorodibenzo-p-dioxin (TCDD). Drug Metab. Rev. 13:387–406.

McDougal, J. N., G. W. Jepson, H. J. Clewell, M. G. MacNaughton, and M. E. Andersen. 1986. A physiological pharmacokinetic model for dermal absorption of vapors in the rat. Toxicol. Appl. Pharmacol. 85:286–294.

Ramsey, J. C., and M. E. Andersen. 1984. A physiological model for the inhalation pharmacokinetics of inhaled styrene monomer in rats and humans. Toxicol. Appl. Pharmacol. 73:159–175.

Ramsey, J. C., and R. H. Reitz. 1981. Pharmacokinetics and threshold concepts. Pp. 239–256 in American Chemical Society Symposium Series 160, Pesticide Chemist and Modern Toxicology, S. K. Bandel, G. J. Marco, L. Goldberg, and M. L. Leng, eds. Washington, D.C.: American Chemical Society.

Sato, A., and T. Nakajima. 1979a. A vial equilibration method to evaluate the drug metabolizing enzyme activity for volatile hydrocarbons. Toxicol. Appl. Pharmacol. 47:41–46.

Sato, A., and T. Nakajima. 1979b. Partition coefficients of some aromatic hydrocarbons and ketones in water, blood and oil. Br. J. Ind. Med. 36:231–234.

Tuey, D. B., and H. B. Matthews. 1980. Distribution and excretion of 2,2′,4,4′,5,5′-hexabromobiphenyl in rats and man: Pharmacokinetic model predictions. Toxicol. Appl. Pharmacol. 53:420–431.

PART IV

Uncertainties: Integration with Risk Assessment and Resources

Dealing with Uncertainty in Pharmacokinetic Models Using SIMUSOLV

Gary E. Blau and *W. Brock Neely*

INTRODUCTION

In the course of investigating the behavior of a chemical, the toxicologist normally doses a mammalian species and follows the movement and distribution of the chemical over a period of time. Having obtained this type of information, it is desirable to have a physiological explanation of the observed results. The usual method of arriving at such an explanation lies in characterizing the concentration-time data by the terms of a mathematical model. If the model is based on known mechanisms, it is possible to predict the behavior of the chemical in other species. The ultimate purpose of these types of studies is to use the results to assess the impact on other species, including man, to help quantify various risk assessment scenarios.

The building of such models is an iterative process requiring a sophisticated mathematical analysis. This paper will discuss the art of model building and demonstrate how the recently developed software package SIMUSOLV (trademark of the Dow Chemical Company) can help the toxicologist analyze pharmacokinetic data in a relatively easy fashion.

MODEL BUILDING

The first step in the building process is to define the problem to be solved. This can range from simply estimating the bioavailability of a chemical for a specific species to the elucidation of the fate and distribution in a variety of mammals, including man. Once the problem has been stated, the next step is to postulate several physiologically meaningful mechanisms to describe the set of data that has been collected. The third step is to use the data to discriminate between candidate models. To help make this choice the maximum likelihood principle will be introduced. Bayesian methods can also be used (Reilly and Blau, 1974). The utility of this later approach, however, can best be demonstrated for parameter estimation rather than model discrimination. Once the best model has been identified, classical statistical procedures can be used to measure the adequacy and to help select additional models.

Quite frequently the existing data set does not contain sufficient information to discriminate rival models. In this case additional experiments must be designed and carried out. The type of new information required is determined by using the models to predict the behavior of the system under a variety of different conditions and to find that point where the predicted behavior is the most different. For example, suppose the problem is to determine whether a chemical follows a monophasic or a biphasic clearance. The key data points to discriminate these models would be at the latest time points that are still detectable by the analytical method being used. In this case the answer of where to look for additional information is obvious, but normally, this is not true. It is a big mistake to try and build models after the data have been collected. Model postulation and analysis should be part of the experimental process. SIMUSOLV has been developed to bring this sophisticated approach to the laboratory so that the experimentalist can quickly examine the data as soon as they are collected. Such instant feedback provides a greater breadth of knowledge with which to plan future experiments.

After the best model has been identified and validated by a residual analysis and/or goodness of fit test, the next step is to determine the uncertainty in the parameter estimates. SIMUSOLV produces linear and nonlinear confidence regions for the parameters. In a similar fashion to the model discrimination discussed above, there may be insufficient data to produce acceptable confidence regions. If this happens, additional experiments must be designed to minimize the region of uncertainty of the parameters (Reilly and Blau, 1974). This entire description of the model-building scenario is summarized in Figure 1.

Having gone through this extensive model-building process, the final task is the obvious one of using the model to solve the problem under consideration. Since model building is based on data, a word of caution is necessary. Data have errors associated with them. This uncertainty is an inherent part of both the model selected and the parameters estimated. Any use of the model must reflect this uncertainty in terms of confidence

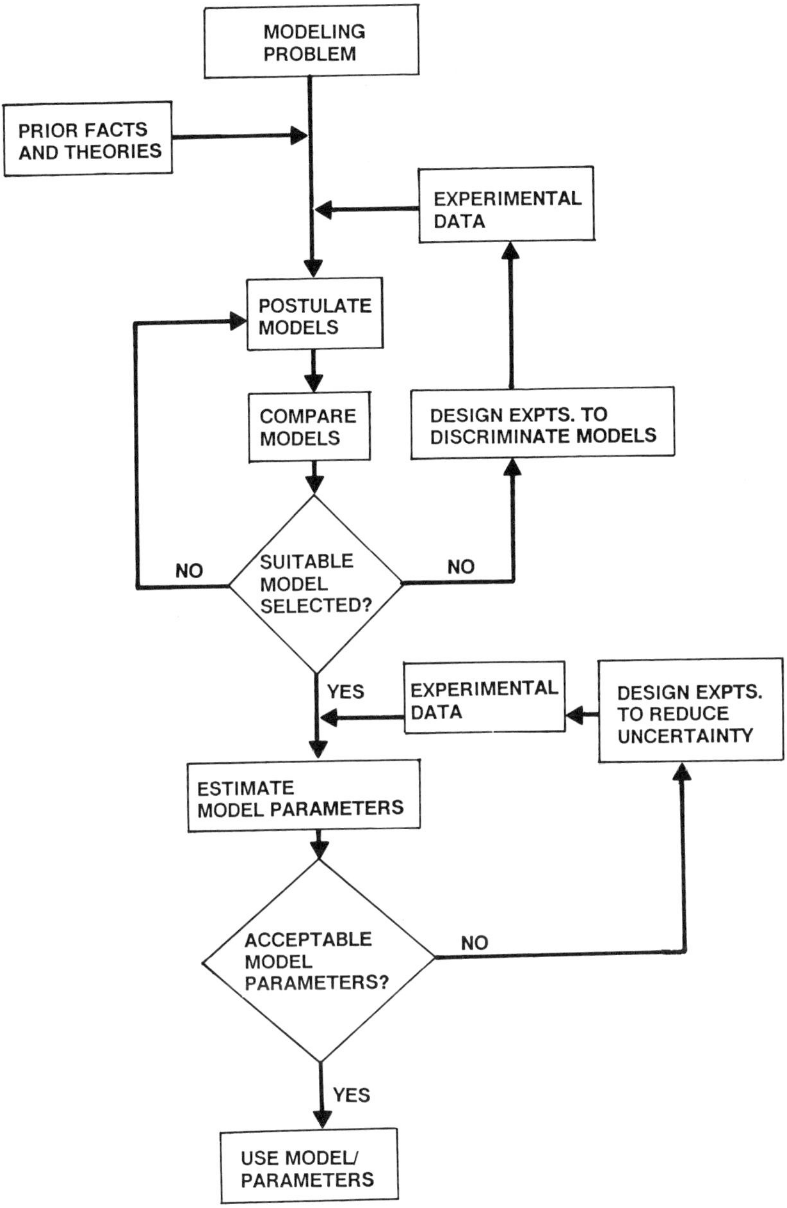

FIGURE 1 Sequential model-building procedure.

limits in the values predicted by the model, as well as confidence regions around the estimated parameters.

Finally, a model is developed to answer a specific question. There is always a great tendency to use the model for something other than its intended purpose. This is fraught with many dangers and should be avoided if at all possible. In the remainder of this paper reduction to practice of these philosophical concepts will be accomplished with SIMUSOLV.

UNCERTAINTIES AND ERROR ANALYSIS IN MODEL BUILDING

To illustrate the concept of model building consider the case of two models (M_1 and M_2) characterized by the following equations:

$$\text{For } M1: \quad y_i = f_1(k_1,k_2,x_i) + \epsilon_i$$

$$\text{For } M2: \quad y_i = f_2(k_1,k_2,k_3,x_i) + \epsilon_i \qquad (1)$$

Each model is a function of two or more parameters, i.e., rate constants for clearance, metabolism, absorption, etc., and a single independent variable (x_i), usually time. In both cases y_i is the dependent variable, such as concentration in the blood, and ϵ_i is the experimental error in the ith measurement.

The first problem in uncertainty is to examine the two models and apply a set of criteria (to be discussed below) to make a decision as to which one is best. In other words, find the most suitable of the models described in Equation 1 which best describes the data. This is done by comparing how well the models fit the data. The comparison is done when the models are examined at their individual best. The actual procedure is accomplished by measuring the probability of generating the observed concentration-time data by each model for a given set of parameters. The method of maximum likelihood accepts it as obvious that the values of the parameters k_1,k_2 for M_1 and k_1,k_2,k_3 for M_2 which maximize this probability are the best. These parameters are called maximum likelihood estimators denoted k_1^*,k_2^* for M_1 and k_1^*,k_2^*,k_3^* for M_2. The function which calculates the probabilities for any set of parameters is called the likelihood function, denoted as $L_1(k_1,k_2)$ for M_1 and $L_2(k_1,k_2,k_3)$ for M_2. To distinguish one model from another, the ratio of likelihood functions is evaluated by using the maximum likelihood estimators. A ratio of 10 is ordinarily taken as showing a difference in plausibility, whereas 100 denotes a strong preference for one model over the other (Reilly and Blau, 1974). For example, if $L_1(k_1^*,k_2^*)/L_2(k_1^*,k_2^*,k_3^*)$ for Models 1 and 2 were 100, there would be a strong preference for M_1 being better at explaining the data than M_2. The above heuristics assume that the number of parameters in each model is the same. If the number is different, then the likelihood ratios must be somewhat higher than 100 to discriminate between them.

The above illustration can be generalized to M models as follows. For M candidate models of the form:

$$y_i = f_n(k_n,t_i) + \epsilon_1 \ (n = 1,2,\ldots M), \tag{2}$$

where k becomes a dimensional vector of parameters for the nth model selected. For each of the M models, SIMUSOLV can be used to calculate the maximum likelihood estimates k_n*. From these estimates likelihood ratios can be determined for all pairs of models. The heuristics can be applied as described above in the two-by-two comparison test, and the most suitable model can be selected. An example of this technique is described in the section "Applications."

The second problem in uncertainty is obtaining the best estimate of the parameters for the model selected. When the method of maximum likelihood is used, this task is embedded within the discrimination problem as described above. In other words, by comparing the models at their optimum the best set of values is automatically generated. The determination of these maximum likelihood estimates is a nontrivial task, except for the special case in which the models in Equation 1 are linear. Unfortunately, this situation does not occur with phenomenologically based models; consequently, it is necessary to apply estimation procedures (Bard, 1974). SIMUSOLV uses both direct and indirect algorithms for parameter estimation.

In the direct method different values of the parameters are selected, and the likelihood function is estimated for each. The set which gives the lowest value of the likelihood function is deleted, and a new point is selected according to a procedure called the flexible polygon method (Bard, 1974). This process of adding and deleting points is continued until the maximum value of the likelihood function is obtained.

Indirect methods not only use values of the likelihood function but also its curvature to direct the search for the maximum value. In SIMUSOLV, the generalized reduced gradient method is used (Lasdon et al., 1978). From an initial guess of the maximum, succeeding estimates are selected along a direction of improved function values. Determination of this direction comes from a knowledge of the gradient, which can be determined numerically or analytically. The former is usually preferred because the user does not need to supply any information. Numerical approximations to the gradient, however, frequently lead to less than adequate directions. This problem has been overcome in SIMUSOLV by solving both the sensitivity and the model equations to yield the gradients directly. This method, called the direct decoupled method, has reduced the computational time over the conventional reduced gradient method by an order of magnitude (Dunker, 1984). In addition, the state-of-the-art numerical integration techniques *L*awrence *S*olver for *O*rdinary *D*ifferential *E*quations (LSODE) are being used to solve all the differential equations associated with the model (Hindmarsh, 1982). Despite the sophistications of the techniques included in SIMUSOLV, every effort has been made to make these capabilities invisible to the user. However, the user must be aware that the true maximum likelihood estimates require this level of technology to ensure efficient use of the computer resources and achievement of the proper solution.

Once point estimates are obtained by maximizing the likelihood function, their uncertainty must be quantified. SIMUSOLV uses several methods to handle this quantification.

The first method, called linearization, represents the region of uncertainty as an ellipse in parameter space (e.g., two-dimensional contour plots for the two parameters k_1 and k_2 in Model 1). Here, the likelihood function is linearized by a Taylor's series expansion about the maximum likelihood estimates. These confidence regions are drawn based on a statistical F distribution (Draper and Smith, 1981). A 95% confidence region means that if the experiment was repeated 100 times, 95 times the maximum likelihood estimates would fall within this region. The linearization method is attractive in that the confidence regions are easily generated by the computer (Reilly, 1976). Unfortunately, linearizing the function distorts the shape of the confidence regions. Rather than ellipses, confidence regions may take on different shapes dependent on the information content of the experimental data.

The second method available in SIMUSOLV to determine these confidence regions, called the nonlinear method, does not linearize the likelihood functions but uses the familiar F statistic to determine the region. In Problem 2 in the section "Applications," it will be shown just how significant these departures from linearity can be in the shape of the regions. The determination of the confidence regions requires a large amount of computational effort because the likelihood function must be evaluated for all points in parameter space. It is also an approximate method because the F statistic, which is being used, is valid only for linear models.

The third method in SIMUSOLV for representing the confidence regions, called the exact method, requires no assumptions. It is based on a subjective interpretation of statistics called Bayesian methods (Reilly and Blau, 1974). By this method the degree of belief in an event is quantified versus the familiar objective or frequency of occurrence interpretation. The Bayesian approach applied to the quantification of parameter uncertainty is expressed by the following theorem: $Pr(k/y) = Pr(k) \cdot Pr(y/k)$, where $Pr(k/y)$ is the posterior probability of the parameter k and is the true value given the set of data y, $Pr(k)$ is the prior probability of the parameter k and is the true value before the data were collected, and $Pr(y/k)$ is the likelihood of generating the data y for the set of parameters k. Although the above likelihood is different in interpretation, it gives the same values as the likelihood function discussed earlier. To apply the method, it is necessary to multiply the believed value of the parameters before the experiment is conducted by the value of the likelihood function calculated after the experiments are performed. These values must be normalized so that the total values of the posterior probabilities add to 1.0. If no prior information is available for the parameters before the experiment is conducted, a uniform, noninformative prior distribution is assumed by SIMUSOLV (Bard, 1974). By using this posterior probability, SIMUSOLV calculates regions of uncertainty by using frequency of occurrence arguments independent of any assumptions in the form of the likelihood function. Once again, an example later in the text will illustrate the effect of the assumptions inherent in the last two methods.

SIMUSOLV allows the user to construct confidence regions by either of these methods. The computational burden associated with the nonlinear and exact methods is prohibitive for large problems (where the number of parameters is greater than four). They must be used, however, if there is any suspicion about the information content of the data. The Bayesian method should be used exclusively if prior information on the parameters is available.

Experimental Error

Before concluding this section, the effect of experimental error on model development must be discussed. Because the method of maximum likelihood is being used, it is necessary to accommodate the structure of the experimental error in the model-building process. To simplify the analysis, it is frequently assumed that the errors in different experiments are (1) independent of one another, (2) uncorrelated, (3) have constant variance, and (4) are normally distributed. If these assumptions are not reasonably valid it may have serious effects on the statistical analysis. To illustrate, suppose that blood plasma concentrations are measured by an analytical procedure in which the error in the measurement is a constant fraction of the quantity being measured. Let the model expressing the blood plasma concentrations as a function of rate parameters be as follows:

$$C_i = f(k_1, k_2, \ldots, t_i) + \epsilon_i, \tag{3}$$

where the k_j's are the individual rate constants, and the error ϵ_i is a constant fraction of the amount being measured. The variance of the error in the observations will not be constant if the blood level extends over any considerable range. On the other hand, if the assumption is made that the variance of the error is constant, the consequences both for parameter estimation and model building could be equally serious. For example, during a long clearance phase from the body, the model will fail to give proper weight to any tailing in the curve from secondary processes. This will be demonstrated more clearly in the next section ("Applications") when the data in Problem 1 are analyzed.

In the past, much effort has been expended in getting models into convenient form for plotting on ordinary or special graph paper. A biological example is the use of Lineweaver-Burk plots for determining kinetic constants in enzyme systems. In most cases this convenience is achieved at the cost of distorting the error structure. When curves are fitted by eye or by simple linear least squares, constant variance is almost invariably assumed. For example, one method of plotting blood clearance data is by using logarithmic graph paper. This assumes that the elimination from the blood follows a first-order process and can be expressed as:

$$dC/dt = -kt, \tag{4}$$

where C is the blood concentration, and k is the elimination rate constant, which on integration yields the following nonlinear equation:

$$C = C_0 \exp(-kt) * \epsilon, \tag{5}$$

where ϵ is the error in measurement and is a multiplicative function of the concentration.

By making the log transform of Equation 4, a linear equation results. In this case the error becomes an additive function of the log of the concentration, as shown in Equation 6:

$$\ln C = -kt + \ln C_0 + \ln \epsilon. \tag{6}$$

Plotting the logarithm of the concentration C versus time yields a straight line. The straight line, however, is achieved by assuming that the scatter in the data is a constant fraction of the quantity being measured when, in fact, it might be that the scatter in the data is constant. If it is the latter, then the width of the error bars as the concentration becomes small will make the interpretation of a straight line from Equation 5 misleading.

It is generally much better to write the model directly in the form of Equation 1 and use SIMUSOLV, which can accommodate the error structure. It is more the rule than the exception that the error, while unpredictable, depends to some extent on the magnitude of the quantity being measured. That is, the absolute value of the error usually tends to be large when larger quantities are measured. In the SIMUSOLV program the error is described by the following model (Reilly et al., 1977):

$$\text{Variance}(y_i) = \text{Variance}(\epsilon_1) = \omega^2[f(k,x_i)]^\gamma, \tag{7}$$

where ω and γ are statistical parameters of the error model. The latter is called the heteroscedasticity parameter. Usually its value will be between 0, in which case the error variance is a constant, and 2, when the variance is proportional to the quantity being measured. Equation 7 is used to weight the measurements to allow for changes in error variance.

Although the maximum likelihood approach is restrictive in requiring explicit assumptions concerning the form of the experimental error, any unknown parameters such as ω and γ appearing in the error model are estimated along with the pharmacokinetic parameters of the model.

APPLICATIONS

The SIMUSOLV computer program was used to analyze the three simulated problems presented below. This is a package that has been developed at the Dow Chemical Co. to aid the nonmathematician in building and solving models containing systems of algebraic and ordinary differential equations. After the program is written, SIMUSOLV allows the investigator to match experimental data with different models. Based on the statistical discussion in the previous section, guidance is provided in deciding which model is "best." If the information content of the experimental data is inadequate to make such decisions, SIMUSOLV will help the user design additional experiments to make the final model selection. Finally, SIMUSOLV contains graphical routines to display simulated results.

Problem 1

Assume that an animal has been given a C^{14}-labeled chemical by means of a single oral dose. At periodic time intervals blood samples are collected and analyzed for total radioactivity. The data in Table 1 are the result of such a hypothetical experiment.

TABLE 1 Simulated Sequential Blood Concentration-Time Data Following an Oral Dose of 45 mg

Time (h)	Blood (mg/liter)
0.5	0.10
1.0	0.11
1.5	0.082
2.0	0.057
3.0	0.035
4.0	0.024
6.0	0.0091
8.0	0.0050
10.0	0.0036

Two possible mechanisms to explain the data in Table 1 are the following:

A. A simple two-compartment model as shown below:

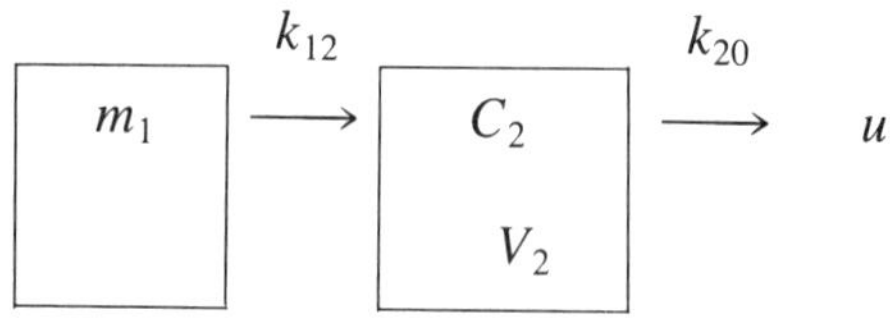

Model A

where m_1 is the mass of chemical in the gastrointestinal tract at time t, V_2 is the volume of distribution for the bloodstream (200 ml), C_2 is the concentration of chemical in the bloodstream, and u is the mass of chemical in the urine.

Assuming a first-order absorption of the oral dose from the gastrointestinal tract into the blood stream followed by a first-order elimination into the urine, the mass-balance equations describing Model A become:

$$
\begin{aligned}
dm_1/dt &= -k_{12}m_1 \\
dC_2/dt &= k_{12}(m_1/V_2) - k_{20}C_2 \\
du/dt &= k_{20}C_2V_2 \ . \qquad (8)
\end{aligned}
$$

B. A more sophisticated model can be postulated with the scheme shown below:

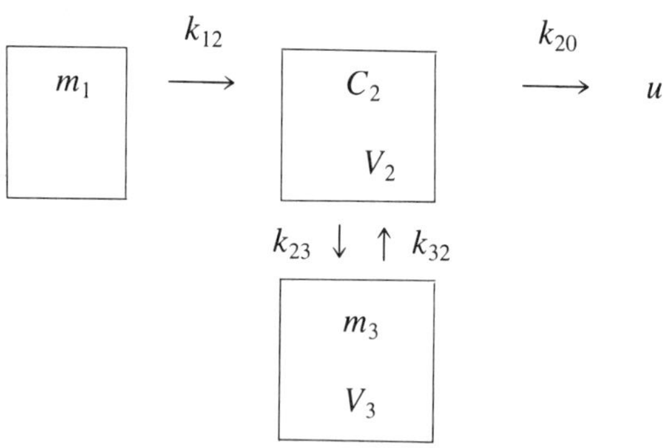

Model B

TABLE 2 Calculations from a SIMUSOLV Anaylsis on the Two Models A and B

Model	Heteroscedasticity Parameter	Likelihood Function	Percent Variance Explained
A	0	1.0×10^{16}	97.66
B	0.89	1.4×10^{20}	99.56

where m_3 is the mass of chemical in compartment V_3 that is not subject to elimination.

In this case the parent compound is capable of moving into a compartment that is not subject to elimination, a so-called deep compartment. Such a movement has the effect of lowering the amount of chemical that can be analyzed in the blood. The slow rate of release (k_{32}) back into the blood generates a long clearance time into the urine pool ($k_{32} \gg k_{20}$).

The following set of differential equations describes this model:

$$\begin{aligned} dm_1/dt &= -k_{12}m_1 \\ dC_2/dt &= k_{12}\, m_1/V_2 + k_{32}m_3/V_2 \\ &\quad - k_{23}C_2 - k_{20}C_2 \\ dm_3/dt &= k_{23}V_2C_2 - k_{32}m_3 \\ du/dt &= k_{20}C_2V_2, \end{aligned} \tag{9}$$

where k_{32} has units of liters per hour and represents the volumetric transfer rate coefficient for the movement of material from the deep compartment (V_3) back into the plasma (V_2).

With a little imagination more models could be postulated to explain the data in Table 1. These two models suffice, however, for our purpose, which is to address the problems associated with discriminating models. By plotting the logarithm of the concentration versus time, a straight line can be drawn by eye through the points, as illustrated in Figure 2.

Without any further consideration, it might be concluded that the simple elimination as described in Model A would explain the results. By using likelihood ratios, however, discrimination between the two models is possible. An example of the complete output from SIMUSOLV for Model A is included in Appendix A. The key statistics used for discrimination are shown in Table 2.

The likelihood ratio of 14,000 indicates that there is a strong preference for Model B over Model A. This conclusion could have been missed if

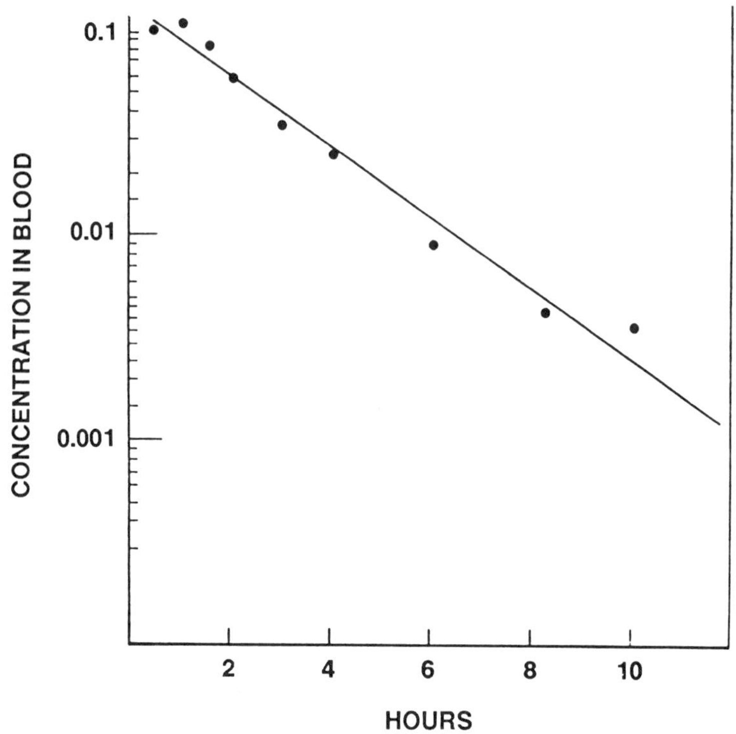

FIGURE 2 Log concentrations versus time for the data in Table 1.

reliance on the simple log plot shown in Figure 2 had been made. The actual comparison of the values predicted by the models using their maximum likelihood estimates and the data are shown in Figure 3. Here, it becomes quite evident that Model A does not describe the blood concentration points at the later time intervals. Thus, SIMUSOLV quickly allows the investigator to model the data; and by examining the key statistics and plots, the investigator can begin the task of eliminating and targeting in on the mechanisms that need to be examined in greater detail.

Problem 2

Problem 2 demonstrates the ability of SIMUSOLV to examine the confidence regions around the rate constants. Metzler (1968) presented data for drug absorption and excretion (Table 3).

The problem was formulated in the manner described for Model A in Equation 8. Figure 4 shows the agreement between the model and the data by using the maximum likelihood estimators. The information content

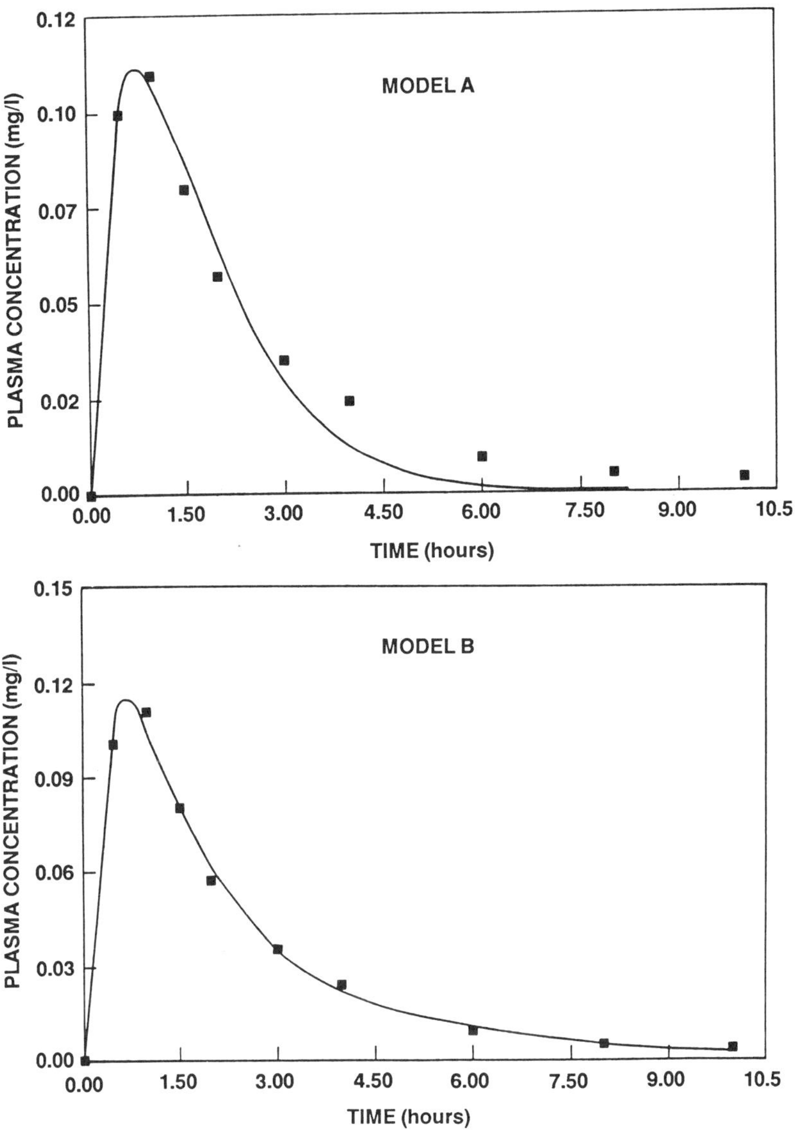

FIGURE 3 Concentration-time data from Table 1 fit to Models A and B.

associated with the data for generating the uptake rate constant is very minimal (one datum point at 1 h from Table 3). This would indicate a strong suspicion that the confidence region around k_1 is very large. By using the linearization approach for quantifying the confidence regions, the contour plot shown in Figure 5 is obtained. Reliance on this approach

TABLE 3 Sequential Blood Concentration-Time Data Following an Oral Dose

Time (h)	Concentration
1	70.2
2	81.9
4	76.4
7	74.1
24	50.3
48	32.4
55	20.5
72	18.3

Source: Metzler (1968).

would lead to the conclusion that the confidence region around the two rate constants is excellent.

The distortion of the region is readily apparent when the exact procedure is used to depict the confidence region (Figure 6).

As suggested by the data, the exact confidence region demonstrates the large uncertainty associated with the uptake rate constant.

Figure 7 indicates the result of taking data points at 0.5 and 0.75 h following administrations.

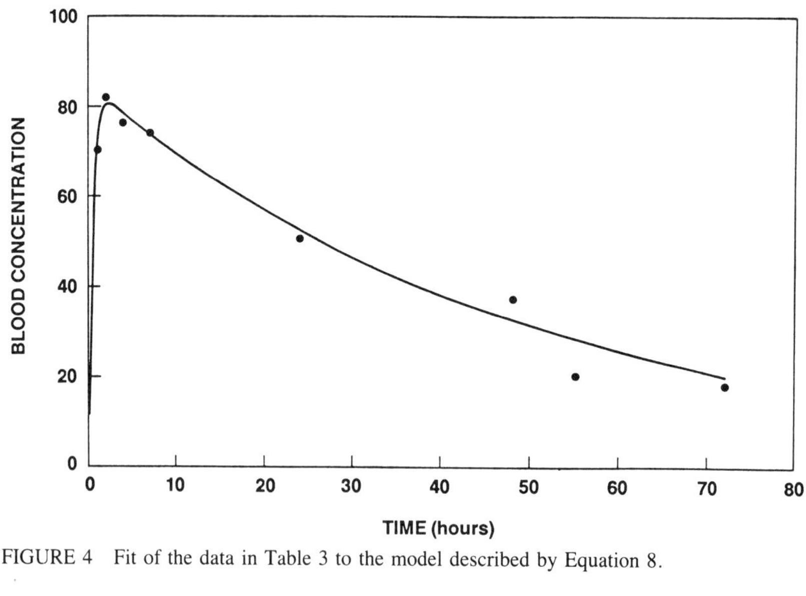

FIGURE 4 Fit of the data in Table 3 to the model described by Equation 8.

TABLE 4 Plasma and Urine Concentrations Resulting from an Oral Dose of 48.15 mg

	Plasma (mg/liters)		Urine (mg)	
Time (h)	Parent	Metabolite	Parent	Metabolite
0.82	0.175	0.822		
1.0			1.87	7.23
1.2	0.166	1.14		
1.4	0.126	1.15		
2.0	0.109	0.860	3.23	15.53
2.4	0.09	0.648		
2.9	0.083	0.601		
3.0			4.02	21.15
3.38	0.070	0.382		
3.92	0.059	0.403		
4.0			4.59	25.88
4.42	0.051	0.304		
5.18	0.035	0.252		
6.0			5.77	32.4
6.35	0.015	0.143		
8.0			6.3	34.9
8.3	0.0081	0.0636		
10	0.0047	0.0332		
12			6.65	37.06
12.4	0.0026	0.0065		
24			6.92	38.7
24.57	0.0009	0.0065		
48			7.3	40.29
72			7.38	40.77

The confidence region has now been reduced, and the new value for k_1 is more reliable. Thus, by using the graphical capabilities of SIMUSOLV, the choice of experiments for improving the quality of the data is greatly improved.

Problem 3

Nichols and Peck (1982) orally dosed an experimental animal. Plasma concentrations of the parent and the metabolite were measured up to 24 h after dosing. Urine collections were made over 72 h, and the cumulative excretion of parent and metabolite in the urine was determined. The data for this experiment are shown in Table 4.

Several models were postulated to explain the data ranging from the simple (Equation 10) to the complex (Equation 13).

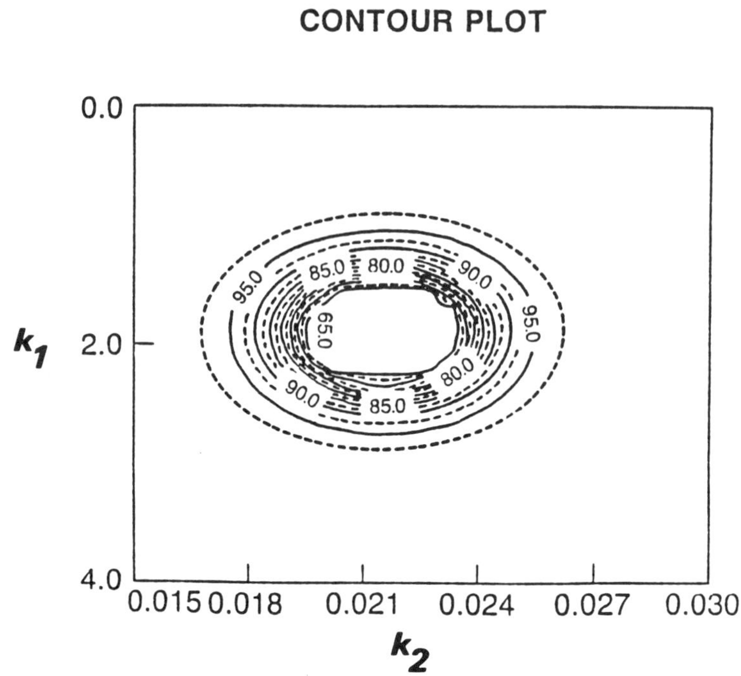

FIGURE 5 Linear confidence regions around k_1 and k_2 for Model A using the data in Table 3.

Model 1

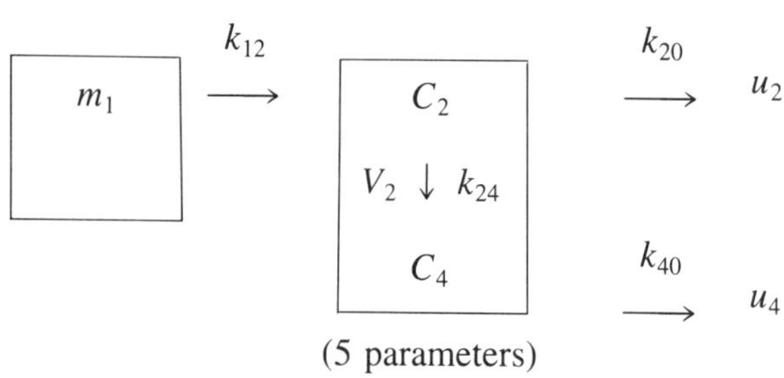

(10)

where m_1 is the oral dose, k_{12} is the absorption rate constant, k_{20} is the excretion constant for parent, k_{40} is the excretion constant for metabolite, C_2 is the concentration of parent in plasma, C_4 is the concentration of metabolite, u_2 is the cumulative amount of parent in urine, u_4 is the cumulative amount of metabolite in the urine, and V_2 is the volume of distribution for plasma. Note that all rate constants are first order.

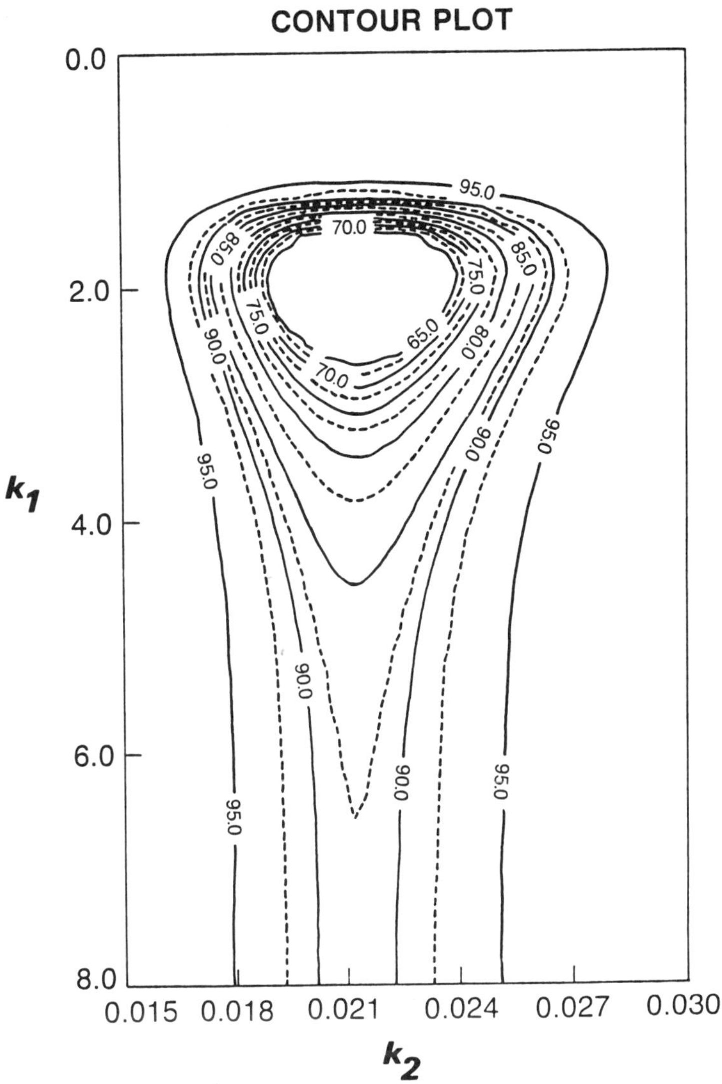

FIGURE 6 Confidence region around k_1 and k_2 for Model A using the data in Table 3. Note the large uncertainty in the confidence around the uptake rate constant.

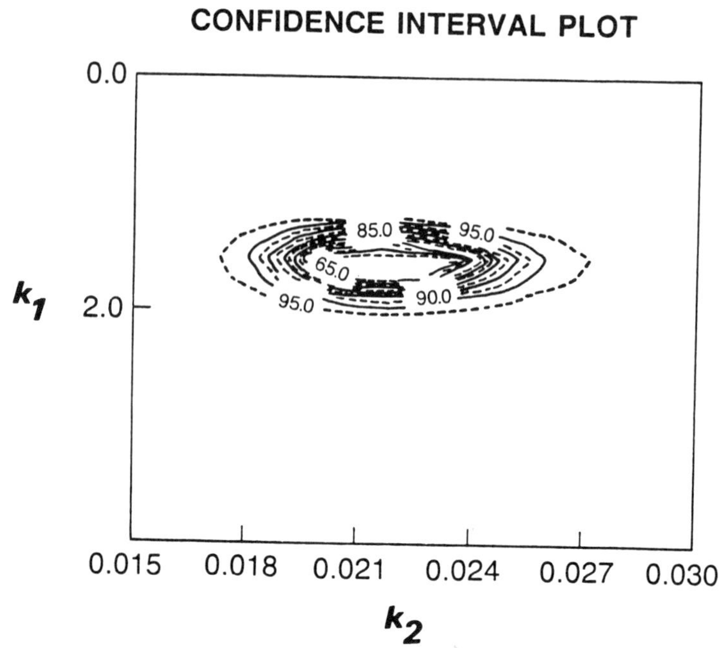

FIGURE 7 Confidence region around k_1 and k_2 for Model A using the data in Table 3 with additional data taken at 0.5 and 0.75 h.

Model 2

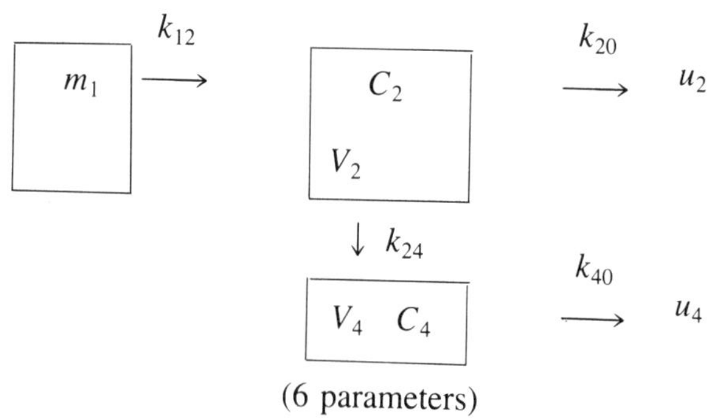

(11)

where all the symbols are similar to those in Equation 10, with the additional parameter of V_4, the volume of distribution of metabolite in the plasma.

Model 3

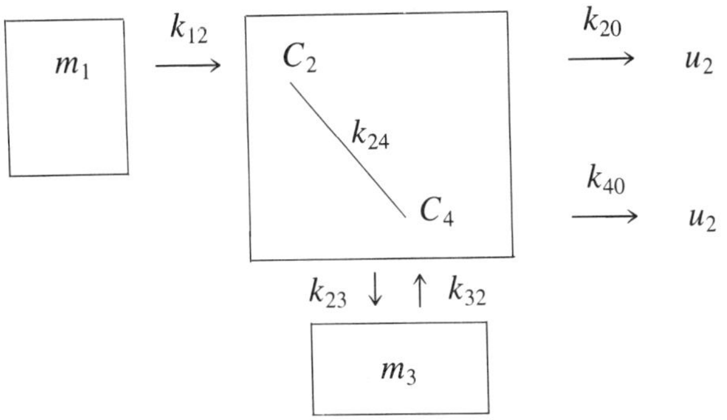

(12)

(7 parameters)

where two additional parameters k_{23} and k_{32} that are similar to those in Model B (Equation 9) are included. Again, these two constants represent the flow of chemical into a deep compartment (V_3) and the slow release back into the plasma, or compartment V_2.

Model 4

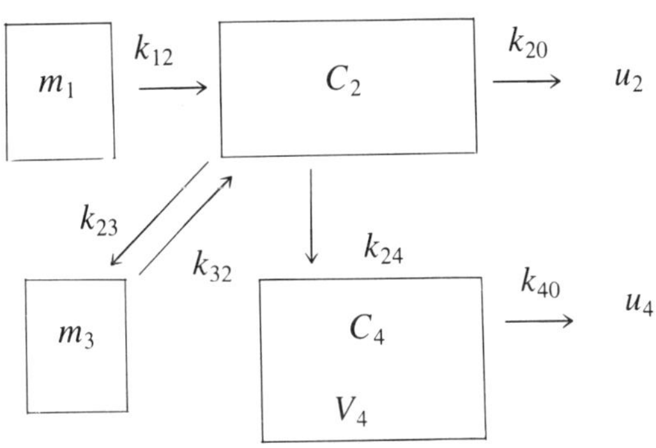

(13)

(8 parameters)

This is the complete model in which the volume of distribution for the metabolite (V_4) has been added to the model represented by Equation 12.

SIMUSOLV was used to determine the maximum likelihood estimates shown in Table 5. Table 5 also shows the progressive improvement in the fit of the model in going from Model 1 to Model 3. By applying likelihood ratios in a pairwise fashion, no discrimination is possible between Model 3 and Model 4. However, Model 2 is slightly better than Model 1, and Model 3 is significantly preferred over Model 2. The lack of discrimination between Models 3 and 4 is also evident when the volume of distributions V_2 and V_4 are examined. In other words, there is no

TABLE 5 Statistical Analysis Using SIMUSOLV on the Four Models Represented by Equations 10–13 and the Data in Table 5

	Model			
	1	2	3	4
Maximum likelihood[a]	57.34	66.28	90.15	91.65
Parameters[b]				
k_{12}	0.362	0.365	0.36	0.34
k_{24}	6.85	5.08	17.37	17.15
k_{20}	1.11	0.83	3.08	3.06
k_{40}	1.37	2.68	2.96	3.14
k_{23}			3.34	2.48
k_{32}			0.069	0.047
V_2	9.8	13.6	3.33	3.41
V_4		4.19		3.19
Percent variation explained	98.79	98.77	98.76	98.75

[a]The maximum likelihood function is expressed in natural logarithms. Thus, the ratio of the function is simply the difference.
[b]All rate constants are in units of reciprocal hours, except for k_{32}, which has units of liters per hour.

difference between the volume of distribution for the parent and the volume of distribution selected for the metabolite.

The big improvement in fitting the data came from the addition of k_{23} and k_{32} to Model 3. The movement of the parent compound into a deep compartment and the slow release back into the plasma allowed for the improved fit at the later time intervals. As can be seen from the plot in Figure 8(A), it is the later time intervals that are not adequately fitted by Model 1. Figure 8(B) represents the fit of Model 3 and is much improved for all time periods.

CONCLUSION

The purpose of this paper was to demonstrate areas of uncertainty which arise in building pharmacokinetic models and the technique required to deal with them. In discriminating rival models and quantifying uncertainty in the model parameters, the inherent error in the actual experimental data must be considered. SIMUSOLV, a computer software package developed at the Dow Chemical Co., provides a tool for the nonmathematician to work with to handle these problems. By using the latest statistical tech-

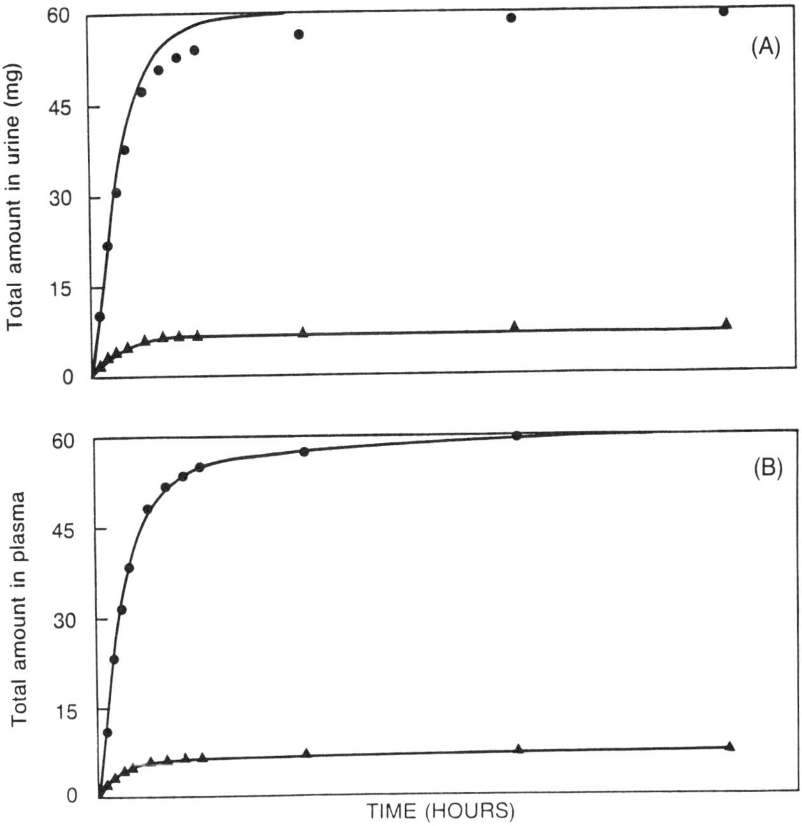

FIGURE 8 A plot of the data in Table 5 fitted to Model 1 (Equation 10). Note the divergence of the lines from the data at the later time periods. The lower curve is the fit of the same data to Model 3 (Equation 12). The incorporation of the additional mechanism has caused a much better fit to the data at all time periods.

niques and employing a graphics package, the user is aided in making the proper choice between models. In addition, the package also provides assistance in the area of experimental design. In other words, where the information in the data is insufficient to make a choice, the program will suggest areas in which additional experiments should be conducted.

REFERENCES

Bard, Y. 1974. Nonlinear Parameter Estimation. New York: Academic Press.

Draper, N. R., and H. Smith. 1981. Applied Regression Analysis, 2nd ed. New York: Wiley-Interscience.

Dunker, A. 1984. The decoupled-direct method for calculating sensitivity coefficients in chemical kinetics. J. Chem. Phys. 81:2085–2093.

Hindmarsh, A. C. 1982. ODePack, A Systematized Collection of ODE Papers. Report no. UCRL-880077, August 1982. Livermore, Calif.: Lawrence Livermore National Laboratory.

Lasdon, L. S., A. D. Waren, and M. Rather. 1978. GRGZ User's Guide. Cleveland: Department of Computer and Information Services, Cleveland State University.

Metzler, C. 1968. Data presented at the Pharmaceutical Science Meeting, November 17–20, Washington, D.C.

Nichols, A. T., and C. C. Peck. 1982. VIII a non-trivial example. Pp. 24–37 in ELSNR-Extended Least Squares Nonlinear Regression Program Users Manual, Technical Report no. 9. Bethesda, Md.: Division of Clinical Pharmacology, Uniformed Services University.

Reilly, P. M. 1976. The numerical computation of posterior distributions in Bayesian statistical inference. J. R. Stat. Soc. Ser. C Appl. Stat. 21:201–209.

Reilly, P. M., and G. E. Blau. 1974. The use of statistical methods to build mathematical models of chemical reacting systems. Can. J. Chem. Eng. 52:289.

Reilly, P. M., R. Bajramovic, G. E. Blau, D. R. Branson, and M. W. Sauerhoff. 1977. Guidelines for the optimal design of experiments to estimate parameters in first order kinetic models. Can. J. Chem. Eng. 55:614.

APPENDIX

The following is an example of the statistical output for Model A and is typical for all SIMUSOLV runs.

Description	Parameter Estimates (Initial)	Parameter Estimates (Final)	Standard Deviation
Objective Function	32.005	36.876	
k_{12}	1.7820	1.5317	0.186
k_{20}	0.7899	0.9693	6.39E−02

Time	C_1 Observed	C_2 Predicted	% Error	Standardized Residual[a]	Residual Plot
0.00E+00	0.0000E+00	0.0000E+00	0.00	0.0000E+00	
0.500	1.0000E−01	9.8685E−02	1.31	1.315E−03	*
1.000	0.1100	0.1067	3.03	3.336E−03	**
1.500	8.0000E−02	8.7028E−02	−8.78	−7.028E−03	****
2.000	5.7000E−02	6.3420E−02	−11.44	−6.520E−03	****
3.000	3.5000E−02	2.9080E−02	16.91	5.920E−03	****
4.000	2.4000E−02	1.2109E−02	49.55	1.189E−02	*******
6.000	9.0000E−03	1.8813E−03	79.10	7.119E−03	******
8.000	5.0000E−03	2.7726E−04	94.45	4.723E−03	***
10.000	3.6000E−03	4.0222E−05	98.88	3.560E−03	**

Statistical Summary

	Maximized Likelihood Function	Weighted Residual Sum of Squares	Weighted Residual Sum	Standard Error of Estimate	Percentage Variation Explained	Weighting Parameters
C_2	36.88	3.668E−04	2.432E−02	6.057E−03	97.667	0.00

Correlation Matrix

	k_{12}	k_{20}
k_{12}	1.000	
k_{20}	0.1293	1.000

Variance-Covariance Matrix

	k_{12}	k_{20}
k_{12}	3.4559E−02	
k_{20}	1.5384E−03	4.0937E−03

[a]This is the difference between the observed and calculated values multiplied by the weighting factor.

Interspecies and Dose-Route Extrapolations

Curtis C. Travis

INTRODUCTION

Risk assessment is a procedure used to synthesize all available data and the best scientific judgment to estimate the risk associated with human exposure to chemicals. Because of gaps in our current scientific understanding of the cancer-causing process, risk assessment requires the use of a series of judgmental decisions on unresolved issues. The major assumptions arise from the necessity to extrapolate experimental results (1) across species from rats or mice to humans and (2) from the high-dose regions to which animals are exposed in the laboratory to the low-dose regions to which humans are exposed in the environment, and (3) across routes of administration. There is a growing awareness of the need for an evaluation of the scientific bases for the assumptions used in the risk assessment process. Pharmacokinetics provides a tool for such an evaluation.

Pharmacokinetics is the study of the absorption, distribution, metabolism, and elimination of chemicals in man and animals. An effective approach for interpreting empirical data relating to pharmacokinetics is the development of predictive physiologically based pharmacokinetic models. These models utilize actual physiological parameters of the experimental animals such as breathing rates, blood flow rates, tissues volumes, etc., to describe the pharmacokinetic process. These physiological parameters are coupled with chemical-specific parameters such as blood/gas coefficients, tissue/blood partition coefficients, and metabolic constants to predict the dynamics of a compound's movement through an animal system. The models have the ability to relate exposure concentrations quantitatively to organ concentrations over a range of exposure intervals. As such, the models allow for prediction of the relationship between inhaled concentrations of a chemical and the concentration found in target tissues. A chief advantage of the physiologically based model is that by simply changing the physiological parameters, the same model can be utilized to describe the dynamics of chemical transport and metabolism in mice, rats, and humans.

DESCRIPTION OF THE MODEL

The pharmacokinetic model used in the present study (Figure 1 and Table 1) is patterned after that developed by Dedrick (1973) and modified by Ramsey and Andersen (1984) for prediction of the behavior of styrene inhalation exposure in humans from behavior observed in rats. The chemicals used as an illustration in this particular study are tetrachloroethylene (perchloroethylene, Perc, PCE), which is used extensively as a dry-cleaning solvent and methylene chloride (dichloromethane, DCM), which is a solvent with wide use in the food industry. Details of the model development are given in Word et al. (in press).

The tissue groups include (1) vessel-rich organs such as brain, kidney, and viscera; (2) vessel-poor organs such as the muscle and skin; (3) slowly perfused fat tissue; and (4) organs with a high capacity to metabolize (principally liver). The model is described mathematically by a set of differential equations which calculate the rate of change of the amount of chemical in each compartment. Metabolism of PCE and DCM, which occurs chiefly in the liver, is described by a combination of a linear metabolic component and a Michaelis-Menten component describing saturable metabolism.

In the modeling of the uptake, distribution, and elimination of PCE and DCM, we used the 0.75 exponential power of the body weight (BW) to scale cardiac output and ventilation rate within species. The scaling formulas are as follows:

$$Q_{alv} = Q_{alvc}\,\mathrm{BW}^{0.75}, \text{ and} \tag{1}$$

$$Q_b = Q_{bc}\,\mathrm{BW}^{0.75}, \tag{2}$$

where the allometric constants Q_{alvc} and Q_{bc} are found in Table 2. The blood flow to a given tissue group is obtained from the total blood flow

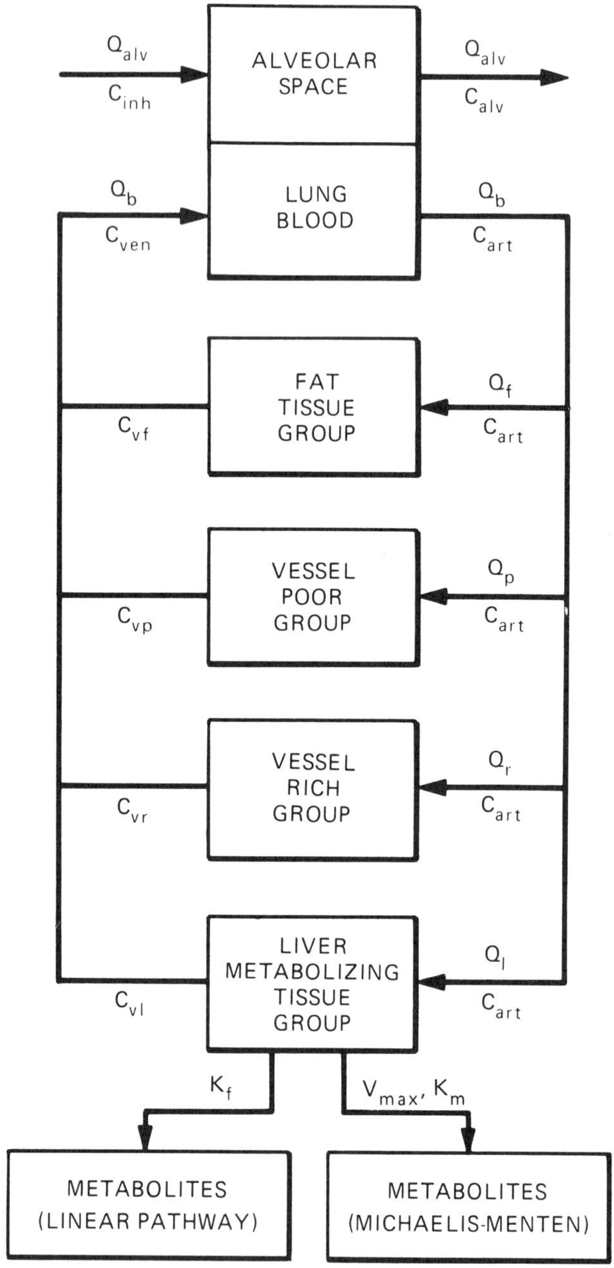

FIGURE 1 Diagram of the pharmacokinetic model used to simulate the behavior of inhaled PCE. The symbols are defined in Table 1, and the parameters used to describe the model are described in Table 3.

TABLE 1 Abbreviations and Symbols Used in Describing a Physiologically Based Pharmacokinetic Model

Symbol	Description
Q_{alv}	Alveolar ventilation rate (liters air/h)
C_{inh}	Concentration in inhaled air (mg/liter air)
C_{alv}	Concentration in alveolar air (mg/liter air)
λ_b	Blood/air partition coefficient (liters air/liters blood)
Q_b	Cardiac output (liters blood/h)
C_{art}	Concentration in arterial blood (mg/liter blood)
C_{ven}	Concentration in mixed venous blood (mg/liter blood)
V_{max}	Michaelis-Menten metabolism rate (mg/h)
K_m	Michaelis constant (mg/liter blood)
K_f	Linear metabolism rate (h^{-1})
A_m	Amount metabolized in the liver (mg)
	Subscripts (i) for tissue groups or compartments
	l Liver (metabolizing tissue group)
	f Fat tissue group
	r Vessel-rich tissue group
	p Vessel-poor tissue group
Q_i	Blood flow rate to tissue group (liters blood/h)
V_i	Volume of tissue group i (liters)
C_i	Concentration in tissue group i (mg/liter)
A_i	Amount in tissue group i (mg)
C_{vi}	Concentration in venous blood leaving tissue group i (mg/liter blood)
λ_i	Tissue/blood partition coefficient for tissue i (liters blood/liter$_i$)
k	Gavage or oral rate constant (h^{-1})
D_0	Total quantity of PCE absorbed via gavage route (mg)

Q_b by multiplying by the fraction of blood flowing to that tissue. The fractions are given in Table 2. Likewise, the volume of a tissue group is determined by multiplying the body weight by the volume fractions given in Table 2. The bone and cartilage volume (9%) is ignored in the model.

The tissue/blood and tissue/air partition coefficients used in our model were made available to us by M. E. Andersen (from D. M. Hetrick, Oak Ridge National Laboratory, Oak Ridge, Tenn., personal communication). The partition coefficients were measured by using a vial equilibration technique (Sato and Nakajima, 1979) in which the chemical was added to a closed vial containing blood or tissue, and the partitioning was determined by estimating the amount that disappeared from the head space after equilibration at 37°C (Gargas et al., 1986). The tissue/blood partition coefficients used in the model (Table 3) are obtained by dividing the tissue/air partition coefficients by the blood/air partition coefficients. Gargas et al. (1986) have also proposed that the metabolic parameters V_{max} and K_f can be scaled using body weight. The Michaelis-Menten metabolic

TABLE 2 Coefficients for Scaling Formulas

	Rats	Humans
Allometric constants		
Ventilation (Q_{alvc}; liters/h)	14	14
Cardiac output (Q_{bc}; liters/h)	14	14
Tissue volume fractions		
Liver	0.04	0.04
Fat	0.07	0.19
VRG[a]	0.05	0.05
VPG[b]	0.75	0.63
Blood flow fractions		
Liver	0.25	0.25
Fat	0.09	0.09
VRG[a]	0.51	0.51
VPG[b]	0.15	0.15

[a]Vessel-rich group.
[b]Vessel-poor group.

rate V_{max} is assumed to be proportional to surface area (Andersen et al., 1984):

$$V_{max} = V_{maxc} \, \mathrm{BW}^{0.7}. \tag{3}$$

The linear metabolic rate constant is scaled as follows:

$$K_f = K_{fc} \, \mathrm{BW}^{-0.3}. \tag{4}$$

Values for the coefficients V_{maxc} and K_{fc} and for the Michaelis-Menten constant K_m are shown in Table 4.

INTERSPECIES EXTRAPOLATION

In this section, it will be determined whether the metabolic scaling factors described above allow for both interspecies extrapolation and dose-route extrapolation. To accomplish this, we will first determine metabolic parameters so that model predictions will reproduce rat data published by Pegg et al. (1979). Then, using scaling parameters, we will attempt to reproduce rat ingestion data published by Pegg et al. (1979) and human data published by Fernandez et al. (1976).

Rat Inhalation

Adult male Sprague-Dawley rats weighing 250 g were exposed to ^{14}C-labeled PCE by inhalation for a duration of 6 h in experiments conducted

TABLE 3 Physiological and Biochemical Parameters Used in Describing the Behavior of PCE in the Pharmacokinetic Model of Figure 1

	Parameter	Rat	Human
Body weight (kg)	BW	0.25	70.0[a]
Alveolar ventilation (liters air/h)	Q_{alv}	5.02	325.0
Blood flow rates (liters blood/h)			
Total blood flow rate	Q_b	5.02	325.0
Blood flow rate in liver	Q_l	1.26	81.3
Blood flow rate in fat	Q_f	0.45	29.3
Blood flow rate in vessel-rich tissues	Q_r	2.56	165.7
Blood flow rate in vessel-poor tissues	Q_p	0.75	48.7
Tissue group volumes (liters)			
Volume in liver	V_l	0.0100	2.8
Volume in fat	V_f	0.0175	13.3
Volume in vessel-rich tissues	V_r	0.0125	3.5
Volume in vessel-poor tissues	V_p	0.1875	44.1
Blood/air partition coefficient	λ_b	18.9	10.3
Tissue/air partition coefficients			
Liver/air partition coefficient	$\lambda_{l/a}$	70.3	70.3
Fat/air partition coefficient	$\lambda_{f/a}$	2,060.0	2,060.0
Vessel-rich/air partition coefficient	$\lambda_{r/a}$	70.3	70.3
Vessel-poor/air partition coefficient	$\lambda_{p/a}$	20.0	20.0

[a]With the exception of Fernandez et al. (1976) (83 kg).

by Pegg et al. (1979). In the 72 h following exposure to 10 ppm, metabolism accounted for 20% of the total radioactivity recovered, while unchanged PCE in expired air accounted for 70%. Pulmonary elimination of PCE had a half-life of about 7 h.

The biological parameters and partition coefficients used to model the empirical data of Pegg et al. (1979) are presented in Table 3. We used Andersen's values for the $\lambda t/a$ (tissue/air) partition coefficients, except for fat. It was necessary to increase the fat/air partition coefficient from Andersen's measured value of 1,638 to 2,060 to account for increased alveolar concentrations of PCE after exposure. Since empirical values for the metabolic parameters are not available, we determined values for these parameters which produced the best fits with the inhalation data from Pegg et al. (1979). Metabolic scaling coefficients which produced the best fit with the empirical data were $V_{max} = 0.068$ mg/h, $K_m = 0.3$ mg/liter blood, and $K_f = 2.73$/h.

In Figure 2, model predictions are compared with the empirical data of Pegg et al. (1979). Figure 2 shows the percentage of PCE recovered in expired air in rats (for a number of different time intervals) following an exposure to 10 ppm of PCE for 6 h. The vertical bars represent the range

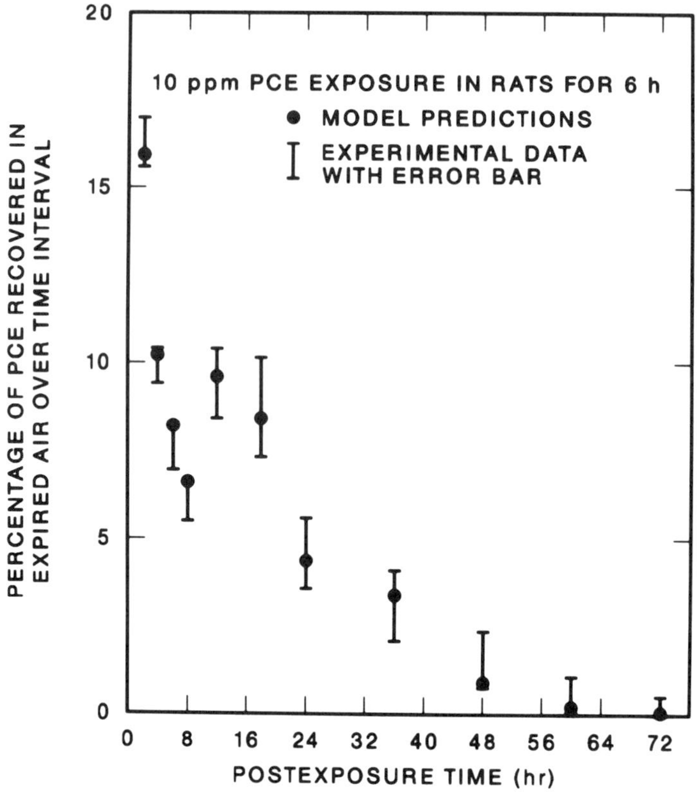

FIGURE 2 Percentage of PCE expired by rats following expsure to 10 ppm in air for 6 h. The percentage plotted at 8 h represented the percentage expired from 6 to 8 h, etc. Note that the time intervals are longer after 8 h. Symbols: ●, model predictions; bars, range of experimental data.Source: Pegg et al. (1979).

of the empirical data. The predictions and empirical data are integrated from the beginning of each time interval to the time at which the points and bars are drawn. Note that the length of the time intervals varies, causing the data points to be nonmonotonic. The model predicted that 68% of the body burden of PCE would be recovered in expired air during the 72 h after exposure, which is in good agreement with the experimentally determined value of 70%.

Rat Ingestion

Adult male Sprague-Dawley rats weighting 250 g were orally administered ^{14}C-labeled PCE in corn oil in experiments conducted by Pegg et al. (1979). In the 72 h following exposure to 500 mg/kg, metabolism accounted for 5% of total radioactivity recovered, while unchanged PCE in expired air accounted for 90%. There was no significant difference in the elimination half-life (approximately 7 h) with dose or route of administration.

In Figure 3, model predictions of expired air concentrations are compared with the empirical data of Pegg et al. (1979). The model, based on parameters obtained from the inhalation study, slightly overpredicted the rate of elimination of PCE. This can be attributed to the effects of corn oil as a carrier vehicle on the pharmacokinetics of PCE. Withey et al. (1983), Withey (1984), and Angelo et al. (1986) have shown that an oil carrier results in a slower elimination pattern for dichloromethane (methylene chloride).

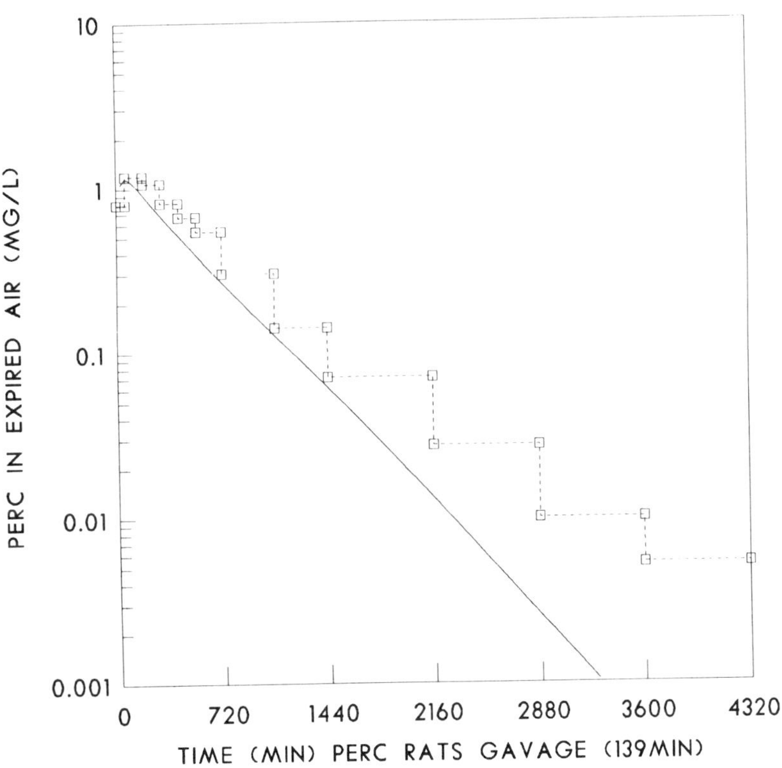

FIGURE 3 Concentration of PCE expired by rats following oral administration of 500 mg of ^{14}C-labeled PCE per kg in corn oil for 6 h. Symbols: solid line, model predictions; □, range of experimental data.

Human Inhalation

Fernandez et al. (1976) exposed 24 subjects to concentrations of 100 ppm of PCE for 1 to 8 h. During the first hours after exposure, the concentration of PCE decreased rapidly in alveolar air; however, more than 2 weeks was necessary to eliminate the PCE retained following an exposure of 100 ppm for 8 h.

The biological parameters and partition coefficients used to model the data of Fernandez et al. (1976) are also presented in Table 3. A body weight of 70 kg was used in all model calculations except for those of Fernandez et al., in which body weight was 83 kg.

Metabolic parameters for humans were obtained using the scaling formulas and the V_{maxc} and K_{fc} determined for rats (Table 4).

Figure 4 is the graph of the data of Fernandez et al. (1976), showing alveolar concentrations resulting from a 100-ppm exposure to PCE for various exposure durations, versus our model results. The model results were calculated assuming 19% fat, and the agreement between the model and the data is very good.

DOSE-ROUTE EXTRAPOLATION

The route of exposure to organic chemicals can significantly affect the quantity of a chemical that reaches a particular target tissue (Angelo et al., 1986; National Research Council, 1986). Pharmacokinetic models provide a tool to quantitatively evaluate the effect of route of administration on dose to the target tissue. This effect must be evaluated on a chemical by chemical basis.

The standard procedures for calculating applied dose following inhalation or ingestion exposures are as follows. For inhalation, the applied dose (in milligrams) is the product of the air concentration (in milligrams per liter) of the chemical, the animal breathing rate (in liters per minute), and the duration of exposure (in minutes). For drinking water ingestion, the applied dose (in milligrams) is the product of the water concentration (in milligrams per liter) of the chemical, the animal drinking water rate

TABLE 4 Metabolic Parameters Used in Describing the Behavior of PCE

	Parameter	Rat	Human
Body weight (kg)	BW	0.25	70.0
Michaelis-Menten metabolic rate (mg/h)	V_{max}	0.068	3.5
Michaelis constant (mg/liter blood)	K_m	0.3	0.3
Linear metabolic rate (h^{-1})	K_f	2.73	0.0

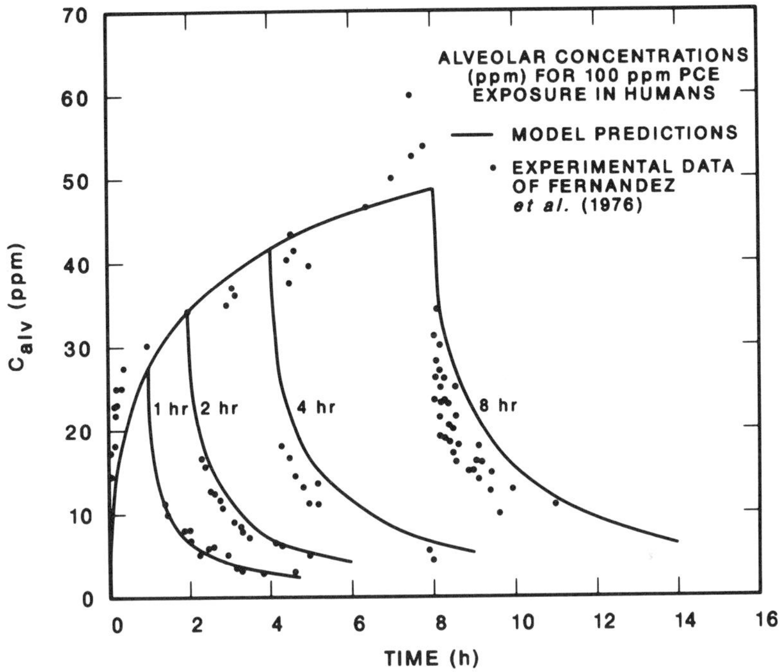

FIGURE 4 PCE concentrations in alveolar air of humans after an exposure concentration of 100 ppm for periods of 1-, 2-, 4-, and 8-h durations. Symbols: solid line, model predictions; ●, experimental data.

(in liters per minute), and the duration of exposure (in minutes). In both of these formulas, 100% absorption into the body has been assumed, a standard assumption in risk analysis when data to the contrary are lacking.

The pharmacokinetic model developed by Andersen et al. (1987) was applied to investigate the dose to target tissues following both inhalation and oral administration of DCM in mice. Figure 5 shows model predictions of total metabolized dose to the mouse liver over a 24-h period. The inhalation exposure was for a duration of 6 h and the ingestion exposure was for 24 h. The applied dose for inhalation exposures was calculated using the alveolar ventilation rate as opposed to the minute volume. Figure 5 indicates fairly good agreement in the effective liver dose for the two routes of administration. The largest difference is a factor of three and occurs in the 500- to 10,000-mg/kg applied dose range.

Figure 6 shows model predictions of total metabolized dose to the mouse lung following inhalation and oral administration. As can be seen, the effective lung dose shows greater dependence on the route of administration. The largest difference is a factor of 5 and, in contrast to the liver, occurs in the low applied dose range.

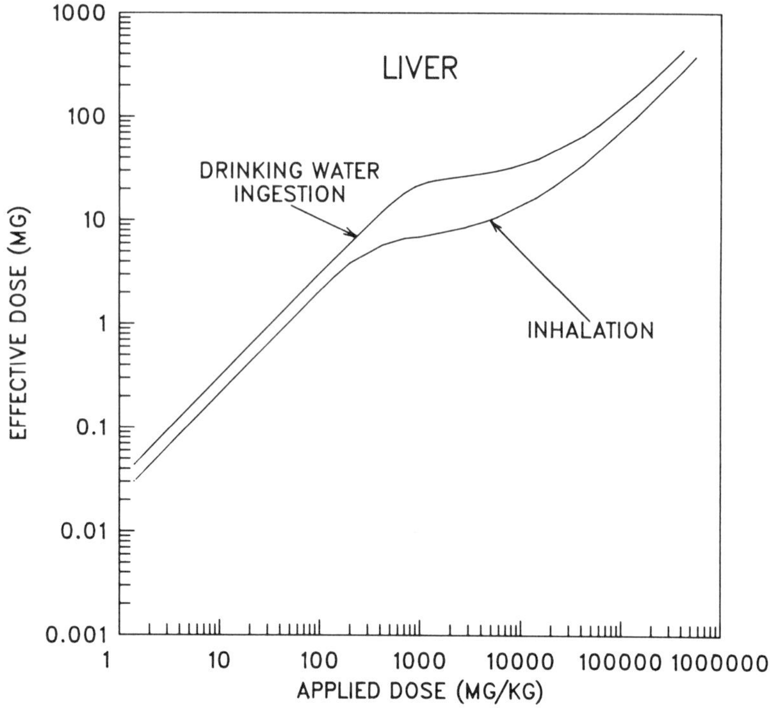

FIGURE 5 Effective dose to the mouse liver from inhalation and oral ingestion of DCM.

DISCUSSION

It has been shown that it is possible to extrapolate pharmacokinetic responses to the chemical PCE by scaling of physiological parameters. Simple scaling of metabolic parameters, originally adjusted to account for the effects in rats, also accounted for the responses in human subjects to doses of PCE administered by inhalation. It has also been shown that it is possible to use pharmacokinetic models to investigate the effect of route of administration on dose to target tissue. Since pharmacokinetic models allow for a quantitative extrapolation of exposure data across species and between routes of administration, they provide a tool to quantitatively evaluate assumptions currently used in the risk assessment process. While the present paper dose not specifically evaluate current assumptions, it does demonstrate that such an evaluation is presently possible.

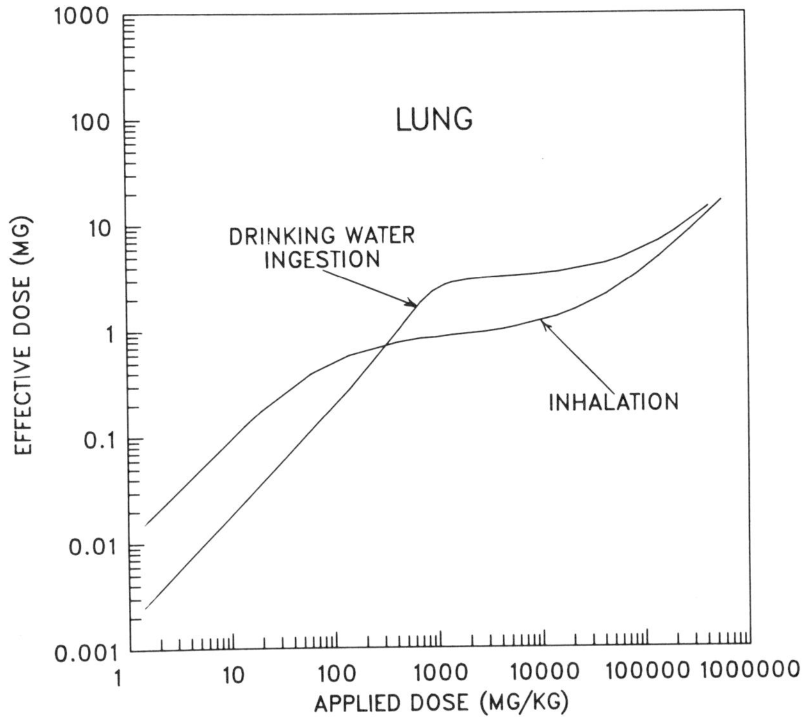

FIGURE 6 Effective dose to the mouse lung from inhalation and oral ingestion of DCM.

REFERENCES

Andersen, M. E., R. L. Archer, H. J. Clewell III, and M. G. MacNaughton. 1987. A physiological model in the rat. *The Toxicologist* 4:111.

Andersen, M. E., H. J. Clewell III, M. L. Gargas, F. A. Smith, and R. H. Reitz. 1987. Physiologically based pharmacokinetics and the risk assessment process for methylene chloride. Toxicol. Appl. Pharmacol. 87:185.

Angelo, M. J., A. B. Pritchard, D. R. Hawkins, A. R. Waller, and A. Roberts. 1986. The pharmacokinetics of dichloromethane I: Disposition in $B_6C_3F_1$ mice following intravenous and oral administrations. Food Chem. Toxicol. 24:965.

Dedrick, R. L. 1973. Animal scale-up. J. Pharmacol. Biopharm. 1:435.

Fernandez, J., G. Guberan, and J. Caperos. 1976. Experimental human exposures to tetrachloroethylene vapor and elimination in breath after inhalation. Am. Indust. Hyg. Assoc. J. 37:143.

Gargas, M. L., H. J. Clewell III, and M. E. Andersen. 1986. Metabolism of inhaled dihalomethanes *in vivo*: Differentiation of kinetic constants for two independent pathways. Toxicol. Appl. Pharmacol. 82:211.

National Research Council. 1986. Drinking Water and Health, Vol. 6. Washington, D.C.: National Academy Press.

Pegg, D. G., J. A. Zempel, W. B. Braun, and P. G. Watanabe. 1979. Disposition of tetrachloro (^{14}C) ethylene following oral and inhalation exposure in rats. Toxicol. Appl. Pharmacol. 51:465.

Ramsey, J. C., and M. E. Andersen. 1984. A physiologically based description of the inhalation pharmacokinetics of styrene in rats and humans. Toxicol. Appl. Pharmacol. 73:159.

Sato, A., and T. Nakajima. 1979. Partition coefficients of some aromatic hydrocarbons and ketones in water, blood and oil. Br. J. Indust. Med. 36:231.

Ward, R. C., C. C. Travis, D. M. Hetrick, M. E. Andersen, and M. L. Gargas. In press. Pharmacokinetics of tetrachloroethylene. Toxicol. Appl. Pharmacol.

Withey, J. R. 1984. Classical pharmacokinetics of methylene chloride-oral administration. In Proceedings of the Food Solvents Workshop I: Methylene Chloride. Washington, D.C.: Nutrition Foundation.

Withey, J. R., B. T. Collins, and P. G. Collins. 1983. Effect of vehicle on the pharmacokinetics and uptake of four halogenated hydrocarbons from the gastrointestinal tract of the rat. J. Appl. Toxicol. 3:249.

Carcinogen-DNA Adducts as a Measure of Biological Dose for Risk Analysis of Carcinogenic Data

Marshall W. Anderson

One of the major problems confronting the regulatory agencies today is the extrapolation of high-dose toxicology data in animals for the assignment of potential risk in the human population. Both low-dose and species-to-species extrapolation of toxicology data are needed for proper risk assessment of human exposure to chemicals.

It is recognized that dose-response relationships for the toxicological response under considerations are of great value in constructing extrapolation procedures. Although it is not practical in most cases to measure toxicological responses, neoplasia, immunosuppression, etc., at relevant environmental or occupational exposure levels, a measurement of dose other than exposure dose may enhance the sensitivity of the risk analysis of high-dose toxicological data. The choice of the biological dose should be based on the mechanisms involved in the toxicological response under consideration. This report will discuss the utility of using carcinogen-DNA adduct levels as a measure of the biological dose in the risk analysis of carcinogenic data.

EVIDENCE FOR USE OF DNA ADDUCTS AS A MEASURE OF BIOLOGICAL DOSE

Considerable evidence has indicated that many mutagens and carcinogens react with cellular DNA either directly or following metabolic formation of reactive products. If DNA replication proceeds on such a modified template before altered bases or nucleotides are removed by enzymatic repair processes, the mutations can be genetically fixed. Thus, the extent of promutagenic damage induced by environmental chemicals and the capacity of the cell to repair such damage may be important factors in both initiation of malignant transformation and tissue specificity of many carcinogens.

Various studies, both *in vivo* and *in vitro*, of carcinogen-DNA adducts in a known target tissue are a good measure of a biological dose for initiation of neoplasia.

1. Data on the mutagenicity of many chemical carcinogens *in vitro* implies that mutation results from the cells' attempt to deal with unexcised DNA adducts at the time of replication (Fahl et al., 1981; McCormick and Maher, 1985; Newbold et al., 1979; van Zeeland et al., 1985; Yang et al., 1980). A similar conclusion was reached by Russell et al. (1982a,b) in a study of ethylnitrosourea-induced transmitted mutations in mice.

2. For equal levels of O^6-ethylguanine in the DNA of V798 cells and in testicular DNA from male mice treated with ethylnitrosourea, the frequency of mutation induction in V79 cells *in vitro* and in the specific-locus assay in the mouse were very similar (van Zeeland et al., 1985).

3. A positive correlation between the carcinogenicity of a series of polycyclic aromatic hydrocarbons and their extent of reaction with DNA has been observed (Brookes and Lawley, 1964; Goshman and Hiedelberger, 1967; Huberman and Sachs, 1967). The binding of 3-propiolactone and other alkylating agents to DNA correlates with their tumor-initiating potency (Colburn and Boutwell, 1968).

4. Several investigators have examined the effect of inhibitors of carcinogenesis on the formation of adducts between carcinogen metabolites and DNA. In general, the degree of inhibition of tumor induction correlates with the degree of inhibition of adduct formation. For example, benzo(*a*)pyrene diol epoxide (BPDE)-DNA adduct formation and benzo(*a*)pyrene-induced neoplasia in mouse lung, forestomach, or skin were inhibited to the same degree by butylated hydroxyanisole (Anderson et al., 1981, 1985; Adriaenssens et al., 1983; Ioannou et al., 1982), inducers of aryl hydrocarbon hydroxylase (Anderson and Bend, 1983; Cohen et al., 1979; Ioannou et al., 1982; Wilson et al., 1981), and a-angelicalactone (Ioannou et al., 1982). Treatment with 2,3,7,8-tetrachlorodibenzo-*p*-dioxin (TCDD) completely inhibited papilloma formation induced by 7,12-dimethylbenz[*a*]-anthracine (DMBA) and DMBA adduct formation (Cohen et al., 1979).

5. Swann et al. (1980) showed that changes in the incidence of dimethylnitrosamine (DMN)-induced kidney tumors produced by changes in the diet and by treatment with benzo(*a*)pyrene correspond to the changes

these treatments produce in the alkylation of the target tissue DNA by DMN.

6. Janss and Benn (1978) found a correlation between the amount of DMBA bound to DNA and the incidence of mammary tumors in rats of different ages.

7. Results from Richardson et al. (1986) demonstrated that interlobe differences in the incidence of diethylnitrosamine-induced hepatocellular carcinomas are due in part to differences in DNA alkylation and cell replication. Thus, the heterogenous tumor response within an organ is quantitatively related to the molecular dosimetry of DNA adducts.

8. Several studies have shown that DNA adducts preferentially accumulate in target cell DNA during continuous exposure to carcinogen. When rats are continuously exposed to hepatocarcinogenic regimens of the alkylating agents dimethylnitrosamine or dimethylhydrazine, O^6-methylguanine accumulates in the target nonparenchymal cells but not in hepatocytes (Bedell et al., 1982; Lindamood et al., 1982). The Clara cell, although accounting for only 1% pulmonary cells in the lungs of rats, was found to possess a much higher concentration of O^6-methylguanine than the other lung cell types, particularly at lower doses (Belinsky et al., 1987). The Clara cell is the putative progenitor cell for lung tumors induced by 4-(*N*-methyl-*N*-nitros amino)-1-(3-pyridyl)-1-butanone (NNK).

9. Recent studies on oncogene activation in rodent tumors have shown that the activating mutations observed in *ras* genes are consistent with the known DNA adduct patterns of the carcinogens (Barbacid, in press).

In general, these observations demonstrate a correlation between specific carcinogen-DNA adduct levels in the target tissue and tumor response. Tissue concentration of carcinogen-DNA adducts, however, may not explain the differences in organ susceptibility and species susceptibility to chemically induced neoplasia. Other aspects in the multistep development of neoplasia may be required to explain these differences. In any case, this does not detract from the use of adduct levels in the target tissue as a measure of biological dose of the carcinogen. The abovementioned results strongly imply that it is biologically more meaningful to relate tumor response to concentrations of specific DNA adducts in the target tissue than it is to relate tumor response to the administered dose of the chemical (Hoel et al., 1983).

FACTORS TO CONSIDER IN CONSTRUCTING A MEASURE OF BIOLOGICAL DOSE FROM CARCINOGEN-DNA ADDUCT LEVELS

Carcinogen-DNA adducts only represent potential promutagenic lesions. Mutations are genetically fixed only if DNA replication proceeds on the modified template before altered bases or nucleotides are removed by DNA repair processes. Thus, the extent of cell replication must be considered in addition to the accumulation of specific DNA adducts. Swenberg et al. (1983, 1985) suggest that the product of (cell replication) × (the concentration of carcinogen-DNA adduct) × (the number of cells at risk) is one possible measure of biological dose. They used this calculation of initiation index to explain the cell specificity in hepatocarcinogenesis during continuous exposure of rats to 1,2-dimethylhydrazine (Bedell et al., 1982) and mice to dimethylnitrosamine (Lindamood et al., 1982).

DETERMINATION OF CARCINOGEN-DNA ADDUCT ACCUMULATION

The accumulation of carcinogen-DNA adducts in a cell is the difference between adduct formation and removal of adducts by DNA repair processes. The dose- and time-dependent accumulation of adducts is required to construct a measure of biological dose. Adduct levels can be determined from direct measurements or, at least in theory, calculated a priori from physiologically based pharmacokinetic models. As shown by several papers in this volume, this type of model can be used to predict blood or tissue levels of the parent compound or metabolites under a variety of experimental conditions. The complexity of adduct accumulation, however, suggests that this same approach may not be feasible to previously predicted adduct levels. Adduct levels comprise a very small percentage of the total administered dose, as small as 10^{-5}% of the total dose (Adriaenssens et al., 1983). Moreover, it may be necessary to determine levels of adducts in selective cell types to obtain adequate measures of biological dose, as illustrated by several reports (Bedell et al., 1982; Belinsky et al., 1987; Lindamood et al., 1982). Prediction of small quantities in individual cell types would probably require very accurate measurements of a large number of metabolic rate constants and, in addition, determination of enzymatic parameters that could determine repair rates of specific carcinogen-DNA adducts. At present, it is probably more feasible to measure directly DNA adduct levels. Sensitive and accurate procedures have been developed to measure adducts at levels of 1 modified base per 10^9 normal nucleotides and potentially at even lower levels of DNA damage (Adamkiewicz et al., 1985; Baan et al., 1985; Balhorn et al., 1985; Fisher et al., 1985; Randerath et al., 1985). Even though the types of pharmacokinetic models constructed to predict blood and tissue levels of parent compounds or metabolites may not predict accumulation of adduct in selective cell types, it may be possible to construct similar types of kinetic models for adduct accumulation.

Chemicals may damage DNA by mechanisms other than forming stable DNA adducts. For example, chemically induced free radical formation may cause DNA strand breaks and/or DNA-DNA or DNA-protein cross-links. Chemically, unstable adducts may generate apurinic/apyrimidinic sites which can be mutagenic (Vousden et al., 1986). More studies are required to evaluate the potential risk from these types of DNA lesions as well as to develop sensitive assays to measure these lesions.

DOSE-RESPONSE RELATIONSHIPS

Dose-response relationships for carcinogen-DNA adducts have been determined for several chemicals. A plot of adduct levels divided by dose versus dose is one way to represent dose-response relationships, and this representation is especially useful for consideration of low-dose extrapolation. The three types of curves illustrated in Figure 1 have been observed. For relationships like that depicted in curve A of Figure 1 (Ioannou et al., 1982), the exposure dose and adduct levels are interchangeable

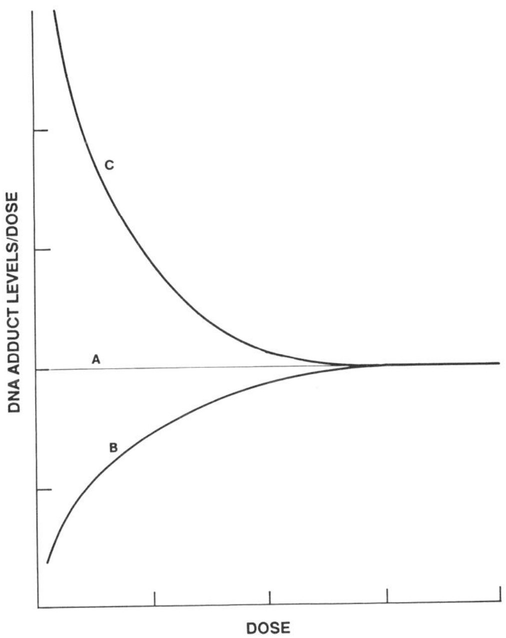

FIGURE 1 Relationships between exposure dose and adduct levels. See the text for an explanation of the three curves.

because a linear relationship exists. For relationships like that depicted in curve B of Figure 1 (Hoel et al., 1983; Ioannou et al., 1982), the adduct levels as a percentage of dose decreases as the dose decreases, and thus, exposure dose would overestimate low-dose risk. For relationships like that depicted in curve C of Figure 1 the adduct levels as a percentage of dose increases as the dose increases (Belinsky et al., 1987), and thus, exposure dose would underestimate low-dose risk.

For continuous exposure to carcinogen, adduct levels must be measured at various time points for each dose. The biological dose would then be related to the area under the time curve for adduct levels. Again, dose-dependent cell replication must be considered in constructing the biological dose. Other modes of action of the chemical related to tumor induction should also be considered when information is available. In any case, for purposes of low-dose extrapolation, adduct levels as a measure of biological dose is superior to exposure levels since for situations like curve C, low-dose risk will not be underestimated.

REFERENCES

Adamkiewicz, J., G. Eberle, N. Huh, P. Nehls, and M. Rajewsky. 1985. Quantitation and visualization of alkyl deoxynucleosides in the DNA of mammalian cells by monoclonal antibodies. Environ. Health Perspect. 62:49–55.

Adriaenssens, P. I., C. M. White, and M. W. Anderson. 1983. Dose-response relationships for the binding of benzo(a)pyrene metabolites to DNA and protein in lung, liver, and forestomach of control and butylated hydroxyanisole-treated mice. Cancer Res. 43:3712–3719.

Anderson, M. W., and J. R. Bend. 1983. In vivo metabolism of benzo-(a)-pyrene: Formation and disappearance of BB-metabolite-DNA adducts and extrahepatic tissues vs. liver. Pp. 459–467 in J. Ragdstrom, J. Montelius, and M. Bengtsson, eds. Extrahepatic Drug Metabolism and Chemical Carcinogenesis. New York: Elsevier Science Publishers.

Anderson, M. W., M. Boroujerdi, and A. G. E. Wilson. 1981. Inhibition in vivo of the formation of adducts between metabolites of benzo(a)pyrene and DNA by butylated hydroxyanisole. Cancer Res. 41:4309–4315.

Anderson, M. W., P. I. Adriaenssens, C. M. White, Y. M. Ioannou, and A. G. E. Wilson. 1985. Effect of the antioxidant butylated hydroxyanisole on in vivo formation of benzo-(a)-pyrene metabolite-DNA adducts. P. 241 in J. W. Finley and D. E. Schwass, eds. Xenobiotic Metabolism: Nutritional Effects. ACS Symposium Series No. 277. Washington, D.C.: American Chemical Society.

Baan, R. A., O. B. Zaalberg, and A. M. J. Fichtinger-Schepman, M. A. Muysken-Schoen, M. J. Lansbergen, and P. H. M. Lohman. 1985. Use of monoclonal and polyclonal antibodies against DNA adducts for the detection of DNA lesions in isolated DNA in single cells. Environ. Health Perspect. 62:81–88.

Balhorn, R., J. A. Mazrimas, and M. Corzett. 1985. Application of HPLC to the isolation of molecular targets in dosimetry studies. Environ. Health Perspect. 62:73–79.

Barbacid, M. In press. Annu. Rev. Biochem.

Bedell, M. A., J. G. Lewis, K. C. Billings, and J. A. Swenberg. 1982. Cell specificity in hepatocarcinogenesis: Preferential accumulation of O^6-methylguanine in target cell

DNA during continuous exposure to rats to 1,2-dimethylhydrazine. Cancer Res. 42:3079–3083.

Belinsky, S. A., C. M. White, J. A. Boucheron, F. C. Richardson, J. A. Swenberg, and M. W. Anderson. 1987. Cell selective alkylation of DNA in rat lung following low dose exposure to the tobacco specific carcinogen 4-(N-methyl-N-nitrosamino)-1-(3-pyridyl)-1-butanone. Cancer Res. 47:1143–1148.

Brookes, P., and P. D. Lawley. 1964. Evidence for the binding of polynuclear aromatic hydrocarbons to nucleic acid of mouse skin; relation between carcinogenic power of hydrocarbons and their binding to deoxyribonucleic acid. Nature 202:781–784.

Cohen, G. M., W. M. Bracken, R. P. Iyer, D. L. Berry, J. K. Selkirk, and T. J. Slaga. 1979. Anticarcinogenic effects of 2,3,7,8-tetrachlorodibenzo-p-dioxin on benzo(a)pyrene and 7,12-dimethylbenz(a)anthracene tumor initiation and its relationship to DNA binding. Cancer Res. 39:4027–4033.

Colburn, N. H., and R. K. Boutwell. 1968. The binding of beta-propiolactone and some related alkylating agents to DNA, RNA, and protein of mouse skin; relation between tumor-initiating power of alkylating agents and their binding to DNA. Cancer Res. 28:653–660.

Fahl, W. E., D. G. Scarpelli, and K. Gill. 1981. Relationship between benzo(a)pyrene-induced DNA base modification and frequency of reverse mutations in mutant strains of *Salmonella typhimurium*. Cancer Res. 41:3400–3406.

Fisher, D. H., J. Adams, and R. W. Giese. 1985. Trace derivatization of cytosine with pentafluorobenzoyl chloride and dimethyl sulfate. Environ. Health Perspect. 62:67–71.

Goshman, L. M., and C. Heidelberger. 1967. Binding of tritium-labeled polycyclic hydrocarbons to DNA of mouse skin. Cancer Res. 27:1678–1688.

Hoel, D. G., N. L. Kaplan, and M. W. Anderson. 1983. Implication of nonlinear kinetics on risk estimation in carcinogenesis. Science 219:1032–1037.

Huberman, E., and L. Sachs. 1967. DNA binding and its relationship to carcinogenesis by different polycyclic hydrocarbons. Int. J. Cancer 19:122–127.

Ioannou, Y. M., A. G. E. Wilson, and M. W. Anderson. 1982. Effect of butylated hydroxyanisole, alpha-angelica lactone, and beta-naphthoflavone on benzo(alpha)pyrene: DNA adduct formation in vivo in the forestomach, lung, and liver of mice. Cancer Res. 42:1199–1204.

Janss, D. H., and T. L. Benn. 1978. Age-related modification of 7,12-demethylbenz(a)anthracine binding to rat mammary gland DNA. J. Natl. Cancer. Inst. 60:173–177.

Lindamood, C., M. A. Bedell, K. C. Billings, and J. A. Swenberg. 1982. Alkylation and de novo synthesis of liver cell DNA from C3H mice during continuous dimethylnitrosamine exposure. Cancer Res. 42:4153–4157.

McCormick, J. J., and V. M. Maher. 1985. Cytotoxic and mutagenic effects of specific carcinogen-DNA adducts in diploid human fibroblasts. Environ. Health Perspect. 62:145–155.

Newbold, R. F., P. Brookes, and R. G. Harvey. 1979. A quantitative comparison of the mutagenicity of carcinogenic polycyclic hydrocarbon derivatives in cultured mammalian cells. Int. J. Cancer 24:203–209.

Randerath, K., E. Randerath, H. P. Agrawal, R. C. Gupta, M. E Schurdak, and M. V. Reddy. 1985. Postlabeling methods for carcinogen-DNA adduct analysis. Environ. Health Perspect. 62:57–65.

Richardson, F. C., J. A. Boucheron, M. C. Dyroff, J. A. Popp, and J. A. Swenberg. 1986. Biochemical and morphologic studies of heterogeneous lobe responses in hepatocarcinogenesis. Carcinogenesis 7:247–251.

Russell, W. L., P. R. Hunsicker, D. A. Carpenter, C. V. Cornett, and G. M. Guinn. 1982a. Effect of dose fractionation on the ethylnitrosourea induction of specific-locus mutations in mouse spermatogonia. Proc. Natl. Acad. Sci. USA 79:3592–3593.

Russell, W. L., P. R. Hunsicker, G. D. Raymer, M. H. Steele, K. F. Stelzner, and H. M. Thompson. 1982b. Dose-response curve for ethylnitrosourea-induced specific-locus mutations in mouse spermatogonia. Proc. Natl. Acad. Sci. USA 79:3589–3591.

Swann, P. F., D. G. Kaufman, P. N. Magee, and R. Mace. 1980. Induction of kidney tumors by a single dose of dimethylnitrosamine: Dose response and influence of diet and benzo(a)pyrene pretreatment. Br. J. Cancer 41:285–294.

Swenberg, J. A., D. E. Rickert, B. L. Baranyi, and J. I. Goodman. 1983. Cell specificity in DNA binding and repair of chemical carcinogens. Environ. Health Perspect. 49:155–163.

Swenberg, J. A., F. C. Richardson, J. A. Boucheron, and M. C. Dyroff. 1985. Relationships between DNA adduct formation and carinogenesis. Environ. Health Perspect. 62:177–183.

van Zeeland, A. A., G. R. Mohn, A. Neuhauser-Klaus, and V. H. Ehling. 1985. Quantitative comparison of genetic effects of ethylating agents on the basis of DNA adduct formation. Use of O^6-ethylguanine as molecular dosimeter for extrapolation from cells in culture to the mouse. Environ. Health Perspect. 62:163–169.

Vousden, K. H., J. L. Bos, C. J. Marshall, and D. H. Phillips. 1986. Mutations activating human c-Ha-ras1 protooncogene (HRAS1) induced by chemical carcinogens and depurination. Proc. Natl. Acad. Sci. USA 83:1222–1226.

Wilson, A. G. E., H.-C. Kung, M. Boroujerdi, and M. W. Anderson. 1981. Inhibition in vivo of the formation of adducts between metabolites of benzo(a)pyrene and DNA by aryl hydrocarbon hydroxylase inducers. Cancer Res. 41:3453–3460.

Yang, L. L., V. M. Maher, and J. J. McCormick. 1980. Error-free excision of the cytotoxic, mutagenic N2-deoxyguanosine DNA adduct formed in human fibroblasts by (+/−)-7 beta, 8 alpha-dihydroxy-9 alpha, 10 alpha-epoxy-7,8,9,10-tetrahydrobenzo(a)pyrene. Proc. Natl. Acad. Sci. 77:5933–5937.

Resources Available for Simulation in Toxicology: Specialized Computers, Generalized Software, and Communication Networks

Daniel B. Menzel, R. L. Wolpert, J. R. Boger III,
and *J. M. Kootsey*

INTRODUCTION

Pharmacokinetics (PK) is a portion of the larger and more general effort of mathematical simulation of physical and biological phenomena. PK simulates certain aspects of the behavior of biological systems, and makes predictions about the future behavior of those systems, by solving systems of algebraic or differential equations.

We describe here some of the general aspects of simulation, the computer facilities needed to carry out simulations, the specialized resources available nationally for simulations with emphasis on PK, computer languages and approaches used in PK, and some projections for the future development of PK and other biological simulations. Our perspective is that of the toxicologist as a user of these resources. We hope to introduce toxicologists to the field of PK modeling, and to inform those already experienced in PK modeling about some national resources available for their use.

GENERAL APPROACHES TO SIMULATION

Simulation is the representation of physical or biological systems by a set of mathematical relations which approximate reality sufficiently well that predictions from the simulation or model closely approximate observations. The emphasis of this conference was on PK. PK models aim to provide an accurate description of the dose of a drug or chemical reaching a specific organ and to predict the evolution of that dose with time. This type of PK model (a dosimetry model) is but the beginning of a larger set of models that predict both the dose and the biological outcome, including estimates of risk. Risk estimates fulfill the ultimate objective of toxicology: predicting the health risks to humans (or other potential target organisms) from exposure to toxic chemicals.

Most PK models describe the dose of a chemical reaching an entire organ with time. Some subdivide the organ into component parts and describe the chemical reactions within the organ or within individual cells as well as the overall dose of chemical in the organ, whereas others aggregate groups of organs into single compartments for the sake of analysis. Regardless of the details, these simulations and PK models all rely on numerical solutions either to partial differential equations or to systems of ordinary differential equations. The problem of numerically solving partial differential equations is usually reduced to that of solving coupled systems of ordinary differential equations by the use of finite-difference methods or, more recently, finite-element methods. Recent progress in numerical analysis has yielded methods that can permit microcomputers to produce useful approximations, even to such complicated systems as the three-dimensional convection of reactive gases in the human lung.

COMPUTER LANGUAGES USED IN PHARMACOKINETIC MODELS

Any computer language can be used to simulate the distribution of chemicals within the body or cells with time. The language chosen depends on the computation equipment available, the experimenter's experience and preferences, and the experimenter's objective for the simulation. Early simulation programs were written in general purpose computer languages, but in recent years special languages have been developed for simulation programs to simplify the process of program development and to reduce the amount of specialized knowledge required.

Stand-alone models developed on large computers have traditionally been written in the general purpose language FORTRAN. Efficient optimizing compilers for FORTRAN are available for nearly all computers. There are widely distributed libraries of FORTRAN subroutines for performing such common operations as solving algebraic and differential equations, solving linear programming problems, performing elementary optimization, and plotting graphs on video screens or flatbed plotters.

Unfortunately, differences among the compilers make it difficult to transport (or "port," in the vernacular) FORTRAN programs and models developed on one machine and operating system to another machine or operating system. In addition, the complete absence in early versions of

FORTRAN of control structures such as "IF . . . THEN . . . ELSE," "DO . . . UNTIL," and "WHILE" loops led to a tradition of writing code which is cryptic and clever and difficult to maintain or upgrade, particularly after the original programmer has left the project. FORTRAN programs require considerable expertise to write and maintain and a major investment in time to develop and debug.

More recent languages such as C and Pascal were designed to enable programmers to write even large systems that are easy to maintain. These languages, both derived from the language Algol, have sophisticated control structures that make a program's logical structure and flow of control stand out. This makes it easier for older programs to be understood and maintained by their authors or by other programmers. These modern languages also encourage modular programming, in which each large task (such as simulating the evolution of the concentrations of a chemical toxicant in each of several human organs) is broken down into smaller tasks (such as reading a data base of organ descriptions, building a data structure for the simulation, approximating the evolution of a single organ for 1 s, stepping through the list of organs, stepping through time, and plotting the results), each of which can be broken down into tasks that are smaller still. Each task is performed by a separate subroutine that can be modified and improved as the simulation model evolves, without requiring that other subroutines be changed. Experience has shown that large programs are much easier to develop and debug when they employ the principles of modular programming.

The languages C and Pascal differ in expected ways, given their origins. Pascal was designed (by the Swiss computer science professor Niklaus Wirth) as a teaching language. It is easy to write well-structured modular programs in Pascal, and very hard not to—the language strictly enforces all sorts of rules and gives informative error messages at the first sign of an error. In contrast, the language C was written at Bell Laboratories (by UNIX pioneer Dennis Ritchie) for expert systems programmers. It gives the programmer more power and flexibility than any other high-level language (and nearly as much as assembler language provides). The price paid for this flexibility is that C is harder to learn than Pascal for novice or even for moderately experienced programmers, and programming errors tend to be caught later (when the program is run) rather than earlier (when the program is complied and when errors are easy to correct). Thus, C is preferred for the experienced computer programmer, and Pascal for the novice to intermediate programmer.

BASIC, the traditional language of microcomputers, is hardly ever used on larger computers because of its relative lack of control structures and because of its traditional implementation as an interpreted (rather than compiled) language. Interpreters for BASIC are widely available for microcomputers, often with good support for on-screen graphics, and the language is not difficult to master. This makes BASIC appealing for novice programmers and adequate for the smallest simulation projects. C and Pascal compilers are also available for microcomputers, minicomputers, and mainframe computers. The dramatic commercial success of Borland's inexpensive "Turbo Pascal®" for International Business Machines (IBM) personal computers (PCs) and compatible computers has made Pascal the most popular structured language used on microcomputers in general and, therefore, in PK models, whereas the success of UNIX in the minicomputer and engineering workstation world has made C the dominant language on mid-sized computers. Language is hardly ever the limiting factor in PK modeling.

SIMULATION LANGUAGES

Simulation languages were developed to provide the constructs and operations required in system simulation. Using such a language, the model developer has only to describe the system to be simulated and does not have to be concerned with writing detailed code for such operations as changing parameters, controlling simulation runs, solving the equations, and printing or plotting the output. Table 1 lists some of the more popular simulation languages; a more detailed listing can be found in the October 1986 issue of the journal *Simulation*. Because engineers have been using simulation longer than biomedical researchers, most of the available simulation languages are designed for engineering applications and are not ideal for biomedical problems.

Many simulation languages, such as CSSL and ACSL, work by translating the simplified model description into a general purpose computer language such as FORTRAN. ACSL is being incorporated into a more

TABLE 1 Popular Simulation Languages

Language	Description
ACSL	A language designed for modeling and evaluating the performance of continuous systems described by time-dependent, nonlinear differential equations.
CSSL IV	A model development language and translator for solving systems of differential equations and a run-time control language and interpreter.
SCoP	An interactive control program for simulation calculations. Used with a C compiler, SCoP greatly simplifies the construction of a simulation program. Available for both IBM PC and VAX systems.
SIMNON	A command-driven interactive program written in FORTRAN for simulation of systems governed by ordinary differential equations and difference equations.
ADSIM	A high-level simulation language providing support for the Applied Dynamics International simulation processor, including run-time commands.

comprehensive program, SIMUSOLV®, which allows even greater flexibility and more complex modeling. Elsewhere in this volume, Gary Blau and Richard Reitz discuss PK models that were programmed using SIMUSOLV.

SCoP (Simulation Control Program) is a simulation package developed specifically for biomedical researchers by the National Biomedical Simulation Resource (NBSR) at Duke University. SCoP uses techniques developed for microcomputers such as menus and screen editing to make it easy for both novice and experienced users to interact with a simulation program. Because biological simulation has not yet developed to the stage where models can be constructed from a set of standard modules, SCoP lets the modeler write equations of any type, e.g. linear or nonlinear algebraic, ordinary, or partial differential equations. To build a simulation program, the modeler makes a copy of one of the template files provided and fills in the model equations in the places specified in the file. For clarity, these equations are written in terms of familiar variable names. The SCoP package includes a library of solvers for several kinds of equations, algebraic and time-dependent, as well as numerous other functions to simplify the description of models and experimental setups. The modeler must also build a data base of parameter and variable information using a spreadsheet-like program provided with SCoP. This data base can later be modified interactively through the simulation program. SCoP provides interactive facilities for changing model parameters; selecting, plotting, printing, and saving output; comparing model output with experimental data visually; and adjusting parameters to fit a model to experimental data. SCoP runs on microcomputers such as the IBM PC/XT/AT and on minicomputers such as the VAX (Digital Equipment Corporation) under the VMS and UNIX® operating systems. Thus, a model developed on a microcomputer can be moved up to a larger machine when it gets too large or too slow.

EQUIPMENT NEEDS FOR PHARMACOKINETIC MODELING

Computational equipment needs vary with the nature of the problem. Complex simulations involving large data sets and multiple equations are best approached using mainframe or minicomputers. Specialized computers provide the necessary resources for solution of these complex programs. Conventional multicompartmental PK models can generally be solved on microcomputers, or microcomputers can be used to access larger shared resources. A microcomputer operating as a smart terminal, as well as a remote data processing unit, is the most desirable approach, providing the flexibility of access to shared resources and local computation and data storage for those models not requiring shared resources.

Specialized computers are available through the Division of Research Resources (National Institutes of Health [NIH]) program in computer resources.

Microcomputers equipped with a modest memory of 256 kilobytes are generally adequate for most PK models. The decrease in cost of hard-disk storage makes the availability of 20–30 megabytes of storage accessible to most laboratories. These microcomputers provide adequate computational ability and storage for the majority of PK models, but speed of computation is sacrificed. The development of math coprocessors (such as the 8087/80287 and 68881 chips) and accelerator boards has made PK model solution on microcomputers an achievable objective.

Graphic output is generally not a major problem with PK models. Statistical and plotting routines are either incorporated into simulation languages or can be used as general utilities once the PK model has been solved and the results of the prediction obtained. Commercially available packages such as SAS/GRAPH® Microsoft Chart or LOTUS 1-2-3® can be used with computer files generated by the PK model program to provide a visual display of the mathematical relationships. A clear graphical representation of pharmacokinetic relationships is important for conveying model predictions to most toxicologists without extensive training in mathematics. Public and lay decision makers also understand graphic displays of information more readily than tabular results or equations. Because of the popular need for graphic displays of data, many commercially available programs on microcomputers and minicomputers afford toxicologists the luxury of dispensing with the plot subroutines once required only a few years ago for computer output.

Microcomputers or terminals equipped with a modem can easily communicate with minicomputers and mainframe computers so that specialized computer facilities are within reach of the toxicologist's laboratory. Commercially available terminal emulator programs allow experimenters to transport input and output data files from microcomputers up to larger computers and back, so that remote storage and manipulation of PK model output is routine. Programs such as Crosstalk, Smartcom, or PC-Talk, for example, allow the experimenter to assemble on one's own microcomputer large data sets for computing on the larger machines and to store the larger computer's output in one's own microcomputer in return. Graphical terminal emulator programs such as PC-Plot III also allow the local microcomputer to display mainframe-generated graphs and charts.

CONVENTIONAL APPROACHES TO MODELING

The traditional approach to modeling has been the construction of a model which stands alone. Generally, these models have been built to describe a complex event and are supported on large general purpose computers (mainframe computers). The regional lung dosimetry model for reactive gases developed by the U.S. Environmental Protection Agency Health Effects Laboratory (Miller et al., 1978, 1985) is an example of such a model. These models are generally written in FORTRAN, C, or Pascal, but rarely in BASIC. Such models can use optimal file formats, data structures, and numerical methods for the needs of a particular model, making them ideal for accommodating large input data sets, complex relationships among the model's state variables, and a need for rapid computation. Mainframe computers running such stand-alone models can provide more memory, speed, and precision than can minicomputers or microcomputers; but most PK models in common use involve few enough equations and are well enough conditioned that the capacity, speed, and precision available on microcomputers are adequate. Especially when microcomputers are used, the user should include diagnostic tests to ensure that rounding errors and other sources of imprecision have not compromised the model's prediction. As discussed elsewhere in this volume by Frederick J. Miller and colleagues, sensitivity analysis of the model is essential to ensure that such computational errors do not bias the model results. The reactive gas lung dosimetry model is now available also in a microcomputer version with little loss in precision compared with that supported on mainframe or minicomputers such as the VAX 11 series (F. J. Miller, U.S. Environmental Protection Agency, personal communication).

AN EXAMPLE OF A MODULAR PROGRAM

The conventional multicompartmental model based on the physiological properties of organs and the chemical and physical properties of chemicals often follows the form shown in Figure 1. This model was developed in our laboratories by Professor R. L. Wolpert and J. R. Boger III. The model program was written in the C language using modular programming techniques. It is a generic model in the sense that it can be used for simulating the time course of chemical concentration in selected organs or parts of organs for different chemicals and for different animal species without reprogramming or recompiling the model program. The model maintains a physiometric data base for each of several animal strains and species, including blood flow rates, organ masses, capillary or plasma volumes, etc., for each animal.

For each new chemical the investigator must choose which organs or parts of organs to represent with separate compartments and which to aggregate together. The rates at which the chemical is introduced, eliminated, or metabolized in each compartment and the extraction ratio re-

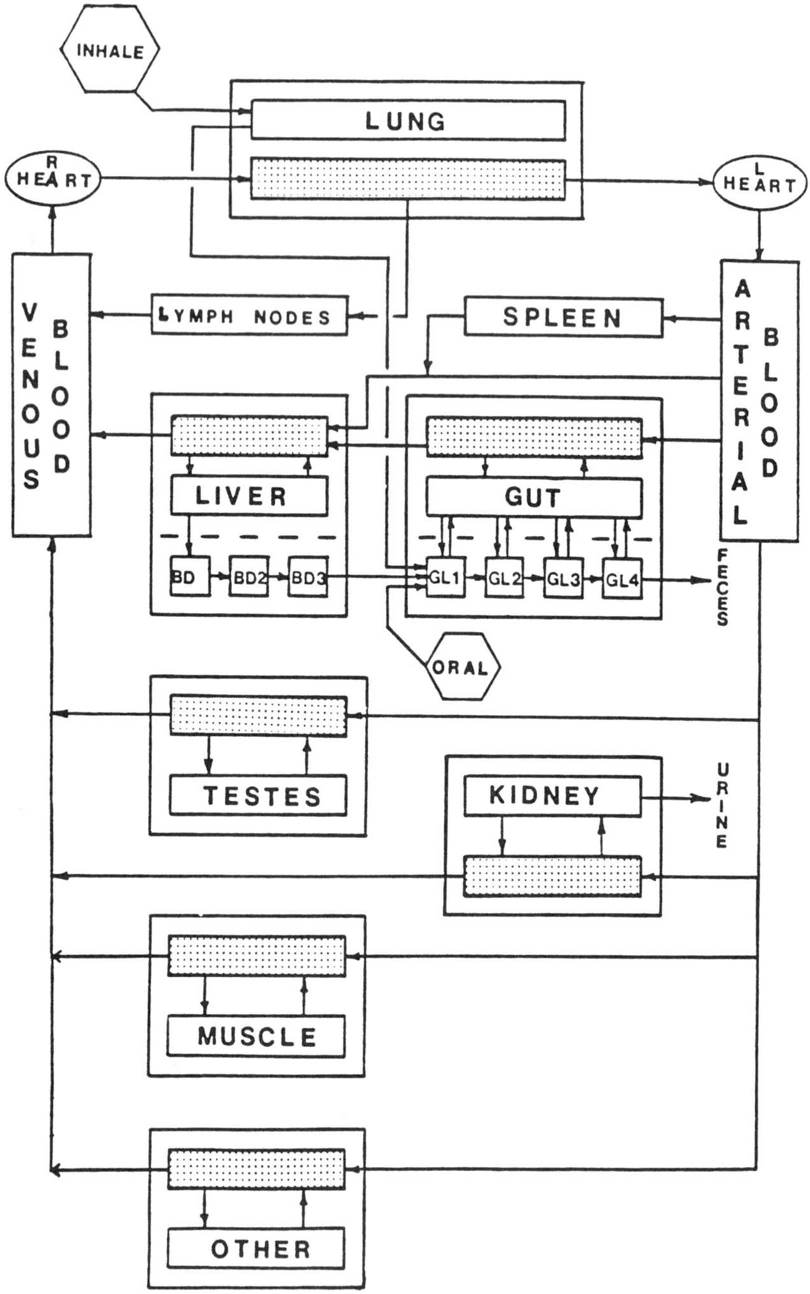

FIGURE 1 A generic multicompartmental model for distribution and metabolism of xenobiotic compounds in mammals.

lating the equilibrium concentrations in the blood or plasma and the tissue of each compartment must also be specified. The model will then plot a color overlay graph showing the time course of chemical concentration in each of a selected number of compartments.

Using this model and measured nickel extraction ratios between each organ and plasma, the kinetics of distribution of nickelous ion in the body have been simulated (Francovitch et al., 1986). Extraction ratios were determined in the whole rat by continuous infusion of radiolabeled nickel chloride until a steady-state concentration was attained in blood. A bolus loading dose was used to accelerate an approach to steady-state concentrations.

This model provides a capillary blood volume within each organ within which the tissue and plasma concentrations are at equilibrium following the transit of blood through the organ. Examples of the simulation of a bolus intravenous injection (shown in Figures 2–5) compare favorably with experimental results (also shown in the figures). Lungs, kidneys, and testes are the three organs of the body known to have a particular affinity for nickel. All of the other organs are lumped together to simplify the interpretation of the results, but the data are not lost for individual organs not reported.

Such a model is generic in the sense that one simulation model can be used for multiple chemicals and can be simplified according to the bio-

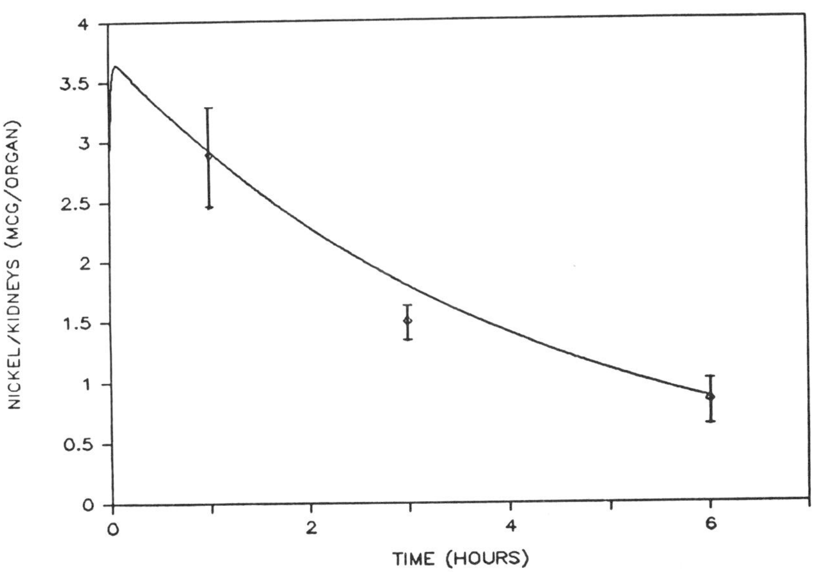

FIGURE 2 Simulated amount of nickel in the kidneys following an intravenous injection of 13 μg. Experimental results are shown as (◇).

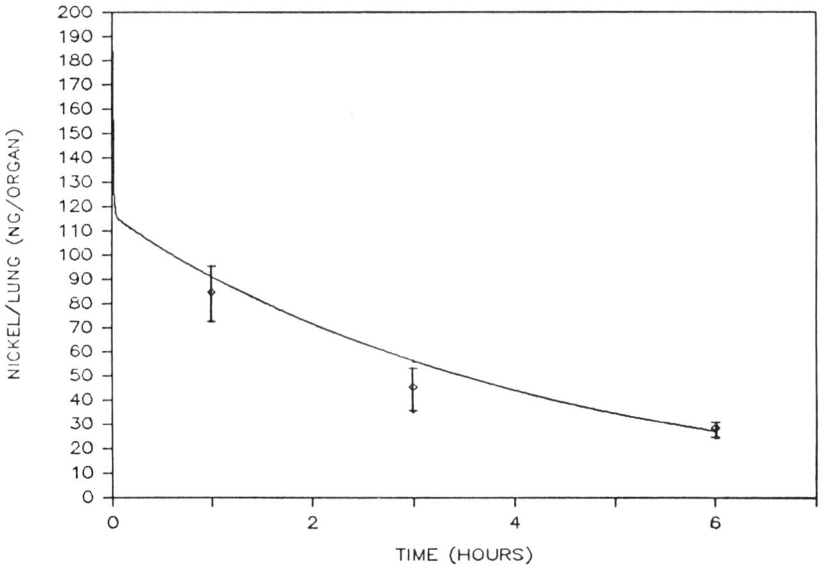

FIGURE 3 Simulated amount of nickel in the lung following an intravenous injection of 13 μg. Experimental results are shown as (◇).

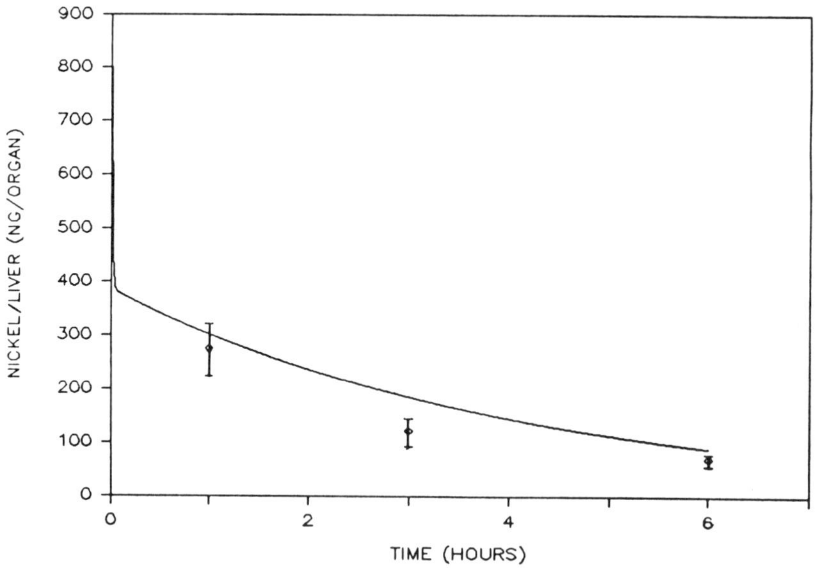

FIGURE 4 Simulated amount of nickel in the liver following an intravenous injection of 13 μg. Experimental results are shown as (◇).

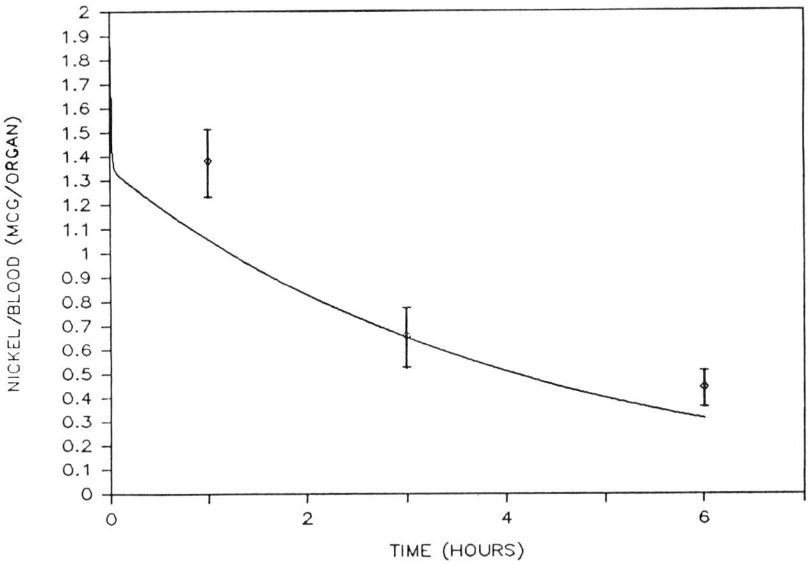

FIGURE 5 Simulated amount of nickel in the blood following an intravenous injection of 13 μg. Experimental results are shown as (◇).

logical activity of the chemical. An individual organ (such as the kidney or lung) can be represented by several compartments if this is necessary to reflect accurately the transport and metabolism of the chemical being studied. There is little point in reporting and studying concentrations in all parts of all organs for all chemicals when biologically each chemical may affect only one or two target organs.

The model can be modified for other species and for different physiological data within a species by simply modifying the input data set with a text processor. Allometric relations have been used to construct tables of organ properties within species and between species where direct information is not available in the literature, so that variations in animal size can be accommodated. The development of a standard data set for a 300-g rat has been completed and is shown in Table 2. Models for children, adult human females, pregnant human females, and senescent humans are projected. Similar models for mice and rats of both sexes and different stages of maturation and senescence are under way.

AN EXAMPLE SCoP PROGRAM

As an example of the application of SCoP to toxicology, a simulation of the effects of sulfite or sulfur dioxide on the covalent reaction of benzo(*a*)pyrene with DNA or its detoxification by conjugation to unreac-

TABLE 2 An Example of Physiological Data Needed for a Generic Model: Male 300-g Sprague-Dawley Rat

A. Anatomical Compartment Dimensions

Compartment	Mass (g)	Capillary Volume (ml)	Extraction Ratio (ml/g)	Elimination Rate (liter/h)
Lung	1.27	0.126	0.86	0.0
Kidney	2.10	0.013	17.4	0.85
Liver	11.71	0.090	0.31	0.025
Gut	4.60	0.016	0.36	0.0
Spleen	0.54	0.0046	0.36	0.0
Testes	2.73	0.0022	0.31	0.0
Muscle	150.0	0.194	0.18	0.0
Carcass	111.46	0.516	0.33	0.0
Venous blood	0.0[a]	12.90	1.0	0.0
Arterial blood	0.0[a]	4.30	1.0	0.0

B. Blood Flow Rates Among Anatomical Compartments

Source	Destination	Flow Rate (ml/h)
Kidney	Venous blood	1,241
Liver	Venous blood	1,522
Gut	Liver	947
Spleen	Liver	237
Testes	Venous blood	61
Carcass	Venous blood	1,795
Arterial blood	Kidney	1,241
Arterial blood	Liver	338
Arterial blood	Gut	947
Arterial blood	Spleen	237
Arterial blood	Testes	61
Arterial blood	Carcass	1,795
Arterial blood	Muscle	846
Muscle	Venous blood	846
Lung	Venous blood	175
Arterial blood	Lung	175
Venous blood	Lung	5,640
Lung	Arterial blood	5,640

[a]Most compartments contain both a tissue (mass) and blood (capillary volume) compartment. However, the blood compartments are unusual in that they contain no tissue.

tive metabolites has been completed (Keller et al., in press). In Figure 6 are shown the chemical reactions involving the production of the putative ultimate carcinogen of benzo(*a*)pyrene, benzo(*a*)pyrene diol epoxide (BPDE), and its reaction with nuclear DNA to form covalent adducts. BPDE is conjugated with glutathione (GSH) by glutathione *S*-transferase

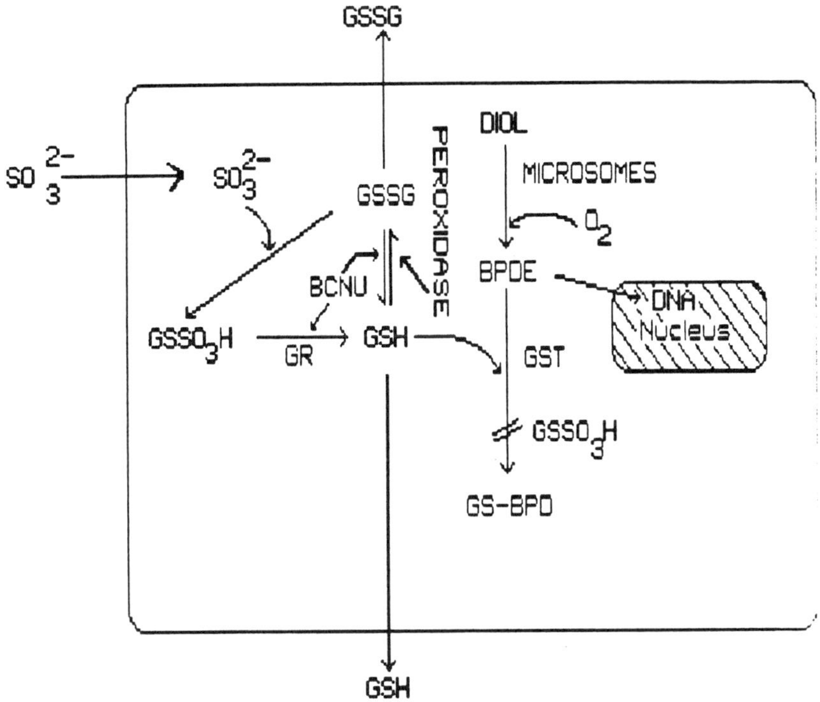

FIGURE 6 The overall scheme of benzo(a)pyrene chemical reactions involving the metabolism of benzo(*a*)pyrene to benzo(*a*)pyrene diol epoxide, and the reaction of benzo(*a*)pyrene diol epoxide with glutathione and nuclear DNA to form covalent adducts. The detoxification of benzo(*a*)pyrene diol epoxide by glutathione is inhibited by glutathione *S*-sulfonate, which is formed by a reaction between sulfite and glutathione disulfide and by added BCNU.

to form the unreactive diol conjugate. This pathway effectively removes BPDE from reaction with DNA. The glutathione *S*-transferase pathway is the rate-limiting step in the detoxification of most polyaromatic hydrocarbon carcinogens. In lung tissue, alternative pathways are limited, making the lung highly dependent on the functioning of the glutathione pathway. Even so, the lung has limited stores of glutathione.

Exposure of cells or the lung to sulfur dioxide or bisulfite results in the formation of glutathione *S*-sulfonate ($GSSO_3H$) from the naturally occurring glutathione disulfide (GSSG). $GSSO_3H$ is a competitive inhibitor of glutathione *S*-transferase (Leung et al., 1985, in press). Exposure of cells to sulfite results in the competitive inhibition of this major detoxification pathway of benzo(*a*)pyrene [B(*a*)P].

$GSSO_3H$ and GSSG are reduced by the same enzyme glutathione reductase (Leung et al., 1985). These two substrates compete with each other for reduction by glutathione reductase. The K_m and V_{max} values for the two substrates differ significantly.

Production of GSSG also depends on the availability of reducing equivalents within the cell. Reduced nicotinamide adenine dinucleotide phosphate (NADPH) is the principal source of reducing equivalent, but $NADP^+$ may be increased by the flow of reducing equivalent through other pathways. NADPH is also consumed by other pathways within the cell. At present, it is only possible to estimate the oxidation of GSH to GSSG in general. Experimental values for GSH and GSSG serve as the basis for choosing the initial values in the cell.

The reactions described are interdependent, and some compete with each other. Thus, the overall fate of B(*a*)P through metabolism, detoxification, and reaction with DNA is not easily predictable from a qualitative or intuitive analysis. The rates of these reactions can be described mathematically, including the efflux of GSSG and GSH from cells by the equations shown in Table 3. The form of the equations suitable for simulation using SCoP are shown in Table 4. The ease of conversion from conventional mathematical form to that needed by the program is a major strength. The names of variables can be made more logical through the use of words descriptive of the variable.

The integrated form of the Michaelis-Menten equation is used to describe the enzymatic reactions. To increase the precision of measurement

TABLE 3 Equations Describing Benzo(*a*)pyrene, Glutathione, and Sulfite Metabolism and DNA Adduct Formation

$$\frac{\partial \mathrm{GSH}}{dt} = k_3[\mathrm{gssg}]\,[\mathrm{SO_3}] + 2[V_{\mathrm{gssg}}] + V_{\mathrm{gsso_3h}} + k_5[\mathrm{SO_3}][\frac{1}{\mathrm{gsh}}] - k_1[\mathrm{gsh}] - V_{\mathrm{ex}} - V_{\mathrm{egsh}}$$

$$\frac{\partial \mathrm{GSSG}}{dt} = k_1[gsh]/2 - k_3[\mathrm{gssg}][\mathrm{SO_3}] - V_{\mathrm{gssg}} - V_{\mathrm{egssg}} + \mathrm{peroxidase}$$

$$\frac{\partial \mathrm{GSSO_3H}}{dt} = k_3[\mathrm{gssg}][\mathrm{SO_3}] - V_{\mathrm{gssc_3h}} - k_{20}[\mathrm{gsso_3h}]$$

$$\frac{\partial \mathrm{EPOXIDE}}{dt} = k_7[\mathrm{diol}] - V_{\mathrm{ex}} - k_2[\mathrm{epoxide}][\mathrm{DNA}]$$

$$\frac{\partial \mathrm{ADDUCT}}{dt} = k_2[\mathrm{epoxide}][\mathrm{DNA}] - k_4[\mathrm{adduct}]$$

$$V_{\mathrm{gssg}} = \frac{V_m\,\mathrm{gssg}\,[\mathrm{gssg}]}{K_m\mathrm{gssg} + [\mathrm{gssg}]}$$

$$V_{\mathrm{gsso_3h}} = \frac{V_m\mathrm{gsso_3h}\,[\mathrm{gsso_3h}]}{K_m\mathrm{gsso_3h} + [\mathrm{gsso_3h}]}$$

$$V_{\mathrm{ex}} = \frac{V_m\mathrm{ex}\,[\mathrm{ex}]}{K_m\mathrm{ex}(1 + [\mathrm{gsso_3h}]/K_i) + [\mathrm{ex}]}$$

$$V_{\mathrm{egsh}} = k_{10}(\mathrm{e}^{(-k_8 * t)} + k_{11}(\mathrm{e}^{(k_9 * t)})$$

$$V_{\mathrm{egssg}} = k_{14}(\mathrm{e}^{(-k_{12} * t)} + k_{15}(\mathrm{e}^{(-k_{13} * t)})$$

TABLE 4 Modifications of Mathematical Equations Describing SO_2-Benzo (*a*)pyrene Interaction in a Form Usable by the Simulation Language SCoP[a]

$$D\_gsh = k3 * gssg * so3 - k1 * gsh + 2 * V\_gssg + V\_gsso3h - V\_ex + k5 * so3 * (1/gsh) - V\_egsh;$$
$$D\_gssg = k1 * gsh / 2 - k3 * gssg * so3 - V\_gssg - V\_egssg + peroxidase;$$
$$D\_gsso3h = k3 * gssg * so3 - V\_gsso3h - k20 * gsso3h;$$
$$D\_ex = k7 * diol - V\_ex - k2 * ex * dna;$$
$$D\_adduct = k2 * ex * dna - k4 * adduct;$$
$$V\_gssg = V mgssg * gssg / (Kmgssg + gssg);$$
$$V\_gsso3h = Vmgsso3h * gsso3h / (Kmgsso3h + gsso3h);$$
$$V\_ex = Vmex * ex / (Kmex * (1 + gsso3h / Ki) + ex);$$
$$V\_egsh = k10 * exp(-(k8 * Time)) + k11 * exp(-(k9 * Time));$$
$$V\_egssg = k14 * exp(-(k12 * Time)) + k15 * exp(-(k13 * Time));$$

[a]See Table 3 for conventional mathematical representation.

of the components of the GSH pathway, the amounts of GSSG and $GSSO_3H$ were increased during the experiment by prior treatment of the cells with the glutathione reductase inhibitor BCNU. By measuring the K_m, K_i, and V_{max} for all of the enzymatic reactions, the simulation can be solved for the condition with or without BCNU inhibition. In other words, the abnormal physiological state of inhibition by BCNU can be used to predict the normal physiological results occurring in the absence of BCNU. Actual experimental values were measured in the absence of BCNU to confirm the validity of the assumptions made for BCNU inhibition.

Differences in chemical reactivity, such as between *syn* and *anti* isomers of BPDE with DNA and glutathione *S*-transferase substrates, can be accounted for also. Experimental values for these rates can be used in the compiled form of the PK model with SCoP and can be modified interactively without having to recompile the model. Interaction between the modeler and the model is made much more rapid by this technique.

An example of the time course formation of the *syn* and *anti* BPDE adducts with DNA over the experiment is shown in Figure 7. The effects of sulfite treatment are shown by the increased B(*a*)P binding to DNA throughout the course of the experiment. The results agree well with the experimental values found by Leung et al. (in press). Increased B(*a*)P-DNA adduct formation in the presence of sulfite or sulfur dioxide is mostly due to the inhibition of the glutathione *S*-transferase pathway by the sulfite metabolite $GSSO_3H$.

Using this model and a simple calculation of the intracellular sulfite concentrations that are likely from exposure to sulfur dioxide concentrations used in three studies of the cocarcinogenicity of sulfur dioxide with

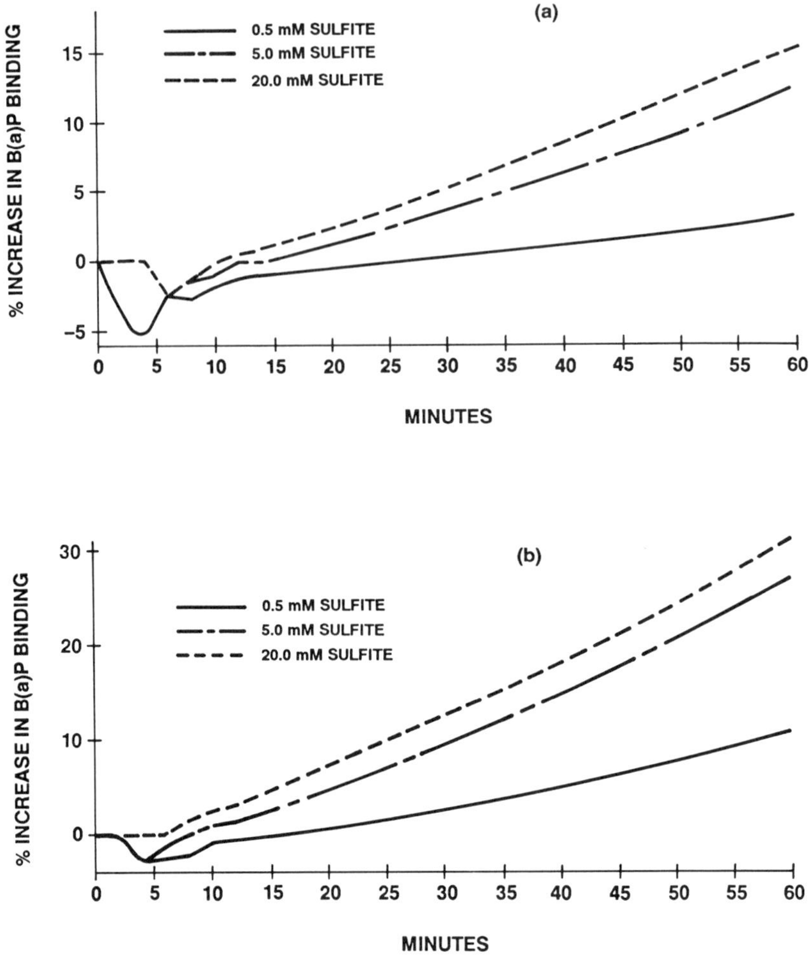

FIGURE 7 Formation of the *anti* (a) and *syn* (b) BPDE adducts with DNA.

polyaromatic hydrocarbons (Laskin et al., 1976; Pauluhn et al., 1985; Pott and Stöber, 1983), we can predict that a marked increase in DNA adducts occurred in the study done by Pauluhn et al. (1985), in which there was a significant increase in the number of lung tumors and shortened time to tumor appearance. Only very small increases of 3–4% in DNA adducts were predicted to have occurred in the other two studies. The changes in numbers and time to tumor for these studies was marginal and of questionable statistical significance (Figure 8).

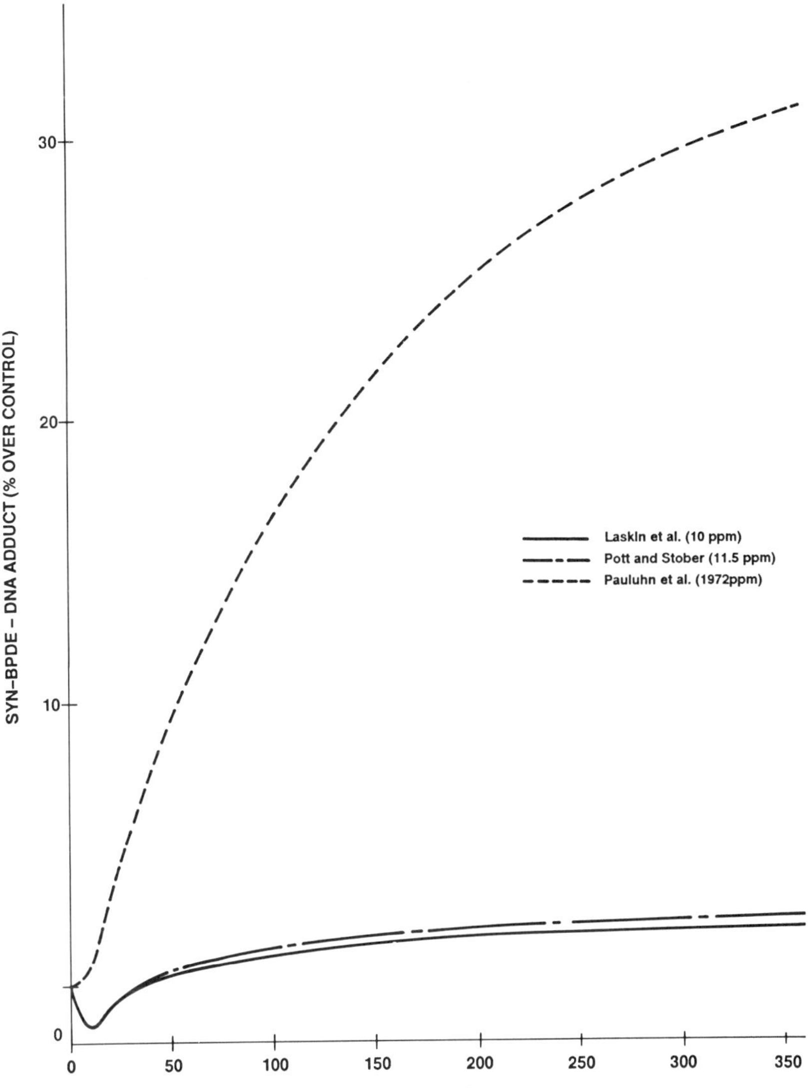

FIGURE 8 Predicted increase in amounts of *syn*-BPDE adduct over control from studies by Laskin et al. (1976), Pott and Stöber (1983), and Pauluhn et al. (1985).

Our model is still only a crude approximation of the actual flux of BAP through the lung cell and the effects of sulfite or sulfur dioxide on these reactions. The relationship between increased DNA binding and tumor rates is unknown at present. The rate of removal of adducts under these conditions has not been measured. The model overestimates the adduct formation as a consequence.

SHARING RESOURCES

As stated at the outset of this paper, the simulation of toxicological events, either pharmacokinetics or pharmacodynamics, is but a special case of the more general methodology of simulation of physical, chemical, and biological events. The same mathematical approaches are applicable to all of these fields, as are the same computational resources. Fortunately for toxicologists much of the developmental work has been accomplished in other fields, particularly engineering, and is now available for application in toxicology.

The sharing of resources, especially computational resources, provides greater power to pharmacokinetics and pharmacodynamics than could be justified by the field alone. The NBSR provides a major computation facility specifically designed for biological simulation (see Table 5).

Training in simulation is available from the NBSR and other NIH simulation resources at two levels. Introductory classes are designed for researchers with little or no background in computer usage or programming. The NBSR presents introductory classes lasting four and a half days several times a year at locations where there is a group of 8 to 15 students and a suitable cluster of microcomputers for teaching. The NBSR introductory class is built around the SCoP software package, and students can take a copy of the software home with them to continue their work after the class is over. Researchers with no prior programming experience can be developing their own simulation programs by the end of the class.

Advanced simulation courses are also available or in preparation by

TABLE 5 Computation Facilities at the National Biomedical Simulation Resource

VAX 11/750
Array Processors (MAP-6430 and Mini-MAP)
Special Parallel Coprocessors for Simulation (AD-10 and AD-100)

NBSR. These courses are for the researcher with some previous experience and cover such topics as the use of special purpose processors, a wider range of simulation languages, and advanced numerical techniques for equation solving.

THE TOXIN CONCEPT

To facilitate the exchange of information in the toxicology community, the Toxicology Information Network (TOXIN) has been established within the NBSR. The goals of TOXIN are to improve communication and understanding among toxicologists about mathematical modeling, to provide access to PK and risk assessment models, to develop consensus on PK and risk assessment models through collaborative use of the same model on the same computer, and to provide access to special data bases needed for modeling. A schematic diagram of the relationship between users and TOXIN is shown in Figure 9.

Users in the United States can interconnect with TOXIN through the commercial data communications network TYMNET. Users enter TYMNET through a local telephone number using regular telephone lines. Microcomputers are encouraged as smart terminals for use with TOXIN, but any terminal can be used. Certain terminals have limited graphics, which may prevent the use of some programs. Depending upon the needs of the user, either a TOXIN account or a regular NBSR collaborators account provides access to the facilities of the NBSR, including the specialized computing facilities. Electronic mail for exchange of information, models, and data files is available.

PK models contributed to TOXIN are placed in common computer files available to every TOXIN user. The models can then be used to run simulations with the same computer code for use on the same computer by different investigators. Problems, improvements, enhancements, and comments on the models can be made by users and sent to the contributor of the model or to the general community of TOXIN users through the electronic mail system. In this manner a consensus can be built about the critical aspects of PK models. The selection of physiological parameters for PK models is a particularly critical area in which comparisons of different values may prove especially useful. As better physiological values are amassed, they can be stored in data tables in a form useful for modeling. Alternative models can be compared and potentially integrated to form second- and third-generation models.

As more PK models are developed and made available to TOXIN users, an archive of PK and pharmacodynamic models can be accumulated. Historical reference will then be possible. At present such an archive does

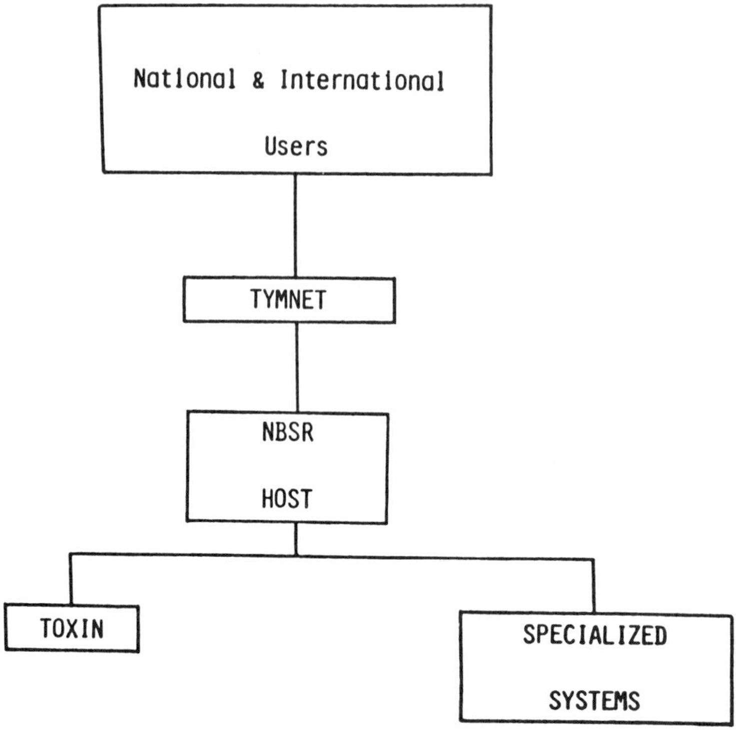

FIGURE 9 Schematic diagram of the relationship between users and TOXIN.

not exist, and original computer codes are inaccessible to investigators beyond the originator. If PK models are to have a major impact on regulatory decision making, the establishment of an archive and the requisite data for risk estimation is essential for the future review and improvement of regulations.

Selection among different PK models is now difficult to make because direct comparisons are generally not possible. Storage of multiple PK models in a form that can be easily used by investigators will provide a means of systematic testing and further validation of PK modeling approaches. Estimation of errors within different PK models can be made more easily, and the methodology of estimating error in PK models (as discussed by Gary E. Blau and W. Brock Neely, this volume) is made more feasible.

TOXIN also provides access to commercial programs that are expensive and difficult to justify for a single laboratory. At present, ACSL, CSSL, SIMNON, ADSIM, SCoP, and SCoPFIT are available as simulation languages on NBSR and are accessible through TOXIN. Should user demand warrant other commerical software, the languages supported by NBSR can be expanded.

FUTURE TRENDS

The rapid development of simulation languages should make simulations of all kinds much more accessible to nonprogrammers. These languages have gone through several revisions and are improving rapidly. Reductions in costs are likely as the programs become more widely used. Versions for microcomputers will provide even greater access. Even those programs not supported by microcomputers should become more available through the use of shared resources and the telecommunications network.

Shared resources provide the opportunity to accumulate libraries of PK models that can be used by multiple users. Consensus can be built and improvements in shared PK models can be accelerated through comparison of predictions for a variety of compounds.

Development of special data bases for physiological parameters can be achieved. Much of our knowledge of organ blood volumes, flow rates, and gross anatomy are gained from tracer techniques. Pharmacokinetic data provide similar data, and the concordance of several experimenters' data increases our confidence in these basic physiological values. Errors are likely to be detected when reference values fail to predict experimental results. Data gaps will be identified and filled as physiologically based PK models become more widely used.

Computational devices are evolving rapidly. The drastically reduced costs of data storage and the ease of transmission of masses of data by electronic means will result in more complex models useful for experimental design and, with hope, long-range planning. Fidelity with human experience should increase as PK models are integrated with pharmacodynamic models, making risk assessment more reliable. Distributed data storage will also improve data utilization and decrease repetition of experiments, lowering costs and animal usage and increasing generalization among classes of chemicals. Digital analog devices and their high-level simulation languages should speed computation to real time rates. Digital equipment reduces the uncertainties and difficulties of conventional wired analog or hybrid computers.

CONCLUSIONS

In summary, there are no major equipment limitations for most PK models in the United States. The specialized computer facilities such as the NBSR at Duke University and other components of the NIH simulation resources provide more than ample power for computation. Most PK models can be solved on microcomputers using commercial or public domain software. The rapid development of simulation software provides an opportunity for toxicologists to use the software firsthand to simulate their experiments, to assist in design of experiments, and to extrapolate from animal experiments to humans.

ACKNOWLEDGMENTS

This work was supported in part by grants from NIH ES01859, ES02916, CA14236, and RP-940-5 from the National Institutes of Health and by a grant from the Electric Power Research Institute. We wish to thank Drs. K. H. Leung and D. A. Keller for sharing their data and models on sulfur dioxide and benzo(*a*)pyrene metabolism and Drs. R. J. Francovitch and C. R. Shoaf for sharing their data and model on nickel metabolism. Messrs. M. I. Tayyeb and J. Sandy and Ms. K. Wilkinson provided excellent technical assistance.

REFERENCES

Francovitch, R. J., C. R. Shoaf, M. I. Tayyeb, and D. B. Menzel. 1986. Modeling nickel dosimetry from kinetic measurements in rats. The Toxicologist 6:528.

Keller, D. A., K. H. Leung, and D. B. Menzel. In press. Glutathione, sulfite, and benzo(a)pyrene interactions: A mathematical model. The Toxicologist.

Laskin, S., M. Kuschner, A. Sellakumar, and G. V. Katz. 1976. Combined carcinogen-irritant animal inhalation studies. Pp. 190–213 in Air Pollution and the Lung, E. F. Aharonson, A. Ben-David, and M. A. Klingberg, eds. New York: John Wiley & Sons.

Leung, K. H., G. B. Post, and D. B. Menzel. 1985. Glutathione S-Sulfonate, a sulfur dioxide metabolite, as a competitive inhibitor of glutathione S-transferase, and its reduction by glutathione reductase. Toxicol. Appl. Pharmacol. 77:388–394.

Leung, K. H., D. A. Keller, and D. B. Menzel. In press. Glutathion S-sulfonate inhibition of glutathione conjugation with benzo(a)pyrene epoxides. Toxicol. Appl. Pharmacol.

Miller, F. J., D. B. Menzel, and D. L. Coffin. 1978. Similarity between man and laboratory animals in regional pulmonary deposition of ozone. Environ. Res. 17:84–101.

Miller, F. J., J. H. Overton, Jr., R. H. Jaskot, and D. B. Menzel. 1985. A model of the regional uptake of gaseous pollutants in the lung. I. The sensitivity of the uptake of ozone in the human lung to lower respiratory tract secretions and exercise. Toxicol. Appl. Pharmacol. 79:11–27.

Pauluhn, J., J. Thyssen, J. Althoff, G. Kimmerle, and U. Mohr. 1985. Long-term inhalation study with benzo(a)pyrene and SO_2 in Syrian golden hamsters. Exp. Pathol. 28:31.

Pott, F., and W. Stöber. 1983. Carcinogenicity of airborne combustion products observed in subcutaneous tissue and lungs of laboratory rodents. Environ. Health Perspect. 47:293–303.

PART V

Poster Session

Introduction

Robert L. Dedrick

During organization of the workshop on which this volume is based, it became clear that additional relevant work and points of view existed that could not be well represented in a limited number of primarily didactic presentations. To enrich the program, a poster session was added in which the presenters were also asked to introduce themselves in the plenary session, give 1- or 2-minute summaries of their work, and then invite general discussion of their topic by workshop participants.

Route-to-Route Extrapolation of Dichloromethane Exposure Using a Physiological Pharmacokinetic Model

Michael J. Angelo and *Alan B. Pritchard*

BACKGROUND

Physiologically based mathematical models are useful devices for exploring the information contained in pharmacokinetic data. Most of the physiological pharmacokinetic models that have appeared in the literature have been developed for pharmacological research. A number of these applications have been reviewed by Gerlowski and Jain (1983), who also present some of the fundamental principles that are used to derive this type of mechanistic model.

There has been increasing interest within the field of toxicology in the use of models to process pharmacokinetic information, especially with regard to exposure assessment. Their value has been recognized for a number of reasons, including their ability to increase a scientist's understanding of physiological factors that control chemical disposition, quantification of *in vivo* metabolism rates, and the extrapolation of pharmacokinetic predictions across dose levels and mammalian species.

This paper illustrates how we have used a previously developed physiological model for the disposition of dichloromethane (DCM; methylene chloride) to compare the pharmacokinetic patterns that result from inhalation and oral administration of DCM in rats. With information generated by computer simulations, we determined the correlations between inhalation doses and the corresponding oral doses of DCM that produce the same level of pharmacokinetic impact. By examining the pharmacokinetic information in this way, we were able to infer how dosing route dependencies can influence the calculation and interpretation of DCM exposure.

RESULTS AND DISCUSSION

The schematic diagram in Figure 1 shows the structure of the physiologically based model that was used to describe the pharmacokinetics of DCM following inhalation and oral administrations. The mathematical details of the model, including parameter values, are given in the Appendix to this paper. Table 1 contains the physiological constants for a 250-g rat that are available in the literature, and Table 2 contains the pharmacokinetic parameters that were determined by using numerical optimization techniques. The kinetics of DCM biotransformation via pathways dependent on the mixed-function oxidase (MFO) system and glutathione (GSH) has been presented recently by Gargas et al. (1986), and it has been incorporated into the mass balance equations.

The model was verified for oral administrations by using data that were collected at the Huntingdon Research Centre in England (Angelo et al., 1986). As an example of this, Figure 2 shows the actual blood concentrations of DCM and model simulations following single oral doses of 50 and 200 mg/kg to rats on days 1 and 14 of a daily gavage dosing regimen.

In developing the correlation between inhalation and oral doses, we first determined that a continuous oral infusion of DCM into the gut lumen

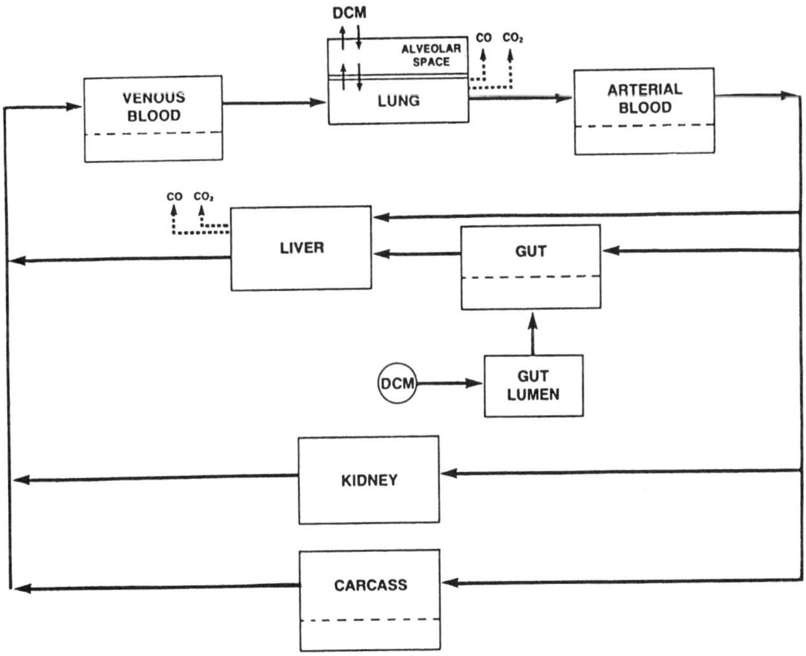

FIGURE 1 Physiological pharmacokinetic model for dichloromethane (DCM).

TABLE 1 Physiological Parameters for a 250-g Rat

Parameter	Abbreviation	Value
Compartment flow rates (ml/min)		
Lung[a]	Q_{Lg}	32.3
Liver	Q_L	16.0
Kidney	Q_K	10.0
Gut	Q_G	9.1
Carcass[b]	Q_C	6.3
Compartment volumes (ml)		
Whole animal (g)	BW	250
Blood	V_B	22.5
Lung	V_{Lg}	2.9
Liver	V_L	10.0
Kidney	V_K	2.3
Gut	V_G	10.1
Gut lumen	V_{Gl}	13.6
Carcass[c]	V_C	183
Alveolar space	V_E	6.25
Alveolar ventilation (ml air/min)	Q_g	90
Blood/air distribution ratio	λ	12.8
Red blood cell/plasma distribution ratio	R_{rbc}	5.4

[a]Sum of venous flows from other compartments.
[b]Volumetric average of muscle, skin, and fat.
[c]Sum of muscle, skin, and fat.

compartment of the model would produce pharmacokinetic profiles similar to those observed during an inhalation exposure. This was done to demonstrate that a constant oral input mimicked the distribution patterns that were achieved during a continuous input to the pulmonary system. Therefore, the correlations between the two methods of administration depended upon the dosing route and not upon the disposition factors that were influenced by the rate of administration. Using the data of McKenna and coworkers (1982), Figure 3 shows that the model simulations of DCM blood concentration during a 6-h oral infusion agree well with the inhalation data. Inhalation simulations at 50, 500, and 1,500 ppm for 6 h produced curves that were virtually identical to those obtained with continuous oral input, indicating that steady infusions of DCM into the lung and gut compartments produced the same blood distribution profiles.

To obtain the inhalation-to-oral dose correlations, we first evaluated delivered doses to target tissues as areas under concentration-time curves (AUCs) for a series of inhalation exposures. Metabolized doses were also

TABLE 2 Pharmacokinetic Parameters for a 250-g Rat

Parameter	Abbreviation	Value
Tissue/plasma distribution ratio		
Blood	R_B	2.8
Lung	R_{Lg}	1
Liver	R_L	1
Kidney	R_K	1
Gut	R_G	4.5
Carcass	R_C	2.4
Gut contents/water distribution ratio	R_{Gl_w}	1
Metabolism parameters		
V_{max} (nmol/ml/min)		40
K_m (nmol/ml)		7
k_2 (min^{-1})		0.016
A_1[a]		0.120
A_2[a]		0.056
f_{CO}		0.60
Alveolar permeability × area product (PA; ml/min)		535
Carbon monoxide clearance (CL_{CO}; ml/min)		0.26
Binding constants for ^{14}CO to Hb		
k_α (nmol/ml)		600
k_β (nmol/ml)		40
Gut absorption (k_a; min^{-1})		0.029

[a]Reitz et al. (1986).

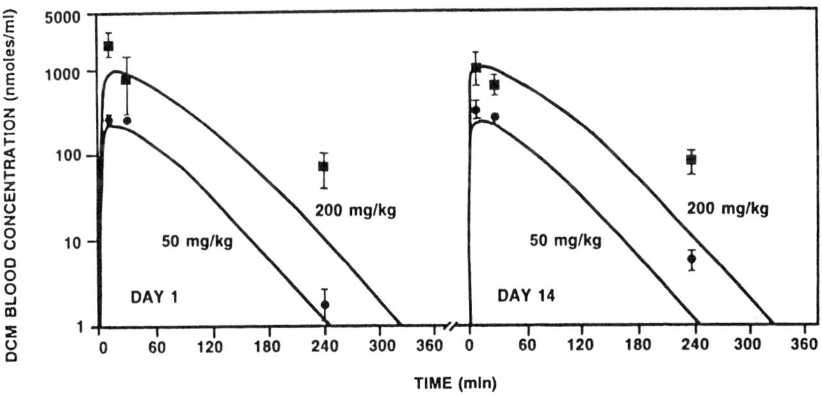

FIGURE 2 Blood concentrations of dichloromethane (DCM) on days 1 and 14 of a repeated oral dosing schedule. Daily gavage doses were administered to rats at 50 (●) and 200 mg/kg (■) in a water vehicle. Data represent the mean ± standard error of the mean for six animals; lines are the predictions from the pharmacokinetic model.

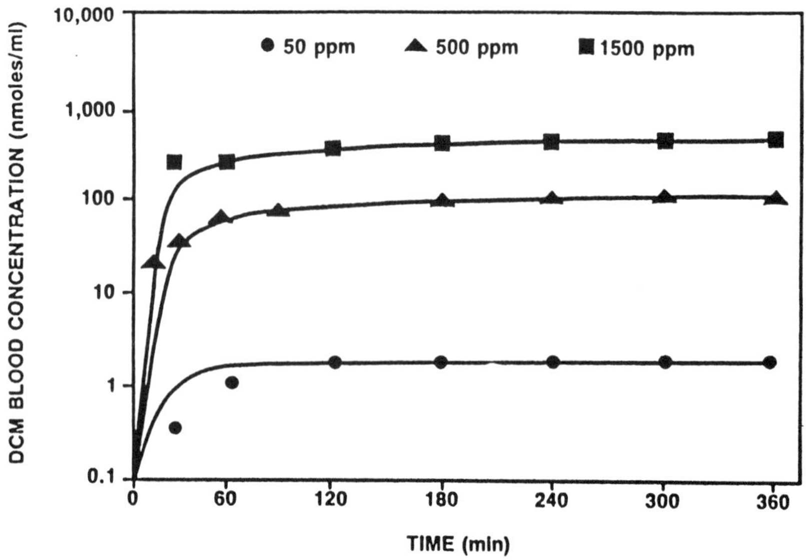

FIGURE 3 Physiological model predictions of dichloromethane (DCM) in blood of rats during 6-h oral infusions. Simulations are compared to data from 6-h inhalation exposures of DCM at 50, 500, and 1,500 ppm. SOURCE: McKenna et al. (1982).

determined in terms of the amount of DCM that was metabolized by each of two biotransformation pathways. A numerical search was then used to find the oral doses that produced the equivalent delivered or metabolized doses to those obtained in the inhalation simulations.

Figure 4 shows the correlations between inhalation exposures and equivalent oral doses of DCM in rats for the lung, liver, and blood compartments. The correlations indicate that at 1,500 ppm, the equivalent oral dose of DCM is between 200 and 300 mg/kg for each tissue.

Figure 5 shows the correlations that were based on the equivalent amounts of metabolized DCM by the two biotransformation pathways. The figure indicates that the correlation for the GSH-dependent pathway is a linear relationship between the oral and inhalation exposures, whereas the MFO pathway exhibited a nonlinear pattern. At 1,500 ppm, the equivalent oral dose was approximately 200 mg/kg for the glutathione pathway and about 100 mg/kg for the MFO pathway, although the latter reached this level between 500 and 750 ppm.

The retention of DCM during steady-state inhalation was less than 15% of the exposure concentration in the 50- to 1,500-ppm treatment range. This was determined by using Equation 11b from the Appendix. Values for the alveolar concentration of DCM (C_a) that were used in the calcu-

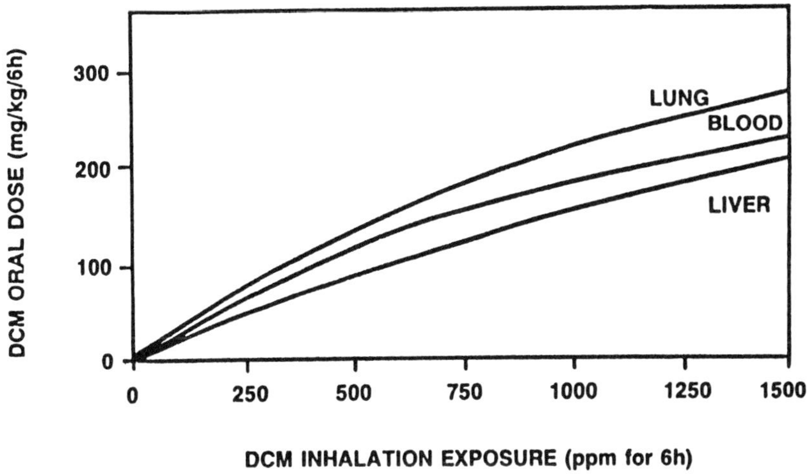

FIGURE 4 Correlations based on equivalent tissue AUC in rats between oral doses (infusions for 6 h) and 6-h inhalation exposures of dichloromethane (DCM).

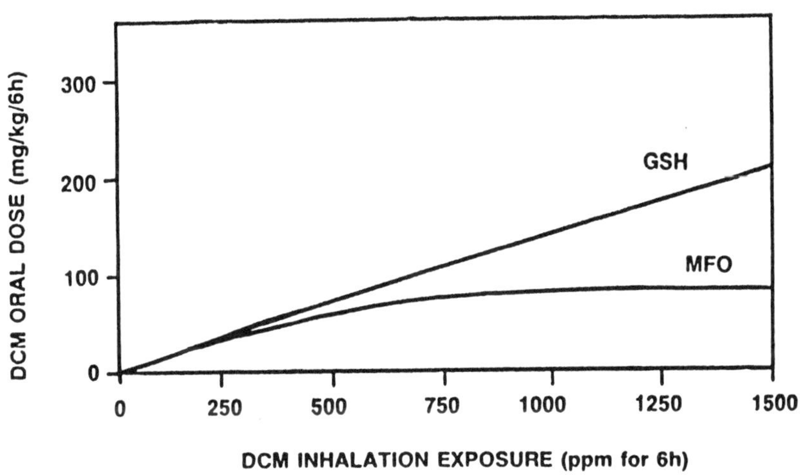

FIGURE 5 Correlations based on equivalent amounts of metabolized dichloromethane (DCM) by MFO and GSH pathways between oral doses (infusions for 6 h) and 6-h inhalation exposures in rats.

lations were obtained as a numerical output from the simulations of inhalation exposures.

Previous studies have shown that the retention of DCM during steady-state inhalation is less than 100% (Fiserova-Bergerova, 1983). In these situations, calculations that are based upon complete absorption of an inhaled dose overestimate the amount of material that enters the systemic portion of the body. Similarly, a sizable hepato-pulmonary first-pass effect during oral absorption of DCM reduces the systemic bioavailability of the compound, even though the retention of an oral dose by the gastrointestinal compartment is usually considered to be 100%. Therefore, quantitative relationships that are based upon 100% retention of the inhaled dose and/or complete absorption of an applied oral dose are misleading if internal challenges to the target tissues or to the metabolic systems are not used as the basis for the correlations.

By their very nature, physiologically based models are designed to describe the physiological phenomena that control the disposition of absorbed chemicals. With respect to DCM, this means that the first-pass effects and incomplete pulmonary retention of an inhaled dose are accounted for in the mass-balance equations that comprise the model. Because of this, the model establishes a theoretical basis for quantifying internal doses of DCM that should be used in the development of route-to-route pharmacokinetic comparisons.

CONCLUSIONS

1. We have established that a physiologically based pharmacokinetic model can be used to compare organ-specific doses between inhalation and oral administrations of dichloromethane.
2. We demonstrated that internal exposures, as measured by the delivered doses to target tissues and by metabolic challenges, are different for a given administered dose.
3. We determined that the retention of DCM during steady-state inhalation was less than 15% over the concentration range tested.
4. The dose of DCM delivered to target organs is overestimated if total absorption of an inhaled dose is assumed and/or the hepato-pulmonary first-pass effect is ignored.
5. The model provides an effective tool for exposure assessment by quantifying the internal doses of DCM that are the appropriate measures to use in route-to-route pharmacokinetic comparisons.

REFERENCES

Angelo, M. J., and A. B. Pritchard. 1984. Simulations of methylene chloride pharmacokinetics using a physiologically based model. Reg. Toxicol. Pharmacol. 4:329–339.

Angelo, M. J., A. B. Pritchard, D. R. Hawkins, E. R. Waller, and A. Roberts. 1986. The pharmacokinetics of dichloromethane. II. Disposition in Fischer 344 rats following intravenous and oral administration. Food Chem. Toxicol. 24:975–980.

Fiserova-Bergerova, V. 1983. Pp. 101–132 in Modeling of Inhalation Exposure to Vapors: Uptake, Distribution and Elimination, Vol. I. Boca Raton, Fla.: CRC Press.

Gargas, M. L., H. J. Clewell III, and M. E. Andersen. 1986. Metabolism of inhaled dihalomethanes *in vivo*: Differentiation of kinetic constants for two independent pathways. Toxicol. Appl. Pharmacol. 82:211–223.

Gerlowski, L. E., and R. K. Jain. 1983. Physiologically based pharmacokinetic modeling: Principles and applications. J. Pharm. Sci. 72:1103–1127.

McKenna, M. J., J. A. Zempel, and W. H. Braun. 1982. The pharmacokinetics of inhaled methylene chloride in rats. Toxicol. Appl. Pharmacol. 65:1–10.

Reitz, R. H., F. A. Smith, M. L. Gargas, H. J. Clewell, and M. E. Andersen. 1986. Physiological modeling and risk assessment: Example, methylene chloride. Toxicologist 6:7.

APPENDIX

The following are the mass-balance equations for each compartment in the physiological pharmacokinetic model for dichloromethane.

Liver

$$V_L \frac{dC_L}{dt} = Q_L\left(C_{B,art} - \frac{R_B}{R_L}C_L\right) + Q_G\left(\frac{R_B}{R_G}C_G - C_{B,art}\right) - V_L[(1-A_1)r1_L + r2_L], \text{ where} \quad (1)$$

$$r1_L = \frac{V_{max}\dfrac{C_L}{R_L}}{K_m + \dfrac{C_L}{R_L}}, \text{ and}$$

$$r2_L = k_2\frac{C_L}{R_L}$$

Gut

$$V_G\frac{dC_G}{dt} = Q_G\left(C_{B,art} - \frac{R_B}{R_G}C_G\right) + D, \text{ where} \quad (2)$$

$$D = \frac{k_aM}{R_{Gl}} \cdot \exp\left(-\frac{k_a}{R_{Gl}}t\right)$$

Lung

$$V_{Lg}\frac{dC_{Lg}}{dt} = Q_{Lg}\left(C_{B,ven} - \frac{R_B}{R_{Lg}}C_{Lg}\right)$$

$$+ PA\left(C_a - \frac{R_B}{R_{Lg}\lambda}C_{Lg}\right)$$

$$- V_{Lg}(A_1 r1_{Lg} + A_2 r2_{Lg}), \text{ where} \qquad (3)$$

$$r1_{Lg} = \frac{V_{max}\frac{C_{Lg}}{R_{Lg}}}{K_m + \frac{C_{Lg}}{R_{Lg}}}, \text{ and}$$

$$r2_{Lg} = k_2\frac{C_{Lg}}{R_{Lg}}$$

Alveolar Space

$$V_a\frac{dC_a}{dt} = PA\left(\frac{R_B}{R_{Lg}\lambda}C_{Lg} - C_a\right) + Q_a(C_{inh} - C_a) \qquad (4)$$

Carcass

$$V_C\frac{dC_C}{dt} = Q_C\left(C_{B,art} - \frac{R_B}{R_C}C_C\right) \qquad (5)$$

Kidney

$$V_K\frac{dC_K}{dt} = Q_K\left(C_{B,art} - \frac{R_B}{R_K}C_K\right) \qquad (6)$$

Arterial Blood

$$V_{B,art}\frac{dC_{B,art}}{dt} = Q_{Lg}\left(\frac{R_B}{R_{Lg}}C_{Lg} - C_{B,art}\right) \qquad (7)$$

Venous Blood

$$V_{B,ven}\frac{dC_{B,ven}}{dt} = Q_L\left(\frac{R_B}{R_L}C_L\right) + Q_K\left(\frac{R_B}{R_K}C_K\right) + Q_C\left(\frac{R_B}{R_C}C_C\right) - Q_{Lg}C_{B,ven} \qquad (8)$$

^{14}CO Balance

$$V_B \frac{[^{14}CO]_t}{dt} = f \cdot [(1 - A_1)V_L r1_L + A_1 V_{Lg} r1_{Lg}] - CL_{CO}[^{14}CO]_f, \text{ where} \tag{9}$$

$$[^{14}CO]_t = [^{14}CO]_f + [Hb^{14}CO], \text{ and}$$

$$[Hb^{14}CO] = \frac{k_\alpha \cdot [^{14}CO]_f}{k_\beta + [^{14}CO]_f}$$

$^{14}CO_2$ Balance

$$\text{Cumulative } ^{14}CO_2 \text{ expired} = \int_0^\infty \{(1 - f) \cdot [(1 - A_1) \cdot V_L r1_L + A_1 V_{Lg} r1_{Lg}] + V_L r2_L + V_{Lg} r2_{Lg}\} dt \tag{10}$$

DCM Balance

$$\text{Cumulative } ^{14}\text{C-DCM expired during oral exposure} = \int_0^\infty Q_a C_a dt \tag{11a}$$

$$\text{Retention of DCM during steady-state inhalation} = \frac{2}{3}\left(\frac{C_{\text{inh}} - C_a}{C_{inh}}\right) \tag{11b}$$

The following nomenclature was used in the model equations.

Compartment symbols

- C_1 Concentration in compartment 1 (nmol/ml)
- Q_1 Blood flow rate compartment 1 (ml/min)
- R_1 Tissue/plasma distribution ratio
- V_1 Volume of compartment 1 (ml)

Subscripts for compartments: B, blood; *art*, arterial; *ven*, venous; L, liver; Lg, lung; G, gut; Gl, gut lumen; C, carcass; K, kidney; a, alveolar space.

Other symbols

- A_1 Distribution of MFO activity between lung and liver
- A_2 Distribution of GSH activity between lung and liver
- C_{inh} Inhaled DCM concentration (nmol/ml of air)
- CL Pulmonary clearance of ^{14}CO (ml/min)
- $[^{14}CO]_t$ Total concentration of ^{14}CO in blood (nmol/ml)

$[^{14}CO]_f$	Free concentration of ^{14}CO in blood (nmol/ml)
f	Stochiometric fraction of CO produced from 1 mol of DCM
K_m	Michaelis constant for reaction $r1$ (nmol/ml)
k_a	Gastrointestinal (GI) absorption rate constant (min^{-1})
k_2	Metabolism rate constant for reaction $r2$ (min^{-1})
k_α, k_β	Binding constants for $Hb^{14}CO$ (nmol/ml)
M	Mass of DCM in oral dose (nmol)
PA	Alveolar permeability $\times$ area product (ml/min)
Q_a	Alveolar ventilation (ml air/min)
R_λ	Blood/air partition coefficient
$r1$	Rate of DCM metabolism via the MFO pathway
$r2$	Rate of DCM metabolism via the GSH pathway
V_{max}	Maximum reaction velocity for $r1$ (nmol/ml/min)

Sensitivity Analysis in Pharmacokinetic Modeling

Murray S. Cohn

INTRODUCTION

On July 15, 1985, R. H. Reitz and M. E. Andersen released a draft paper entitled "Physiologically-Based Pharmacokinetics and the Risk Assessment Process for Methylene Chloride," which was provided to the federal health and safety regulatory agencies. The model used various assumptions and experimental data to predict internal concentrations of "toxiphors," a term used by these authors to describe presumed toxic chemical species resulting from biodegradation of methylene chloride by certain biochemical pathways. The results of the modeling were used to demonstrate that risk assessments done without consideration of the pharmacokinetic approach overestimate predicted risk to humans. The pharmacokinetic model was applied to estimate internal concentrations of toxiphors in the lungs and livers of mice, rats, hamsters, and humans, because these were the two sites of response in the National Toxicology Program's (NTP) inhalation bioassay of methylene chloride in mice (NTP, 1986). Of the many input parameters to the model, some of the most important determinations critical to its output are the values of the kinetic

The material found within this paper contains the view of the author, who is an employee of the U.S. Consumer Product Safety Commission, and does not necessarily reflect official opinions or policies of the U.S. Consumer Product Safety Commission.

constants (K_f, V_{max}, K_m) for the two proposed pathways. According to Reitz and Andersen, K_f is the constant relating to a toxic nonsaturable pathway, that involving glutathione *S*-transferase (GST). V_{max} and K_m, on the other hand, are standard Michaelis-Menten kinetic constants that apply to a nontoxic, saturable oxidative pathway involving monooxygenase activity.

In the original July 15, 1985, paper, these kinetic constants were based on the specific activities of GST and monooxygenase measured in samples of lung and liver tissue from all four species by Lorenz et al. (1984) using 1-chloro-2,4-dinitrobenzene (for GST) and 7-ethoxycoumarin (for monooxygenase) as substrates. Because these critical constants were not based upon experimentation using methylene chloride itself as a substrate, the assumption that the data from the study by Lorentz et al. (1984) served as an appropriate substitute led to one source of uncertainty in the use of the Reitz-Andersen model.

The use of the data of Lorenz et al. (1984), however, was restricted by Reitz and Andersen in a prepublication version of the same paper dated January 19, 1986 (Andersen et al., in press), for the purpose of apportioning the amount of total monooxygenase and GST between lung and liver in the four species. In this new version, a curve-fitting exercise was used instead to obtain the whole-body values of the three kinetic constants in animals. With all other physiological and biological parameters kept constant, the values of K_m, V_{max}, and K_f were varied in an intricate computer optimization procedure, and the pharmacokinetic model was used to get the best approximations of actual experimental chamber data in which disappearance of methylene chloride over time was monitored. The chamber data were obtained for rats, mice, and hamsters by using five initial concentrations and recording the decrease in the chamber concentration of methylene chloride over a period of up to 6 h. The values of the three kinetic constants giving the best approximations to these five experimental concentration versus time curves for the three species were then used in other versions of the pharmacokinetic model to predict toxiphor levels.

For humans, values of K_m (0.58 mg/liter) and V_{max} (119 mg/h) for the oxidative pathway were estimated from the human experimental data of Nolan and McKenna (in press) in which methylene chloride concentrations in expired air were measured following exposure by inhalation. For the critical GST pathway, however, no human data are available upon which an estimate of K_f can be based. Reitz and Andersen noted, however, that allometric scaling based on the concept of clearance served to adequately relate the mouse, hamster, and rat data, using a factor of body weight to the 0.7 power. Using this scaling procedure, they obtained a K_f value of 0.53 h^{-1} for humans, which is similar to the value of K_f for humans used by them in their original 1985 paper. They further noted that the human values of K_m and V_{max} given above, when compared with the optimized animal values, also scale allometrically.

This exercise examines the sensitivity of the Reitz-Andersen optimization procedure to see if K_f, K_m, and V_{max} can be varied to produce an alternative, yet reasonable, fit in the experimental concentration versus time curves. As the mouse was the species that responded in the NTP's inhalation bioassay on methylene chloride at the sites modeled by Reitz and Andersen (lung and liver), the data on mice were chosen for this analysis.

RESULTS

The optimized values of K_m, V_{max}, and K_f for the mice in the chamber experiments were 0.369 mg/liter, 0.8884 mg/h, and 4.3238 h^{-1}, respectively. With all other parameters, as determined by Reitz and Andersen in their original experiment and Andersen et al. (in press), held constant, it was decided to set the value of K_f equal to 1.00 h^{-1} and vary the other two to obtain the best approximation of the experimental data (the value of 1.00 h^{-1} was chosen to replace 4.3238 h^{-1} because use of the pharmacokinetic model as described by Reitz and Andersen makes an approximate fourfold difference in effective human dose). The Reitz-Andersen model was programmed for use on a microcomputer using the Livermore Solver for Ordinary Differential Equations (LSODE; Hindmarsh, 1980). The microcomputer version used was verified by using modeled output data furnished by Reitz and Andersen.

Using trial and error and the value of 1.00 h^{-1} for K_f, values of 0.7 mg/liter and 1.15 mg/h were found for K_m and V_{max}, respectively, which led to the curves labeled "Variant" in Figures 1 to 5. In these figures "RA" refers to the Reitz-Andersen optimized fit, and "Chamber" refers to the actual laboratory data. Each figure displays the time course of disappearance of methylene chloride (in parts per million) versus time (in hours); the five graphs represent five different initial concentrations (490, 960, 2,000, 3,200, and 10,000 ppm in Figures 1 to 5, respectively). The alternative (variant) choices for the three kinetic constants leads to curves which approximate the actual chamber data about as well as the Reitz-Andersen optimized choices, with the exception of the 3,200-ppm initial concentration chamber experiment (Figure 4). This is not disturbing, however, because there was a transient increase in methylene chloride concentration between 2.5 and 4.5 h, and before this transient increase occurred, the chamber data followed the variant curve. A numerical description of the correlation of the variant fit to the chamber data, as compared with the RA fit, was not determined because of the transient increase in the 3,200-ppm data (Figure 4).

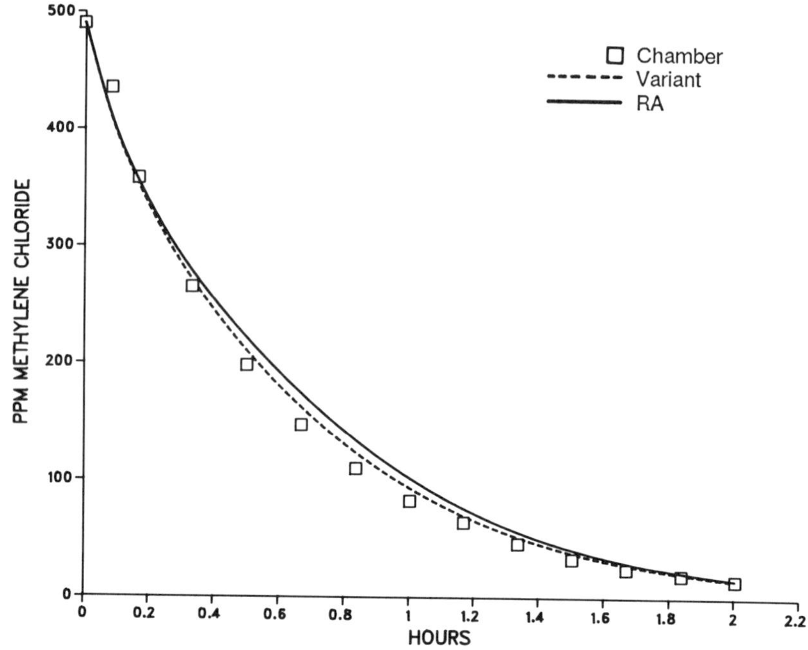

FIGURE 1 Modeled versus experimental concentrations (490 ppm at time zero).

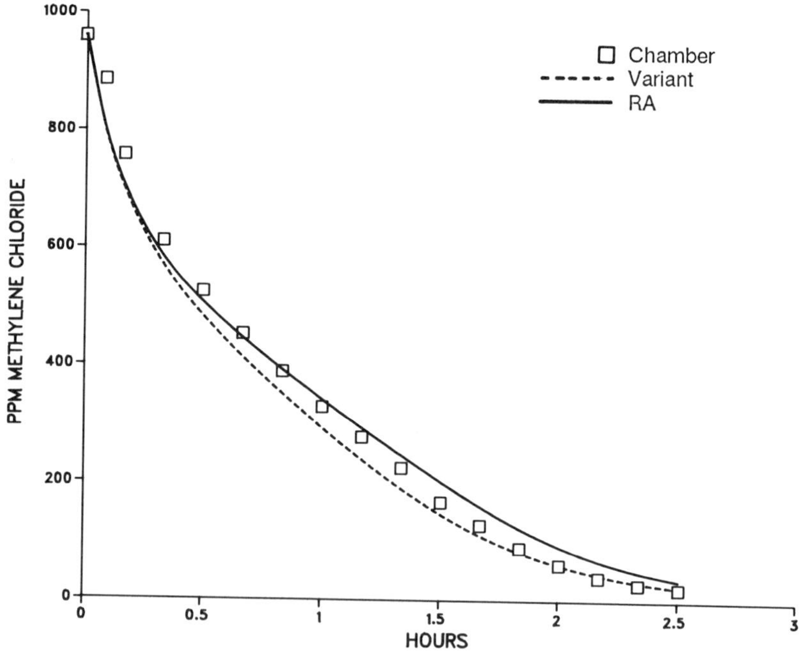

FIGURE 2 Modeled versus experimental concentrations (960 ppm at time zero).

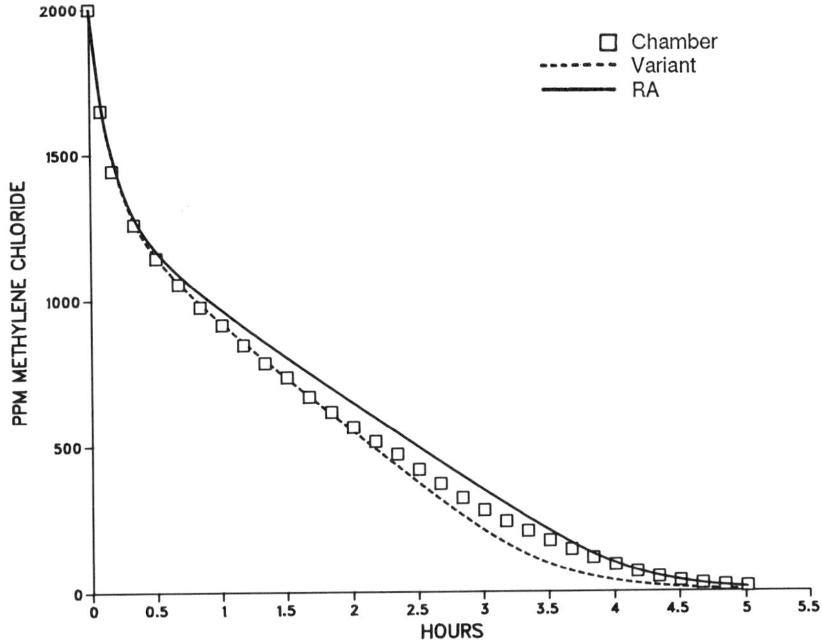

FIGURE 3 Modeled versus experimental concentrations (2,000 ppm at time zero).

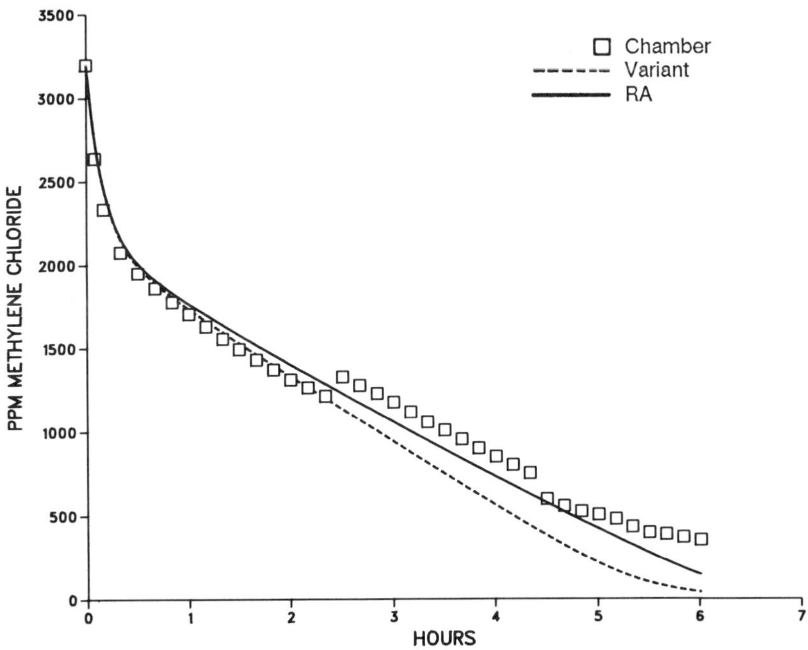

FIGURE 4 Modeled versus experimental concentrations (3,200 ppm at time zero).

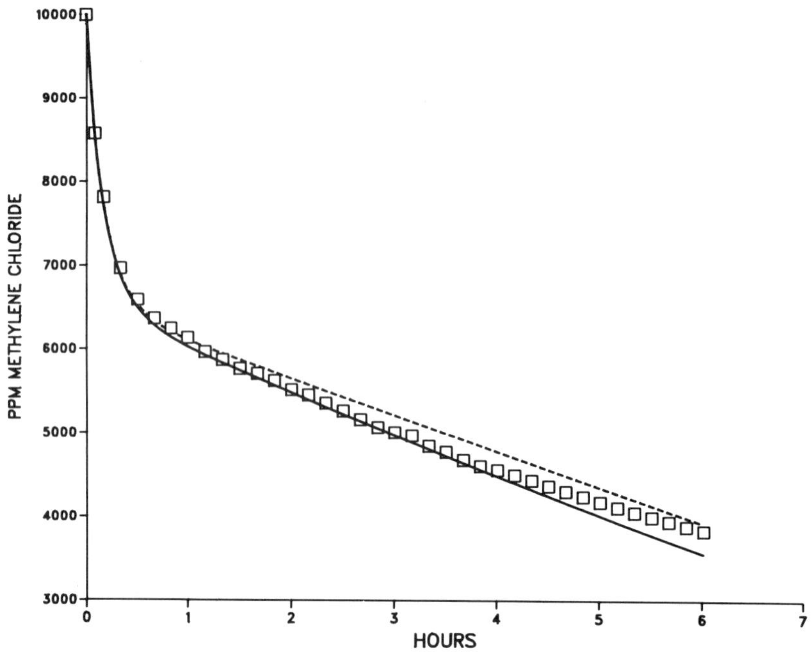

FIGURE 5 Modeled versus experimental concentrations (10,000 ppm at time zero).

The scaling approach of Reitz and Andersen, based on the computer optimization fit to the experimental animal chamber data, is shown in Table 1.

By the Reitz-Andersen approach, if K_f scaled allometrically among the three animal species, the intrinsic clearance values should be similar. This is the reasoning for setting human intrinsic clearance equal to 60 and back-calculating. If 1.00 is used instead of 4.32 for the K_f for mice, however, the clearance for mice becomes 1.08, and the intrinsic clearance becomes 13.5 (Table 1). When compared with 33.5 for hamsters and 57.1 for rats, the mouse no longer scales allometrically. Therefore, the value of a key input parameter, the human K_f, that is needed for application of the pharmacokinetic model to humans is yet to be determined.

CONCLUSIONS

It is concluded that although the Reitz-Andersen optimization procedure may lead to kinetic constants that give an optimal fit, by use of the pharmacokinetic model, to the experimental chamber data, there are alternative combinations (one of which was determined in this analysis) of

TABLE 1 Reitz-Andersen Scaling Approach to Obtain K_f for Humans

Species	Liver Volume (ml)	Body Weight (kg)	K_f (h^{-1})	Clearance (ml/h)[a]	Intrinsic Clearance (ml/h)[a]
Mouse	1.08	0.027	4.32	4.67	58.5
Hamster	5.60	0.140	1.51	8.46	33.5
Rat	9.32	0.233	2.21	20.6	57.1
Human	2,215	70.0	0.53[b]	1,174	60.0[b]
Mouse variant	1.08	0.027	1.00	1.08	13.5

[a]Clearance is defined by Reitz and Andersen as K_f times liver volume, and intrinsic clearance is calculated as clearance divided by body weight to the 0.7 power.
[b]Human K_f is calculated by setting intrinsic clearance equal to 60 and back-calculating.

the three kinetic constants that also can reasonably fit these data. This may be due, perhaps, to the fact that the optimum fit may have correlation coefficients that only trivially exceed those from another fit that has substantial differences in the underlying constants. In other words, there are multiple adequate solutions to the problem (redundancy). On this basis, the scaling approach suggested by Reitz and Andersen for the determination of a human K_f value, a most critical value according to these authors with regard to the prediction of methylene chloride carcinogenic risk to humans, is subject to uncertainty. Thus, the scaling approach should still be considered preliminary, and more research and validation is required before it can be accepted as a reasonable basis for pharmacokinetic modeling in the case of methylene chloride.

Most pharmacokinetic modeling is dependent upon the values of many input parameters. Such modeling can also rely upon the process of solving for those input parameters that are unable to be determined otherwise. The purpose of the exercise provided here is to encourage those who employ pharmacokinetic models as a tool to improve quantitative risk assessment to examine the sensitivity of such models. Small changes in input data or in the goodness of fit to a set of experimental data can greatly alter the effect that inclusion of pharmacokinetic modeling can have on estimates of human risk.

REFERENCES

Andersen, M. E., H. J. Clewell III, M. L. Gargas, F. A. Smith, and R. H. Reitz. In press. Physiologically-based pharmacokinetics and the risk assessment process for methylene chloride. Toxicol. Appl. Pharmacol.

Hindmarsh, A. C. 1980. LSODE and LSODI, two new initial value ordinary differential equations solvers. ACM-Signum Newsletter 15(4):10–11.

Lorenz, J., H. R. Glatt, R. Fleischmann, R. Ferlinz, and F. Oesch. 1984. Drug metabolism in man and its relationship to that in three rodent species: Monooxygenase, epoxide hydrolase, and glutathione S-transferase activities in subcellular fractions of lung and liver. Biochem. Med. 32:43–56.

National Toxicology Program (NTP). 1986. Toxicology and Carcinogenesis Studies of Dichloromethane (Methylene Chloride) (CAS No. 75-09-2) in F344/N Rats and $B6C3F_1$ Mice (Inhalation Studies). Technical Report Series No. 306, NIH Publication No. 86-2562. Washington, D.C.: U.S. Department of Health and Human Services.

Nolan, R. J., and M. J. McKenna. In press. Pharmacokinetics of methylene chloride in human volunteers.

Mutation Accumulation: A Biologically Based Mathematical Model of Chronic Cytotoxicant Exposure

Rory B. Conolly, Richard H. Reitz, and *Melvin E. Andersen*

INTRODUCTION

This report describes a computer-based biological model suggesting why nongenotoxic chemicals such as chloroform test positively when carcinogenicity bioassays use doses that may cause transient cytotoxicity (Heywood et al., 1979; Jorgensen et al., 1985; National Cancer Institute, 1976; Palmer et al., 1979; Roe et al., 1979). The model is roughly isomorphic with actual mammalian physiological and biochemical systems. It describes the mathematical relationships among:

1. *In vivo* cytotoxicant exposure and tissue dose of parent compound and metabolites.
2. Tissue dose and cell death.
3. Cell death and regenerative hyperplasia.
4. Regenerative hyperplasia and normally scheduled replication and the accumulation of genetic mutations.

The model described here is a combination of three discrete models describing pharmacokinetics, cytotoxicity, and carcinogenesis. These discrete models are (1) the physiologically based pharmacokinetic (PB-PK) model of Ramsey and Andersen (1984), (2) a cytotoxicity model, and (3) the two-stage carcinogenicity model of Moolgavkar and Knudson (1981).

The PB-PK model was used as a pharmacokinetic driver for the cytotoxicity model to provide accurate linkage between *in vivo* toxicant exposure and tissue doses of parent compounds and reactive metabolites.

This PB-PK model has been validated for several volatile halogenated compounds and solvents (Clewell and Andersen, 1985; Gargas et al., 1986; Ramsey and Andersen, 1984).

The cytotoxicity component of the model contains equations describing depletion of a generic macromolecule (MM) by a reactive metabolite of the parent compound. The MM is subsequently resynthesized by a feedback-controlled enzymatic system. Depletion of MM leads to cell death, which is followed by regenerative hyperplasia. The description of MM depletion was based on a model of hepatic glutathione depletion and resynthesis that has been previously described (Andersen et al., 1986). A model developed by Reitz (1986) of cytotoxicity mediated by parent chemical bioactivation and macromolecule depletion also contributed to this work.

In Moolgavkar and Knudson's (1981) carcinogenicity model, cellular replication incurs a risk of error leading to mutations in daughter cells. We have used cytotoxicity, as described above, to force regenerative hyperplasia and thereby increase the rate of mutation accumulation. Simulated data are presented showing how cells containing one and two mutations accumulate as a function of time and toxicant exposure.

METHODS

Computer Hardware and Software

The mathematical model was written in ACSL (Advanced Continuous Simulation Language; Mitchell & Gauthier Associates, Inc., Concord, Mass.) and run on an IBM-AT (IBM Co., Boca Raton, Fla.).

Modeling of Cytotoxicity

The liver in Ramsey and Andersen's (1984) PB-PK model was used as the target organ for cytotoxic effects (Figure 1). This choice is not important per se for the results and conclusions reported here and was used only as a matter of convenience. Model parameters were scaled for a 250-g rat, and the liver was modeled as a population of 10^8 cells with basal death and birth rates of 10^{-3}/h. Liver cells contained a generic MM, which is essential for cell viability. (In a model validated for a specific cytotoxicant whose mechanism of action involved macromolecular binding, the generic MM would be replaced with a description of the actual macromolecule(s) involved. Validation experiments using chloroform as the model cytotoxicant are planned.) The rate of hepatic metabolism of parent compound was dependent on the rate of toxicant delivery to the liver and the Michaelis constants for metabolism. Depletion of MM was

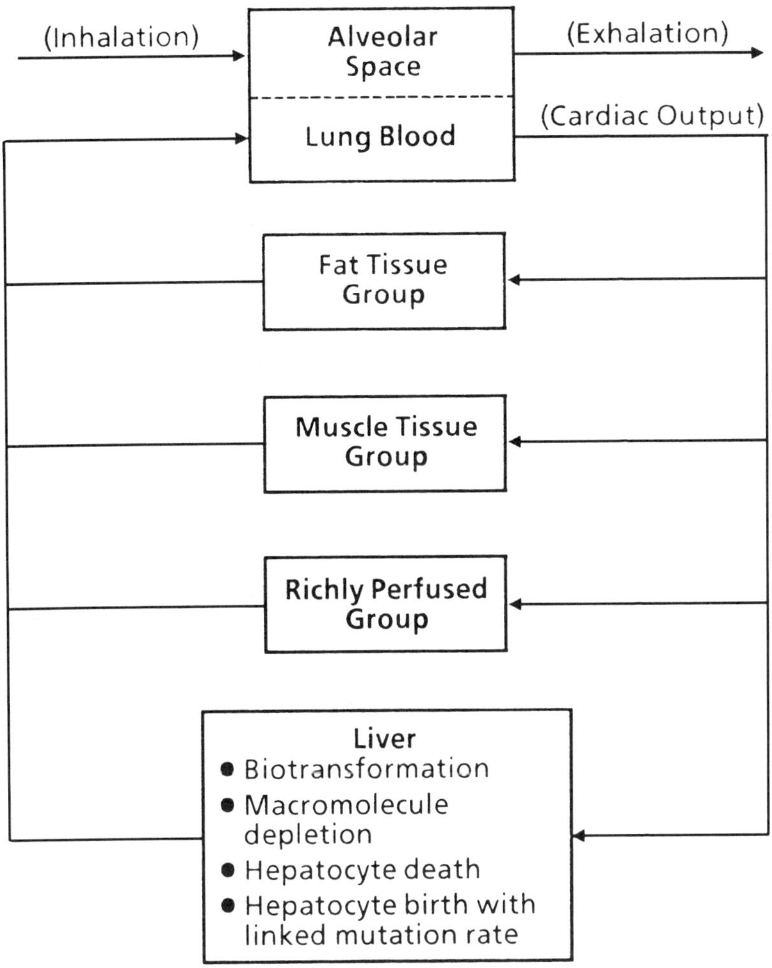

FIGURE 1 Schematic of the biologically based pharmacokinetic model of Ramsey and Andersen (1984) as adapted to describe hepatic cytotoxicity and mutation accumulation. Each compartment is defined by biologically realistic parameters.

modeled by describing a second-order reaction of the parent compound metabolite with MM. Cell death was lined to MM depletion by a normal curve (Hastings, 1955), in which >99.999% of the area under the curve lay between the maximum and minimum possible levels of an MM. The cumulative percentage of the area under the curve, corresponding to the degree of MM depletion, defined the number of hepatocytes committed to die. Actual death took place 1 h later. The liver responded to unscheduled cell death, i.e., cell death advanced by the toxicant, by increasing the rate of cell division, until the steady-state level of 10^8 cells was regained. This compensatory increase in birth rate occurred 8 h after the unscheduled death.

Modeling of mutation accumulation in hepatocytes is described below. Hepatocytes with no mutations (N1 cells) and those with one mutation (N2 cells) were modeled as having the same basal death and birth rates, as being equally susceptible to cytotoxicant and as responding to unscheduled cell death, i.e., cytotoxicity, in the same manner. Cells with two mutations (N3 cells) were considered to be insensitive to cytotoxicant.

The cytotoxicity model did not describe the maximum amount of cell death that could be tolerated without death of the animal. This could easily be done but was not considered necessary for this exercise.

Figures 1, 2, and 3 illustrate various aspects of the cytotoxicity model. Figure 4 is useful for studying model function.

"Liver"

- 10^8 Cells
- Normal Rate Cell Death 10^{-3} HR^{-1}
- Normal Rate Cell Birth 10^{-3} HR^{-1}
- Chance of Mutation Per Cell Birth 10^{-7}

$$\text{Toxicant} \xrightarrow[(K_m,\ V_{MAX})]{\text{(Biotransformation)}} \text{Metabolite}$$

As ↑ [Metabolite], ↓ [Macromolecule]

As ↓ [Macromolecule], ↑ Probability of Cell Death*

As ↑ Cell Death, ↓ Population of Normal (N1) Cells

As ↓ N1, ↑ Birth Rate Until Back to 10^8 Cells.

* Sensitivity of cells to macromolecule depletion is normally distributed about a set level of macromolecule.

FIGURE 2 Detail for the cytotoxicity component of the model. The choice of liver was one of convenience. This description of cytotoxicity could be applied equally well to other tissues.

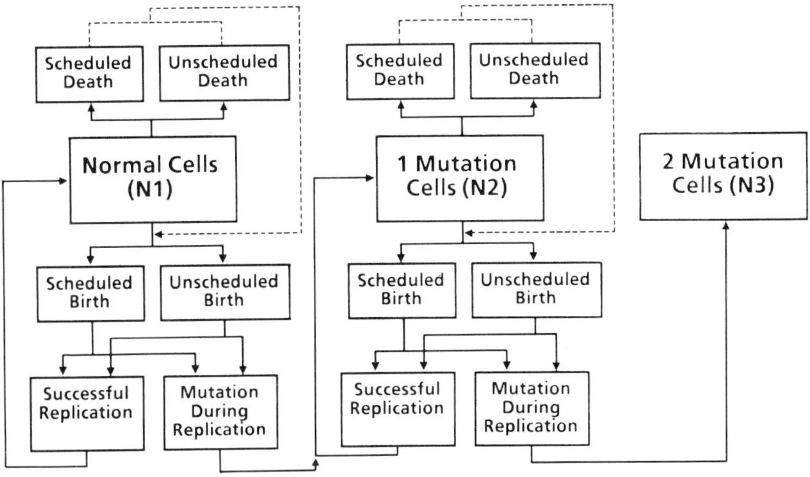

FIGURE 3 Adaptation of the Moolgavkar and Knudson (1981) two-stage model for carcinogenicity to the pharmacokinetic/cytotoxicity model. N1 and N2 cells have the same basal birth and death rates. N3 cells are immortal. Unscheduled events are those due to cytotoxicity.

Modeling the Relationship of Cell Birth to Mutation Accumulation

A premise of the model is that regenerative hyperplasia consequent to cytotoxicity increases the rate at which mutations accumulate in the target organ and that some of these mutations influence tumor development. This premise is based on the assumption that a small fraction of cellular replications "go wrong," resulting in mutant progeny. Normal hepatocytes (N1 cells) were modeled as having a 10^{-7} chance of suffering a mutation during replication. Cells containing one mutation (N2 cells) were modeled as replicating with the same kinetics as N1 cells and also as having a 10^{-7} chance of suffering a mutation during replication. Cells with two mutations (N3 cells) were assumed to be tumorigenic without further mutation. These relationships are illustrated in Figure 3.

Generic Nature of Cytotoxicity Model

We know of no body of data that is suitable for mathematical modeling of the relationships among reactive metabolite production, depletion of critical cellular macromolecules, rates of cell death and regeneration, and the accumulation of mutations. We therefore defined a generic MM and estimated reasonable quantitative relationships among the amount of parent compound metabolized, depletion of MM, cell death and birth, and mutation accumulation. Because the definition of MM is not linked to a

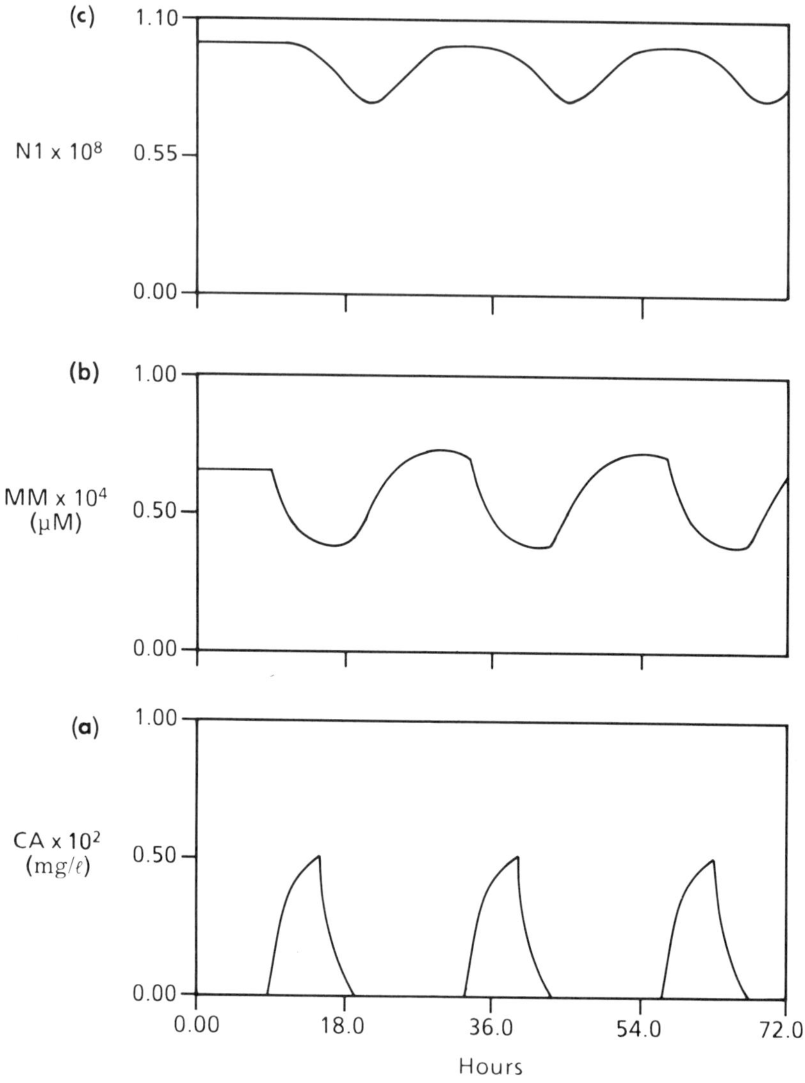

FIGURE 4 Temporal relationships of *in vivo* toxicant exposure, macromolecule depletion, and number of normal (N1) cells. In (a) daily 6-h exposure to parent toxicant is tracked by the spikes in C_A. In (b) depletion of MM by a metabolite of the parent and resynthesis of MM as CA declines are shown. In (c) N1 death as MM is depleted and consequent regenerative hyperplasia are illustrated.

biochemical mechanism of toxicity for a particular chemical, the model's descriptions of cytotoxicity and mutation accumulation are also generic. For these reasons, even though the PB-PK model is well-validated for a number of compounds, we present the simulations of cytotoxicity and mutation accumulation in terms of exposure to a model cytotoxicant rather than to any particular chemical. (For the curious, the model cytotoxicant used has the pharmacokinetic behavior of 1,2-dichloroethane.)

RESULTS

Figure 4 illustrates the qualitative relationships among cytotoxicant exposure, depletion of MM, cell death, and regenerative hyperplasia. Simulation of repeated, 6-h/day inhalation exposure to parent compound results in corresponding spikes in its arterial concentration (C_A). As C_A increases, MM starts to fall. This fall in MM is due to its destruction by the cytotoxic metabolite of the parent compound. As the concentration of cytotoxic metabolite falls, MM is resynthesized and its concentration increases. The model simulates an overshoot or rebound effect in which the level of MM rises above its preexposure concentration. This type of rebound effect is seen after toxicant-induced depletion of hepatic glutathione (GSH) (Wong and Klaassen, 1981), which we have modeled for related projects. Although not shown in Figure 4, the MM concentration would return to the preexposure basal level between 24 and 48 h after a single 6-h exposure. A decreasing concentration of MM is followed by a corresponding decline in N1, the number of normal hepatocytes. N1 returns to its basal level (approximately 10^8 cells) after MM is resynthesized. In all, Figure 4 illustrates several cycles of toxicant-driven cell death and regenerative hyperplasia.

Figure 5 simulates the accumulation of cells with one mutation (N2 cells) over time in toxicant-exposed and control groups. Note the linear nature of N2 accumulation with time and the demonstration that toxicant exposure can greatly increase the rate of N2 accumulation.

Simulated rates of accumulation of cells with two mutations (N3 cells) are shown in Figure 6. Notice that Figure 6 uses log-log scaling and that the lines depicting N3 accumulation with time are described by a power function.

In Figure 7, in which arithmetic scaling is used, the numbers of N1, N2, and N3 cells occurring during the simulation of 1 year of 6-h/day toxicant exposure are shown.

The model calculates a cumulative expectation describing the prevalence of N2 and N3 cells. Figure 6 indicates that about 10^{-4} of one N3 cell was produced during 90 days of simulated exposure, causing 20% cell

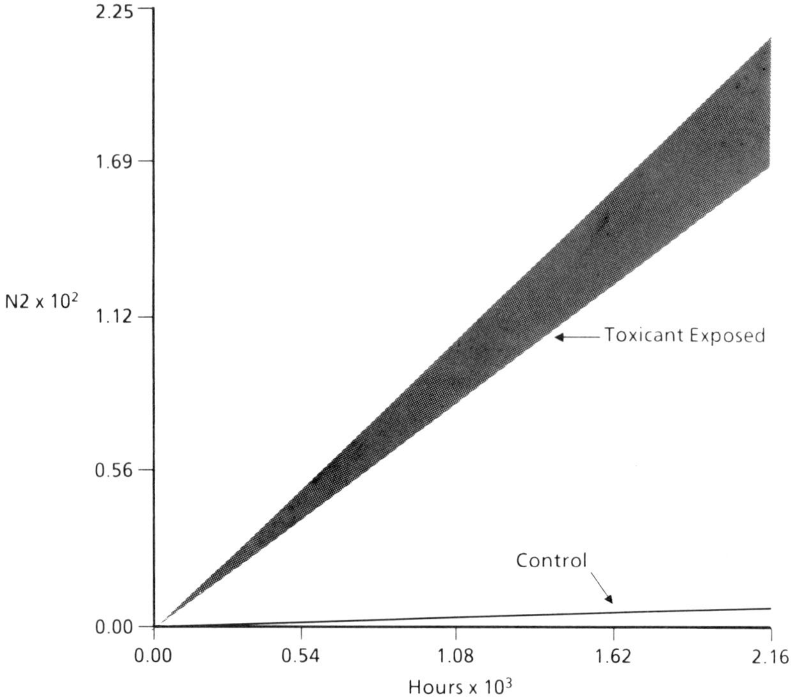

FIGURE 5 Simulated 90-day accumulation of cells with one mutation (N2) in control and toxicant-exposed groups. The daily fluctuation in the number of N2 cells in the toxicant-exposed group reflects the cycle of cell death and regenerative hyperplasia caused by each 6-h exposure to toxicant.

death/day. In this instance the 10^{-4} actually represents the average prevalence for the population of individuals at risk at that point in time.

DISCUSSION

The model described here incorporates some straightforward assumptions about relationships among the tissue dose of a toxicant, depletion of a critical cellular macromolecule, the likelihood of cell death, and the probability of mutations occurring during regenerative hyperplasia. Depletion of hepatic GSH is known to be linked to increased probability of cell death (Docks and Krishna, 1976; Mitchell et al., 1973; Wells et al., 1980). The use of a cumulative probability function to describe this relationship means that the marginal increase in the probability of cell death as MM is depleted is not monotonically increasing, as might be expected. The correct mathematical description of this relationship is not known at

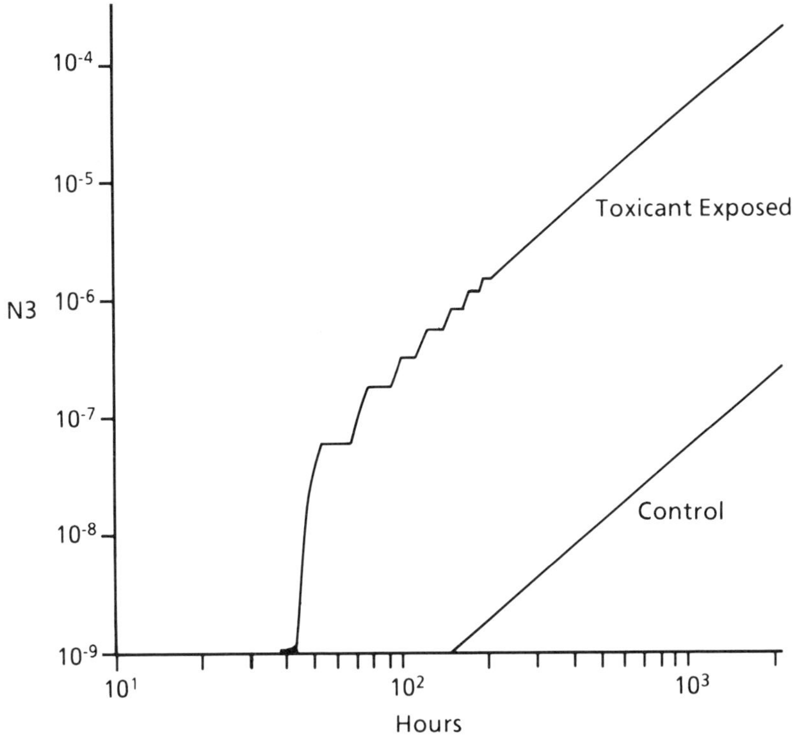

FIGURE 6 Simulated 90-day accumulation of cells with two mutations (N3) in control and toxicant-exposed groups. The toxicant-exposed group suffered about 20% cell death/day, which was completely replaced within 24 h by regenerative hyperplasia. This regenerative hyperplasia leads to the difference between the control and toxicant-exposed groups. In both the control and exposed groups, N3 accumulation is proportional to time^2.

this time. However, over the range of MM depletion used in the simulations reported here (0 to 20%), the cumulative probability did increase monotonically.

A number of workers have studied mutation rates in replicating cell systems, and it is generally accepted that mutation does occur as a consequence of mistakes during cellular replication (Crawford et al., 1983; Tsutsui et al., 1978, 1981).

The model suggests that repeated episodes of cytotoxicity increase the rate at which mutations accumulate in a target tissue. Moreover, cells containing two mutations and that would presumably be more likely to be tumorigenic (Moolgavkar and Knudson, 1981) accumulate with the square of time, whereas the accumulation of cells with one mutation is linear.

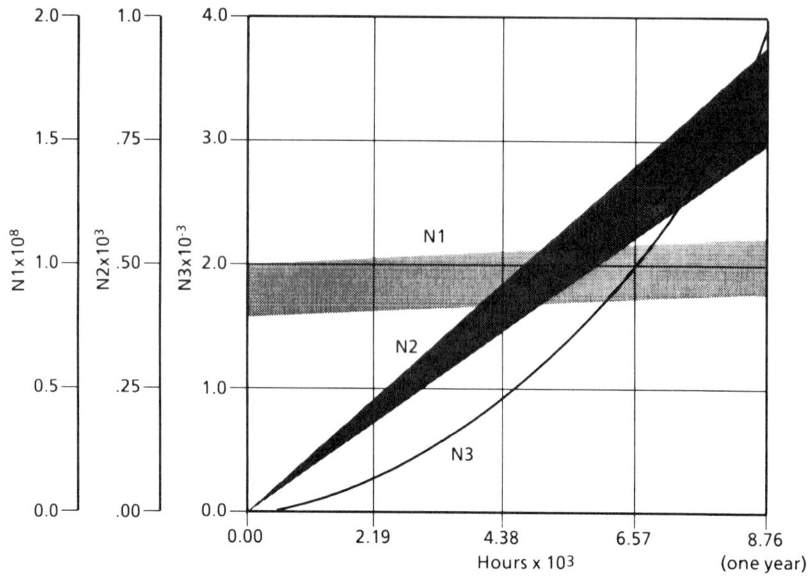

FIGURE 7 Simulation of 1 year of daily 6-h inhalation exposure to a model cytotoxicant. Daily fluctuations in N1 and N2 cells are due to cytotoxicity. Note that the rate of accumulation of cells with one mutation (N2) is linear with time, whereas accumulation of cells with two mutations (N3) depends on time^2.

A significant advantage of this model is that the tissue dose of cytotoxic metabolite is related to *in vivo* parent compound exposure by the well-validated PB-PK model (Clewell and Andersen, 1985; Gargas et al., 1986; Ramsey and Andersen, 1984). Modeling of cytotoxicity and other target organ consequences of *in vivo* exposure has little meaning if the tissue dose of the ultimate toxicant is not well-characterized.

Limitations of the Model—What Is Modeled and What Is Not

Several processes that would affect cytotoxicity and mutation accumulation have not been specifically described in this model. For example, the model was exercised repetitively in order to simulate chronic exposures of up to 1 year (Figures 5, 6, and 7). For these long-term simulations it may be necessary to model the aging process in the rat. Changes in the size of the fat compartment and in the capacity for parent compound metabolism with age would affect pharmacodynamic behavior. The possibility that N2 cells have growth advantages, i.e., altered death and birth rates and different sensitivity to MM depletion relative to N1 cells, may be important and could give rise to N2 birth rates different from those simulated here. Immune surveillance of phenotypically altered cells probably affects the N2 and N3 populations, and modeling of this parameter may be necessary for a comprehensive description. The model does not estimate the biological significance of the simulated mutations. Because not all mutations would be expected to predispose the cell to clonal growth, the accumulation of N3 cells is probably more rapid than the accumulation of tumorigenic cells. These factors may all be important for accurate simulation of rates of cytotoxicity and mutation accumulation. However, the relative shapes of the curves describing the change in N1, N2, and N3 populations with time (Figure 7) should not differ between a model specifically describing these factors and the one described here.

Model Validation

As noted above, the PB-PK model of Ramsey and Andersen (1984) which is used as a pharmacokinetic driver for the cytotoxicity model has been well-validated. We are in the process of obtaining actual data on the quantitative relationships among $CHCl_3$ biotransformation, cell death, cell regeneration, and the consequent accumulation of mutations. These data will be used to develop a $CHCl_3$ cytotoxicity model that will provide a strong test of the hypothesis that $CHCl_3$ cytotoxicity is responsible in whole or in part for the positive outcomes seen in $CHCl_3$ carcinogenicity bioassays (Heywood et al., 1979; Jorgensen et al., 1985; National Cancer Institute, 1976; Palmer et al., 1979; Roe et al., 1979). At the present stage of its development, the generic model (see "Methods") can be used to examine aspects of the quantitative relationships among depletion of critical cellular macromolecules, cytotoxicity, cell death and replication, and the accumulation of mutations.

It should be noted that the model does not describe the accumulation of genetic mutations caused by direct interactions of parent compound or metabolites with DNA. However, the model does provide a framework which could readily be modified to describe these factors.

In conclusion, this presentation illustrates the potential utility of using computer-based models of biological systems to study the relationships of cytotoxicity, cell replication, and the accumulation of genetic damage. A particularly interesting result generated by the model described here is the dependence of the birth rate of cells containing two mutations on the square of time.

ACKNOWLEDGMENTS

Our thanks to Maj. Harvey Clewell and Dr. Thomas B. Starr for their help.

REFERENCES

Andersen, M. E., H. J. Clewell III, M. L. Gargas, and R. B. Conolly. 1986. A physiological pharmacokinetic model for hepatic glutathione (GSH) depletion by inhaled halogenated hydrocarbons. The Toxicologist 6:148. (Abstract 598.)

Clewell, H. J., III, and M. E. Andersen. 1985. Risk assessment extrapolations and physiological modeling. Toxicol. Ind. Health 1:111–131.

Crawford, B. D., J. C. Barrett, and P. O. P. Ts'o. 1983. Neoplastic conversion of preneoplastic Syrian hamster cells: Rate estimation by fluctuation analysis. Mol. Cell. Biol. 3:931–945.

Docks, E. L., and G. Krishna. 1976. The role of glutathione in chloroform-induced hepatotoxicity. Exp. Mol. Pathol. 24:13–22.

Gargas, M. L., H. J. Clewell III, and M. E. Andersen. 1986. Metabolism of inhaled dichloromethanes in vivo: Differentiation of kinetic constants for two independent pathways. Toxicol. Appl. Pharmacol. 82:211–223.

Hastings, C. 1955. P. 26 in Approximations for Digital Computers. Princeton, N.J.: Princeton University Press.

Heywood, R., R. J. Sortwood, P. R. B. Noel, A. E. Street, D. E. Prentice, F. J. C. Roe, P. F. Wadsworth, A. N. Worden, and N. J. Van Abbe. 1979. Safety evaluation of toothpaste containing chloroform. III. Long term studies in beagle dogs. J. Environ. Pathol. Toxicol. 2:835–851.

Jorgensen, T. A., E. F. Meierhenry, C. J. Rushbrook, R. J. Bull, and M. Robinson. 1985. Carcinogenicity of chloroform in drinking water to male Osborne Mendel rats and female B6C3F1 mice. Fund. Appl. Toxicol. 5:760–769.

Mitchell, J. R., D. J. Jollow, W. Z. Potter, J. R. Gillette, and B. B. Brodie. 1973. Acetaminophen-induced hepatic necrosis. IV. Protective role of glutathione. J. Pharmacol. Exo. Ther. 187:211–217.

Moolgavkar, S. H., and A. G. Knudson, Jr. 1981. Mutation and cancer: A model for human carcinogenesis. J. Natl. Cancer Inst. 66:1037–1052.

National Cancer Institute. 1976. Carcinogenesis bioassay of chloroform. NTIS No. PB246018/AS. Bethesda, Md.: National Cancer Institute.

Palmer, A. K., A. E. Street, F. J. C. Roe, A. N. Worden, and N. J. Van Abbe. 1979. Safety evaluation of toothpaste containing chloroform. II. Long term studies in rats. J. Environ. Pathol. Toxicol. 2:821–833.

Ramsey, J. C., and M. E. Andersen. 1984. A physiologically-based description of the inhalation pharmacokinetics of styrene in rats and humans. Toxicol. Appl. Pharmacol. 73:159–175.

Reitz, R. H. 1987. The role of cytotoxicity in the carcinogenic process. In Nongenotoxic Mechanisms in Carcinogenesis, Banbury Report 25. Cold Spring Harbor, N.Y.: Cold Spring Harbor Laboratory.

Roe, F. J. C., A. K. Palmer, A. N. Worden, and N. J. Van Abbe. 1979. Safety evaluation of toothpaste containing chloroform. I. Long term studies in mice. J. Environ. Pathol. Toxicol. 2:799–819.

Tsutsui, T., J. C. Barrett, and P. O. P. Ts'o. 1978. Induction of 6-thioguanine- and oubain-resistant mutations in synchronized Syrian hamster cell cultures during different periods of the S phase. Mutat. Res. 52:255–264.

Tsutsui, T., B. D. Crawford, and P. O. P. Ts'o. 1981. Comparison between mutagenesis in normal and transformed Syrian hamster fibroblasts. Difference in the temporal order of HPRT gene replication. Mutat. Res. 80:357–371.

Wells, P. G., R. C. Boerth, J. A. Oates, and R. D. Harbison. 1980. Toxicologic enhancement by a combination of drugs which deplete hepatic glutathione: Acetaminophen and doxorubicin (adriamycin). Toxicol. Appl. Pharmacol. 54:197–209.

Wong, K.-L., and C. D. Klaassen. 1981. Relationship between liver and kidney levels of glutathione and metallothionein in rats. Toxicology 19:39–47.

Physiologically Based Pharmacokinetic Model for Ethylene Dichloride and Its Application in Risk Assessment

Richard W. D'Souza, William R. Francis, Robert D. Bruce, and *Melvin E. Andersen*

INTRODUCTION

Ethylene dichloride (EDC; 1,2-dichloroethane) is a large-volume chemical extensively used in industry. EDC is presently being used as an alternative to ethylene dibromide as a fumigant for the treatment of food grains to control rodents, insects, and soil nematodes. In 1978, a study conducted by the National Cancer Institute (NCI) demonstrated treatment-related tumors in rats and mice dosed with 75 and 150 mg/kg EDC as a corn oil gavage (NCI, 1978). At about the same time the Bologna Tumor Center in Italy reported that no treatment-related tumors were seen in rats or mice chronically exposed to EDC by inhalation 7 h daily at the maximum tolerated dose of 150 ppm (Maltoni et al., 1980). These apparently contradictory results of the two bioassays have not been satisfactorily resolved to date.

Pharmacokinetic studies employing compartmental models have been conducted on EDC (Reitz et al., 1982; Spreafico et al., 1980). The results of these studies suggest that the differences in the two bioassays may be because of pharmacokinetic differences; that is, target organ exposure to EDC may have been greater after the gavage dosing then after a 7-h inhalation exposure. Unfortunately, the compartmental modeling employed in these studies could not quantitate target organ exposures with different doses and routes of exposure, nor extrapolate the relevance of the observations to human exposure situations for assessing human risk.

The objectives of this study were to develop a physiologically based pharmacokinetic (PB-PK) model for EDC and its reactive metabolites, to demonstrate the use of information derived from this model in cancer risk assessment, and to use the model to help explain the route-of-exposure differences observed in the two EDC bioassays.

METHODS

PB-PK Model Development

PB-PK models are mathematical models that mimic the way the body handles chemicals. These models are physiologically realistic and utilize all available anatomical, physiological, and physiocochemical information.

A schematic representation of the PB-PK model developed for EDC is shown in Figure 1. For the distribution of EDC within the body, organs

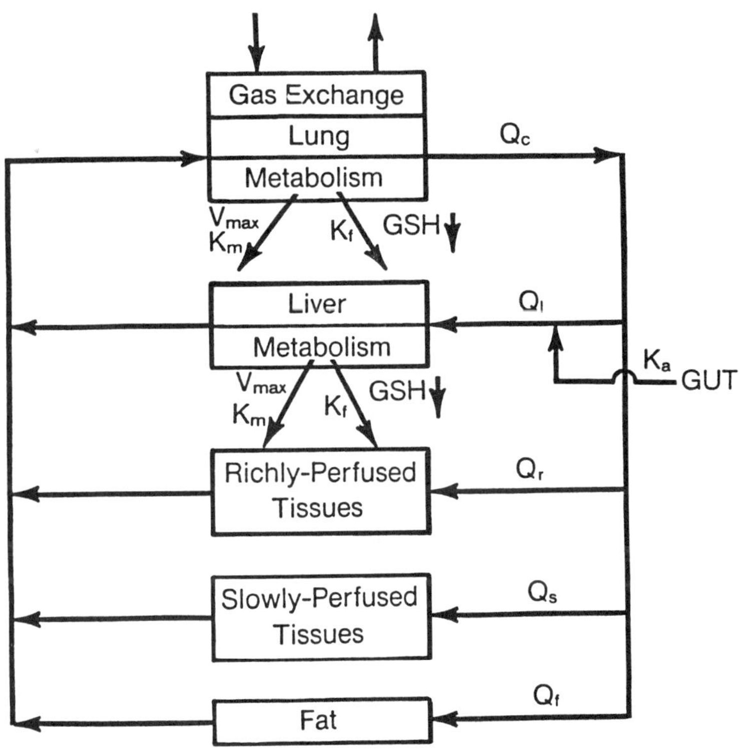

FIGURE 1 Schematic representation of the pharmacokinetic model developed for EDC. The biodistribution of EDC was modeled by dividing body tissues into three compartments based on their blood flow and relative ability to accumulate EDC. The liver and lung were considered the only metabolizing organs. V_{max} and K_m are the rate constants for the oxidation pathway, and K_f is the rate constant for the GSH pathway. GSH ↓ represents a glutathione depletion model.

TABLE 1A Partition Coefficients Obtained for EDC

Ratio	Sprague-Dawley Rat	Fisher 344 Rat	B6C3F1 Mouse	Human
Blood:air	27.6	30.4	29.7	21.1
Richly perfused:blood	1.1	1.2	1.0	—
Slowly perfused:blood	0.8	0.8	0.8	—
Fat:blood	12.2	11.4	12.1	—

were lumped together into three compartments, based on their blood flow and ability to accumulate EDC. For the purposes of this demonstration, the liver and the lung were considered to be the only metabolizing organs. All organs were connected to the circulatory system in an anatomically accurate fashion, that is, the lung was modeled as receiving the entire cardiac output, whereas other organs received a fraction of cardiac output. Because EDC is a small, lipophilic, and readily diffusible molecule, the PB-PK model was set up as a flow-limited model; that is, there were no diffusion barriers to the distribution of EDC. The parameters used in constructing this model are given below.

Partition Coefficient

The steady-state distribution ratio or partition coefficient of EDC was calculated by the method of Sato and Nakajima (1979). These partition coefficients were calculated for the B6C3F1 mouse and Fischer 344 rat and Sprague-Dawley rat. The blood:air partition coefficient was also obtained for human blood. The results are shown in Table 1A. The partition coefficients of many tissues were similar, and for modeling purposes these tissues were grouped together. The richly perfused group included such tissues as the kidney and spleen, the slowly perfused comprised muscle and skin, and the fat compartment represented body fat. It was noted that the human blood:air partition coefficient was somewhat smaller than that for the rat or the mouse. This lower value was, therefore, used in our model when scaling up from the rodent to the human.

Metabolism Rate Constant

EDC is metabolized by two competing pathways, a saturable pathway that involves P-450 oxidation, and by direct conjugation with glutathione (GSH). A simplified metabolism schematic, as reported by Anders and Livesey (1980), is shown in Figure 2.

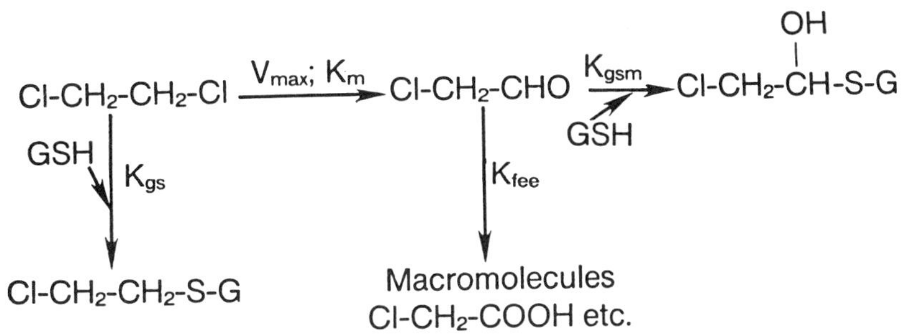

FIGURE 2 EDC metabolism pathway. EDC is metabolized by a saturable oxidation pathway and by conjugation with GSH.

Estimates of metabolism rates for the two pathways were obtained from gas uptake measurements (Gargas et al., 1986). The results of the gas uptake chamber runs are shown in Figure 3. As with similar halogenated hydrocarbons, we observed that the oxidative pathway for EDC is saturated at relatively low concentrations and is best described by a Michaelis-Menten type equation. The GSH pathway is first-order until high EDC exposure concentrations, where we observed a shift from first-order kinetics. This shift from first-order behavior for the GSH pathway was due to depletion of GSH, so that not enough GSH was available to react with EDC. In the recirculating chamber experiments that we (M.E.A.) have conducted with the halogenated hydrocarbons, this behavior has only been observed with EDC and allyl chloride. Metabolism rate constants for the two pathways are shown in Table 1B. At exposure concentrations where GSH is not depleted, a first-order rate constant (K_f) was used in the PB-PK model. At higher exposure concentrations, as used in bioassays, this first-order constant did not provide an adequate description of EDC metabolism, and the model therefore underestimated EDC concentrations. To account for this behavior, a GSH depletion model (D'Souza et al., 1986) had to be incorporated as part of the PB-PK description of EDC. This model kept track of liver and lung GSH levels, which were necessary to compute the amount of GC metabolite. The gas uptake studies provided rate constants for total metabolism. Metabolism rates between the liver and lung were split by using data from the literature on relative enzymatic

TABLE 1B Metabolism Rate Constants Obtained for EDC

EDC	V_{max} = 3.25 mg/h/kg	K_m = 0.25 mg/liter	K_f = 9.0 h^{-1} kg^{-1}
GSH model	K_{gs} = 0.0014 h^{-1} kg^{-1}	K_{fee} = 4,500 h^{-1} kg^{-1}	K_{gsm} = 0.14 h^{-1} kg^{-1}

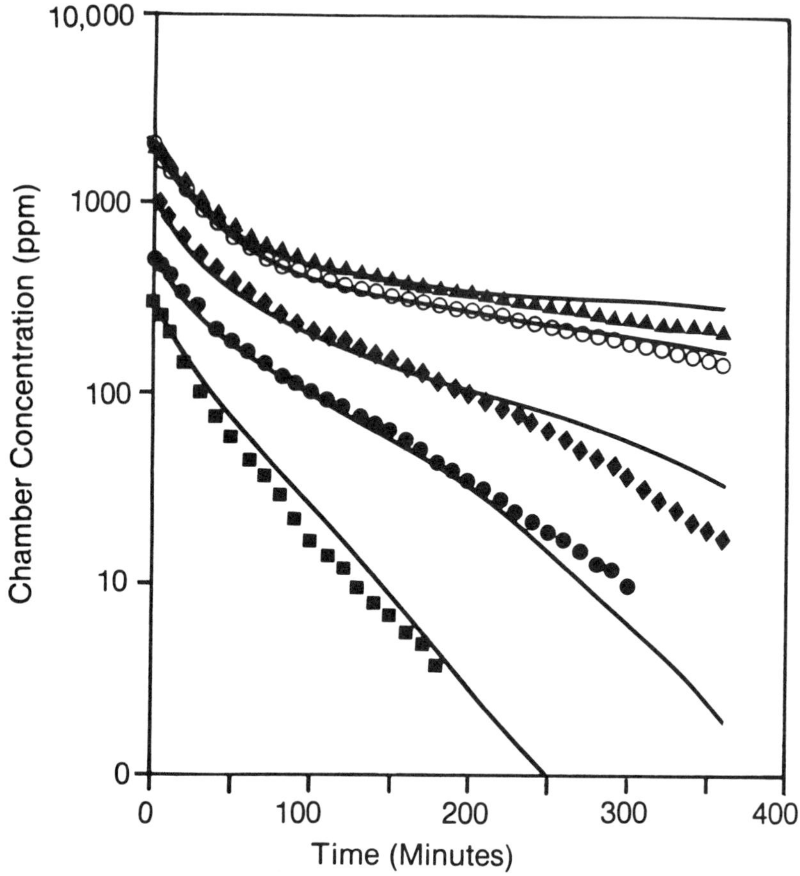

FIGURE 3 Gas uptake runs for EDC. The initial chamber concentrations used were 300 (■), 500 (●), 1,000 (♦), and 2,000 (○) ppm. One group of animals exposed to 2,000 ppm was pretreated with 2,3-epoxy propanol (▲). The lines represent model predictions that were generated by defining a PB-PK model of rats in a closed chamber.

activities between these two organs, as was accomplished for the methylene chloride model (Andersen et al., 1987). Scaleup of metabolism rates for different species was performed by using allometric scaling (Adolph, 1949; Dedrick, 1973; Lindstedt and Calder, 1981).

Physiological Parameters

The physiological parameters that were used to construct the model are well established in the literature (Caster et al., 1956). In most cases allometric equations were used to scale these parameters for different species.

Mass-balance differential equations were written for each compartment of the PB-PK model, describing the inflow, outflow, binding, and metabolism of EDC and the formation and detoxification of the intermediate metabolites. These equations were similar to those used by Andersen et al. (1987), except for the GSH model. The ACSL (Advanced Continuous Simulations Language; Mitchell & Gauthier Associates, Inc., Concord, Mass.) computer program was used to solve this set of simultaneous differential equations by numerical integration, employing Gear's algorithm for stiff systems.

RESULTS

The PB-PK model was validated by measuring EDC and GSH concentrations in the rat and mouse. Because the short-lived reactive metabolites cannot be measured readily, GSH depletion was used as an indirect validation. Figure 4 depicts PB-PK model predictions and experimentally determined EDC blood concentrations in the rat. Model predictions were in good agreement with observed data at the dose range tested.

Blood concentrations were similarly determined for the mouse and are shown in Figure 5. Again, experimental data were in close agreement with model predictions. Our predictions were also compared with data in

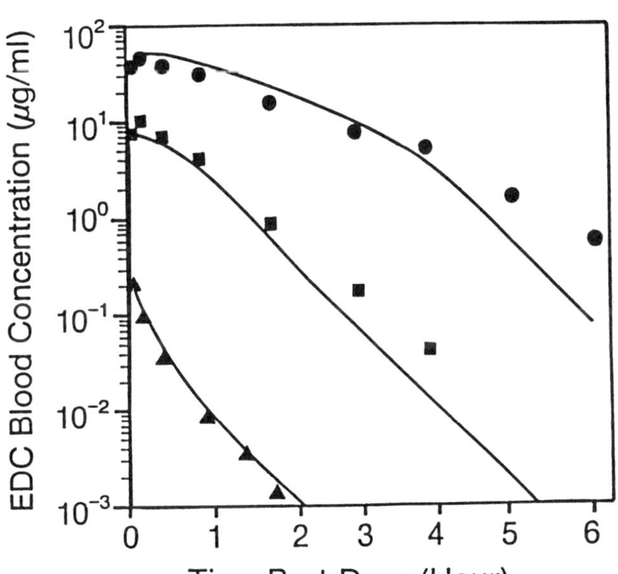

FIGURE 4 Model predicted (—) and experimentally determined EDC blood concentrations in the rat after corn oil gavage doses of 1.5 (▲), 25 (■), and 150 (●) mg/kg.

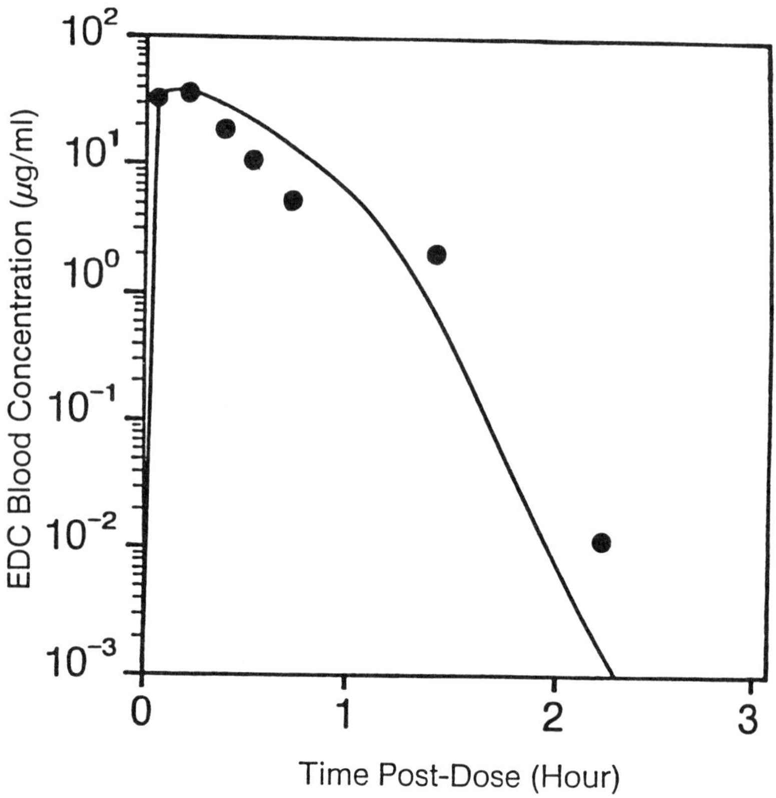

FIGURE 5 Model predicted (—) and experimentally determined (●) EDC blood concentrations in the mouse after a corn oil gavage dose of 150 mg/kg.

the literature and were found to be in excellent agreement. Spreafico et al. (1980) measured EDC levels in blood and tissues in the rat at several dose levels after oral, intravenous, and inhalation exposures. The results of their studies are very predictable by our model, both for blood and tissue concentrations. EDC blood concentration data of Reitz et al. (1982) after both oral and inhalation exposures in the rat were also in close agreement with model predictions.

GSH concentrations for both liver and lung of the rat and mouse after oral doses of 150 mg/kg were in close agreement with model predictions, as shown in Figures 6 and 7, respectively.

The strains of rats used in the studies cited from the literature and ours were different, but there were no apparent differences in the pharmacokinetics of EDC because of this difference. Spreafico et al. (1980) used Sprague-Dawley rats, and Reitz et al. (1982) used Osborne Mendel rats.

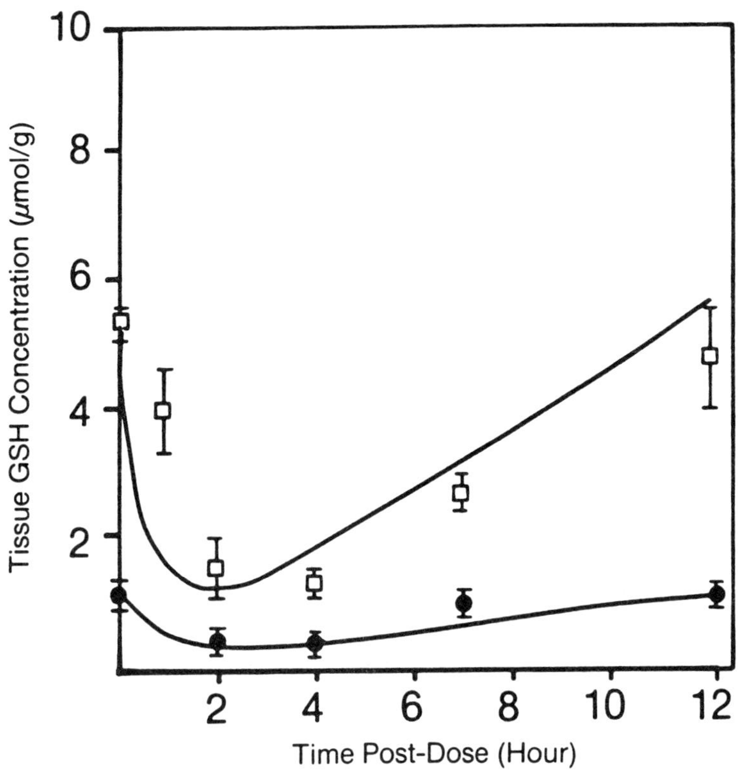

FIGURE 6 GSH concentrations in the liver (□) and lung (●) of the rat following an EDC dose of 150 mg/kg. Symbols represent mean ± standard deviation of four rats. The curves are model predictions.

We employed Fisher 344 rats and B6C3F1 mice in our studies. It appears that strain differences do not significantly affect the biodistribution or metabolism of compounds like EDC.

From the close agreement between model predictions and observed concentrations both from our studies and data reported in the literature, it is clear that the PB-PK model developed has strong predictive powers. This has been seen with different dose levels, routes of exposure, and two species.

Dose Surrogates

Several reports in the literature have concluded that the GSH conjugate formed in the metabolism of halogenated hydrocarbons like EDC and ethylene dibromide (EDB) is the carcinogenic moiety and not the parent

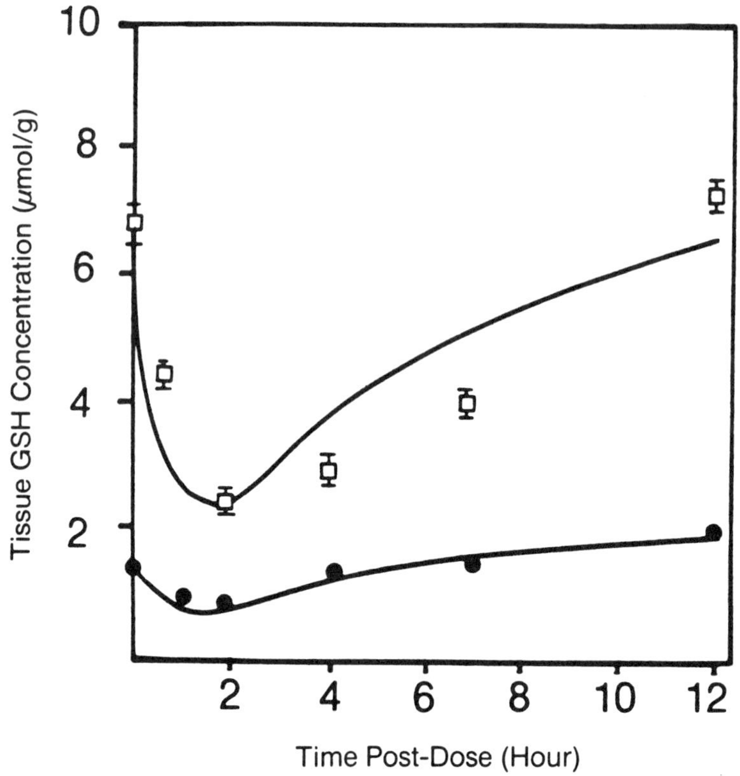

FIGURE 7 GSH concentrations in the liver (□) and lung (●) of the mouse following an EDC dose of 150 mg/kg. Symbols for the liver represent mean ± standard deviation of four mice. The lungs of four mice were pooled together and analyzed as a single sample. The curves are model predictions.

compound. For instance, White et al. (1983) compared the amount of DNA single-strand breaks produced by EDB and its tetradeuterated derivative (EDB-d_4), and found that EDB-d_4 was more genotoxic than EDB. Because the deuterated compound would preferentially be metabolized via the GSH pathway, these results demonstrated that the GSH pathway and not the oxidative pathway was the source of genotoxic metabolite(s). Similarly, Storer and Conolly (1985) studied hepatic DNA damage by EDC after pretreating mice with either piperonyl butoxide to block microsomal oxidation or diethyl maleate to deplete GSH, and found that the GSH pathway was responsible for DNA damage and not the oxidative pathway. With methylene chloride, Andersen et al. (1987) related the number of tumors at each dose level with the PB-PK model predictions of the amount of GSH and oxidative metabolite produced at that dose level and noted that tumors related quite well with GSH metabolite formed. We have, therefore, used the total amount of the parent compound-glutathione conjugate (GC) produced in a target tissue as a surrogate for dose.

PB-PK Model and Risk Assessment

Because of the nonlinear metabolism pathways for EDC, the relationship between administered EDC dose, either as inhalation or oral exposure, and the amount of GC metabolite produced in a target tissue is complex. Figure 8 illustrates the relationship between different oral gavage doses of EDC and the resulting GC metabolite that would be produced in the liver for the B6C3F1 mouse. It can be seen that at low doses, because of first-order metabolism of EDC, the relationship is 1:1. As the exposure dose is increased, the amount of GC produced is proportionally much

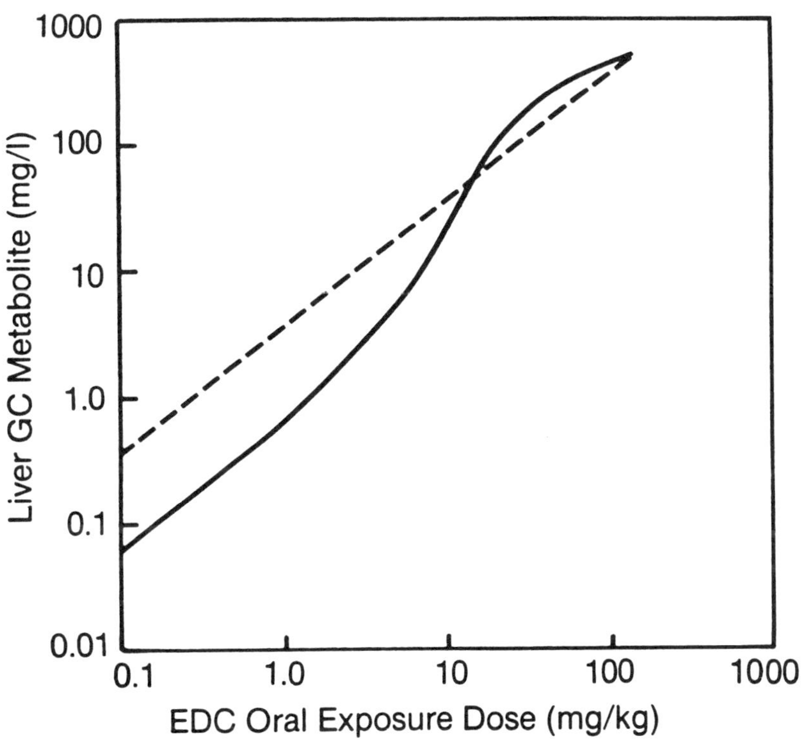

FIGURE 8 Relationship between EDC administered dose and the dose surrogate (the amount of liver GC metabolite) in the mouse. The curves represent the complex relationship predicted by the model. The dashed line is a direct extrapolation from the 150-mg/kg dose, assuming a 1:1 relationship. The curve predicted for the human is virtually superimposable on the mouse curve and is, therefore, not shown in the figure (see text for detailed explanation).

greater than the administered dose. This nonlinearity is due to the fact that the oxidative pathway for EDC metabolism is saturated, resulting in increasing EDC levels and, therefore, increasing GC levels. At exaggerated doses, as employed in bioassays, the capacity of the GSH pathway is also overwhelmed due to depleted GSH concentrations, and the amount of GC produced becomes proportionally lower than the administered dose. The dashed line (Figure 8) is a direct linear extrapolation of the GC metabolite from the 150-mg/kg dose to low doses, ignoring any nonlinearities. It can be readily seen from this type of plot that by not taking into account the nonlinear metabolism of EDC, the risk of liver cancer to the mouse, based on results of a 150-mg/kg dose, can be easily overestimated by an order of magnitude. The PB-PK model was also used to compute this relationship for the human. Liver GC in the human was predicted to be very similar to that in the mouse and is therefore not shown in Figure 8.

Figure 9 depicts the relationship between administered dose and lung GC metabolite for the B6C3F1 mouse. A similar relationship is observed for the lung, except that the human GC exposure doses are about 2.5 times smaller. This suggests that by ignoring the nonlinear metabolism of EDC, lung cancer risk for the mouse can be overestimated by a factor of 10, while for the human it can be overestimated by 25-fold. Administered dose versus GC exposure plots were also constructed for inhalation exposure for the mouse and for oral and inhalation exposures for the rat, with similar findings (plots not shown here).

PB-PK Model and Virtually Safe Doses

As part of the conventional cancer risk assessment process, the number of tumors in each dose group of the bioassay is correlated with the administered dose. Statistical models are used to fit the data to extrapolate these findings to low exposure levels. These models estimate the dose required for 1 out of 1 million animals to develop a certain type of tumor during lifetime dosing. This dose is commonly called the virtually safe dose (VSD). The VSD for humans is obtained by dividing the mouse VSD by a factor of 12.7, which is the relative difference in body surface area between the mouse and man. Additional safety factors are also built into the risk assessments, including the use of confidence intervals and upper limits on estimated exposure of humans. The VSD is then compared with anticipated human exposures to prepare a risk assessment.

The PB-PK model was used to generate GC amounts with different EDC exposure doses in the mouse, rat, and human (as in Figures 8 and 9). The multistage model (Crump, 1982) was then used to relate the number of tumors produced in each dose group with GC produced at that

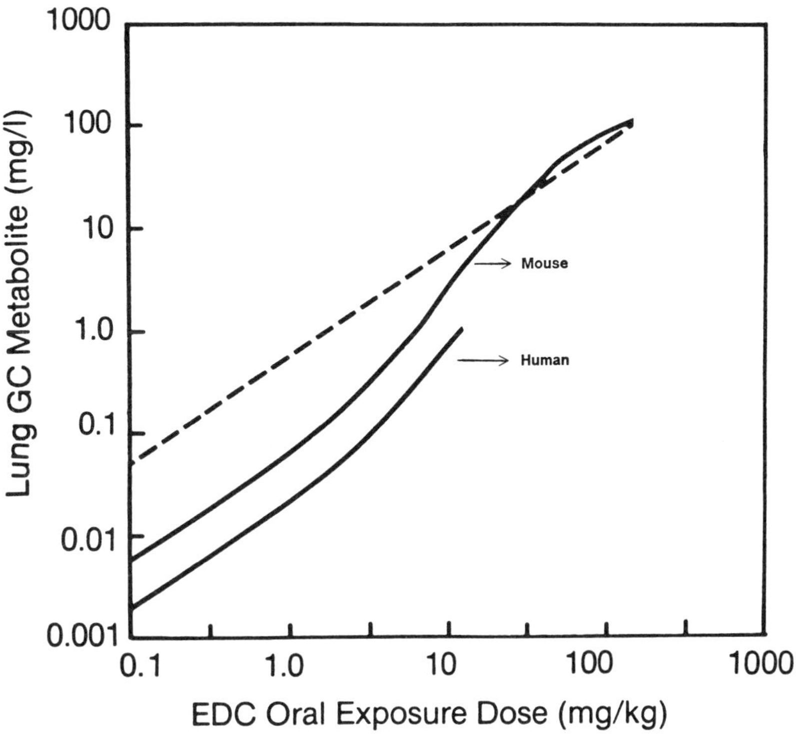

FIGURE 9 Relationship between EDC administered dose and lung GC metabolite for the mouse and human. The dashed line is a direct extrapolation from the 150-mg/kg dose, assuming a 1:1 relationship.

tumor site. That is, instead of correlating tumors with the administered dose, or the external dose, the dose surrogate at the target site, or the internal dose, was employed. There was no additional justification for choosing the multistage model over several commonly used low-dose extrapolation models other than the fact that it is one of the models used most often by regulatory agencies. The VSD obtained by using this internalized dose concept was converted to the corresponding EDC exposure concentration both for the mouse as well as the human employing Figures 8 and 9. That is, the extrapolation from mouse to human was made by using the PB-PK model and not a surface area correction, as would be done in a conventional risk assessment.

The results of this analysis are shown in Table 2. For comparison purposes, VSDs were calculated both by the conventional method and by the PB-PK approach. Because regulatory agencies commonly use the lower confidence limit (LCL) of dose, the comparisons employed here are similarly restricted to the LCLs and not to the fitted lines. When comparing

TABLE 2 VSDs Calculated for EDC by Employing the Conventional and PB-PK Approach

		Conventional			PB-PK		
Organ	Estimate Type	Mouse EDC (μg/kg/day)	Human EDC (μg/kg/day)	GC Metabolite (μg/liter/day)	Mouse EDC (μg/kg/day)	Human EDC (μg/kg/day)	Ratio, PB-PK/Conventional
Liver	Best estimate	3.1	0.24	150	240	230	960
	95% LCL	0.46	0.036	2.3	3.8	3.6	100
Lung	Best estimate	270	21	245	4,100	7,500	360
	95% LCL	0.99	0.078	0.91	15	39	500

NOTE: In the conventional method tumors were correlated with EDC administered dose to obtain a VSD in the mouse (column 3). VSD for the human was obtained by dividing the mouse VSD by 12.7 (column 4). In the PB-PK approach tumors were correlated with the amount of GC metabolite produced in a target organ, and a VSD for this dose surrogate was obtained (column 5). This dose surrogate was then converted into the corresponding EDC administered dose for the mouse (column 6) and human (column 7) by employing the PB-PK model. Column 8 compares the results of these two approaches.

the VSD for liver tumors by the two methods, it can be seen that the VSD for liver tumors obtained by using the pharmacokinetic approach is about 2 orders of magnitude greater than that obtained by the conventional method. VSD for lung tumors, however, shows about a 500-fold difference. These data indicate that by not considering the pharmacokinetics of EDC in the mouse and human, conventional methods may overestimate cancer risk due to low-level EDC exposure by 2 or even 3 orders of magnitude.

PB-PK Model and Route-of-Exposure Differences

To address the route-of-exposure differences in the two EDC bioassays, the amount of GC metabolite produced in the liver and lung was computed for oral gavage doses of 75 and 150 mg/kg and compared with those that would be produced by a 7-h inhalation exposure at 150 ppm. These results are shown in Table 3. It is interesting to note that lower amounts of the dose surrogates are predicted at the maximum tolerated inhalation dose of 150 ppm, compared with the other two oral doses. This observation may be a possible reason that treatment-related tumors were not seen in the inhalation bioassay, but were observed both at 75 and 150 mg/kg for the oral bioassay.

DISCUSSION

PB-PK models are mechanistic models that attempt to account for the important processes that take place from the time a chemical is present

TABLE 3 Amount of GC Metabolite Formed in the Liver and the Lung of the Rat Following Oral and Inhalation Exposures

	GC Metabolite	
Dose and Route	Liver	Lung
Corn oil gavage		
150 mg/kg	630	131
75 mg/kg	372	71
Inhalation		
150 ppm (7 h)	230	64

at an absorption site for potential inhalation or oral or dermal uptake into the body, to the interaction that takes place between the chemical or its metabolites and some body tissue. The amount of chemical that is present at an absorption site may, therefore, have a very complex and indirect relationship with the amount of that chemical or its metabolites that is present at a target site and elicits a biological response. The chemical must first get absorbed; it may then bind to blood components that may limit its further distribution, it may undergo detoxification or may become more toxic, or it may be eliminated rapidly from the body. An understanding of these processes, as well as the quantitation of the time course in the way that these processes operate and interact, may play a very important role in determining the potential risk from exposure to a certain concentration of a chemical.

Once it is known whether the parent chemical or its metabolite(s) is responsible for a toxic or carcinogenic response, a PB-PK model can be developed to quantitate the amount of exposure and the time course of exposure to this moiety at a target site in laboratory animals. Using the model, this information can then be scaled up to the human to assess target organ exposure and, therefore, the risk of exposure to humans. The model does not provide insight into the mechanism of cancer, nor does it predict sensitivity of one target organ over another or one species over another. The model is simply a tool to quantitate target organ exposure to the relevant chemical species.

One of the major problems in assessing toxicity or cancer risk is route-of-exposure extrapolations. Often, data are only available from one route of intake, while a risk assessment must be conducted for exposure via another route. Sometimes, like with the EDC and DCM bioassays, the results are positive from one route and negative via another route of exposure, which makes the results difficult to interpret. Unfortunately, there has been no reliable method so far for any extrapolation between routes of intake. PB-PK models, however, have the potential to quantitate the amount of chemical and its metabolites that is taken in from different exposure sites and that reaches a target tissue to elicit a response.

Another factor that is not readily resolved without these kinds of models is species differences in the biodistribution of chemicals. Although PB-PK models do not predict sensitivity of a biological response of one species over another, they can reliably quantitate target organ doses between species. Differences in the physiology of different species, like organ blood flow or pulmonary ventilation, are taken into account in these models. Anatomical differences, like organ sizes, are appropriately corrected when extrapolating results from one species to another. Also, physicochemical parameters of the chemical, like blood and tissue binding, are taken into consideration. The power and flexibility of these models lies in the fact that they are physiologically realistic and not mathematical "black boxes." As more information is gained about the behavior of a chemical, it can be added to the model without changing its basic structure, thus improving its predictability.

The PB-PK model developed for EDC is in reality quite basic, because the model only computes gross tissue exposures of the relevant chemical species. Further work in this area should concentrate on cellular distribution, as well as molecular dosimetry aspects like DNA binding and repair processes. All of these processes, as well as the processes that take place from the time that an unrepaired DNA adduct is formed to the time that a tumor is produced, are expected to be complex and potentially nonlinear with exposure dose. As information is gained about the rates of these processes, they can readily be added on to the PB-PK model.

REFERENCES

Adolph, E. H. 1949. Quantitative relationships in physiological constituents of mammals. Science 109:579–585.

Anders, M. W., and J. C. Livesey. 1980. Metabolism of dihaloethanes. Pp. 331–341 in Ethylene Diochloride: A Potential Health Risk?, Banbury Report 5, B. Ames, P. Infante, and R. Reitz, eds. Cold Spring Harbor, N.Y.: Cold Spring Harbor Laboratory.

Andersen, M. E., H. J. Clewell III, M. L. Gargas, F. A. Smith, and R. H. Reitz. 1987. Physiologically-based pharmacokinetics and the risk assessment process for methylene chloride. Toxicol. Appl. Pharmacol. 87:185–205.

Caster, W. O., J. Poncelet, A. B. Simon, and W. D. Armstrong. 1956. Tissue weights of rats. I. Normal values determined by dissection and chemical methods. Proc. Soc. Exp. Biol. Med. 91:122–126.

Crump, K. S. 1982. An improved method for low dose carcinogenic risk assessment from animal data. J. Environ. Pathol. Toxicol. 5(2):675–684.

Dedrick, R. L. 1973. Animal scale-up. J. Pharmacokinet. Biopharm. 1(5):435–461.

D'Souza, R. W., W. R. Francis, and M. E. Andersen. 1986. A mathematical model for glutathione depletion and increased resynthesis following ethylene dichloride exposure. Pharm. Res. 3(5):137S.

Gargas, M. L., H. J. Clewell, and M. E. Andersen. 1986. Metabolism of inhaled dihalomethanes in vivo: Differentiation of kinetic constants for two independent pathways. Toxicol. Appl. Pharmacol. 82:211–223.

Lindstedt, S. L., and W. A. Calder III. 1981. Body size, physiologic time, and longevity of homeothermal animals. Q. Rev. Biol. 56:1–16.

Maltoni, C., L. Valgimigli, and C. Scarnato. 1980. Long term carcinogenic bioassays of ethylene dichloride administered by inhalation to rats and mice. Pp. 3–29 in Ethylene Dichloride: A Potential Health Risk?, Banbury Report 5, B. Ames, P. Infante, and R. Reitz, eds. Cold Spring Harbor, N.Y.: Cold Spring Harbor Laboratory.

NCI (National Cancer Institute). 1978. Bioassay of 1,2-Dichloroethane for Possible Carcinogenicity, NCI Carcinogenesis Technical Report, Series No. 55, DHEW Publication No. (NIH) 78-1361. Washington, D.C.: U.S. Government Printing Office.

Reitz, R. H., T. R. Fox, J. C. Ramsey, J. F. Quast, P. W. Langvardt, and P. G. Watanabe. 1982. Pharmacokinetics and macromolecular interactions of ethylene dichloride in rats after inhalation or gavage. Toxicol. Appl. Pharmacol. 62:190–204.

Sato, A., and T. Nakajima. 1979. Partition coefficients of some aromatic hydrocarbons and ketones in water, blood and oil. Br. J. Ind. Med. 36:231–234.

Spreafico, F., E. Zuccato, M. Marcucci, M. Sironi, S. Paglialunga, M. Madonna, and E. Mussinin. 1980. Pharmacokinetics of ethylene dichloride in rats treated by different routes and its long-term inhalatory toxicity. Pp. 107–129 in Ethylene Dichloride: A Potential Health Risk?, Banbury Report 5, B. Ames, P. Infante, and R. Reitz, eds. Cold Spring Harbor, N.Y.: Cold Spring Harbor Laboratory.

Storer, R. D., and R. B. Conolly. 1985. An investigation of the role of microsomal oxidation metabolism in the *in vivo* genotoxicity of 1,2-dichloroethane. Toxicol. Appl. Pharmacol. 77:36–46.

White, R. D., A. J. Gandolfi, G. T. Bowden, and I. G. Sipes. 1983. Deuterium isotope effect on the metabolism and toxicity of 1,2-dibromoethane. Toxicol. Appl. Pharmacol. 69:170–178.

APPENDIX: NOMENCLATURE

GC	Metabolite formed by conjugation of EDC and GSH.
K_f	First-order rate constant for EDC metabolism via GSH pathway (h^{-1})
K_{fee}	First-order rate constant for reaction of chloroacetaldehyde with everything else besides GSH (h^{-1})
K_{gsm}	First-order rate constant for reaction of GSH with chloroacetaldehyde (h^{-1})
K_m	Michaelis-Menten constant for oxidation pathway (mg/liter)
Q_c	Cardiac output
Q_f	Fat blood flow
Q_l	Liver blood flow
Q_{lu}	Lung blood flow
Q_p	Alveolar ventilation
Q_r	Blood flow to richly perfused tissues
Q_s	Blood flow to slowly perfused tissues
V_{max}	Maximum capacity of oxidation pathway (mg/h)

Mathematical Modeling of Ozone Absorption in the Lower Respiratory Tract

John H. Overton, Jr., Richard C. Graham, and
Frederick J. Miller

INTRODUCTION

Environmental toxicologists are confronted with the difficult task of interpreting the manifold results from human clinical, epidemiological, and animal studies on air pollutants and assessing their implications and relevance concerning pollutant levels to which man is exposed. To accomplish this objective, mathematical models that predict respiratory tract dose patterns are essential. A purpose of these models is to extrapolate quantitatively effective pollutant concentrations between animals and man. In so doing, the existing animal data base will have increased direct relevance to national ambient air quality standards, and future studies can be designed to permit improved correlation with human effects.

As part of a program to accomplish this objective, we have developed a mathematical dosimetry model that simulates the uptake and distribution of obsorbed O_3 in the respiratory tract (RT) of humans and laboratory animals (Miller et al., 1985). Originally, the dosimetry model was developed to simulate the local absorption of O_3 only in the lower respiratory tract (LRT). However, the present model can be used with other gaseous

The research described in this article has been reviewed by the Health Effects Research Laboratory, U.S. Environmental Protection Agency, and approved for publication. Approval does not signify that the contents necessarily reflect the views and policies of the Agency nor does mention of trade names or commercial products constitute endorsement or recommendation for use.

pollutants, such as NO_2, if reactions with biological constituents can be modeled the same way as the O_3 reactions. Absorption in the entire respiratory tract can be modeled also by adding appropriate pretracheal "generations" to a given LRT anatomical model. (Although the upper respiratory tract [URT] anatomical features can be accounted for correctly, the treatment of transport and chemical reactions, which is the same as that in the tracheobronchial region, may be inadequate; future studies are required.) The present dosimetry model includes use of LRT anatomy, ventilatory parameters, and varying airway dimensions during the breathing cycle. The model also accounts for the processes of transport in the lumen and air spaces and transport and irreversible chemical reactions in the liquid lining and in the underlying tissue and capillaries.

Plans for future modifications to the model include (1) an improved treatment of URT transport and absorption; (2) taking into account metabolism in lung fluids and tissues, as well as a more general treatment of chemistry, including reversible reactions and reaction orders greater than 1; and (3) interfacing the RT model with pharmacokinetic models to allow a more complete description of the absorption, distribution, and metabolism of inhaled gases in the body than is presently available.

This paper describes the present O_3 dosimetry model and illustrates the use of the model with the results of two investigations of (1) the sensitivity of predictions to anatomical models of the guinea pig and rat, and (2) the effect of exercise on O_3 absorption in man.

METHODS

Given the O_3 concentration at the entrance (e.g., nose, mouth, or trachea), the model simulates, during one or more breathing periods, the transport and absorption of O_3 in airways and alveolar air spaces of each generation or segment of a respiratory tract anatomical model.

Species lung dimensions are taken into account by making use of anatomical or airway models, as illustrated in Figure 1, a stylized diagram. In these models, airways of the LRT are represented by a sequence of sets of right circular cylinders. All cylinders or airways in a set that correspond to a particular generation are the same size. The URT region (if used) consists of pretracheal generations or sequential segments. For each generation, the simulation model requires the specification of the number of airways or segments and their diameters and lengths. Additionally, for the pulmonary region, the alveolar volume and the surface area for each generation are needed. Similarly, the surface area of each URT segment is required.

During the simulated breathing cycle, LRT linear dimensions are varied isotropically (linear dimensions are proportional to the one-third power of

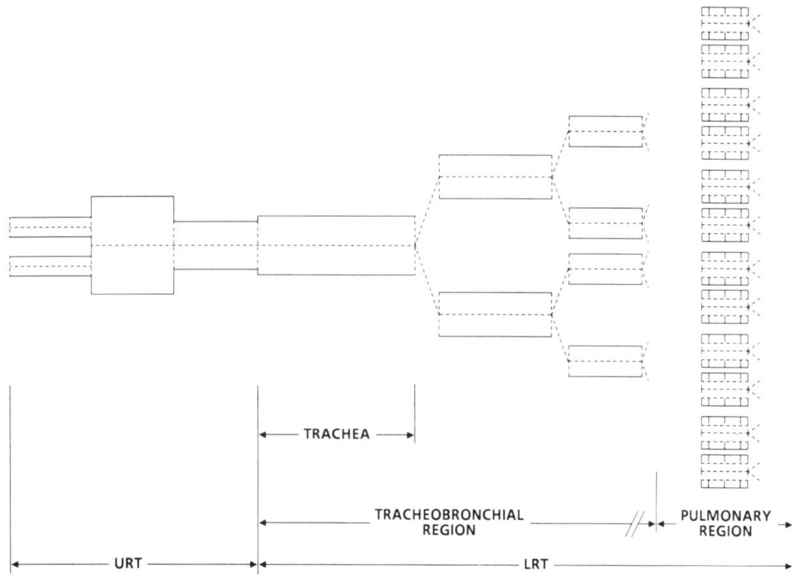

FIGURE 1 A stylized diagram of the type of respiratory tract anatomical models used with the dosimetry model.

the ratio of the expanded volume of the generation to the volume of the generation at FRC) to account for changes in lung volume. The simulation of volume flow rate during a breath can be based on experimental data (e.g., plethysmograph studies) or on an assumed time-dependent function such as the sine function.

The simulation model accounts for transport in the lumen and air spaces a well as for chemical reactions and transport in the liquid lining, tissue, and blood. Although O_3 reacts with many different chemical species, for modeling purposes one effective reaction is assumed,

$$O_3 + HC \xrightarrow{k} \text{products},$$

where k is the effective second-order rate constant and HC represents the chemical components that react with O_3.

The processes of transport and chemical reactions are described in terms of partial differential equations (Figure 2). In the liquid lining, tissue, and blood compartments, where only the processes of molecular diffusion and chemical reactions are considered, the form of the equation is the same (see Miller et al., 1985, for justification of assumptions). In the lumen of the airway and in the alveolar air spaces, axial dispersion, axial convection, the loss of O_3 to the liquid lining, and lung expansion and contraction are taken into account.

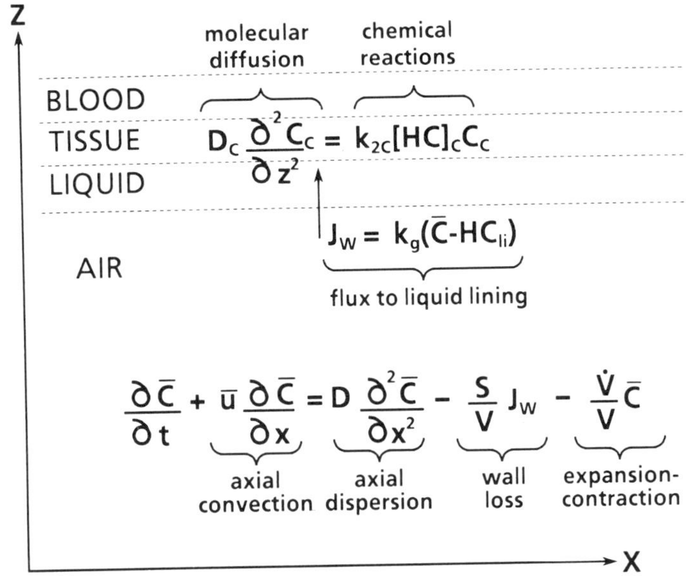

FIGURE 2 Equations used to describe the transport and chemical reactions of O_3 in the compartments of the respiratory tract. Definition of terms includes: t = time; x = distance along airway path; z = distance within compartments; $\bar{C}$ = average cross-sectional O_3 concentration; C_c = pointwise O_3 concentration in compartment; D = effective dispersion coefficient of O_3 in lumen; D_c = molecular diffusion coefficient of O_3 in a compartment; $\bar{u}$ = average air velocity in lumen; S = surface area; V = volume of airway; $\dot{V}$ = time rate of change of airway volume; J_w = radial flux to wall; k_g = gas phase mass transfer coefficient; k_{2c} = effective second-order chemical rate constants; $[HC]_c$ = concentration of components that react with O_3; H = Henry's Law constant; C_{li} = liquid layer O_3 concentration at the liquid-air interface.

Given boundary conditions; initial conditions; and values for the physical, chemical, and biological parameters, the equations can be solved to give simulated O_3 dose and dose patterns.

RESULTS AND DISCUSSION

Effect of Anatomical Model on the Distribution of Simulated Absorbed O_3 in the LRT

Overton et al. (1987) investigated the effects of different anatomical models on the uptake and distribution of absorbed O_3 in the LRT of guinea pigs and rats. Figure 3, which is based on results of the investigation, displays simulated net and tissue dose profiles for the two species when two anatomical models are used for each species. The net and tissue doses

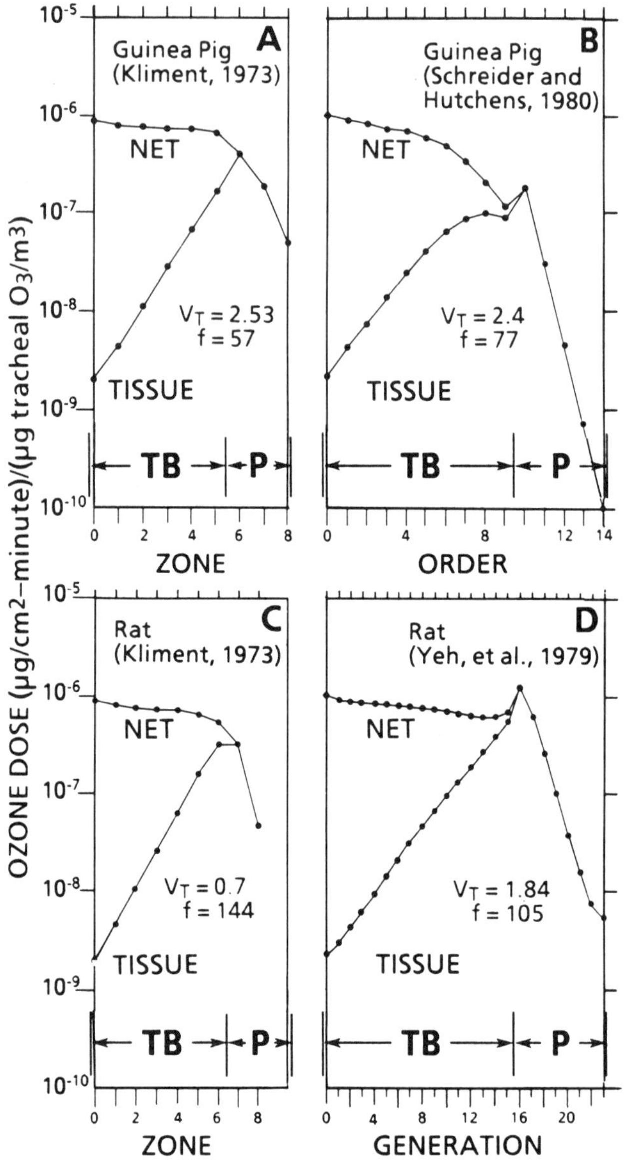

FIGURE 3 Effect of four different lower respiratory tract anatomical models on predicted net and tissue O_3 dose profiles of guinea pigs and rats (based on Overton et al., 1987). Dose is the quantity of O_3 reacting with biological constituents per unit surface area of a zone, order, or generation per unit time per tracheal O_3 concentration. Net dose is the sum of the liquid lining, tissue, and blood compartment doses. The source of the anatomical model used is given at the top of each graph. V_T = tidal volume (in milliliters); f = breaths per minute; TB = tracheobronchial region; P = pulmonary region.

are plotted versus zone, order, or generation (depending on the anatomical model). Dose is defined as the quantity of O_3 reacting with biological constituents per unit surface area of airway per unit time per tracheal O_3 concentration. Net dose is the sum of the blood, tissue, and liquid compartment doses. On each graph the species, source of the anatomical model, and ventilatory parameters are indicated.

Regardless of the ventilatory parameters, the plots are typical for each anatomical model and illustrate qualitatively the differences and similarities resulting from the simulations using the different anatomical models. Generally, in the trachea the tissue dose is relatively low; the dose increases distally to a maximum in the vicinity of the first pulmonary airway unit and then rapidly decreases. Net dose is much larger in the trachea, however, and decreases distally throughout the tracheobronchial region. In the pulmonary region, the net and tissue doses are essentially the same.

The curves, based on the anatomical models of the guinea pig and rat of Kliment (1973) (Figures 3A and 3C), are similar. The difference between the shapes of the two tissue dose curves at zones 6 and 7 is due to our choice of where the pulmonary or surfactant-lined region begins, not to differences in anatomical models. For the rat, this region begins at zone 7, and for the guinea pig it begins at zone 6. Quantitatively, the total and pulmonary uptakes for the Kliment guinea pig model (Figure 3A) are 92% and 85%, respectively. For the Kliment rat model (Figure 3C), these uptakes are 87% and 81%, respectively.

Based on the results with the Kliment zone model, one might conclude that the uptake properties of rats and guinea pigs are very similar with respect to how the absorbed O_3 is distributed in the LRT. However, Figures 3B (guinea pig; Schreider and Hutchens, 1980) and 3D (rat; Yeh et al., 1979) suggest that the LRT deposition pattern for the two animals is dissimilar. For example, the net trachea dose is ~7 times the net dose of the first alveolated airways (order 10, Figure 3B) for the guinea pig. On the other hand, the net dose in the trachea of the rat model of Yeh et al. is less than the net dose of the first alveolated generation (generation 16, Figure 3D).

Results with the Kliment (1973) rat and the Yeh et al. (1979) rat anatomical models were further investigated by Overton et al. (1987) by varying the breathing frequency and keeping the minute volume constant. The two anatomical models were found to result in significantly different LRT total and pulmonary percent uptakes: 20%–40% (depending on breathing frequency) higher uptakes for the Kliment rat model than for the Yeh et al. rat model. Further, the Kliment rat was less sensitive to breathing frequency changes: The Kliment rat total and pulmonary percent uptakes decreased by ~6% when the breathing frequency was increased from 80 to 140 breaths per minute, whereas the percent uptakes of the Yeh et al. rat decreased by ~20% over the same frequency range.

Effect of Exercise on Predicted Uptake and Distribution of Absorbed O_3 in the LRT of Man

The effect of exercise on the uptake and the distribution of absorbed O_3 in man was investigated by Miller et al. (1985). Figure 4 and Table 1 illustrate the results. The anatomical model used was based on that of Weibel (1963) for man, but it was modified so that the lung volume at the end of exhalation was 2,650 ml (based on data obtained from William F. McDonnell, Environmental Protection Agency, Research Triangle Park, N.C., personal communication; the data [FRC versus TLC] were extrapolated to the TLC value that corresponded to the Weibel [1963] anatomical model). Lung expansion was not taken into account for these simulations. All calculations were based on specifying the concentration at the entrance to the trachea.

The effect of exercise on predicted tissue dose is illustrated in Figure 4 for several levels of exercise. The level of exercise increases with increasing curve number, with curve 1 being for normal respiration and curve 4 for heavy exercise. The simulations show that exercise has very little effect on O_3 tissue dose in the tracheobronchial region and a very large effect on this dose in the pulmonary region. In addition, the maximum tissue dose not only shifts distally from generation 17 (the first pulmonary generation) at rest to generation 20 at the heaviest exercise, but it also increases by a factor of 3.3.

Table 1 illustrates the effect of exercise on the distribution of absorbed O_3 in the regions and compartments of the LRT of man. For each level of exercise, Table 1 gives the percent uptake of absorbed O_3 in the regions

TABLE 1 Simulated Effect of Exercise on Regional and Compartmental Uptake of O_3 in Man[a]

					Tracheobronchial Region Uptake (%)		Pulmonary Region Uptake (%)[d]	
ID[b]	V_T[c] (liters)	f[c] (BPM)	$\dot{V}_E$[c] (liters/min)	Total Uptake (%)	Liquid Lining	Tissue	Tissue	Blood
1	0.5	15.0	7.5	89.2	18.4	8.3	60.3	2.0
2	1.0	15.0	15.0	95.8 (2.2)[e]	10.8 (1.2)	5.2 (1.3)	77.1 (2.6)	2.6 (2.6)
3	1.75	20.3	35.5	98.5 (5.2)	5.0 (1.3)	2.5 (1.4)	87.9 (6.9)	3.0 (7.1)
4	2.25	30.0	67.5	99.4 (10.0)	2.8 (1.4)	1.4 (1.5)	92.0 (13.7)	3.1 (14.0)

[a]Uptake is based on the quantity of O_3 inhaled at the trachea.

[b]Simulation identification of curve number (see Figure 4).

[c]V_T = tidal volume; f = breaths per minute; $\dot{V}_E = V_T \times f$ = minute volume.

[d]The liquid lining absorbs less than 0.3%.

[e]Values in parentheses are the ratio of mass absorbed to the mass absorbed at normal respiration (for the first row of data, ID = 1).

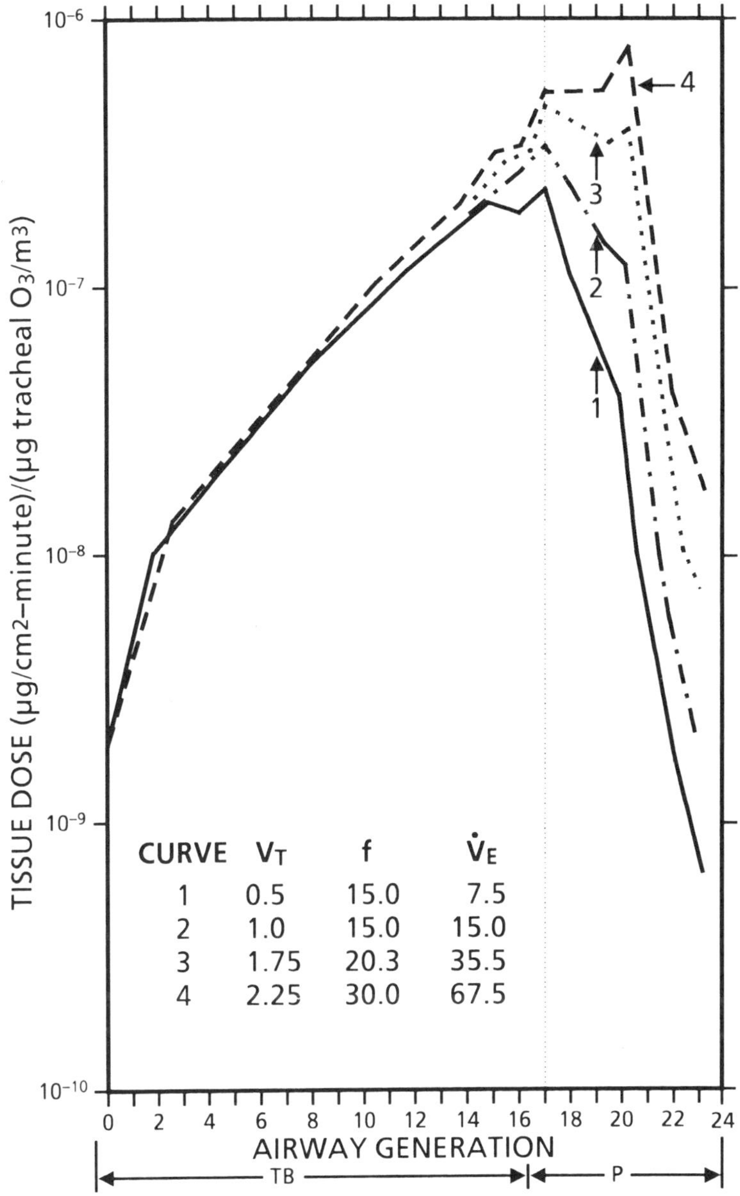

FIGURE 4 Effect of O_3 exercise on predicted tissue O_3 dose in the lower respiratory tract of man (based on data from Miller et al., 1985). Dose is the quantity of O_3 reacting with biological constituents per unit surface area of a zone, order, or generation per unit time per tracheal O_3 concentration. The level of exercise and its corresponding tissue dose curve are indicated on the figure. V_T = tidal volume (liters); f = breaths per minute; $\dot{V}_E$ = minute volume; TB = tracheobronchial region; P = pulmonary region.

and compartments of the LRT. Also presented for each compartment and level of exercise is the ratio of the absorbed mass to the absorbed mass at normal respiration. The first column identifies the simulations with the curves in Figure 4.

Although the percent total uptake does not change much (from ~89% to 99%), the percent uptake in the regions and compartments can vary considerably as the level of exercise increases. With increasing minute volume the tracheobronchial uptake decreases from ~27% to 4% whereas the pulmonary uptake increases from ~62% to 95%. A better indicator of the effect of exercise is the change in the mass of the absorbed O_3. For example, the mass absorbed by the tracheobronchial region increases to ~1.4 times the amount absorbed during the same time of normal respiration, even though the percent uptake decreases. On the other hand, pulmonary absorption increases by a factor of about 14. A discussion of the significance of these results can be found in Miller et al. (1985).

SUMMARY

A brief description of a mathematical dosimetry model has been given. The model was developed to simulate the uptake and distribution of irreversible chemically reacting toxic gases, such as O_3 and NO_2, in the LRT of man and laboratory animals and takes into account species LRT anatomy and ventilatory characteristics, transport in the lumen of the airways and in alveolar air spaces, and transport and chemical reactions in the liquid lining and the underlying tissue and blood compartments.

Model predictions were illustrated with the results of two investigations. In the first, two anatomical models for both the guinea pig and rat were used to investigate the effects of different anatomical models on predictions of O_3 uptake. This investigation indicates the importance of reliable anatomical models for a given animal. The second illustration of O_3 dosimetry model results was a discussion of an investigation into the effects of exercise on predicted O_3 dose and distribution in the LRT of man. As exercise increased, the model predicted a moderate increase in percent total uptake, with very little effect on the amount of O_3 absorbed by the tracheobronchial tissue. However, a pronounced increase in pulmonary absorbed O_3 was predicted.

REFERENCES

Kliment, V. 1973. Similarity and dimensional analysis, evaluation of aerosol deposition in the lungs of laboratory animals and man. Folia Morphol. 21:59–64.

Miller, F. J., J. H. Overton, Jr., R. H. Jaskot, and D. B. Menzel. 1985. A model of the regional uptake of gaseous pollutants in the lung. I. The sensitivity of the uptake of ozone in the human lung to lower respiratory tract secretions and exercise. Toxicol. Appl. Pharmacol. 79:11–27.

Overton, J. H., R. C. Graham, and F. J. Miller. 1987. A model of the regional uptake of gaseous pollutants in the lung. II. The sensitivity of ozone uptake in the laboratory animal lungs to anatomical and ventilatory parameters. Toxicol. Appl. Pharmacol. 88:418–432.

Schreider, J. P., and J. O. Hutchens. 1980. Morphology of the guinea pig respiratory tract. Anat. Rec. 196:313–321.

Weibel, E. R. 1963. Morphometry of the Human Lung. New York: Academic Press.

Yeh, H. C., G. M. Schum, and M. T. Duggan. 1979. Anatomic models of the tracheobronchial and pulmonary region of the rat. Anat. Rec. 195:483–492.

Development of a Physiologically Based Pharmacokinetic Model for Multiday Inhalation of Carbon Tetrachloride

Dennis J. Paustenbach, Harvey J. Clewell III, Michael L. Gargas, and *Melvin E. Andersen*

INTRODUCTION

The extrapolation of animal data to humans has generally been considered a nonquantitative exercise that is affected by numerous unknown biological factors. It has been recognized that this process, often called scaleup or extrapolation, should be influenced by differences in metabolism, body size, and pharmacokinetics between the test species and man. In the past, because a mathematical method for describing these differences was not clearly defined, safety factors were used in an attempt to account for these differences between species (Calabrese, 1983; Dourson and Stara, 1983; Gaylor, 1983; Krewski et al., 1984; National Research Council, 1975; Weil, 1972; Zielhuis and van der Kreek, 1979 a,b).

Physiologically based pharmacokinetic (PB-PK) modeling offers a promising approach for scaling up animal data to predict kinetic behavior in humans (Andersen, 1981a,b; Dedrick and Bischoff, 1980; Himmelstein and Lutz, 1979). These models have been developed for styrene (Ramsey and Andersen, 1984), polybrominated biphenyls (Dedrick, 1973), tetrachlorinated dibenzofurans (King et al., 1983), hepatic glutathione depletion (D'Souza et al., 1987), and methylene chloride (Andersen et al., 1987a). In general, these models quantitatively describe the kinetic behavior of the parent molecule and its metabolites in the test species, and ideally, they successfully predict the metabolism and elimination of that molecule in humans (Clewell and Andersen, 1985). The underpinnings of physiological models have been reviewed by Ramsey and Andersen (1984).

In this paper, we develop and apply a PB-PK model for carbon tetrachloride (CCl_4). Because CCl_4 has been of considerable interest in toxicology, there were sufficient animal and human data to construct and validate the model (Cornish, 1980; Recknagel, 1983). Specifically, the distribution and/or elimination of CCl_4 in the rat have been studied by Dambrauskas and Cornish (1970), Shimizu et al. (1973), Paustenbach et al. (1986a,b), and Uemitsu et al. (1985). The kinetic behavior in the mouse (Bergman, 1979), dog (Robbins, 1929), and monkey (McCollister et al., 1951) has also been evaluated. In those species studied, about 45% of the inhaled $^{14}CCl_4$ was eliminated unchanged in the breath, 45% of the ^{14}C activity was in the feces, and about 2% was in the breath as $^{14}CO_2$. The remaining ^{14}C activity, approximately 7%, was eliminated in the urine. Paustenbach et al. (1986a,b) studied the effects of repeated exposure as well as unusually long periods of exposure. Because of the thoroughness of the data collection, these data were used to validate this model.

In this paper, the PB-PK model was used to predict the behavior of CCl_4 in Sprague-Dawley rats for repeated exposure to schedules of 8 and 11.5 h/day, and the results were compared against actual laboratory data. The behavior of inhaled CCl_4 in humans was also predicted, and the results were compared with previously published data describing the elimination of CCl_4 in human volunteers. Lastly, the model was used to study the potential for day-to-day accumulation of CCl_4 in the adipose tissue of rats and humans following repeated inhalation exposure to 5 ppm CCl_4 (the current American Conference of Governmental Industrial Hygienists [ACGIH] threshold limit value [TLV]) for periods of 8 and 12 h/day.

METHODS

Data Base

Paustenbach et al. (1986a,b) published time course data obtained in male Sprague-Dawley rats following repeated inhalation exposure (average body weight, about 400 g) to CCl_4 at 100 ppm for 8 and 11.5 h/day for up to 10 of 14 days. At a number of serial sacrifices, the concentration of ^{14}C activity in seven tissues and the blood was determined. Samples of fat, liver, adrenal, kidney, brain, heart, spleen, and blood were collected after 1, 2, 3, 4, 5, 7, and 10 days of exposure. A sufficient number of samples was collected to describe the elimination of $^{14}CCl_4$ and $^{14}CO_2$ in the expired air and ^{14}C activity in the urine and feces (Paustenbach et al., 1983).

Elimination

In the study by Paustenbach et al. (1986a,b), the elimination of ^{14}C activity was measured in the expired air, urine, and feces for up to 100 h following exposure, and various half-lives of elimination were determined. Following 2 weeks of exposure to the 8-h/day schedule. $^{14}CCl_4$ in the breath and ^{14}C activity in the feces comprised 45% and 48% of the total ^{14}C excreted. Following 2 weeks of exposure to the 11.5-h/day schedule, the values were 32% and 62%, indicating that repeated exposure to the longer schedule altered the elimination of CCl_4. Regardless of the period of exposure, less than 8% of the inhaled $^{14}CCl_4$ was excreted in the urine and less than 2% was exhaled in the breath as the $^{14}CO_2$ metabolite (Paustenbach et al., 1986a).

Humans

Stewart et al. (1961) exposed human volunteers to 49 or 10 ppm for periods of 70 and 180 mins, respectively. The parent CCL_4 measured in the expired air of these volunteers showed a biphasic elimination that was not unlike that collected in the rat (Paustenbach et al., 1986a), monkey (McCollister et al., 1951), or mouse (Bergman, 1979). We estimated that the α and β half-lives were approximately 50 and 240 min, respectively.

Physiological Modeling

The data of Paustenbach et al. (1986a,b), together with the results of gas-uptake data and handbook data on physiological parameters, were sufficient for developing a PB-PK model. The data of Stewart et al. (1961) provided the mechanism for validating the model's ability to scale up animal data to predict the response in humans.

Ramsey and Andersen (1984) described a physiologically based model for examining the kinetic behavior of inhaled gases and vapors that are essentially nonirritating to the respiratory tract. In this approach, which is similar to that developed by Riggs (1970) and Fiserova-Bergerova and Holaday (1979), the body is lumped into tissue groups analogous to (1) highly perfused organs, excluding the liver; (2) muscle and skin; (3) fat; and (4) organs with a high capacity to metabolize the inhaled chemical. The physiological parameters of the metabolizing tissue groups are essentially those for the liver, although for some chemicals it could represent the kidney, lung, or skin. Blood flows and organ volumes are generally based on literature values for these parameters (Altman and Dittmer, 1979; Snyder, 1975). Organ partition coefficients and metabolic constants for each chemical are determined by appropriate experimentation (Gargas et al., 1986b). These various constants are used in the four mass-balance differential equations that describe the time-dependent changes of tissue concentrations in each of the compartments. Simulations of expected behavior are conducted with a commercial software package (Advanced Continuous Simulations Package, Mitchell & Gauthier Associates, Inc., Cambridge, Mass.) on a CDC 6700 computer (Agin and Blau, 1982). The PB-PK model for CCl_4 has been thoroughly described elsewhere (Paustenbach et al., in press).

Metabolites

The physiological model for volatile compounds used previously (Ramsey and Andersen, 1984) was modified to account for the proposed metabolism and elimination model for CCl_4. In the study by Paustenbach et al. (1986a), the chemical structures of the radioactive metabolites found in the urine and feces were not identified; therefore, kinetic constants were estimated for the model based on the laboratory data. In the schematic of the CCl_4 model (Figure 1), inhaled CCl_4 is either exhaled unchanged or converted to a metabolite. The model accounts for metabolism and excretion of these metabolites via the urine, feces, and breath (as CO_2). Using the actual K_m and V_{max} for the rat, as determined in gas-uptake studies, it was possible to predict the total quantity of metabolites that would be produced.

In the metabolite portion of the model, total metabolite was apportioned to three pools: material excreted in the feces (the major portion of the metabolized CCl_4), material excreted in the urine, and material eliminated as exhaled CO_2. Elimination of these various metabolite pools was assumed to follow first-order behavior for all the material in the compartment. The rate constants for elimination of metabolites via urinary, fecal, and carbon dioxide metabolites were K_2, K_3, and K_1, respectively. The total amount of CCl_4 converted to a metabolite and subsequently eliminated is given the value of 1.0. This is equivalent to the ^{14}C activity excreted in the urine (A_2), as $^{14}CO_2$ in the breath (A_1), and ^{14}C in the feces (A_3).

To develop a model that could simulate the rat data of Paustenbach et al. (1986a) the standard physiological model of Ramsey and Andersen (1984) and Andersen et al. (1987b) was modified to account for four routes of elimination, repeated exposure, varying times of exposure, and the formation of a second or a third metabolite, which may have been eliminated in the feces and/or the urine.

Model Parameters

To develop any physiological model, data need to be either collected in the laboratory or obtained from the literature. Physiological descriptions of most test animals have been developed (Altman and Dittmer, 1979;

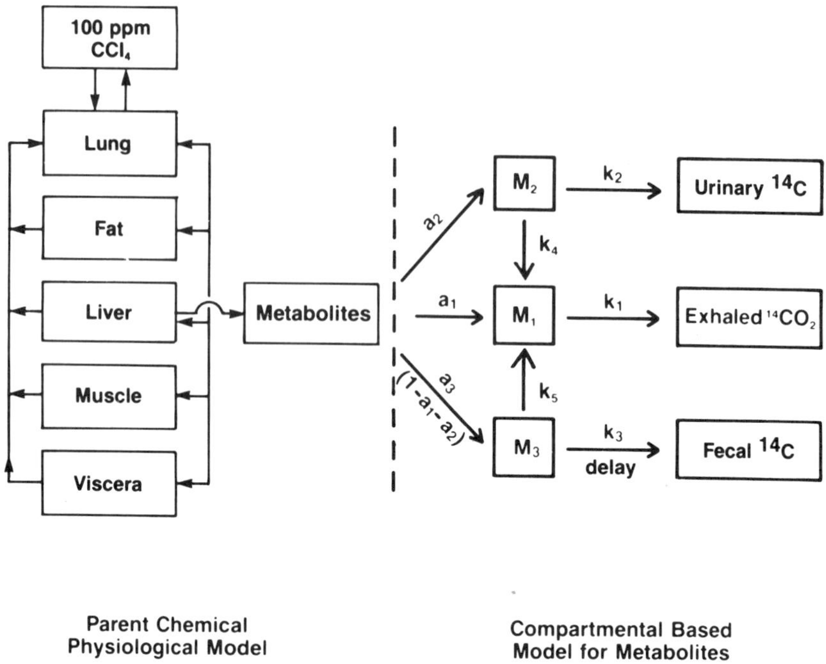

FIGURE 1 Schematic of the physiologically based pharmacokinetic (PB-PK) model used to describe the absorption, metabolism, and excretion of $^{14}CCl_4$ in exposed rats. a_1, a_2, and a_3 describe the amounts of ^{14}C in the elimination compartments; and K_1, K_2, and K_3, respectively, are the rate constants used to describe that elimination. In this model, transport between compartments M_2 and M_1 is accounted for by using the rate constant K_4 and between M_3 and M_1 by using K_5. This later transport accounts for the slow oxidative metabolism of metabolites to CO_2.

Dorato et al., 1983). The tissue volume, perfusion rate, ventilation rate, and percentage of body fat for any given body weight of test animal are estimated (based on literature values). Because tissue volumes do not always scale up in proportion to body weight (e.g., liver, kidney, and bone), some investigators have suggested that an alternative method involving uniformity of perfusion between species should be used (Boxenbaum, 1980, 1982). For humans, values for most anatomical and physiological parameters are also available (Snyder, 1975). The partition coefficients for various tissues and the blood, as well as the kinetic constants for important metabolic pathways, also need to be determined or estimated for volatile chemicals; the latter are usually determined using gas-uptake procedures (Gargas et al., 1986a,b).

Partition Coefficients

Partition coefficients for CCl_4 were determined by using a vial equilibration technique (Sato and Nakajima, 1979) in which CCl_4 was added to a closed vial containing test liquid. Partitioning was determined by estimating the amount of disappearance from the headspace after 1 or 2 h of incubation at 37°C. These are reported in the legend to Figure 2.

Biochemical Constants

Carbon tetrachloride is metabolized via an oxidative pathway involving cytochrome P-450. Lipid peroxidation, presumably initiated by a free radical metabolite of CCl_4 (Butler, 1961), appears to be the most important factor in CCl_4-induced liver injury (Glende, 1972; Recknagel and Glende, 1973). In this work, we estimated the values of V_{max} and K_m using the gas-uptake simulation approach of Gargas et al. (1986b). This is consistent with that reported by Uemitsu (1986).

Validation

The model was validated by comparison with published data. Specifically, the model was used to predict the concentration of ^{14}C activity in the exhaled breath, urine, feces, and adipose tissue, as well as $^{14}CO_2$ in expired air following inhalation of 100 ppm for exposure periods of 8 and 11.5 h/day in the rat (Paustenbach et al., 1986a,b). In addition, a description of the kinetic behavior following repeated exposures to both schedules, for up to 10 of 14 days, was attempted. The ability of this model to accurately describe the elimination of $^{14}CCl_4$ in the breath of humans exposed to 70 ppm for 10 min (Stewart et al., 1961) was also attempted. A more thorough description of the CCl_4 model is described elsewhere (Paustenbach et al., 1988).

RESULTS

Validity of the PB-PK Model

We compared data describing the elimination of ^{14}C activity in the Sprague-Dawley rat for all four excretory pathways with the kinetic behavior (uptake, metabolism, and elimination) predicted by the CCl_4 model. The model (Figure 2A) accurately predicted accumulation of ^{14}C activity in the adipose tissue of test animals following repeated exposure, illustrating that in the rat there is a rapid attainment of steady-state levels in fat following repeated exposure to 100 ppm for 8 h/day.

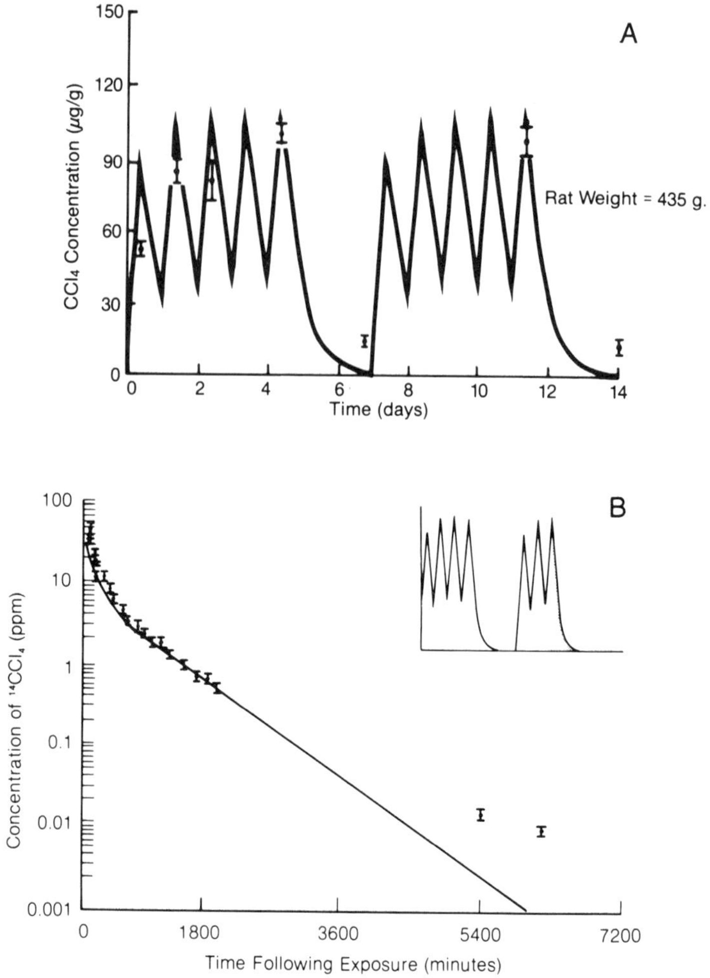

FIGURE 2 Comparison of the model predicted and actual concentrations of ^{14}C activity (expressed as CCl_4) in the adipose tissue of rats exposed to 100 ppm of $^{14}CCl_4$ for 8 h/day, 5 + 5 days (A). Concentration of CCl_4 in the exhaled breath of rats following exposure to 100 ppm of $^{14}CCl_4$ for 11.5 h/day for 7 days (4 of 7, plus 3 of 7 days) (B). The solid line represents the computer-predicted concentration whereas the actual data are shown for the various time points (± standard error). These simulations were based on the actual weight of the rat (as shown). Physiological parameters for the rat (0.42 kg) used in the PB-PK model were: cardiac output, 8.15 liters/h; alveolar ventilation, 7.9 liters/h. Tissue volumes (as percentage of total) were: liver, 4%; fat, 25%; muscle, 7.5%; richly perfused organs, 57%. Blood flows (as percentage of total) were: liver, 25%; fat, 6%; muscle, 15%; richly perfused tissue, 54%. For the rat, the blood:air partition was 4.52, fat:blood was 79.4, liver:blood was 3.13, and muscle: blood was 2.00.

The model (Figure 2B) also predicted the elimination of parent $^{14}CCl_4$ in the expired air of rats following repeated exposure to 100 ppm. This plot shows the actual versus model predicted elimination following the day 7 (4 + 3 days) of exposure to the 11.5-h/day schedule. These simulations were especially sensitive to changes in body weight because CCl_4 is lipophilic and a larger animal (over 350 g) has a greater proportion of fat than does a smaller animal. This produces storage of larger amounts of CCl_4 in fat during exposure and prolonged metabolism of CCl_4 after cessation of exposure.

Elimination via $^{14}CO_2$ and Urine

The behavior of the metabolites in the urine and breath (as $^{14}CO_2$) was more difficult to describe than that of the parent compound in the breath. Interestingly, even with our simple description of the kinetics of the elimination of radioactivity, there was reasonably good agreement between the elimination of $^{14}CO_2$ in the breath with that predicted by the model (Figure 3A). Similarly, adequate results were obtained for ^{14}C activity eliminated in the urine and feces (Figures 3B and C). In both cases, the biological data and model simulation accurately described exposures of 11.5 h/day for 7 of 14 days (4 of 7 days, followed by 3 of 7 days).

One advantage of the PB-PK approach is that when biological behavior is not well predicted by the model (e.g., CO_2 and urine), there is a suspicion that other important factors were omitted from the model description. In this example, we assumed that CO_2 elimination was fast and monoexponential. This model failed to accurately reproduce $^{14}CO_2$ elimination because it did not account for the formation of CO_2 from other compartments. Accordingly, the movement of radiolabel from the fecal and urinary compartments to the CO_2 compartment was included to improve the description. K_4 was the most important constant, and it accounted for the movement of long-lived fixed (bound) radioactivity in tissues and its release as CO_2 over time.

Using the published data of Stewart et al. (1961), we scaled the PB-PK model for the rat to humans. In their work, healthy volunteers were exposed to either 10 or 49 ppm for 10 or 70 min, respectively. Expired air was collected for 77 h after exposure. As shown (Figure 4A), there is good agreement between the actual human data and what was predicted by the model. This agreement is especially impressive in that the model was based on data obtained only in rats at exposures of 100 ppm and the use of a different duration of exposure.

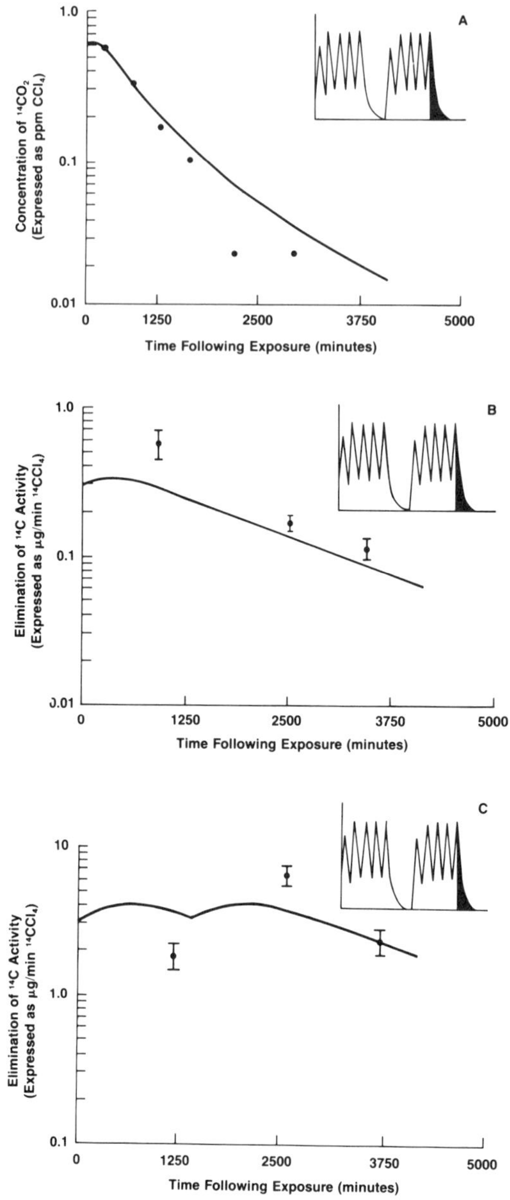

FIGURE 3 Comparison of the actual versus predicted concentration of $^{14}CO_2$ in the expired air of rats exposed to 100 ppm of CCl_4 for 8 h/day for 10 days (5 of 7, plus 5 of 7 days) (A), the actual versus predicted concentration of ^{14}C activity eliminated in the urine (B), and ^{14}C activity in the feces (C) of rats exposed to the same schedule. The model was initially unable to accurately describe the formation and elimination of CO_2 for periods after 3,600 min. This was later corrected by the addition of K_4 to the model.

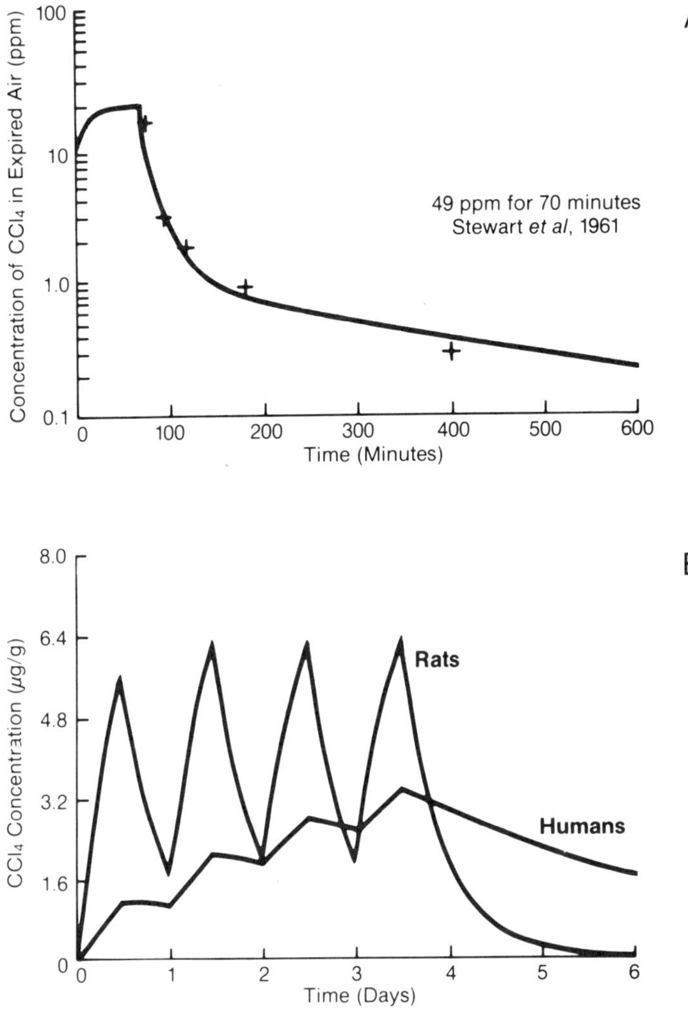

FIGURE 4 Comparison of the actual versus predicted concentration of CCl_4 in the expired breath of humans exposed to 49 ppm of CCl_4 for 70 min (A). The experimental data for humans was obtained from Stewart et al. (1961). This simulation of the predicted human response was based solely on the scale up of data collected in the Sprague-Dawley rat. The model was used to predict behavior for repeated inhalation exposure to 5 ppm of CCl_4 for both the rat and human (B). The PB-PK model predicted that rats exposed to 5 ppm for 8 h/day will not accumulate CCl_4 or its metabolites in adipose tissue with repeated exposure. In contrast, humans exposed to 5 ppm of CCl_4 for 8 h/day (the current ACGIH TLV concentration) would be expected to have day-to-day increases in adipose tissue levels. Physiological parameters for the human (70 kg) used in the PB-PK model were: cardiac output, 256 liters/h; alveolar ventilation, 254 liters/h. Tissue volumes (as percentage of total) were: liver, 4%; fat, 20%; muscle, 62%; richly perfused tissue, 5%. Blood flows (as percentage of total) were: liver, 25%; fat, 6%; muscle, 18%; and richly perfused organs, 51%. For humans, the blood:air partition was 2.7.

A validated PB-PK model can also be used to examine the behavior of a chemical in humans under a number of different exposure scenarios (Figure 4B). In this example, the model shows that rats repeatedly exposed to 5 ppm would not demonstrate CCl_4 accumulation with repeated exposure. In contrast, when humans are repeatedly exposed to 5 ppm of CCl_4 (the current ACGIH-TLV), there is a day-to-day accumulation in adipose tissue. Due to degassing after a weekend away from exposure, it is not anticipated that this degree of accumulation poses a measurable increased risk to exposed persons. Additional development of PB-PK models for volatile chemicals should be very useful in allowing us to better understand the various margins of safety presently implicit in the more than 1,000 occupational exposure limits that have been established for workers (Cook, 1987; WHO, 1986).

DISCUSSION

A PB-PK model that accurately describes the behavior of CCl_4 in rats was developed by using fundamental biological data and was validated by comparison with rat and human kinetic data. The model was capable of describing CCl_4 pharmacokinetics for a variety of exposure regimens and for concentrations between 10 and 100 ppm. We obtained an excellent description of the elimination of parent CCl_4 in the exhaled breath and were able to predict the concentration of CCl_4 in the adipose tissue of rats at any time following exposure to either an 8- or 11.5-h/day dosing regimen. By making some basic assumptions, we also predicted the time course of formation and elimination of CCl_4 metabolites in the urine, the breath (CO_2), and the feces. The usefulness of applying the PB-PK approach to solving industrial hygiene questions, such as the modification of exposure limits for long work shifts, has been discussed (Paustenbach, 1985).

It was difficult to simulate adequately the elimination of ^{14}C activity in feces and CO_2 by using a model involving just one metabolite. It appears likely that two or more metabolites are formed and then eliminated in the feces and that perhaps this is responsible for the latent production of $^{14}CO_2$ seen in the exhaled breath. Whenever this occurs, plausible biological reasons for these deviations can be developed; these can be incorporated into the model (Figure 1). We have found that one benefit of developing a PB-PK model is to better understand the potential mechanism of action of the test chemical, and from this one can identify the optimal laboratory experiments to be used to examine the kinetics of toxicity of a particular chemical.

While there is always the danger of overinterpreting disposition studies which monitor only radioactivity, the combination of information on parent chemical, particular metabolites ($CHCl_3$ and CO_2), and total radioactivity, allow some conclusions to be made regarding the pathways of metabolism of CCl_4 and the persistence of bound metabolites derived from this chemical. The V_{max} of CCl_4 is very low, about 0.28 mg/h (1.81 μmol/h) in a 300-g rat. In contrast, the V_{max} for trichloroethylene is 37.6 μmol/h (Andersen et al., 1987c). With CCl_4 the vast majority of the metabolized material becomes bound in compartments that lead to slow elimination in the feces and urine. Thus, a_1 plus a_2 is 93.5% of total metabolism. Only 6% is metabolized to CO_2 with a rapid elimination rate constant. Paustenbach et al. (1986) reported that some 3% of the exhaled material trapped on charcoal was $CHCl_3$ and not CCl_4. This amount of radioactivity represents only 2–3% of the total metabolized dose. Thus, the sum of CO_2 and $CHCl_3$ is somewhat less than 10% of the amount metabolized. The CO_2 produced and eliminated with a rapid rate process is most likely a secondary product of $CHCl_3$ metabolism.

Lipid peroxidation initiated by CCl_4 is generally regarded to represent abstraction of a proton from polyunsaturated fatty acids. This labelizes the fatty acid structure to further oxidation and produces $CHCl_3$. From the result of modeling CCl_4 metabolism, this type of peroxidative process is clearly only a minor pathway for CCl_4 metabolism, accounting, in this study at 100 ppm, for only 0.028 mg/h of total CCl_4 metabolism.

Importantly, we found that the PB-PK model could predict the elimination of unchanged CCl_4 in the breath of exposed humans based solely on the partition data collected in human blood, pharmacokinetic data collected in rats, and typical scaleup of parameters. The ability to predict the behavior of chemicals at various concentrations through models that account for the metabolism and pharmacokinetics of a chemical in lower mammalian species should be a powerful tool for assessing the human risk of exposure to these agents. Already, these models have been used to estimate better occupational exposure limits for persons who are exposed to airborne chemicals during shifts as long as 10, 12, or 14 h/day (Andersen et al., 1987b). PB-PK models have also been used to improve the quantitative risk assessment process for methylene chloride (Andersen et al., 1987a), and this appears to be an important new use for this approach.

In this model, we were successful in describing metabolism by assuming a single, time-independent, oxidative pathway for CCl_4 metabolism. It is well-documented, however, that CCl_4 can cause destruction of the cytochrome P-450 enzyme system under various *in vivo* exposure conditions. The success of this present model may be fortuitous, and there may well be time-dependent changes in V_{max} during exposure of rats to a concentration of 100 ppm. In this case, the basal V_{max} level would be larger than that estimated in this study, and the residual V_{max} level at the end of exposure would be lower than the 0.4 mg/h/kg estimated by gas-uptake methods. Andersen et al. (1987a) have described techniques for examining suicide enzyme inhibition by reactive metabolites *in vivo* by using various exposure conditions in the gas-uptake chamber.

One such approach is coexposure to binary mixtures of structurally related chemicals, where one chemical is very well metabolized and the other is suspected of causing destruction of the major oxidative enzymes responsible for the metabolism of the two chemicals. The chemical suspected of inhibiting metabolism is added first, and after a variable preexposure period, the second well-metabolized chemical is added to the chamber atmosphere. The experiment then consists of observing the loss of metabolizing capacity for the second substrate. These experiments are now under way in our laboratory in Dayton, Ohio, with CCl_4 and $CHCl_3$. It would be premature to use this present model for risk assessment calculations until these types of inhibition experiments are completed and fully interpreted.

REFERENCES

Agin, G. L., and G. E. Blau. 1982. Application of DACSL (Dow Advanced Continuous Simulation Language) to the design and analysis of chemical reactors. AICHE Symposium Series No. 214. 78:108–118.

Altman, P. L., and D. S. Dittmer, eds. 1979. P. 1582 in Biology Data Book, 2nd ed. Vol. II. Bethesda, Md.: Federation of American Societies for Experimental Biology.

American Conference of Governmental Industrial Hygienists, Threshold Limits Committee. 1986. Threshold limit values for chemical substances in the workroom air adopted by ACGIH for 1986–87. Cincinnati, Ohio: American Conference of Governmental Industrial Hygienists.

Andersen, M. E. 1981a. A physiologically based toxicokinetic description of the metabolism of inhaled gases and vapors: Analysis at steady state. Toxicol. Appl. Pharmacol. 60:509–526.

Andersen, M. E. 1981b. Pharmacokinetics of inhaled gases and vapors. Neurobehav. Toxicol. Teratol. 3:383–389.

Andersen, M. E., H. J. Clewell III, M. L. Gargas, F. A. Smith, and R. H. Reitz. 1987a. Physiologically-based pharmacokinetics and the risk assessment process for methylene chloride. Toxicol. Appl. Pharmacol. 87:185–205.

Andersen, M. E., M. G. MacNaughton, H. J. Clewell III, and D. J. Paustenbach. 1987b. Adjusting exposure limits for long and short exposure periods using a physiological pharmacokinetic model. Am. Ind. Hyg. Assoc. J. 48:335–341.

Bergman, K. 1979. Whole-body autoradiography and allied tracer techniques in distribution and elimination studies of some organic solvents (including carbon tetrachloride). Scand. J. Work Environ. Health 59:1–263.

Boxenbaum, H. 1980. Interspecies variation in liver weight, hepatic blood flow, and antipyrine intrinsic clearance: Extrapolation of data to benzodiazepines and phenytoin. J. Pharmacokinet. Biopharm. 8:165–176.

Boxenbaum, H. 1982. Interspecies scaling, allometry, physiological time, and the ground plan of pharmacokinetics. J. Pharmacokinet. Biopharm. 10:201–227.

Butler, T. C. 1961. Reduction of carbon tetrachloride in-vivo and reduction of carbon tetrachloride and chloroform in-vitro by tissues and tissue constituents. J. Pharmacol. Exp. Ther. 134:311–319.

Calabrese, E. J. 1983. Principles of Animal Extrapolation. New York: John Wiley & Sons.

Clewell, H. J., III, and M. E. Andersen. 1985. Use of physiologic pharmacokinetics in risk assessment. J. Toxicol. Ind. Health 1:111–122.

Cook, W. A. 1987. A Compendium of World-Wide Occupational Exposure Limits. Akron, Ohio: American Industrial Hygiene Association.

Cornish, H. H. 1980. Solvents and vapors. In Casarett and Doull's Toxicology, 2nd ed., J. Doull, C. D. Klaassen, and M. O. Amdur, eds. New York: Macmillan.

Dambrauskas, T., and H. H. Cornish. 1970. Effect of pretreatment of rats with carbon tetrachloride on tolerance development. Toxicol. Appl. Pharmacol. 17:83–97.

Dedrick, R. L. 1973. Animal scale-up. J. Pharmacokinet. Biopharm. 1:435–461.

Dedrick, R. L., and K. B. Bischoff. 1980. Species similarities in pharmacokinetics. Fed. Proc. 39:54–59.

Dorato, M. A., K. H. Carlson, and D. L. Copple. 1983. Pulmonary mechanics in conscious Fischer 344 rats: Multiple evaluations using nonsurgical techniques. Toxicol. Appl. Pharmacol. 68:344–353.

Dourson, M. L., and J. F. Stara. 1983. Regulatory history and experimental support of uncertainty (safety) factors. Reg. Toxicol. Pharm. 3:224–238.

D'Souza, R. W., W. R. Francis, and M. E. Andersen. 1987. A physiologic model for tissue, glutathione depletion and increased resynthesis following ethylene dichloride exposure. J. Pharmacol. Exp. Ther.

Fiserova-Bergerova, V., and D. A. Holaday. 1979. Uptake and clearance of inhalation anesthetics in man. Drug Metab. Rev. 9(1):43–60.

Gargas, M. L., H. J. Clewell III, and M. E. Andersen. 1986a. Metabolism of inhaled dihalomethanes *in vivo*: Differentiation of kinetic constants for two independent pathways. Toxicol. Appl. Pharmacol. 82:211–223.

Gargas, M. L., H. J. Clewell III, and M. E. Andersen. 1986b. A physiologically-based simulation approach for determining metabolic constants from gas uptake data. Toxicol. Appl. Pharmacol. 86:341–352.

Gaylor, D. W. 1983. The use of safety factors for controlling risk. J. Toxicol. Environ. Health. 11:329–336.

Glende, E. A., Jr. 1972. On the mechanism of carbon tetrachloride toxicity—coincidence of loss of drug metabolizing activity with perioxidation of microsomal lipid. Biochem. Pharmacol. 21:2131–2138.

Himmelstein, K. J., and R. J. Lutz. 1979. A review of the applications of physiologically based pharmacokinetic modeling. J. Pharmacokinet. Biopharm. 7:127–145.

King, F. G., R. L. Dedrick, J. M. Collins, H. B. Matthews, and L. S. Birnbaum. 1983. A physiological model for the pharmacokinetics of 2,3,7,8-tetrachlorodibenzofuran in several species. Toxicol. Appl. Pharmacol. 67:390–400.

Krewski, D., C. Brown, and D. Murdoch. 1984. Determining "safe" levels of exposure: Safety factors or mathematical models. Fund. Appl. Toxicol. 4:S383–S384.

McCollister, D. D., W. H. Beamer, G. J. Atchison, and H. C. Spencer. 1951. Absorption, distribution and elimination of radioactive carbon tetrachloride by monkeys upon exposure to low concentrations of vapor. J. Pharmacol. Exp. Ther. 102:112–120.

National Research Council. 1975. Principles for Evaluating Chemicals in the Environment. Washington, D.C.: National Academy of Sciences. 454 pp.

Paustenbach, D. J. 1985. Occupational exposure limits, pharmacokinetics, and unusual work schedules. Pp. 111–277 in Patty's Industrial Hygiene and Toxicology, Vol. IIIA, L. J. Cralley and L. V. Cralley, eds. New York: John Wiley & Sons.

Paustenbach, D. J., G. P. Carlson, G. S. Born, J. E. Christian, and J. E. Rausch. 1983. A dynamic closed-loop recirculating inhalation chamber for conducting pharmacokinetic and short-term toxicity studies. Fund. Appl. Toxicol. 21:128–132.

Paustenbach, D. J., G. P. Carlson, J. E. Christian, and G. S. Born. 1986a. A comparative study of the pharmacokinetics of carbon tetrachloride in the rat following repeated inhalation exposures of 8 and 11.5 hr/day. Fund. Appl. Toxicol. 6:484–497.

Paustenbach, D. J., J. E. Christian, G. P. Carlson, and G. S. Born. 1986b. The effect of an 11.5 hr/day exposure schedule on the distribution and toxicity of inhaled carbon tetrachloride in the rat. Fund. Appl. Toxicol. 6:472–483.

Paustenbach, D. J., M. E. Andersen, H. J. Clewell III, and M. L. Gargas. 1988, submitted. A physiologically-based pharmacokinetic model for inhaled carbon tetrachloride in the rat. Toxicol. Appl. Pharmacol.

Ramsey, J. C., and M. E. Andersen. 1984. A physiologically based description of the inhalation pharmacokinetics of styrene in rats and man. Toxicol. Appl. Pharmacol. 73:159–175.

Recknagel, R. O. 1983. A new direction in the story of carbon tetrachloride hepatotoxicity. Life Sci. 33:401–408.

Recknagel, R. O., and E. N. Glende. 1973. Carbon tetrachloride hepatotoxicity: An example of lethal cleavage. Crit. Rev. Toxicol. 2:263–297.

Riggs, M. 1970. Chapt. 13 in The Mathematical Approach to Physiological Problems. Cambridge, Mass.: MIT Press.

Robbins, B. H. 1929. The absorption, distribution and excretion of carbon tetrachloride in dogs under various conditions. J. Pharmacol. 37:203–216.

Sato, A., and T. Nakajima. 1979. Partition coefficients of some aromatic hydrocarbons and ketones in water, blood and oil. Br. J. Ind. Med. 36:231–234.

Shimizu, Y., C. Nagase, and K. Kawai. 1973. Accumulation and toxicity of carbon tetrachloride after repeated inhalation in rats. Ind. Health 11:48–54.

Snyder, W. S. 1975. Report of the Task Group on Reference Man. International Commission on Radiological Protection, Report No. 23. Elmsford, N.Y.: Pergamon.

Stewart, R. D., H. H. Gay, D. S. Erley, C. L. Hake, and J. E. Peterson. 1961. Human exposure to carbon tetrachloride vapor—relationship of expired air concentration to exposure and toxicity. J. Occup. Med. 3:586–590.

Uemitsu, N. 1986. Inhalation pharmacokinetics of carbon tetrachloride in rats based on arterial blood:inhaled air concentration ratios. Toxicol. Appl. Pharmacol. 83:20–29.

Uemitsu, N., Y. Minobe, and H. Nakoyoshi. 1985. Concentration-time-response relationship under conditions of single inhalation of carbon tetrachloride. Toxicol. Appl. Pharmacol. 77:260–266.

Weil, C. S. 1972. Statistics versus safety factors and scientific judgement in the evaluation of safety for man. Toxicol. Appl. Pharmacol. 21:454–463.

World Health Organization (WHO). 1987. Occupational exposure limits. Geneva, Switzerland.

Zielhuis, R. L., and F. van der Kreek. 1979a. Calculations of a safety factor in setting health based permissible levels for occupational exposure. A proposal I. Int. Arch. Occup. Environ. Health 42:191–201.

Zielhuis, R. L., and F. W. van der Kreek. 1979b. Calculations of a safety factor in setting health based permissible levels for occupational exposure. A proposal II, comparison of extrapolated and published permissible levels. Int. Arch. Occup. Environ. Health 42:203–215.

The Delivered/Administered Dose Relationship and Its Impact on Formaldehyde Risk Estimates

Thomas B. Starr

LIMITATIONS OF CONVENTIONAL LOW-DOSE RISK EXTRAPOLATION

It is now well-established that the typical chronic bioassay lacks sufficient power to discriminate among the different mathematical dose-response models that are commonly employed for low-dose extrapolation purposes (Krewski et al., 1983). Indeed, even the so-called megamouse or ED_{01} study of 2-acetylaminofluorine, in which over 24,000 mice were used, was inadequate to this task (Brown and Hoel, 1983a, b; SOT Task Force, 1981). Generally, several dose-response models each provide an adequate fit to tumor incidence data in the observable response range, and yet the predicted risks at exposure levels below this range differ from one another by orders of magnitude. Low-dose extrapolations of risk employing data from the Chemical Industry Institute of Toxicology formaldehyde bioassay (Kerns et al., 1983) provide an excellent illustration of this phenomenon (Starr and Buck, 1984).

It is also known that optimization of the experimental design of chronic bioassays (in terms of the number of treatment groups, their relative size, and their placement relative to the maximum tolerated dose) so as to minimize the uncertainty in predicted risks at low exposure levels does not significantly improve the situation (Portier and Hoel, 1983). In essence, knowledge of the risk at high exposure levels is by itself insufficient to predict accurately the risks at low exposure levels.

Also clear is the fact that the various mathematical models of carcinogenesis utilized for low-dose extrapolation were conceptualized initially and formulated in terms of interactions between the biologically active forms of chemicals agents (i.e., the pharmacokinetic delivered dose) and cellular macromolecules in target tissues (Brown, 1976; Cornfield, 1977; Crump, 1979; Hoel et al., 1983). These models lack the structure necessary to characterize the many physiologic and pharmacokinetic factors that are likely to govern the complex relationship between the dose delivered to various target tissues and the dose administered to whole animals. Furthermore, information on internal measures of exposure is not routinely available. Rather, the externally administered dose, e.g., milligrams per kilogram in feed, micrograms per liter in water, or parts per million in chamber air, is usually the only measure of exposure provided by a bioassay.

Consequently, a critical assumption is made when low-dose risk extrapolation is performed in the absence of data regarding internal measures of exposure. This assumption is that the dose administered in a bioassay is a valid linear proxy for the biologically active dose delivered to specific target tissues (EPA, 1986; Starr and Buck, 1984). This is equivalent to assuming that the kinetics of distribution and disposition of chemicals within whole animals are entirely and exactly linear. It is additionally assumed that a simple scaling of the administered dose (e.g., on a relative body weight or body surface area basis, depending on the exposure route) is all that is necessary to convert predicted risks for one species to those for another (EPA, 1986). Indeed, for exposure via inhalation, different species exposed to the same airborne concentration of a chemical for the same fraction of their lifetimes are presumed to experience the same risk (EPA, 1986).

EVIDENCE THAT LINEAR PROPORTIONALITY DOES NOT HOLD FOR FORMALDEHYDE

Subchronic mechanistic studies of formaldehyde have provided strong direct evidence that the dose delivered to the DNA of replicating cells in the respiratory epithelium of the rat nasal cavity is nonlinearly related to the airborne formaldehyde concentration (Casanova-Schmitz and Heck, 1985; Casanova-Schmitz et al., 1984). Specifically, as is shown in Figure 1, significantly less formaldehyde (by approximately a factor of 3) is covalently bound to respiratory mucosal DNA at low airborne concentrations than is predicted by downward linear extrapolation from the amounts of such binding observed at high concentrations.

Related studies have demonstrated that exposure of rats to formaldehyde via inhalation induces the respiratory depression reflex (Chang et al.,

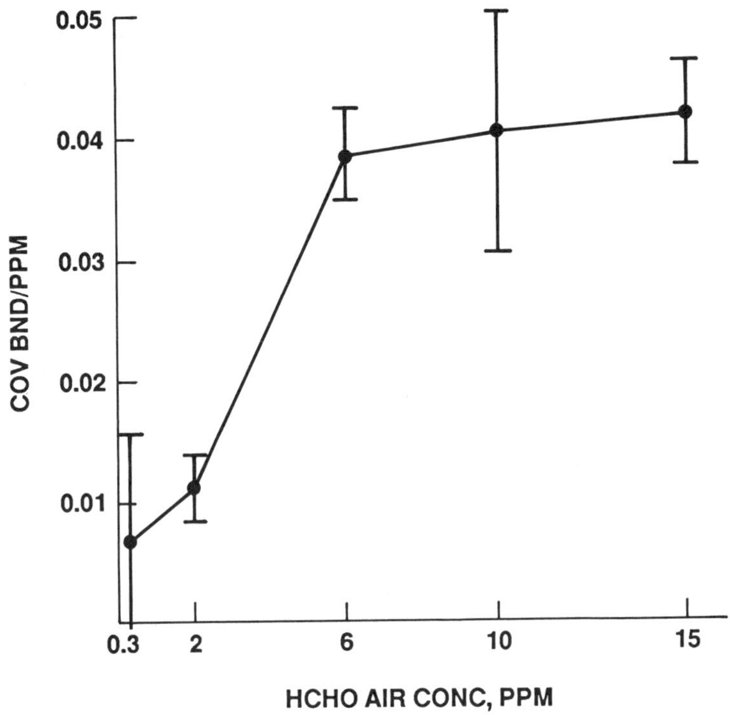

FIGURE 1 Amount of formaldehyde (HCHO) covalently bound (COV BND) to respiratory mucosal DNA of Fischer 344 rats normalized by airborne concentration (i.e., expressed as nanomoles covalently bound/mg of DNA/ppm of HCHO) plotted versus the airborne HCHO concentration to which the animals were exposed. Adapted from Casanova-Schmitz et al. (1984).

1983), an inhibition of mucociliary clearance (Morgan, 1983; Morgan et al., 1983) and intracellular metabolism of formaldehyde (Casanova-Schmitz and Heck, 1985), and stimulation of cell proliferation (Chang et al., 1983; Swenberg et al., 1983, 1986), all as nonlinear functions of the airborne formaldehyde concentration. Because each of these phenomena appears to be an important controlling factor in the relationship between administered and delivered doses (Starr and Gibson, 1985), the nonlinear relationship between covalent binding to DNA and airborne concentration is not unexpected.

The respiratory depression reflex mediates the inhaled dose. Mucociliary clearance mediates the fraction of the inhaled dose that penetrates the mucus layer covering underlying epithelial cells in the nasal cavity. Intracellular metabolism mediates the fraction of formaldehyde entering these cells that remains free to bind with cellular macromolecules, including DNA. Important in this regard is the fact that formaldehyde is an essential biochemical that is normally present in all living cells. it is thus not surprising that efficient metabolic pathways exist for its detoxication. Finally, the rate of cell replication mediates the fraction of DNA that is single stranded, and it is known that formaldehyde binds covalently only to single-stranded DNA (Lukashin et al., 1976; von Hippel and Wong, 1971).

COMPARISON OF RISK ESTIMATES: ADMINISTERED VERSUS DELIVERED DOSE

The implications of these studies for risk assessment have been explored and elucidated (Starr and Buck, 1984). Use of the amount of formaldehyde covalently bound to respiratory mucosal DNA (rather than airborne concentration) as the measure of exposure leads to lower point and upper bound estimates of risk at low doses, *irrespective* of the mathematical dose-response model employed. For illustration, point and upper bound estimates corresponding to an airborne concentration of 1 ppm of formaldehyde are displayed in Tables 1 and 2. It is apparent that risk assessments that do not explicitly incorporate the data regarding covalent binding of formaldehyde to target tissue DNA are likely to overestimate (at least relatively) the cancer risk associated with formaldehyde exposure.

QUESTIONS STILL TO BE RESOLVED

Related mechanistic studies have also provided strong evidence that dramatic *interspecies* differences in the inhaled dose are likely to exist even when the species are exposed identically to the same airborne formaldehyde concentration (Chang et al., 1983). Such differences would presumably also be manifest in the resulting delivered dose, and the re-

TABLE 1 Maximum Likelihood Risk Estimates[a] for 1.0 ppm of Airborne Formaldehyde

Dose Measure	Probit	Logit	Weibull	Multistage
ADM	2.65(−11)[b]	2.87(−6)	5.94(−6)	2.51(−4)
DEL	4.00(−20)	2.15(−8)	7.13(−8)	4.70(−6)
Reduction factor	6.60(+8)	133.0	83.4	53.4

[a]Predicted excess nasal cancer risk for rats exposed to 1 ppm of formaldehyde, 6 h/day, 5 days/week, for up to 24 months.

[b]Numbers in parentheses represent powers of 10.

TABLE 2 Upper 95% Confidence Bounds on Risk for 1.0 ppm of Airborne Formaldehyde

Dose Measure	Probit	Logit	Weibull	Multistage
ADM	2.58(−10)[a]	1.24(−5)	2.54(−5)	1.80(−3)
DEL	7.09(−19)	1.22(−7)	3.98(−7)	6.24(−4)
Reduction factor	3.63(+8)	101.8	63.8	2.9

[a]Numbers in parentheses represent powers of 10.

markable 50-fold disparity in tumor incidence in rats and mice identically exposed to 14.3 ppm of formaldehyde in the CIIT bioassay (Kerns et al., 1983) may be attributable to this phenomenon.

Clearly, interspecies differences in anatomy and physiology, metabolism, and the rates of cell proliferation and repair of DNA damage may each play a critical role in the accurate assessment of the risk to humans from formaldehyde exposure. If the cancer risk in two species as similar as rats and mice can differ by a factor of 50, even though they are identically exposed to the same airborne concentration, can we have much confidence in the extrapolation of risks from rodents to the very different human species under very different exposure conditions?

Additional research is clearly desirable on many aspects of this problem. For example, the above-mentioned phenomena can be studied intensively in nonhuman primates so that the potential effects (on DNA binding and tumor incidence) of differences in anatomy and physiology of the upper respiratory tract can be established. These phenomena also need to be studied following more extended periods of exposure to establish whether or not the dramatic nonlinear effects observed in the short term are representative of effects that occur at later times during the course of a standard 2-year rodent bioassay. Identification of the specific DNA adducts that form as a result of interaction with formaldehyde can also be pursued. Studies that clarify the potential role of these adducts as causal agents in the carcinogenic process must also be conducted. Finally, mechanistic mathematical models that describe quantitatively dependences of these phenomena on airborne concentration must also be developed.

It has been argued by some that pharmacokinetic data cannot be employed in regulatory assessments of carcinogenic risk until uncertainties such as those mentioned above are resolved (cf., Cohn et al., 1985). However, the opposite view, namely, that such data cannot in good conscience be ignored, seems reasonable. At the very least, the comparison of risk estimates based upon pharmacokinetic data with similar estimates derived without such data can expose some of the real uncertainties that would otherwise remain hidden in the many unverified assumptions that are employed in the quantitative risk assessment process.

REFERENCES

Brown, C. C. 1976. Mathematical aspects of dose-response studies in carcinogenesis—the concept of thresholds. Oncology 33:62–65.

Brown, K. G., and D. G. Hoel. 1983a. Modeling time-to-tumor data: Analysis of the ED_{01} study. Fund. Appl. Toxicol. 3:458–469.

Brown, K. G., and D. G. Hoel. 1983b. Multistage prediction of cancer in serially dosed animals with application to the ED_{01} study. Fund. Appl. Toxicol. 3:470–477.

Casanova-Schmitz, M., and H. Heck. 1985. DNA-protein cross-linking induced by formaldehyde (FA) in the rat respiratory mucosa: Dependence on FA concentration in normal rats and in rats depleted of glutathione (GSH). Toxicologist 5:128.

Casanova-Schmitz, M, T. B. Starr, and H. Heck. 1984. Differentiation between metabolic incorporation and covalent binding in the labeling of macromolecules in the rat nasal mucosa and bone marrow by inhaled [^{14}C]- and [^{3}H] formaldehyde. Toxicol. Appl. Pharmacol. 76:26–44.

Chang, J. C. F., E. A. Gross, J. A. Swenberg, and C. S. Barrow. 1983. Nasal cavity deposition, histopathology, and cell proliferation after single or repeated formaldehyde exposure in B6C3F1 mice and F-344 rats. Toxicol. Appl. Pharmacol. 68:161–176.

Cohn, M. S., F. J. DiCarlo, and A. Turturro. 1985. Letter. Toxicol. Appl. Pharmacol. 77:365–368.

Cornfield, J. 1977. Carcinogenic risk assessment. Science 198:693–699.

Crump, K. S. 1979. Dose response problems in carcinogenesis. Biometrics 35:157–167.

EPA. 1986. (U.S. Environmental Protection Agency). Guidelines for carcinogen risk assessment. Fed. Regist. 51:33992–34003.

Hoel, D. G., N. L. Kaplan, and M. W. Anderson. 1983. Implication of nonlinear kinetics on risk estimation in carcinogenesis. Science 219:1032–1037.

Kerns, W. D., K. L. Pavkov, D. J. Donofrio, E. J. Gralla, and J. A. Swenberg. 1983. Carcinogenicity of formaldehyde in rats and mice after long-term inhalation exposure. Cancer Res. 43:4382–4392.

Krewski, D., K. S. Crump, J. Farmer, D. W. Gaylor, R. Howe, C. Portier, D. Salsburg, R. L. Sielken, and J. Van Ryzin. 1983. A comparison of statistical methods for low dose extrapolation utilizing time-to-tumor data. Fund. Appl. Toxicol. 3:140–160.

Lukashin, A. V., A. V. Vologodskii, M. D. Frank-Kamenetskii, Y. L. Lyubchenko. 1976. Fluctuational opening of the double helix as revealed by theoretical and experimental study of DNA interaction with formaldehyde. J. Mol. Biol. 108:665–682.

Morgan, K. T. 1983. Localization of areas of inhibition of nasal mucociliary function in rats following in vivo exposure to formaldehyde. Am. Rev. Respir. Dis. 127:166.

Morgan, K. T., D. L. Patterson, and E. A. Gross. 1983. Formaldehyde and the nasal mucociliary apparatus. Pp. 193–210 in Formaldehyde Toxicology, Epidemiology, and Mechanisms, J. J. Clary, J. E. Gibson, and R. S. Waritz, eds. New York: Dekker.

Portier, C., and D. G. Hoel. 1983. Low-dose-rate extrapolation using the multistage model. Biometrics 39:897–906.

SOT Task Force. 1981. Re-examination of the ED_{01} study: Overview. Fund. Appl. Toxicol. 1:28–63.

Starr, T. B., and R. D. Buck. 1984. The importance of delivered dose in estimating low-dose cancer risk from inhalation exposure to formaldehyde. Fund. Appl. Toxicol. 4:740–753.

Starr, T. B. and J. E. Gibson. 1985. The mechanistic toxicology of formaldehyde and its implications for quantitative risk estimation. Annu. Rev. Pharmacol. Toxicol. 25:745–767.

Swenberg, J. A., E. A. Gross, H. W. Randall, and C. S. Barrow. 1983. The effect of formaldehyde exposure on cytotoxicity and cell proliferation. Pp. 225–236 in Formaldehyde Toxicology, Epidemiology, and Mechanisms, J. J. Clary, J. E. Gibson, and R. S. Waritz, eds. New York: Marcel Dekker.

Swenberg, J. A., E. A. Gross, and H. W. Randall. 1986. Localization and quantitation of cell proliferation following exposure to nasal irritants. Pp. 291–300 in Toxicology of the Nasal Passages, C. S. Barrow, ed. Washington, D.C.: Hemisphere.

von Hippel, P. H., and K. Y. Wong 1971. Dynamic aspects of native DNA structure: Kinetics of the formaldehyde reaction with calf thymus DNA. J. Mol. Biol. 61:587–613.

Pharmacokinetic Simulation as an Adjunct to Experimental Data in Risk Assessment: Predicting Exposure of the Bladder Epithelium in Dogs to Urinary *N*-Hydroxy Metabolites of Carcinogenic Arylamines

John F. Young and *Fred F. Kadlubar*

INTRODUCTION

An analog-digital hybrid computer (Figure 1) (Pearce and Young, 1981; Young et al., 1981) and LOTUS 1-2-3 on an International Business Machines (IBM)-AT personal computer was used to predict the extent of urinary bladder exposure to *N*-hydroxy (*N*-OH) arylamines, which form adducts with urothelial DNA and are believed to serve as ultimate carcinogenic metabolites (Kadlubar et al., 1981). A three-compartment model (Figures 2 and 3) (Young and Kadlubar, 1982) was used to fit the data obtained from rats given an intraurethral instillation of *N*-OH-2-naphthylamine and its acid-hydrolyzable *N*-glucuronide conjugate (Oglesby et al., 1981). This rat model was then assumed to be valid for other species, and simulations were conducted under varying conditions of urinary pH and voiding intervals. These simulations were then used to aid in the data interpretation of experiments with several dogs to which were administered orally the arylamine carcinogen 4-aminobiphenyl. This predictive approach is used to assess the role of urinary pH and voiding interval in the release of the free *N*-OH arylamines in the bladder lumen and the subsequent formation of arylamine-DNA adducts in the urothelium and in the levels of 4-aminobiphenyl-hemoglobin adducts in blood, the latter of which is being used as a potential biological marker of carcinogen exposure (Green et al., 1984).

FIGURE 1 Photograph of the EAI Pacer 500 analog-digital hybrid computer. The large wiring layout is the analog patch panel, and the small wiring layout is the logic panel. All control function inputs and graphic outputs were accomplished via the cathode-ray tube in the right-hand lower corner.

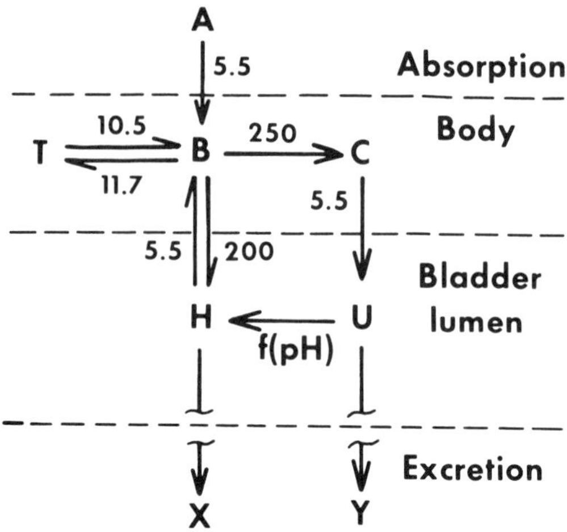

FIGURE 2 Three-compartment (B, T, and H) pharmacokinetic model for the distribution of *N*-hydroxyarylamine (NHAA). A = NHAA to be absorbed; B = NHAA in blood; T = NHAA in tissue; H = NHAA in bladder; X = NHAA excreted; C = NHAA metabolite in blood; U = NHAA metabolite in bladder; and Y = NHAA metabolized and excreted. Values are half-lives in minutes.)

METHODS

The analog-digital hybrid computer (Pacer 500; Electronics Associates, Inc., West Long Branch, N.J.) was used to generate amount-time curves for each component of the model. A set of differential equations was written that described the model (Figure 2) and were wired directly onto the analog patch panel (Figure 1). The schematic diagram of the differential equations describing the model is presented in Figure 3 (Young et al., 1981). The analog portion of the hybrid computer solved all of the differential equations simultaneously; the digital portion sampled each analog curve 100 times during the 1-s solution and created an amount-time matrix. This matrix was then used in iterative schemes for optimization of parameters and statistical analyses of the data. The digital portion of the hybrid computer controlled the setting of the rate constants, optimization schemes, graphics, printouts, and many other special features of this hybrid configuration (Pearce and Young, 1981). All operator interactions with the system were through a cathode ray tube (CRT) input/output terminal. The spreadsheet software package LOTUS 1-2-3 was used extensively to manipulate the exposure data obtained from the hybrid computer-generated curves. Values for integral H (*N*-OH arylamine bladder exposure), integral U (*N*-OH arylamine conjugate bladder exposure), sum X (total *N*-OH arylamine excreted into the urine), and sum Y (total *N*-OH arylamine conjugate excreted into the urine) as a function of voiding

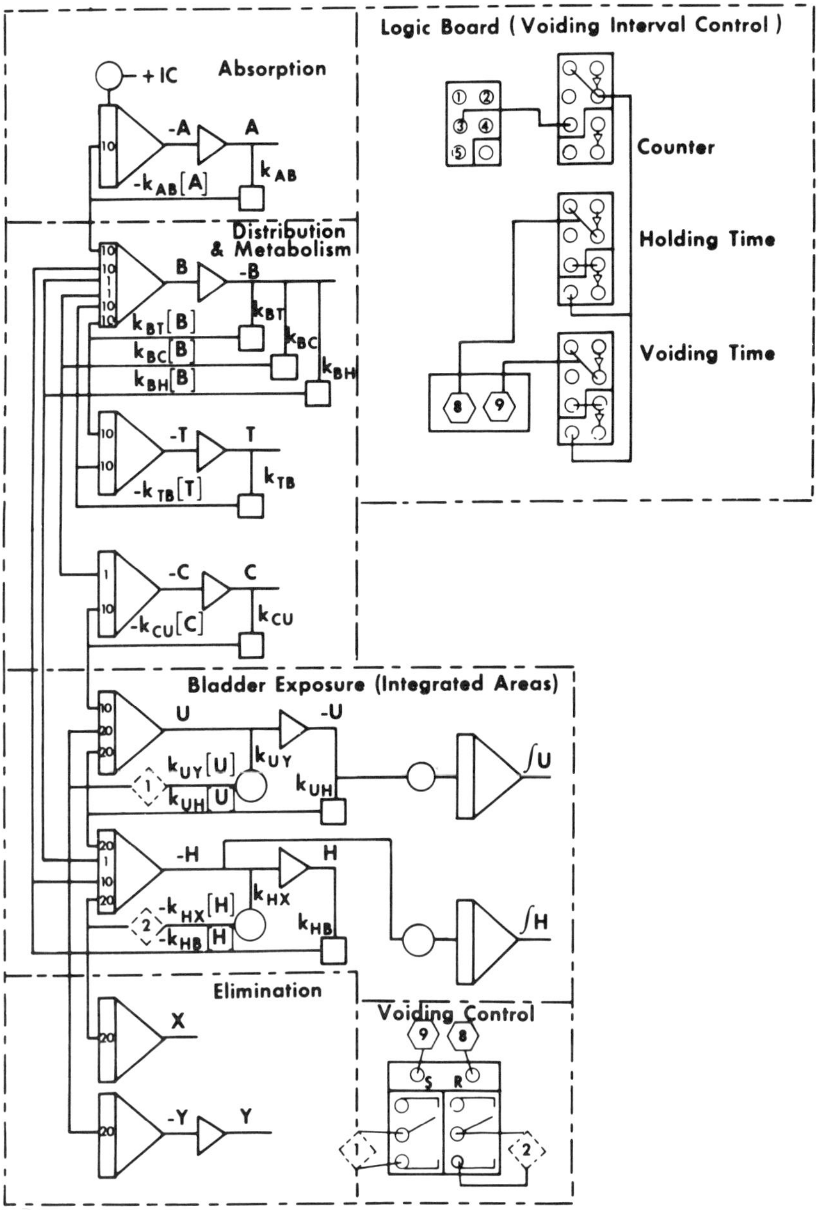

FIGURE 3 Analog and logic schematic diagram of the model. The analog schematic represents the differential equations that are written directly from the model. Holding time → Voiding interval; K_{UH} → controls pH.

interval and urinary bladder pH were obtained from the hybrid simulation and entered into the LOTUS spreadsheet. These data were plotted individually as well as with various combinations of the data to observe the relationships among the components of the model. This type of data manipulation and graphic visualization is extremely easy, fast, and accurate when the spreadsheet approach is used.

The three-compartment model was designed and tested by using data from the bladder instillation experiments of *N*-OH-2-naphthylamine in rats described by Oglesby et al. (1981). The bladder lumen portion of the model was included to be consistent with physiological reality because independent *in vitro* studies had determined the rate of *N*-OH arylamine *N*-glucuronide hydrolysis to increase as a function of decreasing pH. The ability to control voiding interval was also included for consistency because release of the free *N*-OH metabolite occurs as a linear, time-dependent process. Both of these latter parts of the model were unique to this application; normally, in pharmacokinetic modeling the excretion of a chemical into the urine is considered an end product that is not available for recycling and is dealt with as a continuous, cumulative function.

[2,2′-3H]-4-Aminobiphenyl (4-ABP; 5 mg/kg; 12 mCi/mmol) was administered orally to three male dogs. Blood and urine samples were taken at various intervals over the next 24 h. Total radioactivity was determined

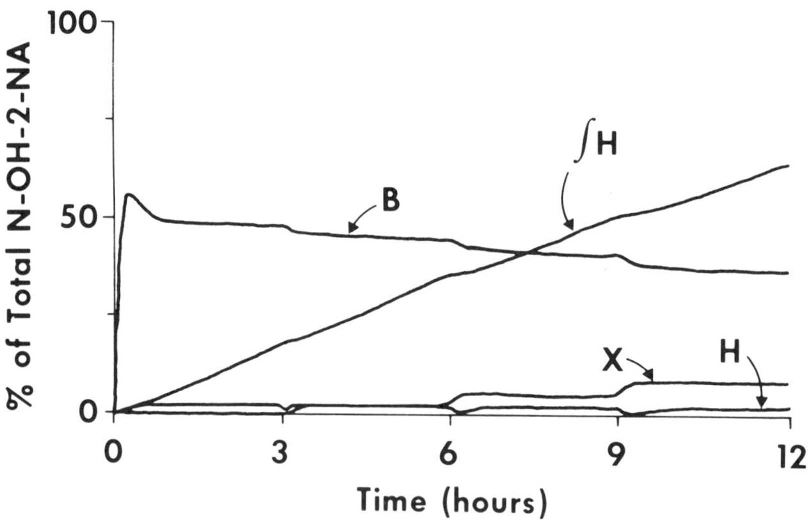

FIGURE 4 Hybrid computer simulation curves for model components B, H, and X and integral H (area under the concentration-time curve [AUC]) for a urine pH of 5 and a voiding interval of 3 h. Plots such as this were generated for each combination of pH and voiding interval.

for whole blood, plasma, and urine samples. Analysis of blood for levels of 4-ABP-hemoglobin adducts at selected intervals were provided by P. Skipper (Massachusetts Institute of Technology, Boston, Mass.). Urine samples were subjected to high-pressure liquid chromatographic analysis to determine the metabolic profile as a function of time (Frederick et al., 1981). The experimental treatment of the dogs differed only by the way in which the urine was collected: dog 2, natural voiding intervals; dog 3, 2 h voiding intervals via catheter for the first 12 h and naturally thereafter; dog 4, continuous voiding via catheter for the first 12 h and naturally thereafter.

RESULTS AND DISCUSSION

A set of amount-time curves was generated by the hybrid computer for each pH and voiding interval. This set was visually simplified by the CRT display of only those model component curves of interest. Figure 4 represents a set of curves for three components (B, H, X) of the model and the integrated area for the H curve under one set of conditions (pH 5; voiding interval, 3 h). The hybrid computer was used to generate a set of data at each of five pHs (4, 5, 5.5, 6, and 7) and eight voiding intervals (1, 2, 3, 4, 5, 6, and 10 h and continuous). These data were entered into the LOTUS 1-2-3 spreadsheet program, which was then used to manipulate the data and generate the plots.

Figure 5 is a plot of integral H (urinary bladder exposure to *N*-OH arylamine) versus voiding interval; each curve represents a different pH value. The average value taken from the literature for pH and voiding interval for four species is superimposed on the plots. The relative bladder exposure values of these species are in the same order as their relative bladder carcinogenic potential as reported in the literature (human > dog > monkey > rat) (Deichmann, 1967; Radomski, 1979; Wynder and Goldsmith, 1977). This correlation suggests that the residence time of the *N*-OH arylamine in the bladder lumen should be directly related to the species sensitivity to urinary bladder cancer.

Another set of curves obtained from the hybrid simulation resulted in a plot of sum X (excretion of *N*-OH arylamine in the urine) versus voiding interval as a function of pH (Figure 6). By dividing the value of the integral H by the value for the sum X at each pH and voiding interval, a matrix of data resulted that is plotted in Figure 7. This relationship in turn can be used to predict the exposure of the urothelium to *N*-OH arylamine metabolites (integral H) by urinary excretion measurements of the *N*-OH arylamine (sum X). Therefore, a noninvasive biological marker for bladder exposure is obtained from the analysis of levels of the *N*-OH arylamine at each urination, given the urine pH and voiding interval.

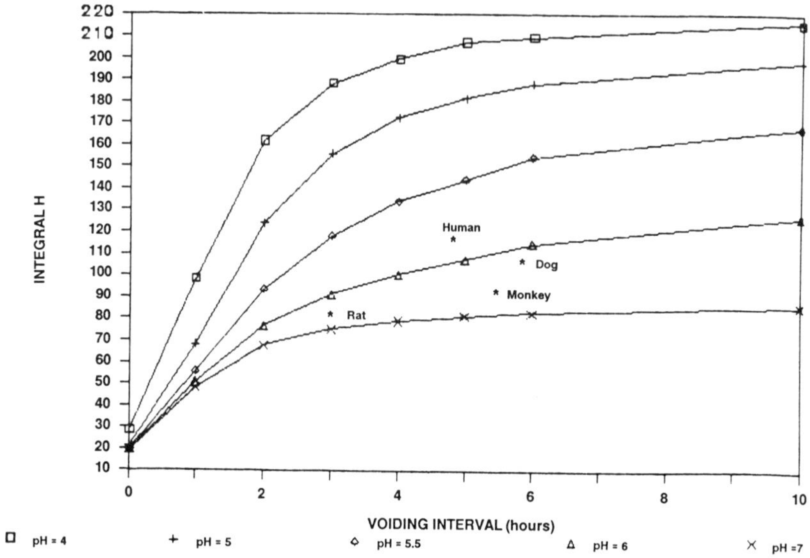

FIGURE 5 Plot of integral H versus voiding interval as a function of pH. The asterisk represents the positions on the graph for each species as determined by average urine pH and voiding interval taken from the literature. Symbols: □, pH 4; +, pH 5; ◇, pH 5.5; △, pH 6; ×, pH 7.

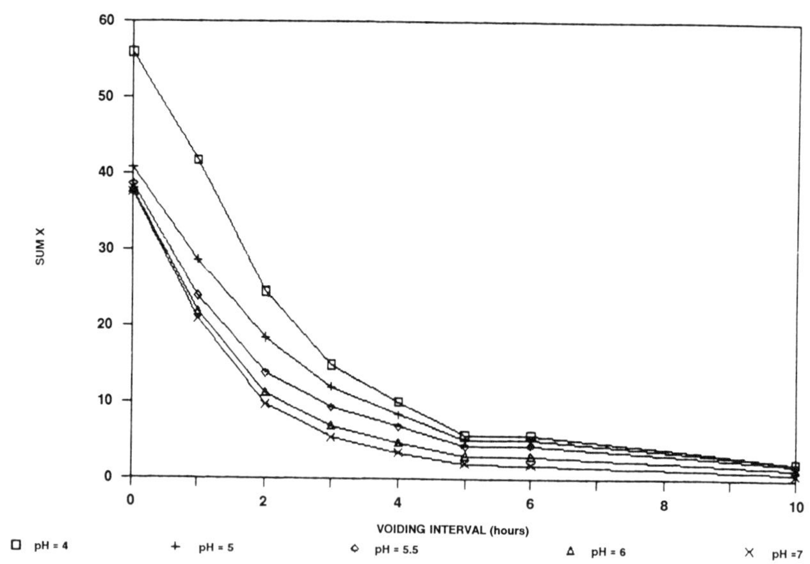

FIGURE 6 Plot of sum X versus voiding interval as a function of pH. For symbol definitions, see Figure 5 legend.

Another relationship to estimate bladder exposure was obtained from the plot of integral U (bladder exposure to the *N*-OH arylamine conjugate) versus voiding interval as a function of pH (Figure 8). The data from Figures 5 and 8 were then combined to generate Figure 9, which is a plot of integral H percentages [(integral H × 100)/(integral H + integral U)] as a function of pH and voiding interval. The data in Figure 9 can then be used to predict bladder exposure (integral H) based on total urinary recovery of the arylamine and its matabolites.

To test our hypothesis that *N*-OH arylamine bladder exposure can be used as a biological marker for assessing carcinogenic potential, [^{3}H]4-ABP has thus far been given orally (5 mg/kg) to three dogs. The only differences experimentally between the three dogs is the manner in which urine was collected: dog 2, natural voiding intervals; dog 3, 2-h voiding intervals for the first 12 h via bladder catheterization and natural voiding thereafter; dog 4, continuous voiding via indwelling bladder catheterization for the first 12 h and natural voiding thereafter.

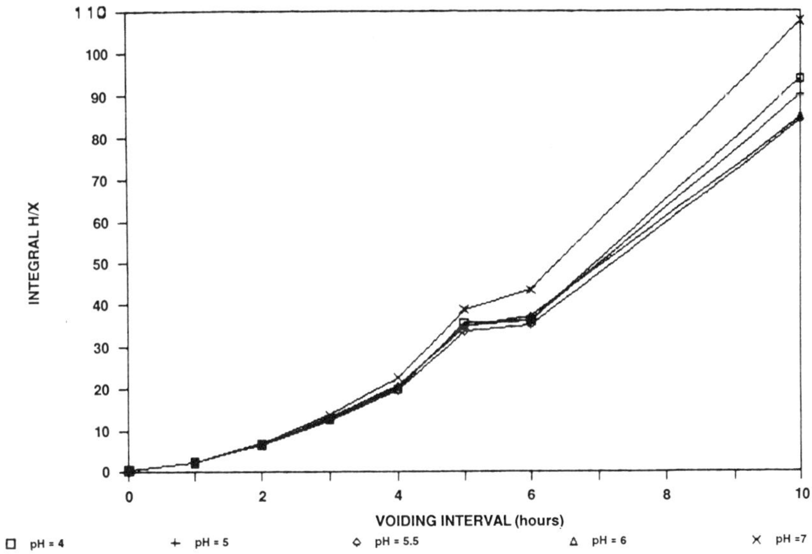

FIGURE 7 Plot of integral H/sum X versus voiding interval as a function of pH. By knowing the amount of *N*-OH arylamine excreted in the urine, the urinary pH, and the voiding interval, an estimate of the bladder exposure to *N*-OH arylamine can be determined (△ symbols in Figure 13). For symbol definitions, see Figure 5 legend.

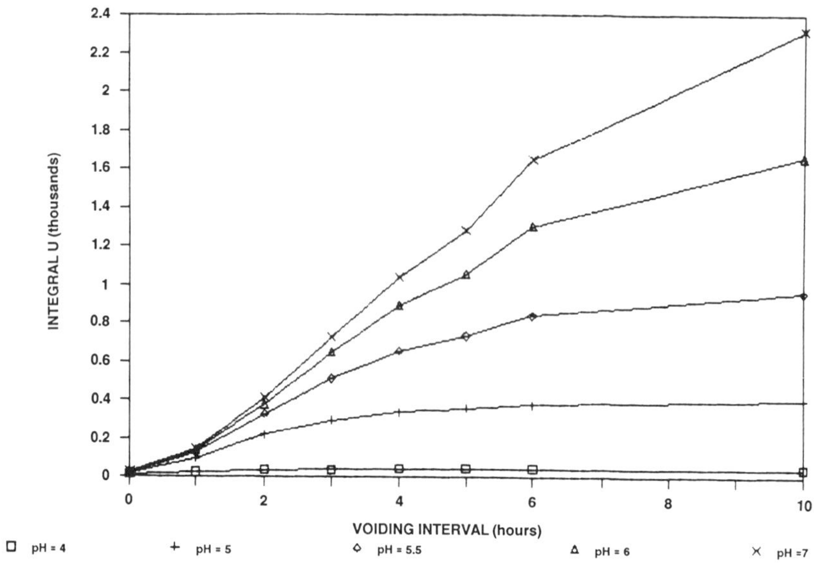

FIGURE 8 Plot of interval U versus voiding integral as a function of pH. For symbol definitions, see Figure 5 legend.

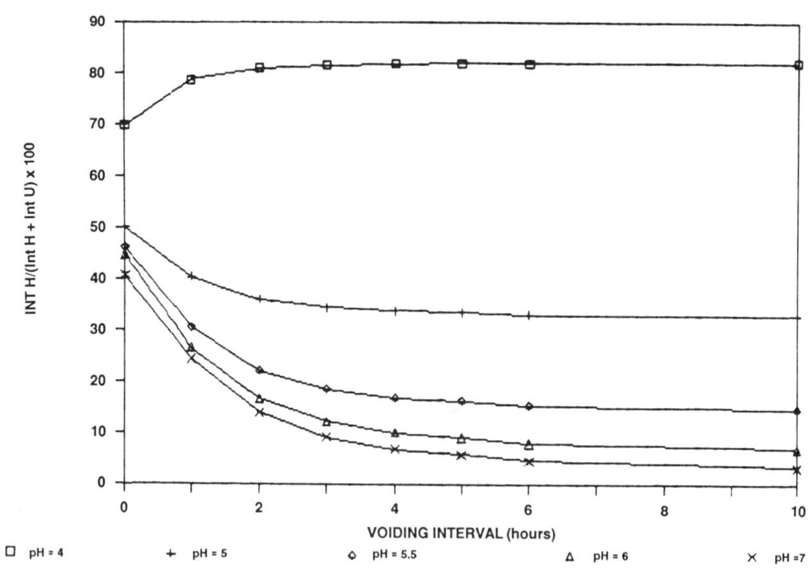

FIGURE 9 Plot of integral H percent versus voiding integral as a function of pH. By knowing the total percentage of dose excreted for a given voiding interval and urinary pH, the bladder exposure to *N*-OH arylamine can be estimated (◇ symbols in Figures 13–15). For symbol definitions, see Figure 5 legend.

The carcinogenic potential of 4-ABP is assumed to be directly related to the amount of DNA adducts formed in the bladder from *N*-OH-4-ABP, which is assumed to be proportional to the residence time of the *N*-OH metabolite in the bladder lumen. The DNA adduct measurements involve the use of highly sensitive and specific immunoassays and are currently in progress.

At present, however, this is not a convenient or readily convertible measurement to be applied to human populations. Therefore, a biological marker is needed to assess exposure to the potentially carcinogenic arylamines. Hemoglobin adducts (Hb-ABP) offer a good possibility because they can be readily measured from a blood sample and may reflect accumulative exposure.

Figures 10–12 are the percentage of dose versus time plots for whole blood, plasma, and hemoglobin–4-ABP adducts (Hb-ABP). The data for the three dogs were very similar, regardless of the manner in which the urine was collected. This was somewhat unexpected because we had anticipated that the Hb-ABP adducts might be formed as a consequence of reabsorption of the *N*-OH metabolite across the bladder wall rather than hepatic metabolism/transfer or direct formation in the blood.

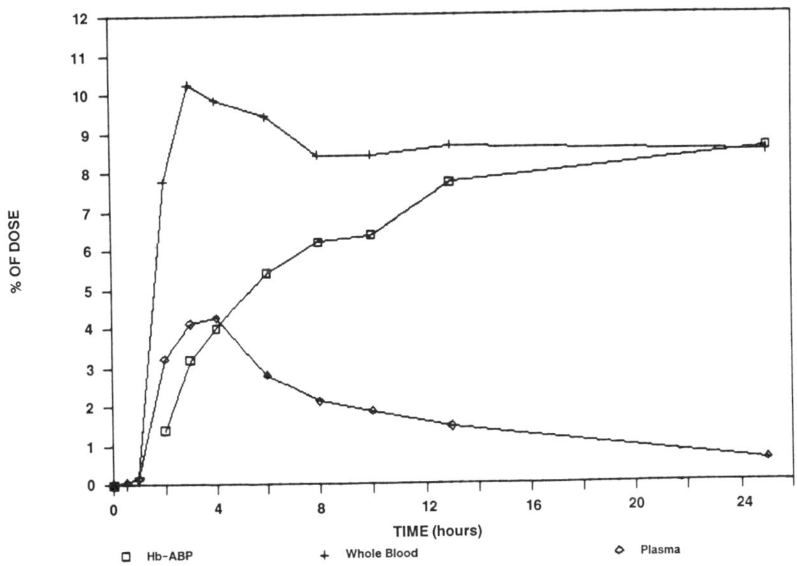

FIGURE 10 Plot of percentage of dose (5 mg/kg) versus time for dog 2. Urine samples were collected as they occurred naturally. Symbols: □, Hb-ABP; +, whole blood; ◇, plasma.

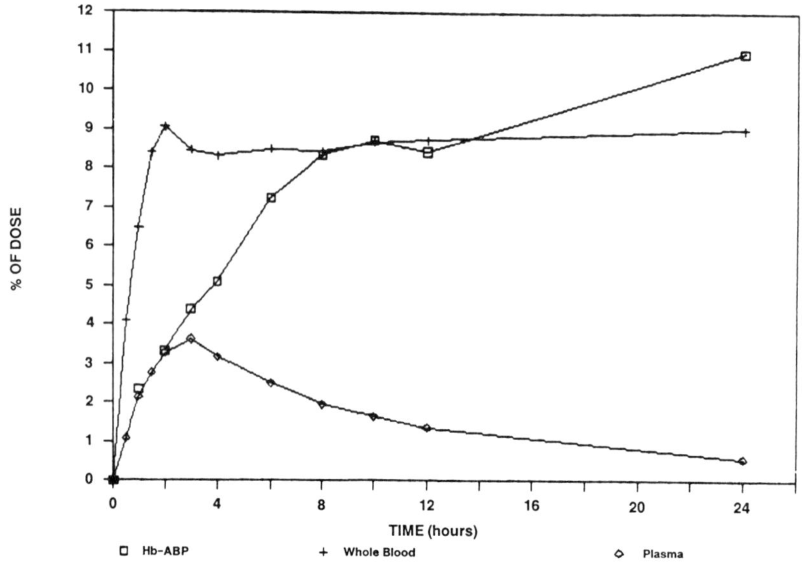

FIGURE 11 Plot of percentage of dose (5 mg/kg) versus time for dog 3. Urine samples were collected every 2 h via a catheter. For symbol definitions, see Figure 10 legend.

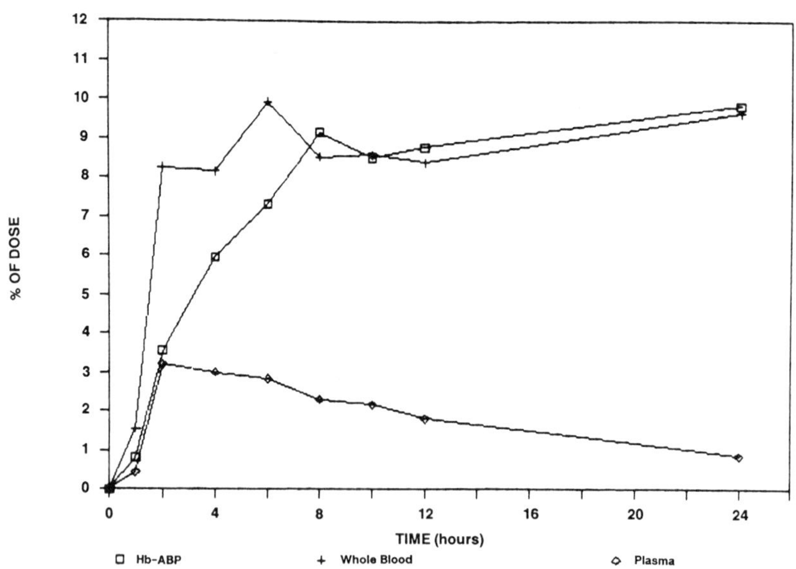

FIGURE 12 Plot of percentage of dose (5 mg/kg) versus time for dog 4. Urine samples were collected continuously via a catheter. For symbol definitions, see Figure 10 legend.

Through pharmacokinetic modeling, three other potential biological markers are being examined. Plasma AUC determination requires multiple blood samples and therefore may not be useful in population studies. Urinary excretion patterns either for total excretion or specifically for the *N*-OH arylamine levels may be easier to obtain but do not reflect multiple exposures over extended times.

Table 1 presents the experimental data and calculated exposure data from the hybrid computer simulation for dog 2. Four biological markers are presented that are potential measures of bladder exposure to the *N*-OH arylamine. The Hb-ABP adduct level was measured directly from the blood sample (column D). The area under the plasma concentration-time curve was calculated by use of the trapezoid rule (column H). Two separate cumulative integral H values were calculated based on total recovery of radioactivity (columns E and I–K) or on excretion of *N*-OH-ABP (columns F and L–N). These latter two values differed by about a factor of 2 but had the same shape as the Hb-ABP curve (Figure 13).

Figures 13–15 are plots of these various exposure measurements versus time for the three dogs. The most complete set of data is for dog 2. The analysis of the rest of the data for dogs 3 and 4 is ongoing.

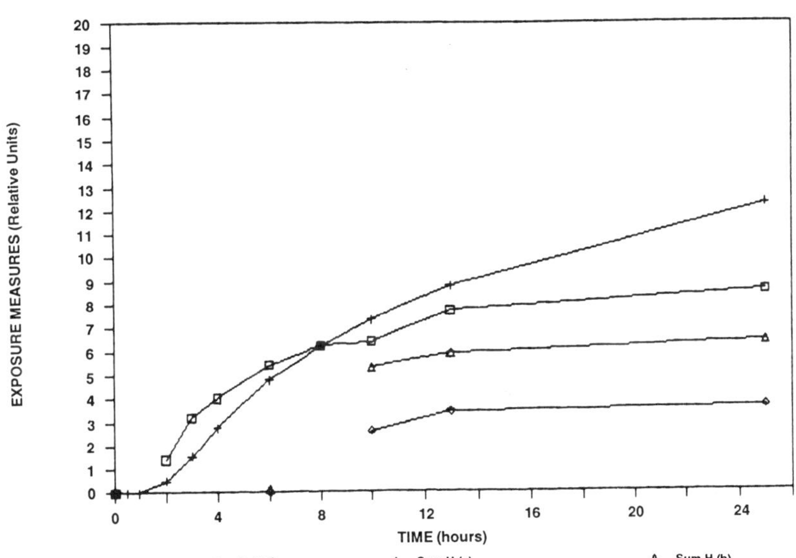

FIGURE 13 Plot of four measures of *N*-OH arylamine bladder exposure as a function of time for dog 2. The dose of 4-ABP was 5 mg/kg. Symbols: □, Hb-ABP; +, p-AUC; ◇, sum H (a); △, sum H (b). (a) Sum of integral H based on total urinary recovery. (b) Sum of integral H based on urinary recovery of *N*-OH arylamine.

TABLE 1 Experimental and Calculated Exposure Data of 4-ABP from the Hybrid Computer Simulation for Dog 2 (Natural Voiding Intervals)

	Measured Data (% of Dose)						Calculated from Total Urine				Calculated from *N*-OH-ABP		
Time (h)	Whole Blood ABP[a]	Plasma ABP[b]	Hb-ABP Adduct[c]	Urine Total	Urine *N*-OH-ABP	Urine pH	Plasma AUC (Adj)[d]	Integral H% (Fig. 9)	Integral H[e]	Sum of Integral H[f]	Integral H/Sum X (Fig. 7)	Integral H[g]	Sum of Integral H[h]
0	0	0	0.				0			0			0
0.5	0.01	0.07	0				0						
1	0.22	0.17	0				0						
2	7.77	3.24	1.40				0.5						
3	10.26	4.13	3.21				1.6						
4	9.85	4.28	4.01				2.8						
6	9.42	2.81	5.41	0.04	0.0029	4.2	4.8	72.3	0.03	0.03	36.1	0.1	0.1
8	8.43	2.12	6.21				6.2						
10	8.43	1.87	6.38	36.58	0.2304	7.1	7.3	7.0	2.56	2.59	22.5	5.2	5.3
13	8.66	1.47	7.73	8.14	0.0445	6.8	8.8	10.0	0.81	3.40	13.6	0.6	5.9
25	8.49	0.58	8.58	4.34	0.0049	6.4	12.3	5.7	0.25	3.65	93.9	0.5	6.4
A	B	C	D	E	F	G	H	I	J	K	L	M	N

[a] + symbol in Figures 10–12.

[b] ◇ symbol in Figures 10–12.

[c] □ symbol in Figures 10–15.

[d] + symbol in Figures 13–15.. Calculated by the trapezoid rule. Adjusted by dividing by 3.5 for ease of comparison to Hb-ABP values.

[e] Column E × column I.

[g] Column F × column L.

[f] ◇ symbol in Figures 13–15.

[h] △ symbol in Figure 13.

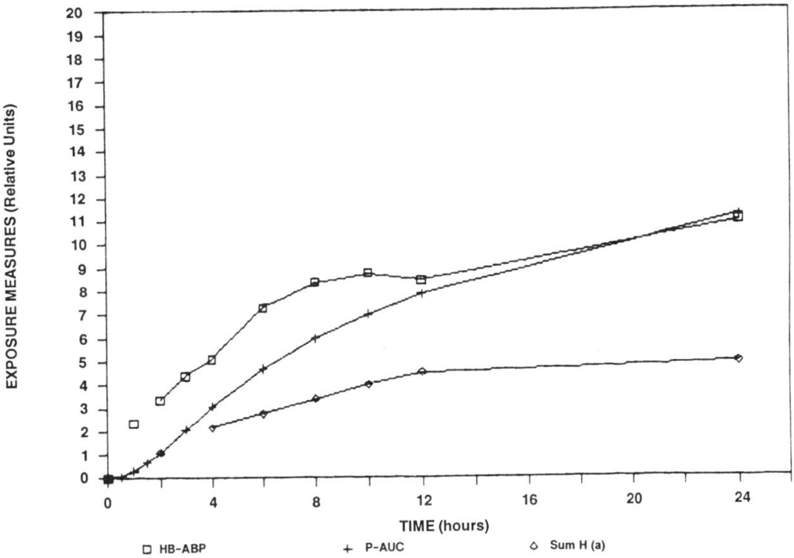

FIGURE 14 Plot of three measures of *N*-OH arylamine bladder exposure as a function of time for dog 3. The dose of 4-ABP was 5 mg/kg. Symbols: □, Hb-ABP; +, P-AUC; ◇, sum H (a). (a) Sum of integral H based on total urinary recovery.

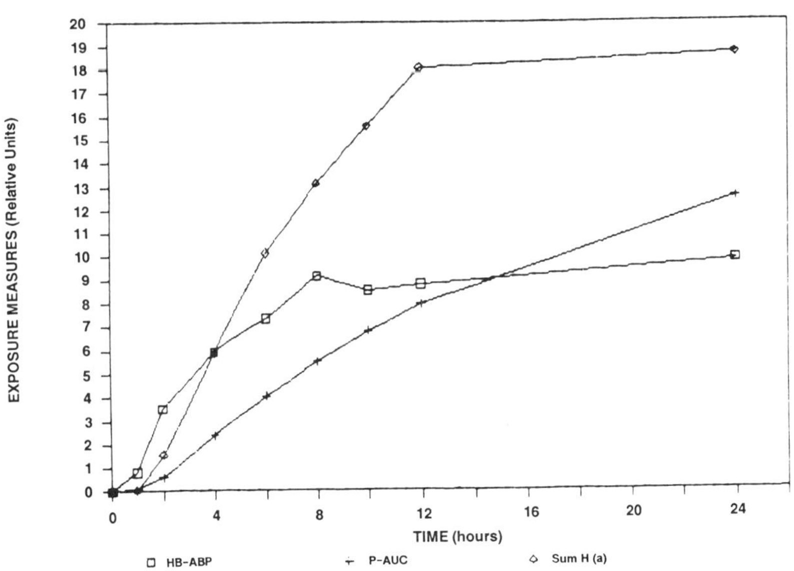

FIGURE 15 Plot of three measures of *N*-OH arylamine bladder exposure as a function of time for dog 4. The dose of 4-ABP was 5 mg/kg. For symbol definitions, see Figure 14 legend. (a) Sum of integral H based on total urinary recovery.

CONCLUSIONS

If carcinogenic potential is related to *N*-OH arylamine bladder exposure and DNA adduct formation, then Hb-ABP measurements may not be indicative of this process because the level of Hb-ABP adduct formation was about the same for all 3 days under three different urinary voiding conditions. We had anticipated that dog 4 (Figure 12) would have had a much lower level of Hb-ABP adducts because the bladder was continuously drained via the indwelling catheter, and therefore, the potential for bladder exposure should have been greatly reduced. Nevertheless, Hb-ABP appears to be an accurate measure of the external dose and hence a potentially useful biological marker.

Prediction of *N*-OH arylamine bladder exposure based on integral H percent calculations from total recovery in the urine also does not seem predictable across various conditions. We would anticipate that continuous urinary excretion would allow the least amount of bladder exposure; however, Figure 15 indicates a high level of exposure potential from this calculation. This might be an artifact of the simulation and data manipulation, or it could be an indication that the experimental conditions were not exact and that the bladder was not entirely empty at all times.

Prediction of *N*-OH arylamine bladder exposure based on integral H calculations from excretion levels of *N*-OH arylamine appears to be most promising, but we must wait for the completion of the urine analyses for 4-ABP metabolites.

On completion of the arylamine-DNA adduct determinations in the urothelium, the endpoint evaluation of our simulations will have more meaning, and new avenues of analysis can then be explored.

ACKNOWLEDGMENTS

The authors wish to thank Mary Ann Butler, Candee Teitel, John Bailey, Ken Dooley, Paul Skipper, and Steven Tannenbaum for the use of their data in this simulation exercise.

REFERENCES

Deichmann, W. G. 1967. Bladder Cancer, A Symposium, K. F. Lampe, ed. Birmingham, Ala.: Aesculapius, p. 30.

Frederick, C. B., J. B. Mays, and F. F. Kadlubar. 1981. A chromatographic technique for the analysis of oxidized metabolites: Application to carcinogenic N-hydroxyarylamines in urine. Anal. Biochem. 118:120–125.

Green, L. C., P. L. Skipper, R. J. Turesky, M. S. Bryant, and S. R. Tannenbaum. 1984. In vivo dosimetry of 4-aminobiphenyl in rats via a cysteine adduct in hemoglobin. Cancer Res. 44:4254–4259.

Kadlubar, F. F., J. F. Anson, K. L. Dooley, and F. A. Beland. 1981. Formation of urothelial and hepatic DNA adducts from the carcinogen 2-naphthylamine. Carcinogenesis 2:467–470.

Oglesby, L. A., T. J. Flammang, D. L. Tullis, and F. F. Kadlubar. 1981. Rapid absorption, distribution, and excretion of carcinogenic *N*-hydroxy-arylamines after direct urethral instillation into the rat urinary bladder. Carcinogenesis 2:15–20.

Pearce, B. A., and J. F. Young. A hybrid computer system for pharmacokinetic modeling. I. Software considerations. Pp. 117–121 in Proceedings of the 1981 Summer Computer Simulation Conference. La Jolla, Calif.: Simulation Council, Inc.

Radomski, J. L. 1979. The primary aromatic amines: Their biological properties and structural-activity relationships. Annu. Rev. Pharmacol. Toxicol. 19:129.

Wynder, E. L., and R. Goldsmith, Jr. 1977. The epidemiology of bladder cancer. Cancer 40:1246.

Young, J. F., and F. F. Kadlubar. 1982. A pharmacokinetic model to predict exposure of the bladder epithelium to urinary N-hydroxyarylamine carcinogens as a function of urine pH, voiding interval, and resorption. Drug Metab. Dispos. 10(6):641–644.

Young, J. F., C. G. White, and B. A. Pearce. 1981. A hybrid computer system for pharmacokinetic modeling. II. Applications. Pp. 122–129 in Proceedings of the 1981 Summer Computer Simulation Conference. La Jolla, Calif.: Simulation Council, Inc.

PART VI

Applications of Mathematical Modeling

Hazard Assessment Using an Integrated Physiologically Based Dosimetry Modeling Appproach: Ozone

Frederick J. Miller, John H. Overton, Jr., Elaine D. Smolko, Richard C. Graham, and *Daniel B. Menzel*

INTRODUCTION

In examining the possible role of pharmacokinetics and pharmacodynamics in risk assessment, the underlying philosophy of the legislative mandate through which the risk assessment is applied must be kept in mind. For example, the ability of the Food and Drug Administration to invoke the Delaney Clause to regulate a substance is different from the risk assessments required for the Environmental Protection Agency's National Ambient Air Quality Standards (NAAQSs). The kind and level of information that is available can vary greatly. Reevaluation of the NAAQSs for ozone (O_3) focuses on whether or not the value of the standard should be changed by as little as 20%, while many carcinogenic risk assessments are trying to establish the level of risk to within one to two orders of magnitude. In any case, the uncertainties identified in risk assessments help to establish areas in which additional research would be useful.

The intent of this paper is to broaden the awareness that pharmacokinetics and mathematical dosimetry models are useful tools in risk assessments of noncarcinogenic as well as carcinogenic effects. While examples

---

The research described in this paper has been reviewed by the Health Effects Research Laboratory, U.S. Environmental Protection Agency, and approved for publication. Approval does not signify that the contents necessarily reflect the views and policies of the Agency nor does mention of trade names or commercial products constitute endorsement or recommendation for use.

of modeling applications have been predominantly related to carcinogenesis, health effects such as emphysema and fibrosis are not to be ignored. This is particularly true for oxidant gases. For example, several animal studies in different species show that long-term exposure to nitrogen dioxide produces emphysema (Fenters et al., 1973; Hyde et al., 1978; Riddick et al., 1968), and O_3 causes pulmonary fibrosis (Fujinaka et al., 1985). Moreover, subchronic exposures at relatively low levels of these gases have been shown to lead to cellular changes (Barry et al., 1985; Chang et al., 1986) that are indicative of a structural remodeling of the lung. The impetus is clear that man must be protected from such effects. But to use the animal toxicological data more quantitatively in setting appropriate NAAQSs for these gases, extrapolation modeling is required.

In this volume the need to incorporate pharmacodynamics in the modeling process has been discussed. Thus far, most risk assessments have assumed a priori that an equivalency of response exists between animals and man. As one proceeds from the molecular level or biochemical event toward injury at the tissue or organ level, that assumption becomes less tenable because of possible species differences in repair processes, levels of antioxidant enzymes, etc. The dose, if it is sufficiently high, can produce damage anywhere from the molecular to the organ level. On the other hand, various host defense systems can interact, and if the damage is not sufficiently severe, they can yield recovery. If the delivered dose overwhelms these defense systems, various disease states can result. Defense systems might conceivably play a role in the etiology of lung disease. All of this is a dynamic situation.

Figure 1 (based on Figure 1-1 of NRC, 1983) summarizes an overview of the research, risk assessment, and risk management processes and their interrelationships that lead to the recommendation of a NAAQS. This paper focuses on the phases of the processes outlined by the dashed rectangle. Here, the area of pharmacokinetics offers great potential for improved risk assessments, particularly when these assessments are required to be more quantitative in nature. The development of dosimetry models that can provide a description of the uptake and distribution of chemical compounds throughout the body and the availability of toxicological data that can be used to establish dose-response relationships are integral to these efforts. The incorporation of laboratory and field observations of adverse health effects, hazard information, and extrapolation methods to yield dose-response assessments is critical to the risk assessment process. The combining of these phases can be facilitated by the integrated physiologically based dosimetry modeling approach, illustrated schematically in Figure 2.

The major components of the approach are the critical toxicity reference (CTR) system (Smolko et al., in press) and the physiologically based

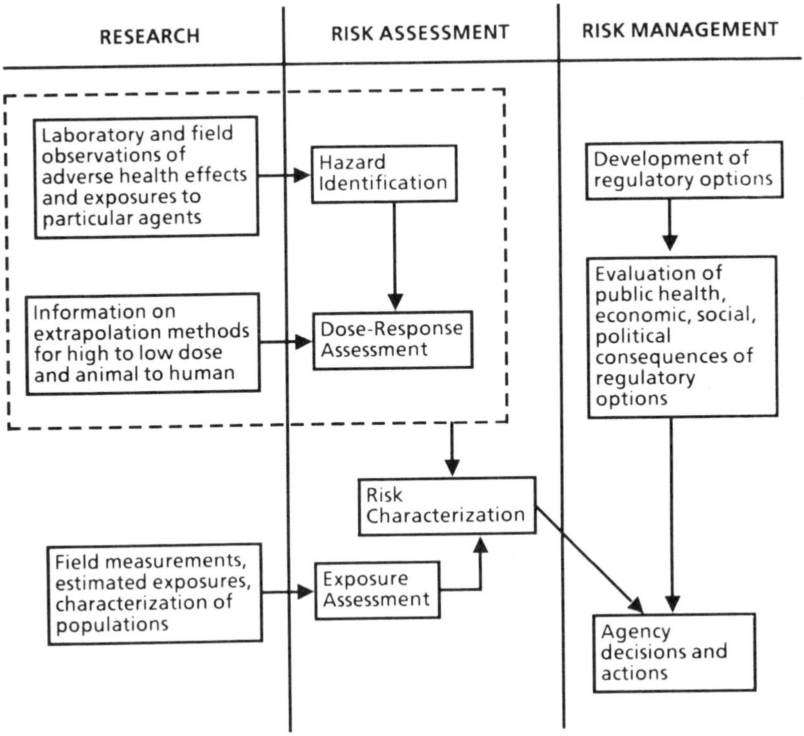

FIGURE 1 Major elements of risk assessment and risk management. The phases of the process that can be addressed by using an integrated physiologically based dosimetry modeling approach are contained within the dashed rectangle. Overall figure schematic is based on Figure 1-1 of NRC (1983).

dosimetry (PBD) model. The CTR system is comprised of four elements: (1) searching the literature for references reporting toxicity data relevant to the toxicant being studied; (2) abstracting information from the selected references in a form appropriate for both quantitative and descriptive extraction; (3) constructing a data base that consists of bibliographic and abstracted information; and (4) compiling concentration-response data by a series of searches of the developed data base. The primary factors considered in developing PBD models are mammalian physiological processes, anatomical characteristics, physicochemical properties of the gas and of relevant biochemical constituents, and mass transport processes. PBD model predictions of dose, appropriate for species with the specific characteristics described in the literature, are combined with the concen-

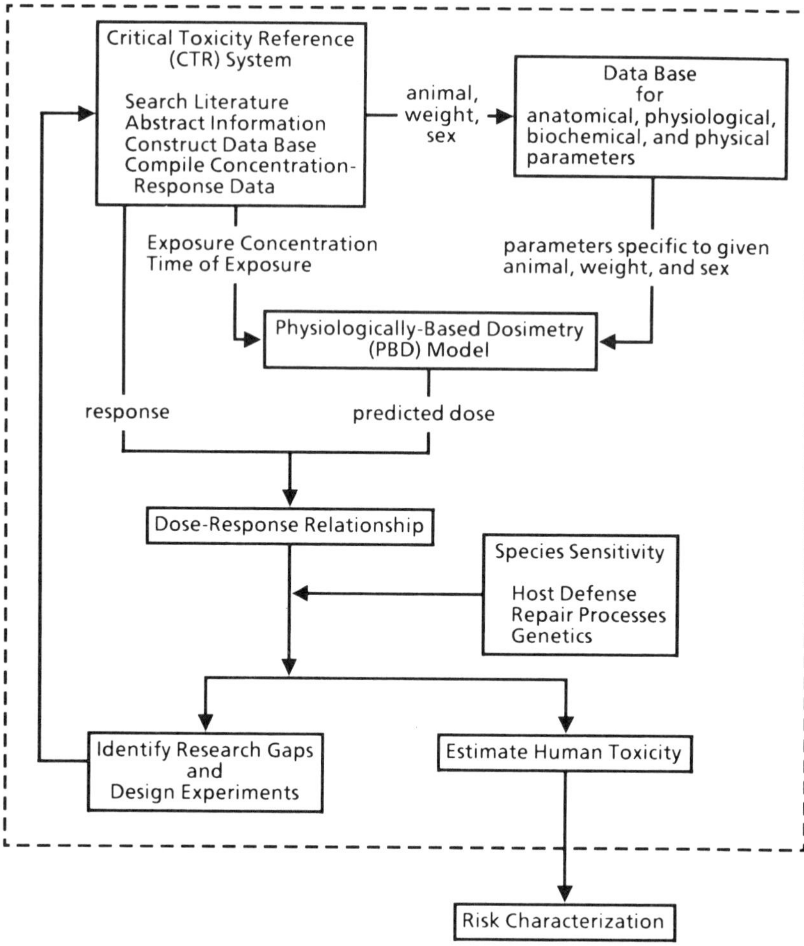

FIGURE 2 A schematic of the elements of an integrated physiologically based dosimetry modeling approach to estimating human toxicity that leads to one of the components of risk characterization. The procedure diagrammed in the dashed rectangle is the proposed approach for implementing the dashed rectangle process shown in Figure 1.

tration-response data that have been collated by species and endpoint. The resulting dose-response relationships can use various expressions of dose to obtain the most appropriate quantitative representation of toxicological effects.

The analyses can incorporate species sensitivity information for consideration of such factors as host defense, repair processes, and genetics. One outcome of this approach is an estimation of human toxicity that can be used as input into the risk assessment and risk management processes.

A second outcome is information to aid in the identification of data gaps and design of experiments. This latter outcome is particularly important because it can eventually yield new data that will be used in the iterative application of the integrated PBD model approach which will strengthen the overall risk assessment and risk management processes.

The remaining sections of this paper illustrate the integrated PBD modeling approach (Figure 2) by applying data from the CTR system and predictions from a PBD model for O_3 to construct dose-response relationships. This will be done first by discussing the formulation, as well as sample simulation results, of a PBD model that predicts the uptake and distribution of absorbed O_3 in the lungs of mammals. Although the focus will be on O_3, the intent is primarily to illustrate the methodology and to provide an understanding of various aspects of delivered dose. The methods discussed will be particularly important when some of these procedures are extended to volatile organic compounds, for which first-pass metabolic effects in the lung are a concern. Also, a brief discussion is provided of a data base management system, CTR, that has been developed to store and retrieve quantitative data in a manner useful for mathematical dosimetry models. An application to ozone toxicological data is presented to illustrate the methodology of using mathematical dosimetry models to examine quantitative dose-response relationships.

LOWER RESPIRATORY TRACT MATHEMATICAL DOSIMETRY MODELING

In the papers presented in this volume, much attention is placed on the use of physiologically based pharmacokinetic models to provide a description of dose distribution following inhalation of a chemical compound. To date, these models predict average levels of the chemical throughout an entire organ (or body compartment). As illustrated below, however, additional complexities are involved when the distribution of inhaled gases is evaluated within the lower respiratory tract (LRT), where dose in various lung regions can vary greatly and consequently have different health effect outcomes.

Model Conceptualization

The major factors affecting the regional uptake of O_3 are morphology of the respiratory tract, the route of breathing, the depth and rate of breathing, gaseous physiocochemical properties, the physical processes governing gas transport, and the physiocochemical properties of the tracheobronchial (TB) liquid lining and of the air-blood barrier in the pulmonary region. All these factors interact in a complex way to determine dose and must be considered in developing a simulation model.

The mathematical model formulation will focus on the lower respiratory tract for several reasons. Upper respiratory tract removal of inhaled gases and particles is amenable to experimental determination of deposition (Corn et al., 1976; Miller et al., 1979; Yokoyama, 1968; Yokoyama and Frank, 1972). The morphology of the upper respiratory tract is quite complex, difficult to measure, and variable between species (Schreider, 1986; Schreider and Raabe, 1981), so mathematical descriptions are difficult to obtain. Further, airflow patterns are also complex (Patra et al., 1986), so proper treatment of gas transport processes is not apparent. Experimental values for upper respiratory tract uptake of the inhaled gas, however, can provide appropriate boundary conditions for mathematically modeling delivered doses to LRT regions.

Aspects of LRT structure and their relationship to model compartments are illustrated in Figure 3. Briefly, anatomical descriptions of the lung are needed such as those available for man (e.g., Weibel, 1963; Yeh and Schum, 1980) and for various animals (e.g., Kliment, 1973; Schreider and Hutchens, 1980; Yeh et al. 1979). The top portion of Figure 3 shows Weibel's (1963) representation of the TB and pulmonary regions of the human lung. For mathematical modeling purposes, we conceptualize this representation into a series of right circular cylinders; and information is needed on the lengths, diameters, and radii of the TB airways. In the pulmonary region, data on the structure of the pulmonary ducts and sacs and on the number of alveoli, their surface areas, and volumes are necessary to apply the model description of gas transport processes.

In the TB or conducting airways, a mucociliary layer protects the underlying ciliated, goblet, brush, and basal cells, etc., from direct insult by the inhaled gas. As can be seen in Figure 3, this layer consists of an epiphase and a hypophase. Lacking definitive data on the thickness and chemical composition of these phases in various portions of the TB region, the current model formulation combines them into one compartment. For ambient exposures to the highly reactive gases, such as O_3, penetration to the bloodstream in the TB region can be ignored (see Miller et al., 1985, for details). Thus, in the TB region, model compartments correspond to the air, liquid lining layer, and underlying tissue. For highly soluble and nonreactive gases, however, a blood compartment and a description of the fate of the gas in other organs and compartments of the body probably would be needed.

In the pulmonary region, the epithelium is chiefly comprised of type I and type II cells over which a very thin layer of a surfactant fluid can be found. Underlying the epithelium are the interstitium and the endothelial cells of the capillary network. Because the alveoli are arranged back to

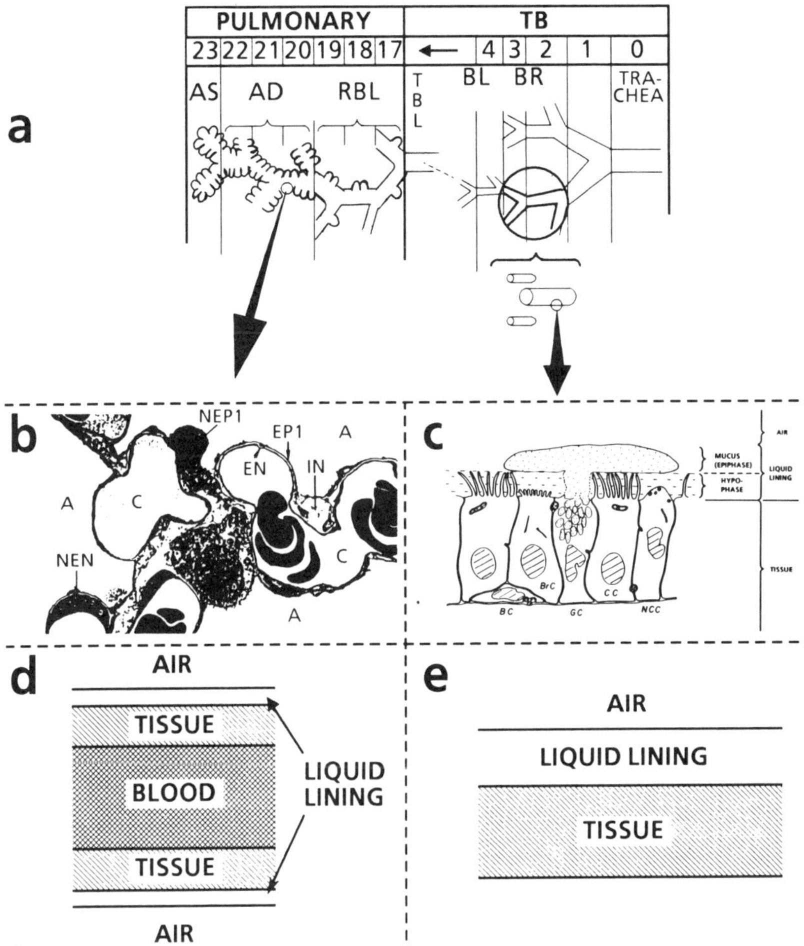

FIGURE 3 The relationships between morphologies and their model representations. Panel a is a schematic of the branching airways of the TB and pulmonary region of man (based on Weibel, 1963). The generations are labeled from the right, beginning with the trachea. BR, BL, and TBL indicate bronchi, bronchioles, and terminal bronchioles, respectively; the respiratory bronchioles, alveolar ducts, and alveolar sacs are indicated by RBL, AD, and AS, respectively. Below the TB portion of the lung schematic are three cylindrical figures that indicate the model representation of the airways. Panel c shows a diagram of the structure of the liquid lining and tissue of the TB region (diagram based on Jeffrey and Reid, 1979). The different cells represented are basal (BC), ciliated (CC), brush (BrC), goblet (GC), and nonciliated serous (NCC). Panel e illustrates the model representation of TB liquid lining and tissue compartments. Panel b is an electron micrograph of the interalveolar septa (based on Gehr et al., 1978). The air spaces (A), capillaries (C), type I cells and their nuclei (EP1 and NEP1), endothelial cells and their nuclei (EN and NEN), and interstitial space (IN) are indicated. Panel d illustrates the model representation of the liquid lining, tissue, and capillaries of the pulmonary region.

back, with capillaries between them, the model formulation for the air, liquid lining, tissue, and blood compartments is assumed to be symmetrical (Figure 3).

A one-dimensional equation of mass transport is used to describe the dynamics of the cross-sectional average concentration. Experimental evidence for this approach is discussed by Overton (1987). The processes of axial convection and axial dispersion, the loss of O_3 to the liquid lining, and lung expansion and contraction are taken into account. Lung expansion and contraction during the breathing cycle become important when pollutant uptake for minute ventilations corresponding to heavy exertion is modeled. When mass transfer in the liquid lining, tissue, and blood compartments is modeled, the transport of ozone is related to molecular diffusion and chemical reaction terms. The reader is referred to Figure 2 of Overton et al. (J. H. Overton, R. C. Graham, and F. J. Miller, this volume) for explicit forms of the partial differential equations used in the model. For a detailed discussion of the model formulation and assumptions, as well as the basis for these assumptions, see Miller et al. (1985) and Overton et al. (1987). Note that when other gases are modeled, modifications of the expression for loss of the gas to airway walls may be necessary. In addition, high solubility or nonreactivity may necessitate changes in some aspects of treating gas transport in the airways.

The information required to model adequately transport in the mucus or surfactant lining liquid, tissue, and blood compartments includes quantitative biochemical data on the constituents as well as on the chemical reactions of O_3 with these constituents. Currently, a pseudo-first-order reaction scheme is assumed when ozone uptake is modeled (see Miller et al., 1985). Estimates of liquid lining, tissue, and blood compartment thickness are also required. At present, this is an area in which most of the available information is from animal studies. Data for man are very limited, and assumptions about the distribution of compartment thickness must be made. Compartmental diffusion coefficients and partitition coefficients, such as those used in physiologically based pharmacokinetic models, are also needed. Much of the above topic is discussed in detail elsewhere (Miller et al., 1985; Overton et al., 1987).

Examples of O_3 Dosimetry Modeling Results

Utilizing the model formulation concepts discussed above, Miller et al. (1985) examined the uptake of O_3 in the LRT of man and performed various sensitivity analyses to illustrate the importance of LRT secretions on delivered dose. Their work shows that the net amount of O_3 removed in the trachea can be several orders of magnitude different from the amount of O_3 that penetrates to the underlying tissue. Proceeding distally from the trachea, this difference diminishes mainly because of the decline in the thickness of the liquid lining. Furthermore, because the gas-exchange region is lined with a very thin fluid that does not contain many constituents that react with O_3, the net amount of O_3 removed and the tissue dose of O_3 are essentially the same. According to Miller et al. (1985), the net O_3 dose curves are much less dependent on the thickness of the TB liquid lining than are the O_3 tissue dose curves.

Another example of sensitivity analysis demonstrating the importance of model parameters is concerned with the values of the liquid lining rate constants. Simulations show that the effect of increased chemical reactivity in the TB liquid lining is to increase the net airway doses in the TB region and to decrease the tissue dose throughout the LRT. Furthermore, tissue dose is found to be much more sensitive to the TB liquid lining reactions than is the net dose. For example, the tracheal tissue dose is predicted to decrease by about 3 orders of magnitude, and the tracheal net dose is predicted to increase by about 1 order of magnitude because of a rate constant change from 0 to twice the reference value. However, pulmonary net and tissue dose values are predicted to change by less than a factor of 2 as a result of the same change in the TB liquid lining rate constant. By contrast, changes in the pulmonary liquid lining rate constant lead to simulation results for LRT net and tissue doses that are essentially unchanged from those obtained with the reference values. These types of analyses show that to determine dose-effect relationships for various cell types or components (e.g., cilia) in the conducting airways, better quantitative data than are presently available on TB liquid lining rate constants and thickness are desirable. On the other hand, the lower sensitivity of predicted pulmonary doses to parameter uncertainties gives a greater confidence in making interspecies comparisons of dose-effect relationships in this region compared with making such comparisons in the TB region.

An example of interspecies dosimetric comparisons is illustrated in Figure 4, in which the solid and dashed lines are O_3 dosimetry simulation data for the rats and man, respectively. Predictions for the rat lung use the Yeh et al. (1979) anatomical model in which the morphometric data are represented in a generational-type model analogous to the Weibel (1963) generational morphometric data for the human lung. Both the net and tissue doses of O_3 are shown for these species. Dose is expressed in Figure 4 as micrograms of O_3 per square centimeter of airway surface area per minute, standardized to a tracheal O_3 value of 1 $\mu g/m^3$. In the mathematical dosimetry model the processes are linear. Thus, dose is proportional to the tracheal concentration. For example, for a tracheal concentration of 1,000 $\mu g/m^3$, the predicted airway doses are 1,000 times the plotted values; for a tracheal concentration of 1 ppm, multiply the plotted values by 1,960 ($\mu g/m^3$ per ppm).

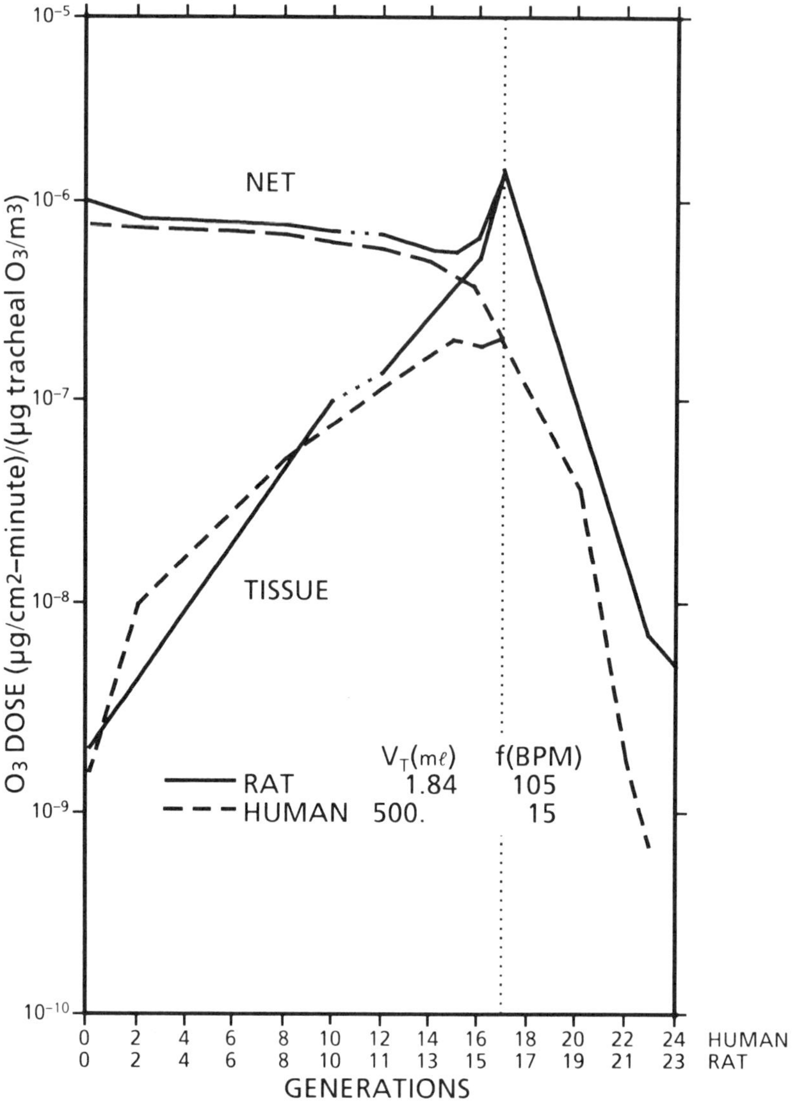

FIGURE 4 An example of a way of comparing species doses by plots of airway dose versus generation for two species, man (dashed curves) and rats (solid curves). The first generations of the pulmonary regions (17 and 16 in the anatomical models for man and rat, respectively) have been matched at the dotted vertical line to allow a better comparison of equivalent regions; this results in the break in the rat generations along the abscissa. The tidal volume (V_T) and breathing frequency (f) are given. Tissue dose is the predicted dose of the epithelial layer in the tracheobronchial region and of the air-blood barrier in the pulmonary region. Net dose is the total absorbed by the liquid lining, tissue, and blood compartments.

As can be seen from Figure 4, the net removal of O_3 is very comparable in the rat and human lungs. To help examine tissue dose patterns, we have matched the first generation of respiratory bronchioles in man with the first alveolated airway in the rat lung. This region is where the histopathological data show the first and the most pronounced lesions in O_3 toxicological studies with rats (Barry et al., 1985; Boorman et al., 1980; Plopper at al., 1979; Stephens et al., 1973). The maximal dose in both man and rat is predicted to occur at this particular region. Almost a 10-fold increase in deposition or removal of O_3 is predicted in this region in the rat compared with that in man according to the dose curves shown in Figure 4, which are standardized to a unit tracheal O_3 value. To make a more definitive interspecies dosimetric comparison at this point, however, nasopharyngeal removal of O_3 in man versus rat must be considered. Nasopharyngeal removal studies are currently being conducted, and when available, they will facilitate translation of the net and tissue dose curves of Figure 4 into delivered dose curves associated with exposure to O_3.

CRITICAL TOXICITY REFERENCE SYSTEM

Data collection for the O_3 analyses was accomplished by the CTR system (Smolko et al., in press). The four basic elements of this system have been briefly described. Sources of references for inclusion in the CTR data base include searches of nationally available data bases (e.g., National Library of Medicine), comprehensive documents, personal communication with colleagues working with the toxicant being studied, and personal review of the relevant literature. Upon establishing an initial base of information from a thorough literature search, the same sources are used to update periodically the CTR data base. A coding strategy was developed to provide a mechanism for retrieving the specific data that are required for dose-response analyses. All information is abstracted onto a coding sheet according to a predetermined format. Abstractors are intensively trained and assist in all aspects of this phase upon completion of their training. The information abstracted onto the coding sheet is entered into the computerized data base. Each reference comprises a unique file and is assigned a unique identification number.

A critical element for dose-response analyses is a quantitative representation of the effect observed and reported in the literature. The last phase of the CTR system, then, compiles the concentration-response data in a form useful for combination with the PBD model predictions. Sub-data bases are generated for easy access to subsets of information or to facilitate future searches for very specific information. For example, a sub-data base could be established for all references reporting concentration-response data. This could subsequently be searched to sort by animal species. Further searches of these sub-data bases could identify studies of particular biological endpoints at specific pollutant concentrations.

After the concentration-response data are compiled, individual studies are examined for inclusion in the dose-response analyses. The basic criteria applied include a quantitative representation of effect, the use of a species appropriate for combination with the PBD model for O_3, an adequate presentation of data, a statistical analysis of data, exposures to O_3 only, a lack of variables that might confound results, and continuous exposures to a single concentration of O_3. Other items to be considered when selecting studies concern both interpretive and technical quality. Interpretive quality criteria address items such as the use of an appropriate experiment to test the hypothesis, while technical quality criteria address items such as providing precise descriptions of experiments and data analysis.

When specific studies appropriate for dose-response analyses are identified, information regarding, for example, weight of the species used is extracted. These data are used with the PBD model to predict the dose of O_3 that reaches specific regions of the lung. Figure 5 shows PBD model

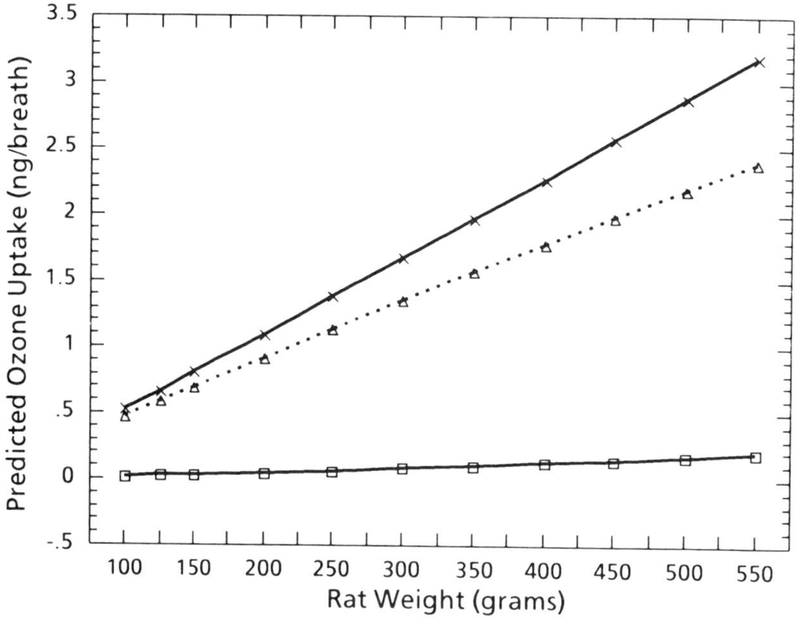

FIGURE 5 Predicted quantity of absorbed O_3 per breath versus rat weight for a 1-ppm exposure concentration with 25% upper respiratory tract removal. The three curves are for total lower respiratory tract (×), tracheobronchial region tissue (□), and pulmonary region tissue (△) absorption. Predictions were made using the PBD model described in the text. Lower respiratory tract anatomical dimensions were based on Yeh et al. (1979) for a 330-g rat. Allometric equations were used to extrapolate the lung dimensions to those corresponding to other weights.

predictions of dose to the mucus-lined tissue, the surfactant-lined tissue, and the total lung for rats (using the anatomical model of Yeh et al., 1979) exposed to 1,960 $\mu g/m^3$ (1 ppm) O_3.

Combining dosimetry data, such as those contained in Figure 5, with biological effects data to yield dose-response curves is illustrated using the formation of edema as an endpoint. Experimental studies (Ichikawa and Yokoyama, 1982; Maitani and Suzuki, 1981; Nambu and Yokoyama, 1981; Suzuki, 1976) were selected in which edema formation was indicated by an increase in lung wet weight. Because nasopharyngeal removal of O_3 in the rat has not been determined experimentally, one analysis was conducted in which 10% removal of O_3 in the head was assumed and another in which 25% removal was assumed. Unconfirmed preliminary experiments (Drs. Jean Wiester and Gary Hatch, personal communication) indicate that the rat may remove 10–25% of the inhaled O_3 in the nasopharyngeal region. Because edema is generally considered to reflect injury to gas-exchange regions, pulmonary tissue dose was used in the dose-response analysis (Figure 6). Data points represent various combinations of exposures to O_3 concentrations, ranging from 1 to 5 ppm for durations ranging from 3 to 12 h. Although a linear fit to the dose-response data is depicted in Figure 6, other curves may be equally plausible for describing the dependence of increased lung wet weight on the dose of O_3 delivered to the pulmonary region of the rat. The next step in the analysis would be to determine, using the PBD model, which O_3 exposure profiles would result in comparable predicted doses of O_3 in the human lung. This would allow better judgments on the reasonableness of extrapolating to man the toxicological data on edema in rats following exposure to O_3.

SUMMARY

Determining tissue dose is a fundamental starting point when making interspecies dose-response comparisons. Because of the complexity and cellular diversity of the lung, toxicologically relevant lung dosimetry needs to be related to specific sites within the respiratory tract to achieve a better understanding of the tissue dose dependence on toxicity of inhaled pollutants. Direct measurement of tissue or cellular dose by radiometric, physical, or chemical means within specific segments of the lung is one approach to determining equivalent doses between species. Such experiments are inherently limited, however, because providing experimental data on all exposure situations of interest for hazard assessment is practically impossible, and obtaining such a broad data base in man is ethically impossible. One approach to this problem is to develop mathematical dosimetry models of the lung that take into account not only the physical and chemical properties of the inhaled toxicant but also the anatomical

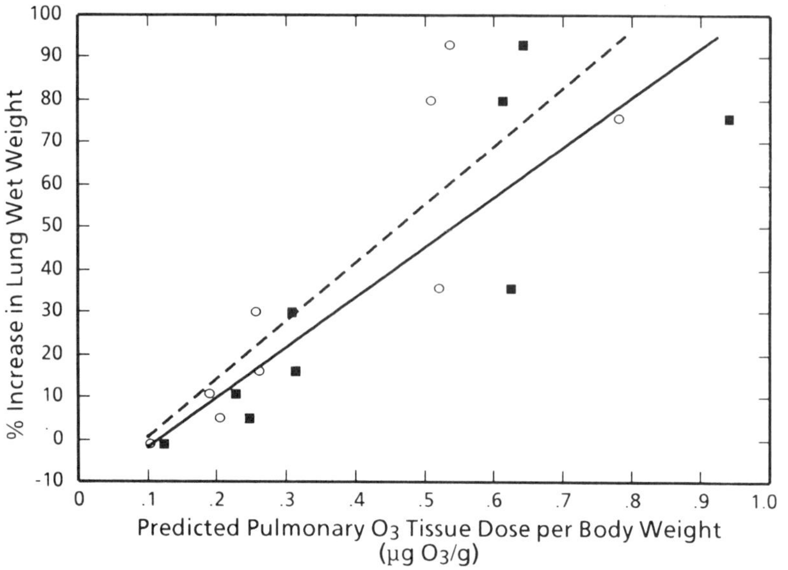

FIGURE 6 Estimated dose-response curves. Percent increased lung weight caused by O_3 exposure versus predicted pulmonary O_3 tissue dose per rat body weight resulting from 25% (○, dashed line) and 10% (■, solid line) upper respiratory tract removal is plotted. Dose is the total quantity of O_3 absorbed per body weight during exposure. Curves are fit by linear regression. Data from the CTR data management system, such as breathing frequency, time of exposure, animal weight, increased lung weight, and exposure concentration, combined with the PBD model predictions, as plotted for 25% head removal in Figure 5 and in other similar figures, were used to construct this figure.

and physical properties of the lung of the exposed animal species or man. This approach, with dose predicted by respiratory tract region or airway generation, provides (1) a guide for future biological experimentation, (2) a framework for further refinement and validation of the assumptions of mathematical dosimetry models, and (3) a basis for quantitative inter- and intraspecies comparisons of toxicological results.

A data base management system is available to store and retrive quantitative data in a manner that is usable as input files in dosimetric models. The effects of O_3 on all studied organ systems have been encoded, and citations reporting health effects are retrievable. Scaling for age, weight, sex, and species for critical characteristics are provided to yield appropriate data to use in conjunction with the PBD model for O_3. Predicted dose and toxicological response are combined to establish dose-response relationships. The methodology was illustrated for pulmonary edema, in which results from several laboratories using different exposure times, concentrations, animal sizes, and experimental methods were consolidated to obtain a single dose-response relationship. Ultimately, when the O_3

dosimetry modeling approach is sufficiently developed, it can provide an improved basis for estimating dose-response relationships for toxic effects observed in experimental animal studies for the purpose of estimating human risk. These assessments, in turn, can serve as an important input to risk management decisions concerning the regulation of O_3.

ACKNOWLEDGMENTS

We are most appreciative of the efforts of Mr. John Boger in the generation of model predictions for use in the analyses. We also wish to express their appreciation to Ms. Carolyn Wheeler for her excellent typing of the manuscript.

This work was supported in part by EPA Contracts 68-02-3809 and 68-02-3869 and by grants RR01693 and CA14236 from the National Institutes of health.

REFERENCES

Barry, B. E., F. J. Miller, and J. D. Crapo. 1985. Effects of inhalation of 0.12 and 0.25 ppm ozone on the proximal alveolar region of juvenile and adult rats. Lab. Invest. 53:692–704.

Boorman, G. A., L. W. Schwartz, and D. L. Dungworth. 1980. Pulmonary effects of prolonged ozone insult in rats. Lab. Invest. 43(2):108–115.

Chang, L., J. A. Graham, F. J. Miller, J. J. Ospital, and J. D. Crapo. 1986. Effects of subchronic inhalation of low concentrations of nitrogen dioxide. Toxicol. Appl. Pharmacol. 83:46–61.

Corn, M., N. Kotsko, and D. Stanton. 1976. Mass-transfer coefficient for sulphur dioxide and nitrogen dioxide removal in cat upper respiratory tract. Ann Occup. Hyg. 19:1.

Fenters, J. D., J. P. Findlay, C. D. Port, R. Ehrlich, and D. L. Coffin. 1973. Chronic exposure to nitrogen dioxide. Arch. Environ. Health 27:85–89.

Fujinaka, L. E., D. M. Hyde, C. G. Plopper, W. S. Tyler, D. L. Dungworth, and L. O. Lollini. 1985. Respiratory bronchiolitis following long-term ozone exposure in bonnet monkeys: A morphometric study. Exp. Lung Res. 8:167–190.

Gehr, P., M. Bachofen, and E. R. Weibel. 1978. The normal human lung: Ultrastructure and morphometric estimation of diffusion capacity. Respir. Physiol. 32:121–140.

Hyde, D., J. Orthoefer, D. Dungworth, W. Tyler, R. Carter, and H. Lum. 1978. Morphometric and morphologic evaluation of pulmonary lesions in beagle dogs chronically exposed to high ambient levels of air pollutants. Lab. Invest. 38:455–469.

Ichikawa, I., and E. Yokoyama. 1982. Effect of short-term exposure of ozone on the lecithin metabolism of rat lung. J. Toxicol. Environ. Health 10:1005–1015.

Jeffrey, P. K., and L. E. Reid. 1979. The respiratory mucous membrane. Pp. 193–245 in Respiratory Defense Mechanism, Part 1, J. D. Brain, D. F. Proctor, and L. M. Reid, eds. New York: Marcel Dekker.

Kliment, V. 1973. Similarity and dimensional analysis, evaluation of aerosol deposition in the lungs of laboratory animals and man. Folia Morphol. 21:59–64.

Maitani, T., and K. T. Suzuki. 1981. Changes to essential metal levels in lungs of rats acutely exposed to ozone. Toxicol. Lett. 8:99–104.

Miller, F. J., C. A. McNeal, J. M. Kirtz, D. E. Gardner, D. L. Coffin, and D. B. Menzel. 1979. Nasopharyngeal removal of ozone in rabbits and guinea pigs. Toxicology 14:273–281.

Miller, F. J., J. H. Overton, Jr., R. H. Jaskot, and D. B. Menzel. 1985. A model of the regional uptake of gaseous pollutants in the lung. I. The sensitivity of the uptake of ozone in the human lung to lower respiratory tract secretions and exercise. Toxicol. Appl. Pharmacol. 79:11–27.

Nambu, Z., and E. Yokoyama. 1981. The effect of age on the ozone-induced pumonary edema and tolerance in rats. Jpn. J. Ind. Health 23:146–150.

NRC (National Research Council, Committee on the Institutional Means for Assessment of Risks to Public Health, Commission on Life Sciences). 1983. P. 91 in Risk Assessment in the Federal Government: Managing the Process. Washington, D.C.: National Academy Press.

Overton, J. H. 1984. Physicochemical processes and the formulation of dosimetry models. J. Toxicol. Environ. Health 13:273–294.

Overton, J. H., R. C. Graham, and F. J. Miller. 1987. A model of the regional uptake of gaseous pollutants in the lung. II. The sensitivity of ozone uptake in the laboratory animal lungs to anatomical and ventilatory parameters. Accepted for publication by Toxicol. Appl. Pharmacol. 88:418–432.

Patra, A. L., A. Gooya, and K. T. Morgan. 1986. Airflow characteristics in a baboon nasal passage. J. Appl. Physiol. 61:1959–1966.

Plopper, C. G., C. K. Chow, D. L. Dungworth, and W. S. Tyler. 1979. Pulmonary alterations in rats exposed to 0.2 and 0.1 ppm ozone: A correlated morphological and biochemical study. Arch. Environ. Health 34:390–395.

Riddick, J. A., K. I. Campbell, and D. L. Coffin. 1968. Histopathologic changes secondary to nitrogen dioxide exposure in dog lungs. Am. J. Clin. Pathol. 49:239. (Abstract)

Schreider, J. 1986. Comparative anatomy and function of the nasal passages. Pp. 1–25 in Toxicology of the Nasal Passages, C. S. Barrow, ed. Washington, D.C.: Hemisphere.

Schreider, J., and O. Raabe. 1981. Anatomy of the nasal pharyngeal airways of experimental animals. Anat. Rec. 200:195–205.

Schreider, J. P., and J. O. Hutchens. 1980. Morphology of the guinea pig respiratory tract. Anat. Rec. 196:313–321.

Smolko, E. D., D. J. McKee, and D. B. Menzel. In press. Critical toxicity reference system. I. An approach for managing quantitative toxicity data. J. Am. College Toxicol.

Stephens, R. J., M. F. Sloan, M. J. Evans, and G. Freeman. 1973. Early response of lung to low levels of ozone. Am. J. Pathol. 74:31–58.

Stephens, R. J., M. F. Sloan, M. J. Evans, and G. Freeman. 1974. Alveolar type 1 cell response to exposure to 0.5 ppm O_3 for short periods. Exp. Mol. Pathol. 20:11–23.

Suzuki, T. 1976. Studies on the acute toxicity of 5-hydroxytryptamine in the rats pre-exposed to ozone. Bull. Tokyo Med. Dent. Univ. 23:217–226.

Weibel, E. R. 1963. Morphometry of the human lung. New York: Academic Press.

Yeh, H. C., and G. M. Schum. 1980. Models of human lung airways and their application to inhaled particle deposition. Bull. Math. Biol. 42:461–480.

Yeh, H. C., G. M. Schum, and M. T. Duggan. 1979. Anatomic models of the tracheobronchial and pulmonary region of the rat. Anat. Rec. 195:483–492.

Yokoyama, E. 1968. Uptake of SO_2 and NO_2 by the isolated airways. Bull. Inst. Public Health 17:302–306.

Yokoyama, E., and R. Frank. 1972. Respiratory uptake of ozone in dogs. Arch. Environ. Health 25:132–138.

Role of Pharmacokinetic Modeling in Risk Assessment: Perchloroethylene as an Example

Chao W. Chen and *Jerry N. Blancato*

INTRODUCTION

The assessment of potential risks to humans from xenobiotic chemicals is a major component of the environmental decision-making process. To ensure high-quality assessments, the risk assessor must consider all available data and perform a wide range of interpretations. All such interpretations should be based on sound and logical scientific thought and should use state-of-the-art methodologies.

Ideally, prediction of human risk because of exposure to environmental agents should be made on the basis of human experience and sound laboratory findings. The available human data are, however, rarely sufficient for this purpose. A common alternative approach is to predict human risk on the basis of experimental animal data. The estimation of human risk from these types of data involves several steps and assumptions. These steps include extrapolation from risk in animals exposed at high doses to risk in animals that might be exposed at low doses, because the available experimental data frequently are the result of high-dose experiments. Thus, even at this first level of extrapolation, there are seldom enough data to permit verification of the assumptions made with regard to the shape of the dose-response relationship at low doses. Other steps in the human risk assessment process include extrapolating from

The views expressed in this paper are those of the authors and do not necessarily reflect the views or policies of the U.S. Environmental Protection Agency.

animal low-dose risk to human low-dose risk (species conversion) and from one route of exposure to another (route-to-route extrapolation). Each of these steps involves certain assumptions; for high- to low-dose extrapolation, usually a parametric dose-response relationship is assumed; for species conversion, the assumption is usually made that animals and humans have equal lifetime risks from exposure when the dose rate is expressed in a particular unit (equivalent dose). The dose units that have been used as equivalent doses include milligrams/body surface area/day, milligram/kilogram of body weight/day, milligram/target tissue volume (or mass)/day, parts per million in air, parts per million in diet, and milligrams/kilogram/lifetime. These conversion factors are presumed to account for all of the differences between animals and humans, including longevity, body size, and metabolic processes and rates. The basis for these conversions and extrapolations is not well documented or proven. For route-to-route extrapolation, the Stokinger and Woodward (1958) approach is usually employed, with the assumption of 100% absorption from air and oral exposure. This assumption can result in significant error in a number of cases. For example, there is little evidence that many substances are fully absorbed after they are inhaled, even at low concentrations. In the case of ingestion by the oral route, less than 100% absorption may occur, depending on the physicochemical nature of the substance, diet contents, species, and other factors.

The use and analysis of pharmacokinetic and metabolic information in risk assessments may reduce some of the ambiguities associated with these three major areas of uncertainty, namely, high- to low-dose extrapolation, species conversion, and route-to-route extrapolation. This is particularly true of those uncertainties associated with route-to-route and, to a lesser extent, high- to low-dose extrapolation. Consideration of metabolism and pharmacokinetics alone cannot completely eliminate uncertainties, however, especially in the case of interspecies extrapolation. It is essential that the carcinogenic processes that are affected by environmental agents be better understood if the uncertainties involved in risk assessment are to be minimized.

Pharmacokinetic analyses can range from very simple arithmetic operations, which are used to calculate the amount of the dose that is converted to some suspected or proven toxin, to complicated physiologically based pharmacokinetic (PB-PK) models. PB-PK models can be used to study the kinetics of absorption, distribution, metabolic processing, and elimination of the parent compound and its metabolites. They may also be used to predict the disposition of a compound at different dosage regimens. In principle, such predictions may be made across species lines, thus greatly reducing the number of different species of animals that must be used to test a given compound. When properly formulated, PB-PK models may give information regarding the concentration of a putative toxin at a target tissue or cell. Most importantly, they can provide a description of the disposition of a compound over time rather than just at some predetermined endpoint. As these models become better tested, a variety of questions pertaining to particular circumstances can be answered. For example, the effects of less-than-lifetime exposure, intermittent exposure, or bursts of exposure can be examined; and differences with regard to various parameters across and within species can be measured and accounted for systematically. As more and more parameters are considered and accurately determined, the uncertainties involved in risk assessment will be correspondingly reduced.

This paper discusses the use of metabolic and pharmacokinetic data on perchloroethylene (PCE) in the formulation of PB-PK models. The results of these models are applied in a cancer risk assessment of PCE. The emphasis here is on methodologies for improving risk assessment, rather than on the risk assessment of PCE per se. The basis for the discussion contained in this paper was work done in preparation of an Addendum to the Health Assessment Document for Tetrachloroethylene (EPA, 1986).

PCE offers an attractive opportunity to evaluate the usefulness of PB-PK modeling in cancer risk assessment for the following reasons:

1. The same tumors (hepatocellular carcinomas) have been observed in both male and female mice exposed to PCE by gavage and inhalation. This information permits the evaluation of uncertainty with regard to route-to-route extrapolation.

2. The metabolism and pharmacokinetics of PCE have been studied extensively in both laboratory animals and humans. The available data are sufficient to enable a PB-PK model to be independently constructed for each of the three species: humans, rats, and mice. These results serve as a useful basis for evaluating certain assumptions about parameters in human models when not enough data exist to permit construction of a PB-PK model for humans.

PCE is biotransformed by microsomal monooxygenases (cytochrome P-450 system) to reactive metabolites, including the putative carcinogen trichloroacetic acid (TCA) and a short-lived, highly reactive epoxide intermediate that can bind covalently to cellular macromolecules. It is generally considered that the hepatic toxicity and carcinogenicity potential of PCE resides in its biologically reactive intermediate metabolites rather than in the parent compound itself. This factor further strengthens the argument for using metabolites, rather than the administered dose, in risk assessment.

METABOLIC DATA PERTINENT TO THE CONDITIONS OF PCE CARCINOGEN BIOASSAYS

Oral Studies

Pegg et al. (1979) and Schumann et al. (1980) administered [^{14}C]PCE in corn oil vehicle as single intragastric doses to rats and mice. ^{14}C-radioactivity was determined for exhaled breath, urine, feces, and carcass for 72 h following dosing. Pulmonary excretion of unchanged PCE was measured. The data from these two studies were collected and collated and then re-expressed in milligram equivalent units (Table 1).

Inhalation Studies

ANIMAL DATA

Pegg et al. (1979) and Schumann et al. (1980) also determined body burdens and metabolism of rats and mice after inhalation exposure to PCE for 6 h. The animals were exposed to [^{14}C]PCE; the radioactivity was determined in urine, feces, expired air, etc.; and unchanged [^{14}C]PCE was determined in expired air for 72 h postexposure. These data are

TABLE 1 Disposition of [^{14}C]PCE Radioactivity for 72 h After Single Oral Doses to Sprague-Dawley Rats and B6C3F1 Mice

	mg Eq./animal		
	Rats[a]		
Drug disposition	1 mg/kg (0.25 mg/animal)	500 mg/kg (125 mg/animal)	Mice[a] (500 mg/kg, 12.25 mg/animal)
Expired unchanged	0.174 (71%)	110.67 (90%)	8.90 (83%)
Metabolized			
$^{14}CO_2$	0.007	0.57	0.14
Urine	0.040	5.72	1.53
Feces	0.015	4.82	0.13
Carcass	0.008	1.41	0.05
	0.070 (29%)	12.52 (10%)	1.85 (17%)
Total recovered	0.244	123.19	10.75

[a]There was an average of three rats and three mice. Data are based on average experimental animal weights: 250 g for rats and 24.5 g for mice.

SOURCE: Adapted from Pegg et al. (1979) and Schumann et al. (1980).

TABLE 2 Disposition of [^{14}C]PCE Radioactivity for 72 h After Inhalation Exposure for 6 h by Sprague-Dawley Rats and B6C3F1 Mice

Drug disposition	mg Eq./animal Rats[a] 10 ppm	 600 ppm	 Mice[a] (10 ppm)
Expired unchanged	1.008 (68%)	68.39 (88%)	0.048 (12%)
Metabolized			
$^{14}CO_2$	0.053	0.54	0.032
Urine	0.275	4.54	0.285
Feces	0.076	2.36	0.027
Carcass	0.063	1.67	0.012
	0.467 (32%)	9.11 (12%)	0.356 (88%)
Total recovered	1.475	77.50	0.404

[a]There was an average of three rats and three mice. Data are based on average experimental animal weights: 250 g for rats and 24.5 g for mice.

SOURCE: Adapted from Pegg et al. (1979) and Schumann et al. (1980).

collated and re-expressed for comparison of rat and mouse metabolism in Table 2.

Metabolic data from oral and inhalation studies have been used to estimate metabolic constants in the PB-PK models for mice and rats. Further details on the use of these data are provided in Appendix A.

HUMAN DATA

Estimates of the extent of PCE metabolism in man by measuring metabolites excreted in urine have been made from several human volunteer studies. A review by Monster (1984) provides useful data for constructing

TABLE 3 Mean Amounts of the PCE Metabolite TCA Excreted in Human Urine over a 72-h Period After a Single Inhalation Exposure to PCE

Exposure concentration (mg/liter)	Duration of exposure (h)	Observed TCA (mg)	No. of subjects	Reference
0.480	4	5.9	6	Monster et al., 1979
0.960	4	11.2	6	Monster et al., 1979
1.029	8	23.5	2	Fernandez et al., 1976

TABLE 4 Mean Concentrations of PCE in Human Alveolar Air After a Single Inhalation Exposure to PCE

Exposure concentration (mg/liter)	Duration of exposure (h)	Postexposure alveolar concentration (mg/liter) at: 0.5 h	1 h	2 h	16 h	64 h
0.686	8	0.137	0.103	0.082	0.027	0.014
1.029	8	0.278	0.216	0.165	0.042	0.020
1.371	8	0.343	0.260	0.192	0.055	0.027

SOURCE: Fernandez et al. (1976).

a PB-PK model for humans. The data presented in Tables 3 and 4 have been used to estimate the metabolic constants in the PB-PK model. These data were selected because they appeared to be obtained from well-conducted studies and were from similarly designed experiments. Additional details on the utilization of these data to estimate metabolic constants in the PB-PK model are given in Appendix A.

The metabolic data presented in Tables 1 through 4 enable us to independently construct a PB-PK model for each of the three species: humans, rats, and mice. These data have been used to optimize the metabolic constants V_m and K_m. The following sections describe the use of these models to assess the cancer risk of PCE.

TABLE 5 Tumor Incidence and the Corresponding Metabolized Dose for B6C3F1 Mice in NCI (1977) Gavage Study, Calculated by PB-PK Model

Administered dose (mg/kg/day)	Daily metabolized dose (mg/kg/day)[a]	Hepatocellular carcinoma incidence[b]
Males (0.03 kg)		
0	0	2/20
536	56.32	32/48
1,072	75.25	27/45
Females (0.025 kg)		
0	0	0/20
386	40.58	19/48
772	57.52	19/45

[a]Daily metabolized dose = Average daily metabolized dose × 78 week/90 week, where the factor 78/90 reflects the fact that mice were exposed to PCE for 78 weeks out of a 90-week lifetime. The average metabolized doses are given in Table 7.

[b]Only animals surviving beyond the first liver tumor death (27 weeks for males and 41 weeks for females) are included in the denominators.

TABLE 6 Tumor Incidence and the Corresponding Metabolized Dose for B6C3F1 Mice in NTP (1985) Inhalation Bioassay, Calculated by PB-PK Model

Administered dose (ppm)	Daily metabolized dose (mg/kg/day)[a]	Hepatocellular carcinoma incidence[b]
Males (0.03 kg)		
0	0	7/49
100	47.23	25/47
200	66.29	26/50
Females (0.025 kg)		
0	0	1/46
100	54.97	13/42
200	76.76	36/47

[a]Taken from Table 8.
[b]Only animals surviving beyond the first death from liver tumor (60 weeks) are included in the denominators.

USE OF PB-PK MODEL TO ESTIMATE METABOLIZED DOSE FOR NCI (GAVAGE) AND NTP (INHALATION) BIOASSAY MICE

As mentioned earlier, PCE experiments in mice offer an attractive opportunity to evaluate the usefulness of PB-PK models in cancer risk assessment because tumors have been induced in the same site by both gavage and inhalation routes of exposure. The dose response data from gavage (NCI, 1977) and inhalation (NTP, 1985) are presented, respectively, in Tables 5 and 6. Under NCI gavage or NTP inhalation dosing patterns, the amount of metabolized dose reaches the steady state during weekdays and is almost totally eliminated during the weekend (nonexposed period). Tables 7 and 8 present daily metabolites (in milligrams/day) for mice in the NCI and NTP bioassays.

CALCULATION OF THE DOSE-RESPONSE RELATIONSHIP

The two-stage model,

$$P(d) = 1 - \exp(-q_0 - q_1 d - q_2 d^2), \qquad (1)$$

is used to calculate the dose-response relationship by using hepatocellular carcinoma incidence rates and the corresponding metabolized dose (in milligrams/kilogram/day) presented in Tables 5 and 6. The incremental cancer risk (for mice) is defined as

$$F(d) = [P(d) - P(0)]/[1 - P(0)] = 1 - \exp[-q_1 d - q_2 d^2]. \qquad (2)$$

The maximum likelihood estimates of parameters q_1 and q_2, along with the 95% upper limit of q_1, denoted q_u, are presented in Table 9. At low doses, the upper-bound risk is predicted by $q_u \times d$ at dose rate d (milligram of metabolites/kilogram/day). It is seen from Table 9 that q_u, calculated on the basis of different data bases, gavage or inhalation, male or female mice, are all comparable, ranging from 6.27×10^{-3} to 1.76×10^{-2} per mg of metabolites/kg/day. the geometric means of q_u are 1.6×10^{-2} and 9.4×10^{-3}, respectively, calculated from gavage and inhalation bioassays. All risk estimates presented in this paper are 95% upper-limit estimates.

To predict human cancer risk caused by exposure to PCE, the assumption must be made as to what constitutes an equivalent dose among species, that is, the dose unit that induces the same magnitude of tumor response in animals and humans. Table 10 presents risk for humans continuously exposed to PCE by inhalation under various equivalent dose assumptions. Table 11 presents parallel risk estimates as in Table 10 when exposure is 8 h/day, 5 days/week. In these calculations, the geometric means of the two upper-bound risks calculated from male and female mice dose-response curves are used.

TABLE 7 Daily Amounts (mg) of Total Metabolites in Male and Female Mice Exposed to PCE by Gavage According to the NTP Bioassay Pattern

	In males (0.035 kg bw)		In females (0.025 kg bw)	
Day	At low dose (536 mg/kg/day)	At high dose (1,072 mg/kg/day)	At low dose (386 mg/kg/day)	At high dose (772 mg/kg/day)
Week 1				
1	2.675	3.545	1.621	2.291
2	2.707	3.583	1.634	2.308
3	2.708	3.584	1.634	2.308
4	2.708	3.584	1.634	2.308
5	2.708	3.584	1.634	2.308
6	0.138	0.349	0.036	0.090
7	0.002	0.006	0.0004	0.0009
Week 2				
1	0.002	3.545	1.621	2.291
2	2.675	3.583	1.634	2.308
3	2.707	3.584	1.634	2.308
4	2.708	3.584	1.634	2.308
5	2.708	3.584	1.634	2.308
6	0.138	0.349	0.036	0.090
7	0.002	0.006	0.0004	0.0009
Average (week 2)	1.949 mg or 64.98 mg/kg/day	2.605 mg or 86.83 mg/kg/day	1.170 mg or 46.82 mg/kg/day	1.659 mg or 66.37 mg/kg/day

TABLE 8 Daily Amounts (mg) of Total Metabolites in Male and Female Mice Exposed to PCE by Inhalation According to the NTP Bioassay Pattern

	In males (0.035 kg bw)		In females (0.03 kg bw)	
Day	At low dose (100 ppm)	At high dose (200 ppm)	At low dose (100 ppm)	At high dose (200 ppm)
Week 1				
1	2.273	3.176	2.285	3.150
2	2.302	3.214	2.312	3.185
3	2.303	3.215	2.312	3.186
4	2.303	3.215	2.312	3.187
5	2.303	3.215	2.312	3.187
6	0.085	0.202	0.087	0.222
7	0.0016	0.004	0.0013	0.003
Week 2				
1	2.274	3.176	2.285	3.150
2	2.303	3.214	2.312	3.185
3	2.303	3.215	2.312	3.186
4	2.303	3.215	2.312	3.187
5	2.303	3.215	2.312	3.187
6	0.085	0.202	0.087	0.222
7	0.0016	0.004	0.001	0.003
Average	1.653 mg or 47.23 mg/kg/day	2.320 mg or 66.29 mg/kg/day	1.649 mg or 54.97 mg/kg/day	2.303 mg or 76.76 mg/kg/day

COMPARISON OF RISK ESTIMATES FOR DRINKING WATER AND INHALATION EXPOSURE, WITH AND WITHOUT THE USE OF PB-PK MODELING

The conventional approach to risk estimation, without the use of PB-PK modeling, would be to assume 100% absorption by either the oral or the inhalation route of exposure. In the following calculations, it is assumed that a person weighs 70 kg, drinks 2 liters of water per day, and inhales 15 m^3 of air per day. The value of 15 m^3/day is used because it is approximately the value used in PB-PK modeling. For convenience, only body surface equivalent dose is used in these calculations, except where parts per million is assumed to be equivalent. When the PB-PK model is used, water intake is assumed to occur uniformly over a 16-h period in a day. Table 12 summarizes cancer risk estimates associated with 1 μg/liter (1 $\mu g/m^3$) of PCE in water (air), calculated with and without PB-PK modeling.

TABLE 9 Parameter Estimates of Two-Stage Model Based on Data from Gavage and Inhalation Bioassays

Data base	q_1[a]	q_2	q_u	Geometric mean
Gavage (NCI, 1977)				
Male mice	1.35×10^{-2}	0	1.76×10^{-2}	
Female mice	1.08×10^{-2}	0	1.40×10^{-2}	1.6×10^{-2}
Inhalation (NTP, 1985)				
Male mice	1.0×10^{-2}	0	1.40×10^{-2}	
Female mice	0	1.87×10^{-4}	6.27×10^{-3}	9.4×10^{-3}

[a]Dose unit = milligrams of metabolite/kilogram/day.

TABLE 10 Predicted Human Cancer Risk Caused by Continuous 24-Hour Exposure to PCE by Inhalation at Different Assumed Equivalent Doses

Concentration (ppm)	Dose metabolized (mg/day)	Assumed equivalent doses[a] mg/$W^{2/3}$/day	mg/kg/day	mg/liver tissue volume/day	ppm[b]
0.01	0.0215	3.8×10^{-5}	2.9×10^{-6}	5.6×10^{-6}	2.6×10^{-4}
0.10	0.215	3.8×10^{-4}	2.9×10^{-5}	5.6×10^{-5}	2.6×10^{-3}
1.00	2.15	3.8×10^{-3}	2.9×10^{-4}	5.6×10^{-4}	2.6×10^{-2}
50	97.56	1.6×10^{-1}	1.3×10^{-2}	2.5×10^{-2}	1.0×10^{0}
100	178.47	2.7×10^{-1}	2.3×10^{-2}	4.6×10^{-2}	1.0×10^{0}

[a]Human body weight (W) is assumed to be 70 kg.
[b]ppm in air is assumed to be equivalent between mice and humans.

TABLE 11 Predicted Human Cancer Risk Caused by Inhalation Exposure (8 h/day, 5 days/week) at Different Assumed Equivalent Doses

Concentration (ppm)	Dose metabolized (mg/day)	Assumed equivalent doses mg/$W^{2/3}$/day	mg/kg/day	mg/liver tissue volume/day	ppm
0.01	5.12×10^{-3}	9.0×10^{-6}	6.9×10^{-7}	1.3×10^{-6}	6.0×10^{-5}
0.10	5.12×10^{-2}	9.0×10^{-5}	6.9×10^{-6}	1.3×10^{-5}	6.0×10^{-4}
1.00	5.12×10^{-1}	9.0×10^{-4}	6.9×10^{-5}	1.3×10^{-4}	6.0×10^{-3}
50	24.55	4.2×10^{-2}	3.3×10^{-3}	6.4×10^{-3}	2.9×10^{-1}
100	47.23	8.0×10^{-2}	6.3×10^{-3}	1.2×10^{-2}	5.6×10^{-1}

TABLE 12 Risk at a Unit Dose of 1 μg/liter in Water or 1 μg/m³ in Air With and Without the Use of PB-PK Modeling

	Without PB-PK modeling			With PB-PK modeling	
		Based on inhalation data			
Medium	Based on gavage data	Method 2a	Method 2b	Based on gavage data	Based on inhalation data
Water (1 μg/liter)	7.4×10^{-7}	NE[a]	NE	1.5×10^{-7}	NE
Air (1 μg/m³)	6.1×10^{-6}	3.8×10^{-6}	8.4×10^{-6}	9.4×10^{-7}	5.5×10^{-7}

[a]NE = Not estimated.

There are several ways to calculate air risk without using a PB-PK model:

1. Indirect estimate from gavage data. On the basis of gavage data, the potency slope (i.e., 95% upper bound of the linear parameter in the two-stage model, under the body surface equivalent dose assumption) is calculated to be 2.8×10^{-2}/mg/kg/day. This is equivalent to 6.1×10^{-6}/μg/m³, using the Stokinger and Woodward (1958) procedure and the assumption of 100% absorption.
2. Direct estimate from inhalation data.
   a. When parts per million is assumed to be equivalent, the potency slope is 2.6×10^{-2}/ppm or, equivalently, 3.8×10^{-6}/μg/m³.
   b. When animal dose in parts per million is converted to milligrams/kilogram/day, assuming 100% absorption, the potency slope is 3.9×10^{-2}/mg/kg/day or, equivalently, 8.4×10^{-6}/μg/m³.

These calculations indicate that estimation of inhalation risk without adjusting for metabolism overestimates risk by about an order of magnitude when compared with that estimated by PB-PK modeling. For drinking water, the conventional approach overestimates risk by about 5 times when compared with that estimated by PB-PK modeling. It should be noted that the variability observed in these estimates is mainly due to metabolism adjustment and not to the use of different dose-response models and/or different assumptions about equivalent dose.

DISCUSSION

In calculating the dose-response relationship for PCE, the amount metabolized per day is considered to be an effective dose. The use of this surrogate effective dose may not eliminate the uncertainty associated with the low-dose extrapolation because the dose actually reaching the receptor sites may not be linearly proportional to the total amount of metabolites, and the shape of the dose-response relationship is still unknown. It seems reasonable to expect, however, that the uncertainty with regard to the low-dose extrapolation would be somewhat reduced by the use of the metabolized dose because the metabolized dose better reflects the dose-response relationship, particularly in the high-dose region.

Ideally, a dose-response function should be derived according to the mechanism of carcinogenic action if the available scientific information is sufficient. A two-event model proposed by Moolgavkar and Venzon (1979) and Greenfield et al. (1984) provides a promising framework for building a biologically based dose-response function. Under this model, the probability of cancer is a function of mitotic rate (M), transition rate (P), birth rate (B), and death rate (D) for cells at different stages: normal, initiated, or transformed. The ultimate goal is to construct a biologically based dose-response function and a physiologically based pharmacokinetic model that estimates target dose under any exposure pattern. For a given target dose d, however, one would not expect mitotic rate $M(d)$, transition rate $P(d)$, birth rate $B(d)$, or death rate $D(d)$ to be identical between animals and humans. If they were identical, one would see much more cancer incidence in humans than in animals, because humans have more cells and a longer life span than animals. Therefore, even if target dose is known, one still faces the problem of dose scaling unless one can measure $M(d)$, $P(d)$, $B(d)$, and $D(d)$ in both humans and animals.

PCE provides a rare opportunity to compare the relationship of the metabolic constants V_{max} and K_m among species. The allometric equations calculated from data from mice, rats, and humans are as follows:

$$V_{max} = 0.016W^{0.89}, \text{ and} \tag{3}$$

$$K_m = 5.0W^{0.44}, \tag{4}$$

where W is body weight, in kilograms. These equations imply:

$$V_{max}(\text{humans})/V_{max}(\text{animals}) = (W_h/W_a)^{0.89}, \text{ and} \tag{5}$$

$$K_m(\text{humans})/K_m(\text{animals}) = (W_h/W_a)^{0.44}, \tag{6}$$

where W_h and W_a are body weights for humans and animals, respectively. These observations are not consistent with assumptions that are usually made in the absence of human data; i.e., the assumptions that V_{max} is related to body weight by the 0.67th power and that K_m is identical between animals and humans (Ramsey and Andersen, 1984). However, one should not make any judgment on the basis of a single observation.

SUMMARY

The overall goal of this paper has been to demonstrate the use of a physiologically based pharmacokinetic model to estimate the risk to humans from various concentrations of PCE.

A relatively simple physiologically based pharmacokinetic model has been used to calculate the amount of PCE metabolized by animals that showed tumor responses in bioassay testing. From the amount metabolized, a multistage model was used to calculate the dose-response relationship for mice. Subsequently, the physiologically based pharmacokinetic model, with appropriate changes in parameters to account for human physiological and biochemical processes, was used to calculate the amounts of PCE that would be metabolized in humans under various ambient exposure concentrations. The amounts, as determined from the model, were then used, together with the dose-response function from the mouse bioassay studies, to calculate human risk.

The metabolized amounts calculated for humans were expressed in a variety of equivalent forms, because some expression of dosimetry is needed to equilibrate the tumor response between species. Because there is no agreed upon standard factor for species conversion, several forms of equivalent dose have been chosen to express the risk calculated from the metabolized dose. The cancer potency per unit of metabolized dose appears to be comparable regardless of whether gavage or inhalation data are used.

Comparisons were also made on the calculation of risk for inhalation based on potencies calculated from gavage, without consideration for route-to-route differences. The results of calculating the human risk by the pharmacokinetic model method show that by using the gavage potency, without accounting for pharmacokinetics, calculating risk for inhalation results in error. Thus, pharmacokinetic considerations must be accounted for when risk is to be calculated from animal studies using a different route of exposure from the route of potential human exposure.

From these PB-PK model analyses, it is observed that the daily formation of metabolites reaches the steady state faster in rodents than in humans. One possible explanation for this is that the percentage of body fat is greater in humans than in rodents.

FUTURE DIRECTIONS

We believe that we have demonstrated here the utility and advantages of using physiologically based pharmacokinetic modeling in exposure and risk assessments. We anticipate that in the future this approach will be used more extensively. As further data become available on the characteristics of the disposition, metabolism, and mechanism of action of other compounds, PB-PK models will become correspondingly more sophisticated. Already, models have been formulated to describe the disposition of foreign chemicals in far greater detail than the model presented here (Bischoff et al., 1971; King et al., 1986; Lutz et al., 1976). The concentrations of the metabolites can be described and predicted not only in the organ of formation but also in other body regions after hemodynamic transport. The sequestering of parent compound and metabolite can also be described (Angelo et al., 1984). This situation can be of particular importance for compounds exhibiting a long biological life, as, for example, in the adipose tissues. It is widely accepted that differences in exposure patterns could have an effect on the true human risk, which is not discernible by the traditional risk calculation approach. As demonstrated in this paper, with the use of these types of models, information can be quickly and easily obtained to help determine those exposure conditions under which risk is greatest, and what types of studies are needed to elucidate more information. As more becomes known about mechanisms of action, risk assessors will benefit from knowing the toxin concentrations at subcellular sites. For example, dose-response functions can be calculated by using the concentration at a particular subcellular site as the dose. Describing and modeling exposure at such sites requires a greater understanding of mechanisms of action but ultimately can result in reducing some uncertainties that are inherent in present risk assessments. Subcellular models have already been formulated for the purpose of exposure assessments (Blancato and Bischoff, 1985; Gehring and Young, 1978). These types of models, and the data that are generated in their formulation, are valuable in providing guidance in the design and implementation of key laboratory studies that help to identify and discern the mechanisms of action of toxic substances.

ACKNOWLEDGMENT

The authors wish to thank members of the Carcinogen Assessment Group and the Exposure Assessment Group, Office of Health and Environmental Assessment, for their contributions to this presentation.

REFERENCES

Angelo, M. J., K. B. Bischoff, A. B. Pritchard, and M. A. Presser. 1984. A physiological model for the pharmacokinetics of methylene chloride in B6C3F1 mice following i.v. administrations. J. Pharmacokinet. Biopharm. 12(4):413–436.

Bischoff, K. B., R. L. Dedrick, D. S. Zaharko, and J. A. Longstreth. 1971. Methotrexate pharmacokinetics. J. Pharm. Sci. 60(8):1128–1133.

Blancato, J. N., and K. B. Bischoff. 1985. Sub-cellular pharmacokinetics of 2,5-hexane dione. Paper presented at North American Symposium on Risk Assessment and the Biological Fate of Xenobiotics, Key Biscayne, Fla., November 17–22, 1985.

EPA (U.S. Environmental Protection Agency). 1986. Addendum to the Health Assessment Document for Tetrachloroethylene (Perchloroethylene): Updated Carcinogenicity Assessment for Tetrachloroethylene (Perchloroethylene, PERC, PCE). EPA/600/8-82/005FA. External Review Draft. (Available from National Technical Information Service, Springfield, Va., as publication no. PB86-174489).

Fernandez, T., E. Guberan, and J. Caperos. 1976. Experimental human exposures to tetrachloroethylene vapor and elimination in breath after inhalation. Am. Ind. Hyg. Assoc. J. 37(March):143–150.

Gehring, P. J., and J. P. Young. 1978. Application of pharmacokinetic principles in practice. Pp. 119–141 in Toxicology: Proceedings of First International Congress on Toxicology, G. L. Plaa and W. H. M. Duncan, eds. New York: Academic Press.

Gehring, P. J., L. Watanabe, and C. Park. 1979. Risk of angiosarcoma in workers exposed to vinyl chloride as predicted from studies from rats. Toxicol. Appl. Pharmacol. 49:15–21.

Greenfield, R. E., L. B. Ellwen, and S. M. Cohen. 1984. A general probabilistic model of carcinogenesis: Analysis of experimental urinary bladder cancer. Carcinogenesis 5(4):437–445.

Hattis, D., S. Tuler, L. Finkelstein, and Z.-Q. Luo. 1986. A pharmacokinetic/mechanism-based analysis of the carcinogenic risk of perchloroethylene. Cambridge, Mass.: Center for Technology, Policy and Industrial Development, Massachusetts Institute of Technology.

King, F. L., R. L. Dedrick, and F. F. Farris. 1986. Physiological pharmacokinetic modeling of cis-dichlorodiamineplatnium (II) (DDP) in several species. J. Pharmacokinet. Biopharm. 14(2):131–155.

Lutz, R. J., R. L. Dedrick, H. B. Matthews, T. E. Eling, and M. W. Anderson. 1976. A preliminary pharmacokinetic model for several chlorinated biphenyls in the rat. Drug Metab. Dispos. 5(4):386–396.

Monster, A. C. 1984. Tetrachloroethylene. Pp. 131–139 in Biological Monitoring and Surveillance of Workers Exposed to Chemicals, A. Aitio, V. Riihimaki, and H. Vainio, eds. New York: McGraw-Hill.

Monster, A. C., G. Boersma, and H. Steenweg. 1979. Kinetics of tetrachloroethylene in volunteers; influence of exposure concentrations and work load. Int. Arch. Occup. Environ. Health 42:303–309.

Moolgavkar, S. H., and D. J. Venzon. 1979. Two-event models for carcinogenesis: Incidence curves for childhood and adult tumors. Math. Biosci. 47:55–77.

NCI (National Cancer Institute). 1977. Bioassay of tetrachloroethylene for possible carcinogenicity. DHEW Pub. No. (NIH) 77-813. Bethesda, Md.: Public Health Service, National Institutes of Health, U.S. Department of Health, Education, and Welfare.

NTP (National Toxicology Program). 1985. Toxicology and carcinogenesis of tetrachloroethylene in F344/N rats and B6C3F1 mice (inhalation studies). Draft technical report.

Ohitsu, T., K. Sato, A. Koisumi, and M. Kumai. 1983. Limited capacity of humans to metabolize tetrachloroethylene. Int. Arch. Occup. Environ. Health 51:381–390.

Pegg, D. G., J. A. Zempel, W. H. Brown, and P. G. Watanabe. 1979. Deposition of tetrachloro(^{14}C)ethylene following oral and inhalation exposure in rats. Toxicol. Appl. Pharmacol. 51:465–474.

Ramsey, T. C., and M. E. Andersen. 1984. A physiologically-based description of the inhalation pharmacokinetics of styrene in rats and humans. Toxicol. Appl. Pharmacol. 79:389–400.

Schumann, A. M., T. F. Quast, and P. G. Watanabe. 1980. The pharmacokinetics of perchloroethylene in mice and rats as related to oncogenicity. Toxicol. Appl. Pharmacol. 55:207–219.

Stokinger, M. E., and R. L. Woodward. 1958. Toxicological methods for establishing drinking water standards. J. Am. Water Works Assoc. 50:517.

APPENDIX: CONSTRUCTION OF PB-PK MODELS

Description of the PB-PK Model for Inhaled PCE

PCE in vapor form in air is readily absorbed through the lungs into blood by first-order diffusion processes. Pulmonary uptake of PCE is largely determined by the ventilation rate, duration and concentration of exposure, solubility in blood and body tissues, and metabolism. PCE is eliminated by pulmonary excretion and metabolism (primarily in the liver). Figure A-1 depicts the PB-PK model used to simulate PCE distribution and metabolism over time. This model was used by Ramsey and Andersen (1984) to study the behavior of inhaled styrene. Abbreviations used in the figure are given in Table A-1.

The model in Figure A-1 is described by a system of differential equations that quantify the rate of change of PCE concentration in each tissue group over time. Pulmonary uptake and elimination are described by the following equation of mass balance of PCE entering and leaving the lungs:

$$Q_a(C_I - C_a) = Q_t(C_{art} - C_{ven}). \qquad \text{(A-1)}$$

This equation assumes a concentration equilibrium between arterial blood and alveolar air. By substituting the relationship $N = C_{art}/C_a$ into the equation above, we have:

$$C_{art} = (Q_aC_I + Q_tC_{ven})/(Q_t + Q_a/N), \qquad \text{(A-2)}$$

where venous blood concentration is given by:

$$C_{ven} = (Q_lC_l/P_l + Q_fC_f/P_f + Q_rC_r/P_r + Q_pC_p/P_p)/Q_t. \qquad \text{(A-3)}$$

The metabolism of PCE occurs mainly in the liver, with a rate of metabolism characterized by the Michaelis-Menten type equation:

$$\frac{dA_m}{dt} = (V_mC_l/P_l)/(K_m + C_l/P_l). \qquad \text{(A-4)}$$

The PCE concentrations in each of the four tissue groups are described by the following equations.

Metabolism tissue group:

$$V_l\frac{dC_l}{dt} = Q_l(C_{art} - C_l/P_l) - \frac{dA_m}{dt}. \qquad \text{(A-5)}$$

Nonmetabolism tissue group:

$$V_i\frac{dC_i}{dt} = Q_i(C_{art} - C_i/P_i), \qquad \text{(A-6)}$$

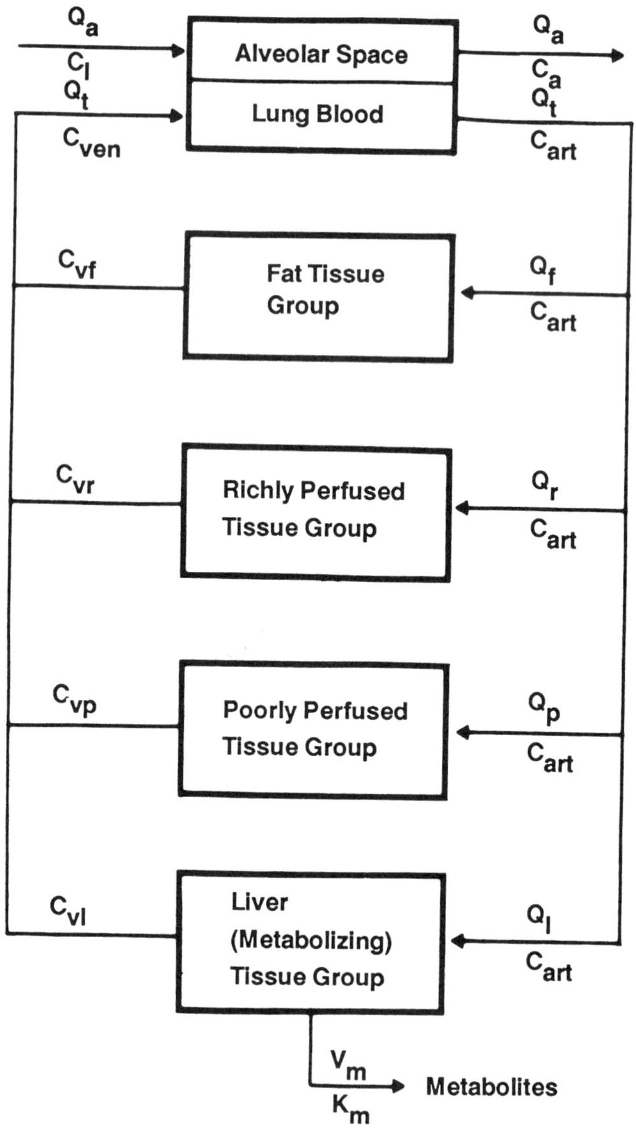

FIGURE A-1 Diagram of the physiologically based pharmacokinetic model. Abbreviations are defined in Table A-1.

Table A-1 Abbreviations Used in Figure A-1

Q_a	Alveolar ventilation rate (liters/min)
Q_t	Cardiac blood output (liters/min)
Q_l	Blood flow rate to liver (metabolizing) tissue group (liters/min)
Q_f	Blood flow rate to fat tissue group (liters/min)
Q_r	Blood flow rate to richly perfused tissue group (liters/min)
Q_p	Blood flow rate to poorly perfused tissue group (liters/min)
V_l, V_f, V_r, and V_p	Volumes of tissue groups (liters) corresponding, respectively, to liver, fat, richly perfused, and poorly perfused tissue groups
N	Blood:air partition coefficient
P_l, P_f, P_r, and P_p	Tissue:blood partition coefficient corresponding, respectively, to liver, fat, richly perfused, and poorly perfused tissue groups
V_m	Maximum velocity of metabolism (mg/min)
K_m	Michaelis constant (mg/liter of blood)
A_m	Amount metabolized (mg)
C_{art}	Concentration in arterial blood (mg/liter)
C_{ven}	Concentration in venous blood (mg/liter)
C_l, C_f, C_r, and C_p	Concentrations in tissue groups corresponding, respectively, to liver, fat, richly perfused, and poorly perfused tissue groups
C_I	Concentration in inhaled air (mg/liter)
C_a	Concentration in alveolar air (mg/liter)

where the subscript i represents fat, richly perfused, and poorly perfused tissue groups.

Parameters Used in the Model

Table A-2 presents the physiological and biochemical parameters used in the model. How these parameters were obtained is described below.

PHYSIOLOGICAL PARAMETERS

These values are taken from the U.S. Environmental Protection Agency (EPA) preliminary draft entitled "Reference Physiological Parameters in Pharmacokinetic Modeling," a contract project headed by Dr. Curtis Travis of the Oak Ridge National Laboratory, Oak Ridge, Tenn. The document has exhaustively reviewed over 100 articles on rodent and human physiological parameters. In our calculation, it is assumed that within a given species the ventilation and blood flow rates are proportional to the two-thirds power of body weight, and the volume of each tissue group is directly proportional to the body weight. For example, the alveolar ventilation rate for a 0.35-kg rat is calculated by the following equation: $Q_a = 0.075 \times (0.35/0.3)^{2/3} = 0.083$ liters/min, where 0.075 liter/min is the ventilation rate for 0.30-kg rat.

TABLE A-2 Physiological and Biochemical Parameters Used in the Model

Parameters	Rats	Mice	Humans
Body weight (kg)	0.35	0.025	70
Alveolar ventilation rate			
Q_a	0.083	0.028	7.500
Blood flow rate			
Q_t	0.104	0.019	6.200
Q_l	0.0389	0.0048	1.550
Q_f	0.00920	0.0017	0.314
Q_r	0.0434	0.0097	2.765
Q_p	0.0126	0.0029	1.570
Tissue volume			
V_l	0.0140	0.0012	1.700
V_f	0.0315	0.0027	14.000
V_r	0.0150	0.0015	3.500
V_p	0.2550	0.022	43.400
Partition coefficient			
N	18.9	16.9	10.3
P_l	3.719	4.159	3.719
P_f	108.994	121.893	108.994
P_r	3.719	4.159	3.719
P_p	1.058	1.183	1.058
V_m	0.00586	0.003	0.703
K_m	2.9378	1.472	32.043

PARTITION COEFFICIENTS

The parameters for rats and mice were provided by Curtis Travis of the Oak Ridge National Laboratory, which is the contractor for the EPA project "Interpretation of Metabolic Data in Exposure Analysis," which is sponsored by the Exposure Assessment Group, Office of Health and Environmental Assessment. It is our understanding that these parameters were obtained by Dr. Melvin E. Andersen of the Air Force Medical Research Laboratory at Wright-Patterson Air Force Base, Ohio, by using the equilibrium vial technique. The tissue:blood concentration ratio was calculated by taking the ratio of two concentration ratios, air/blood and air/homogenate tissue. The tissue:blood coefficients for humans are assumed to be identical to those of rats.

METABOLISM RATE CONSTANTS, V_m AND K_m

Rats

The constants V_m and K_m were estimated by least-square optimization by using the system of differential equations and data from a metabolism study by Pegg et al. (1979). As presented in Table 2 in the text, the amounts metabolized over 72 h for a 0.25-kg rat exposed to PCE at 10 and 600 ppm for 6 h were recorded to be 0.467 and 9.11 mg, respectively. These two data points (72 h, 0.467 mg and 72 h, 9.11 mg) were used to fit the system of differential equations, described previously, with V_m and K_m as unknown parameters. The resultant estimates are $V_m = 0.00468$ mg/min and $K_m = 2.937$ mg/liter. For the NTP (1985) bioassay rats, which had approximately 0.35 kg of body weight, the estimates are $V_m = 0.00468 \times (0.35/0.25)^{2/3} = 0.00586$ mg/min, with K_m being the same, independent of body weight.

Mice

As indicated in Tables 1 and 2 of the text, there are two metabolic data points: one from the gavage study and another from the inhalation study. The metabolic constants are estimated to be $V_m = 0.003$ mg/min and $K_m = 1.472$ mg/liter by using two metabolic data points to optimize the system of differential equations. When PCE is given by gavage, the equation involving liver tissue is given by:

$$V_l \frac{dC_l}{dt} = Q_l(C_{art} - C_l/P_l) - \frac{dA_m}{dt} + Dke^{-kt}, \qquad \text{(A-7)}$$

where D (in milligrams) is the administered dose by gavage, and $k = 11$ (min^{-1}) is estimated from gavage data from rats (Table 1). Varying the value of k within 1 order of magnitude does not appear to affect the predicted metabolism over 72 h. (Obviously, varying k would affect metabolic time pattern over a very short period.)

Humans

The constants V_m and K_m are estimated by using the urinary metabolites, as presented in Table 3 of the text, assuming that TCA in urine accounts for 30% of total metabolites. The value of 30% is selected because it fits the data in Table 4 better than other assumptions about the percentage of urinary TCA in total metabolites. The resultant estimates are $V_m = 0.703$ mg/min and $K_m = 32.043$ mg/liter. The predicted (observed) alveolar concentrations for the three subjects (Table 4) are given in Table A-3.

It is interesting to compare the metabolic constants calculated above and those calculated by other investigators. On the basis of the data of Ohitsu et al. (1983), Hattis et al. (1986) estimated 0.784 mmol/day of urinary metabolites after 8 h/day of occupation exposure. The V_{max} was estimated as follows:

$$V_{max} = \frac{0.784 \text{ mmol/day}}{24 \text{ h/day} \times 60 \text{ min/h}} = 5.00 \times 10^{-4} \text{ mmol/min.} \tag{A-8}$$

Hattis et al. (1986) assumed that the urinary metabolites account for 60% of the total metabolites, and adjusted the V_{max} accordingly to a value of 5.00×10^{-4} mmol/min/0.6 = 8.0×10^{-4} mmol/min, which can be converted to 0.13 mg/min by 8.0×10^{-4} mmol/min $\times$ 164 mg/mmol.

In actual risk calculation, Hattis et al. (1986) adjusted the V_{max}, for better fit, to 9.07×10^{-4} mmol/min, which would be about one-fifth the V_{max} estimated in this paper. We, however, propose two changes from the method used by Hattis et al. (1986).

1. As stipulated by Gehring et al. (1979) for rapidly metabolized compounds, the metabolic period should be 8 h (the exposed period) and not 24 h.
2. Our calculations indicate that human urinary metabolites account for only 30% rather than 60% of the total metabolites. Taking the value of 0.784 mmol/day of urinary output and calculating in the same manner as Hattis et al. (1986), but with the two changes discussed, the V_{max} = 0.869 mg/min, a value very close to our V_{max} of 0.7 mg/min.

As for the value of K_m, Hattis et al. (1986) calculated, on the basis of the data of Ohitsu et al. (1983), a value of 3.4×10^{-4} mmol/liter of blood. Using the value of V_{max} = 0.131 mg/min, the ratio of V_{max} to K_m is 0.0235. Assuming that this ratio is correct, and using the corrected V_{max} of 0.869 mg/min, the K_m would be equal to 37 mg/liter of blood, a value nearly equal to our value of 32 mg/liter of blood.

TABLE A-3 Predicted (Observed) Alveolar Concentrations for Three Subjects

8-Hour Exposure Concentration (mg/liter)	Predicted (Observed) Alveolar Concentration at the Following Postexposure Periods:				
	0.5 h	1 h	2 h	16 h	64 h
0.686	0.24 (0.14)	0.17 (0.10)	0.07 (0.08)	0.01 (0.03)	0.008 (0.010)
1.029	0.37 (0.28)	0.26 (0.22)	0.10 (0.17)	0.02 (0.04)	0.01 (0.02)
1.371	0.49 (0.34)	0.34 (0.26)	0.14 (0.19)	0.02 (0.05)	0.02 (0.03)

Development of Multispecies, Multiroute Pharmacokinetic Models for Methylene Chloride and 1,1,1-Trichloroethane (Methyl Chloroform)

Richard H. Reitz, Richard J. Nolan, and *Alan M. Schumann*

INTRODUCTION

Chronic toxicity tests in rodents are typically conducted with homogeneous populations of young, healthy animals. For practical reasons, only small numbers of animals and limited numbers of doses can be evaluated. Consequently, doses are usually set as high as possible, in the hope of providing maximum sensitivity for the test.

The animals used in these tests develop large numbers of certain types of tumors in the absence of any chemical treatment (e.g., liver tumors in B6C3F1 mice, leukemia in Fischer 344 rats), and have a normal life span of approximately 2 years. The test agent is usually administered by the most experimentally convenient route.

When these studies are completed, they are used to estimate the risk that tumors will be produced in human populations exposed to the same agents. In most cases, large numbers of heterogeneous humans of varying ages and health status are involved. The human exposures are usually orders of magnitude lower than the doses employed in the animal studies. The average lifetime of the human populations is much greater than that of the rodent species, and the spontaneous incidence of specific tumor types is usually much lower. Furthermore, the primary route(s) of exposure in human populations may differ from the route used in the animal study.

Furthermore, there appear to be at least two general types of chemical agents that are capable of affecting the tumor incidences in chronic bioassays: those interacting directly with genetic material (genotoxic agents), and those which do not interact directly with genetic material (nongenotoxic agents). There are theoretical reasons for believing that the dose dependency of the two types of agents may be quite different.

Considering all these factors, it is clear that risk estimations based upon animal data contain large areas of uncertainty. Physiologically based pharmacokinetic (PB-PK) analysis offers us a method for eliminating *some* of the uncertainty in these risk estimates, and may also provide a tool for testing our theories of chemical carcinogenesis. PB-PK analysis cannot, however, eliminate *all* sources of uncertainty from risk estimation, and for optimum use, PB-PK analysis must be integrated with other disciplines.

In the examples that follow, it is shown how the techniques of PB-PK are used to aid in estimating the internal doses associated with exposures to methylene chloride ($MeCl_2$), methylchloroform (MC), and perchloroethylene (tetrachloroethylene; PERC). Risk estimations based on internal dose at the target organ are considerably more reliable than risk estimations that fail to consider such factors as nonlinear metabolic processes and physiological differences between species.

ROLE OF METABOLISM ($MeCl_2$)

Bioassay Results

$MeCl_2$ is a solvent with a variety of uses, including paint removal, metal degreasing, aerosol applications, and food processing. Several studies of the chronic toxicity of $MeCl_2$ have been conducted. Inhalation studies in the Syrian golden hamster, exposed to concentrations up to 3,500 ppm for 6 h/day, did not show a tumorigenic response at any site (Burek et al., 1984). Similarly, a drinking water study sponsored by the National Coffee Association failed to reveal a dose-related statistically significant increase in tumors in the B6C3F1 hybrid mouse strain (Serota et al., 1984a) or the Fischer 344 rat (Serota et al., 1984b).

Two inhalation studies of $MeCl_2$ in the Sprague-Dawley rat (at doses up to 3,500 and 500 ppm, respectively) revealed treatment-related effects on the number of spontaneously occurring benign mammary tumors per tumor-bearing rat. Low incidences of tumors in the ventral neck region in and around the salivary gland of male rats were noted in one study (Burek et al., 1984) but were not noted in the other two rat inhalation studies (Burek et al., 1984; NTP, 1985).

Significant increases in spontaneously occurring lung and liver tumors were noted when B6C3F1 mice were exposed to 2,000 or 4,000 ppm of $MeCl_2$. In contrast to the studies mentioned above, in which $MeCl_2$ either failed to affect the tumor incidence or affected primarily benign tumors, the incidence of *both* benign and malignant tumors was elevated in this study (NTP, 1985).

Obviously, it would be useful to understand the reason for the different results obtained in the different studies. PB-PK analysis can be used in conjunction with biochemical studies to provide a single hypothesis that is consistent with all of these results.

Electrophilic Intermediates

Miller and Miller (1966) noted that most of the potent chemical carcinogens either were directly electrophilic or could be converted to electrophilic species by metabolic activation. Miller and Miller hypothesized that these electrophilic species would be capable of covalently binding to DNA and inducing mutations and/or cancer.

The generation of reactive intermediates also appears to play a role in the tumorigenicity of agents which do *not* react directly with DNA. For example, PERC is inactive in almost all of the short-term tests for genotoxicity (direct reaction with DNA), but the toxicity/carcinogenicity of PERC appears to be strongly correlated with the rate of metabolic activation (Buben and O'Flaherty, 1985; Schumann, 1984; Schumann et al., 1980).

The chemical reactivity of $MeCl_2$ itself is very low, and it is unlikely that it undergoes any direct reaction with macromolecules. Therefore, it appears likely that metabolism would play a role in the toxicity/carcinogenicity of $MeCl_2$.

Metabolic Pathways

$MeCl_2$ is metabolized via two pathways: (1) an oxidative pathway (Kubic et al., 1974) that appears to yield CO as well as considerable amounts of CO_2 (Gargas et al., 1986; Reitz et al., 1986) and (2) a glutathione-dependent pathway (Ahmed and Anders, 1978) that produces CO_2 but no CO (Figure 1). Both pathways release 2 mol of halide ion per mol of dihalomethane consumed. The oxidative pathway (MFO) is readily saturated at concentrations of a few hundred parts per million, but the glutathione-*S*-transferase (GSH) pathway showed no indication of saturation at inhaled concentrations up to 10,000 ppm (Gargas et al., 1986). Reactive, potentially toxic intermediates are formed in both pathways (Figure 1). The question is which of these pathway(s) is related to the *in vivo* tumorigenic activity of $MeCl_2$?

There are several reasons for believing that metabolism by MFO is not involved in the tumorigenicity of $MeCl_2$. First, it is clear that the incidence of most of the treatment-related tumors increases as the exposure concentration is raised from 2,000 to 4,000 ppm (Table 1). However, it has been demonstrated that HbCO, a major product of MFO activity, does

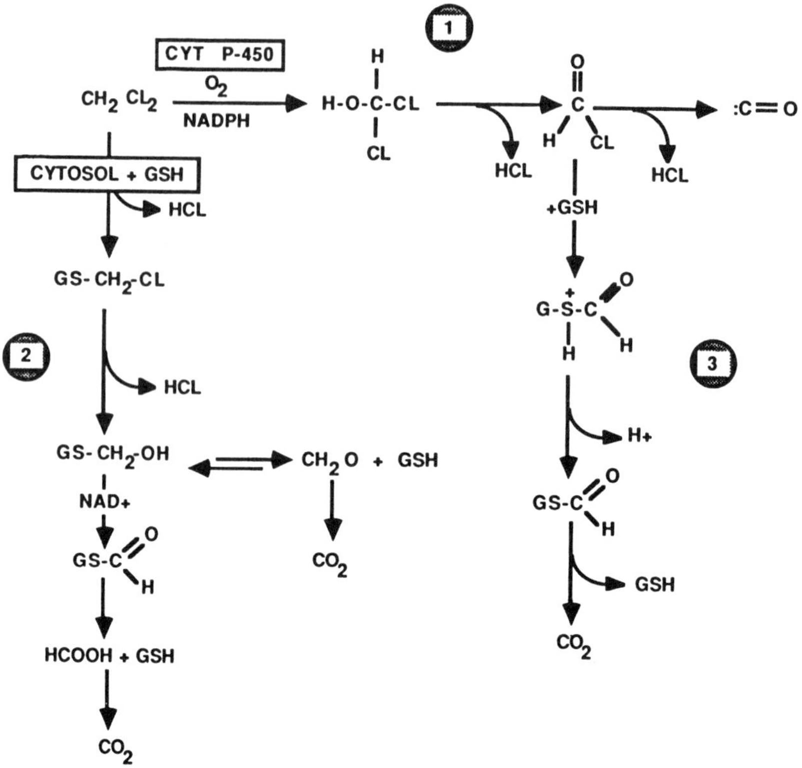

FIGURE 1 Proposed metabolic pathways for methylene chloride metabolism (based on Ahmed and Anders, 1978, and Kubic et al., 1974). Potentially reactive intermediates are formed in each pathway: formyl chloride in the cytochrome P-450 (MFO) pathway, and chloromethyl glutathione in the cytosolic (GSH) pathway. Either metabolic pathway can produce carbon dioxide in this scheme, but only the MFO pathway yields carbon monoxide and carboxyhemoglobin.

not increase in rats or hamsters as exposure levels of $MeCl_2$ are raised above 500 ppm (Burek et al., 1984; McKenna et al., 1982).

In addition, Green (1983) has separated enzymes from the MFO and GSH pathways and studied their ability to metabolically activate $MeCl_2$ *in vitro*. He found that cytosolic preparations from rat liver (containing GSH transferase) significantly increased the yield of bacterial mutations in *Salmonella typhimurium* exposed to DCM. However, rat liver microsomes rich in MFO failed to show this effect.

We have developed a PB-PK model (Andersen et al., 1987) that allowed us to investigate the question further by making direct comparisons between a National Coffee Association drinking water study in B6C3F1 mice and the NTP inhalation study in B6C3F1 mice. The model allowed

TABLE 1 Comparison of Average Daily Values of Dose Surrogates[a] and Tumor Incidences in Lung and Liver Tissues of Female B6C3F1 Mice in Two Chronic Bioassays of Methylene Chloride

Tissue	Control	Inhaled (2,000 ppm)	Inhaled (4,000 ppm)	Drink (250 mg/kg/day)
Liver				
MFO Pathway	0.0	3,575.0	3,701.0	5,197.0
GSH Pathway	0.0	851.0	1,811.0	15.1
Tumors (%)	6.0	33.0	83.0	6.0
Lung				
MFO Pathway	0.0	1,531.0	1,583.0	1,227.0
GSH Pathway	0.0	123.0	256.0	1.0
Tumors (%)	6.0	63.0	85.0	8.0

[a]Dose surrogate is calculated as the average amount of metabolite (milligram equivalents of $MeCl_2$) formed by each pathway per day per liter of organ.

us to calculate the levels of reactive intermediates formed by each pathway in the various tissues. Because these intermediates were presumed to be too reactive to leave the site of their formation, the average amounts of metabolite formed per day per liter of tissue have been calculated as dose surrogates for each organ. The model predictions for these dose surrogates are summarized in Table 1.

Reference to Table 1 indicates that the dose surrogates for the MFO pathway in the liver or lung tissue of B6C3F1 mice are virtually identical in each dose of the inhalation as well as the drinking water study. This is inconsistent with the observed tumor incidences. Tumors in the inhalation study increased with dose and were significantly different than the control, but tumor incidences in the drinking water study were not significantly different than those in controls (Table 1).

Tumor incidences were consistent with the dose surrogates produced by the GSH pathway, however, supporting the hypothesis that the tumorigenicity of $MeCl_2$ is related to the production of GSH metabolites in the target organ. This conclusion has significant implications for the estimation of human risk. For example, the model indicates that exposure to high concentrations of $MeCl_2$ (which saturate the MFO system) produces disproportionately high levels of the GSH dose surrogates. In addition, the model suggests that when humans and rodents are exposed to equivalent concentrations of $MeCl_2$ in the atmosphere or drinking water, the rodents produce higher levels of the GSH dose surrogates because of the relatively higher rates of metabolism in the rodent species.

Future Research

Another important outcome of the PB-PK analysis is that it suggests important directions for future research. For example, it is obvious that the values chosen for the human metabolic rate constants affect the levels of the dose surrogates calculated by the model. The rate constants for the MFO pathway in humans were estimated from existing *in vivo* human data as described elsewhere (Andersen et al., 1987). Because *in vivo* determinations of the activity of the GSH pathway in humans would involve exposing humans to high concentrations of $MeCl_2$, however, the *in vivo* activity of the GSH pathway in humans was estimated by allometric scaling of the rodent data.

Use of allometric scaling is supported by the observation that allometric scaling of the GSH pathway worked reasonably well in the three rodent species; intrinsic clearances (scaled to body weight to the 0.3 power) for this pathway were nearly identical in mice, hamsters, and rats (Andersen et al., 1987). Nevertheless, measurement of *in vitro* GSH activities (with $MeCl_2$ as the substrate) would be very useful in verifying the predictions of this model. We have subsequently conducted such studies, and they are in reasonable agreement with the predictions of this model (Reitz et al., manuscript in preparation, 1987).

EXTRAPOLATION BETWEEN SPECIES (MC)

1,1,1-Trichloroethane (methylchloroform, MC) is a volatile chlorinated hydrocarbon. It is widely used as a metal-degreasing solvent, and is also present in commercial aerosols and adhesive products. Several studies of the absorption, distribution, and elimination of MC have been conducted in humans as well as different species of laboratory animals (Nolan et al., 1984; Monster et al., 1979, Schumann et al., 1982a,b). However, a comprehensive model capable of describing a variety of pharmacokinetic endpoints in multiple species after administration by different routes has not been reported.

We have developed a PB-PK model for MC based on that of Ramsey and Andersen (1984). Metabolic parameters for the rat were determined from the ^{14}C balance study of Schumann et al. (1982a), and metabolic parameters for the B6C3F1 mice and human volunteers were calculated by scaling body weight to the 0.7 power as described by Ramsey and Andersen (1984). Partition coefficients were determined by M. Gargas (Armstrong Aerospace Medical Research Laboratory, Wright-Patterson Air Force Base, Ohio) using a vial equilibration technique similar to that of Sato and Nakajima (1979). Physiological parameters were taken from the scientific literature (Caster et al., 1956; Davis and Mapleson, 1981; ICRP, 1975). Physiological and biochemical parameters used in the model are summarized in Table 2. The objective of these studies was to see whether a PB-PK model developed from one species could be employed to predict data gathered in other species.

The PB-PK model was used to predict time course data for MC in venous blood for all species. In addition, several other types of data were collected either from the ^{14}C balance studies in animals or from clinical pharmacokinetic studies in human volunteers. The additional parameters evaluated included (1) time/course data for MC in expired air, (2) end-exposure body burdens, (3) MC concentrations in fat, (4) MC concentra-

TABLE 2 Parameters Used in the Physiologically Based Pharmacokinetic Model for Methylchloroform

Parameter	Human	Rat 1	Rat 2	Mouse 1	Mouse 2
Body weight (kg)	83.0	0.215	0.468	0.029	0.038
			Percentage of Body Weight		
Liver	2.6	4.0	4.0	4.0	4.0
Rapidly perfused	6.4	5.0	5.0	5.0	5.0
Slowly perfused	63.5	76.0	54.7	79.0	67.0
Fat	19.5	7.0	28.3	4.0	16.0
Flow (liters/h)					
Alveolar ventilation	330.7	5.11	8.81	1.26	1.52
Cardiac output	330.7	5.11	8.81	1.26	1.52
			Percentage of Cardiac Output		
Liver	24	24	24	24	24
Rapidly perfused	53	53	53	53	53
Slowly perfused	14	18	18	21	21
Fat	9	5	5	2	2
Partition coefficients					
Blood/air	2.53[a]	5.76	5.76	10.8[a]	10.8[a]
Liver/blood	3.40	1.49	1.49	0.796	0.796
Rapidly perfused/blood	3.40	1.49	1.49	0.796	0.796
Slowly perfused/blood	1.25	0.547	0.547	0.292	0.292
Fat/blood	104.0	45.7	45.7	24.4	24.4
Metabolic constants					
V_{max} (mg/h)	5.84	0.0904	0.156	0.0222	0.0269
K_m (mg/liter)	6.43	6.43	6.43	6.43	6.43

NOTE: Abbreviations: Rat 1, young rats (2–4 months); Rat 2, old rats (18.5 months); Mouse 1, young mice; Mouse 2, old mice. Parameters used for drinking water simulation, gavage, and intravenous administration in rats are very close to those for Rat 1 and are not shown.

[a]Value of partition coefficient from M. L. Gargas and M. E. Andersen (personal communication, 1985).

tions in liver, and (5) total amounts of MC metabolized. The model predictions were tested by comparison with actual data sets for humans (Nolan et al., 1984), rats, and mice (Schumann et al., 1982a).

Rats

The first parameter was the time course of the venous blood concentrations of MC in rats after exposure to 150 or 1,500 ppm of MC. Experimental data collected by Schumann et al. (1982a) were used to check the predictions. The data showed good agreement with the model predictions over the period studied (Figure 2).

Comparisons between the model predictions and actual data for several other parameters are summarized in Table 3. The model was reasonably successful in predicting values for these additional parameters. The ratio of predicted to actual data ranged from a low of 0.73 to a high of 1.89, and the mean ratio was 1.17 ± 0.44 for 150 ppm and 1.28 ± 0.38 for 1,500 ppm (Table 3).

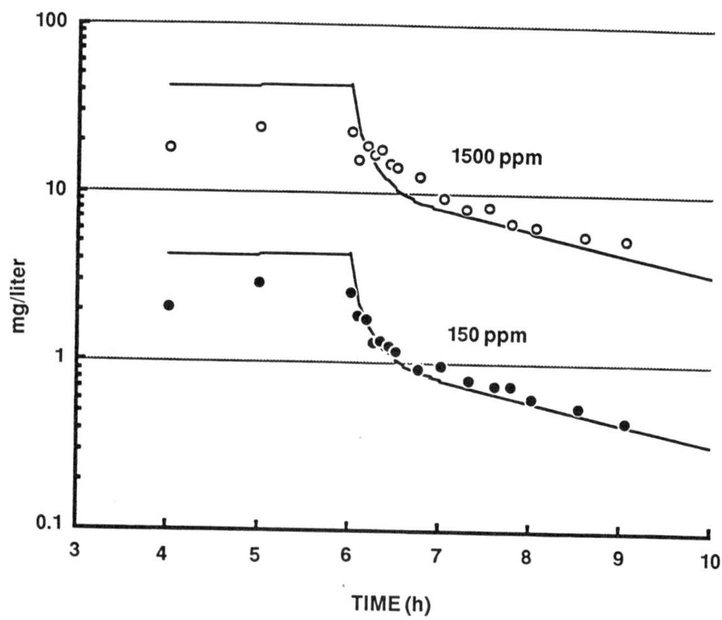

FIGURE 2 Simulated and observed values of venous blood concentration in Fischer 344 male rats during and following a 6-h inhalation exposure to 150 or 1,500 ppm of methyl chloroform (MC). Computer predictions are shown as solid lines, while actual data are given as open circles (1,500 ppm) or closed circles (150 ppm). Data are milligram equivalents of MC/liter of blood.

TABLE 3 Comparison of Values for Selected Parameters in Young Rats and Young Mice as Predicted by the Models and Observed by Schumann et al. (1982a)

Parameter	Model Prediction	Actual Data	Ratio[a]
Young rat (150-ppm inhalation exposure)			
End expt. blood level (mg/liter)	4.43	2.81	1.58
Body burden at 6 h (μmol)	25.10	33.03	0.76
Amount metabolized (μmol)	1.90	1.90	1.00
Conc. in fat[b] (nmol/g)	1,304.00	724.00	1.80
Conc. in liver[b] (nmol/g)	49.60	68.20	0.73
Young rat (1,500-ppm inhalation exposure)			
End expt. blood level (mg/liter)	44.70	23.7	1.89
Body burden at 6 h (μmol)	241.30	264.28	0.91
Amount metabolized (μmol)	5.62	5.53	1.02
Conc. in fat (nmol/g)	13,126.00	8,403.00	1.56
Conc. in liver (nmol/g)	502.00	504.00	1.00
Young mouse (150-ppm inhalation exposure)			
End expt. blood level (mg/liter)	8.70	9.27	0.94
Body burden at 6 h (μmol)	2.96	4.97	0.60
Amount metabolized (μmol)	0.62	0.65	0.95
Conc. in fat (nmol/g)	1,575.00	1,329.00	1.19
Conc. in liver (nmol/g)	51.70	76.00	0.68
Young mouse (1,500-ppm inhalation exposure)			
End expt. blood level (mg/liter)	87.90	111.60	0.79
Body burden at 6 h (μmol)	25.20	39.90	0.63
Amount metabolized (μmol)	1.23	1.19	1.03
Conc. in fat (nmol/g)	15,915.00	16,198.00	0.98
Conc. in liver (nmol/g)	525.00	631.00	0.83

[a]The ratio is equal to the model prediction column divided by the actual data column.
[b]Concentrations in fat and liver were those observed at the end of the 6-h exposure.

Mice

The model was then used to describe the pharmacokinetic behavior of MC after inhalation exposure in male B6C3F1 mice. To convert from rat simulations to mouse simulations, the appropriate physiological and biochemical parameters were adjusted as outlined in Table 2. Metabolic constants for the mouse were obtained by setting the Michaelis constant (K_m) for the mouse equal to that of the rat, and scaling the value of V_{max} to the 0.7 power of body weight. The size of the body fat compartment in the young mice was reduced from 7% of body weight to 4% of body weight, and the percentage of cardiac output directed to the fat tissue was reduced from 5% to 2%.

The time course of MC in venous blood (collected at the orbital sinus) was accurately described by the model for two different exposure concentrations: 150 and 1,500 ppm (Figure 3).

The physiologically based model predicted that MC would be eliminated from the mouse much more rapidly than from the rat. This is consistent with the data of Schumann et al. (1982a), who reported that initial elimination half-lives for MC in mice were 5- to 10-fold less than the corresponding half-lives in rats.

As outlined for rats, five additional parameters simulated by the model were checked against experimental data collected by Schumann et al. (1982a) in the mouse. The model was also reasonably successful in describing these data (Table 3). The ratio of predicted to actual data ranged from a low of 0.60 to a high of 1.19, and the mean ratio was 0.87 ± 0.21 for 150 ppm and 0.85 ± 0.14 for 1,500 ppm (Table 3).

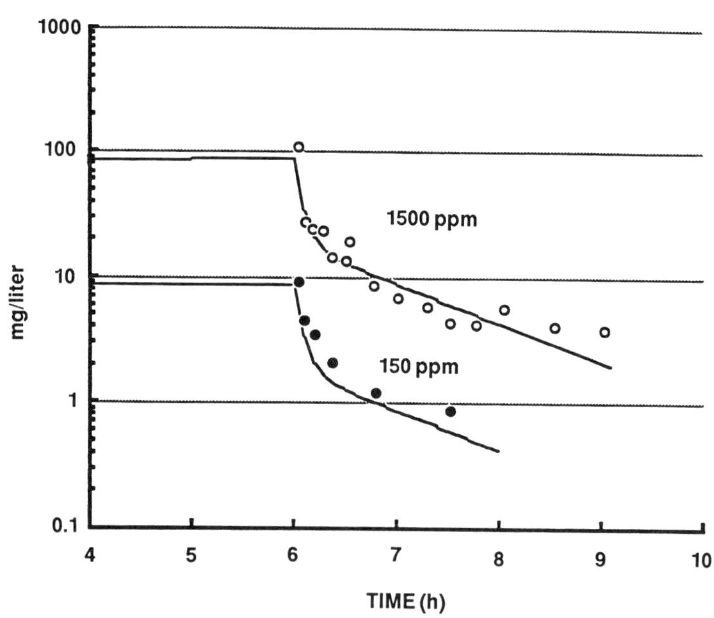

FIGURE 3 Simulated and observed values of venous blood concentration in B6C3F1 male mice during and following a 6-h inhalation exposure to 150 or 1,500 ppm of methyl chloroform (MC). Computer predictions are shown as solid lines, while actual data are given as open circles (1,500 ppm) or closed circles (150 ppm). Data are milligram equivalents of MC/liter of blood.

from a low of 0.60 to a high of 1.19, and the mean ratio was 0.87 ± 0.21 for 150 ppm and 0.85 ± 0.14 for 1,500 ppm (Table 3).

Human Data

After adjustment of model parameters, as described above for mice, the time courses of MC in expired air (Figure 4a) and venous blood (Figure 4b) during the following human exposures to 350 or 35 ppm MC were simulated with the model. Elimination of MC was triphasic in nature, and the model clearly reflected this (Figures 4a and 4b). The model predicted that expired air and venous blood concentrations should be proportional to exposure concentration throughout the range of 35–350 ppm MC. These predictions were also in good agreement with the experimental data of Nolan et al. (1984) (Figures 4a and 4b).

The model was then used to estimate the amount of MC metabolized by humans during the two exposures. The model predicted that 2.47 mg equivalents of MC would be metabolized during the 240 h after a 6-h exposure to 35 ppm of MC, and 4.32 mg equivalents of MC metabolites were actually recovered (Nolan et al., 1984). Similarly, the model predicted that 18.6 mg equivalents of MC metabolites would be formed during the 240 h following a 6-h exposure to 350 ppm of MC, and 32.9 mg equivalents were actually recovered.

Was the Simulation Successful?

In this exercise, a single basic PB-PK model has been used to predict a variety of pharmacokinetic data from three different species. Only those parameters known to differ between species or measured in the laboratory were varied between species. In general, the model predictions and animal data were within a factor of 2, and often they were substantially closer. In view of the fact that apparently identical animal studies of toxicity and/or carcinogenicity often differ between themselves by more than a factor of 2, the consistency of the PB-PK model seems remarkable. Furthermore, the fact that this consistency extends across a variety of endpoints, including end-exposure blood levels, total amounts metabolized, and tissue concentrations, suggests that this technique offers considerable promise in understanding and predicting interspecies differences in the delivery of internal dose to various organs.

MODELING DRINKING WATER EXPOSURES (MC, PERC)

Halogenated hydrocarbons such as $MeCl_2$, MC, and PERC have limited solubilities in water. Consequently, the doses of these materals that can

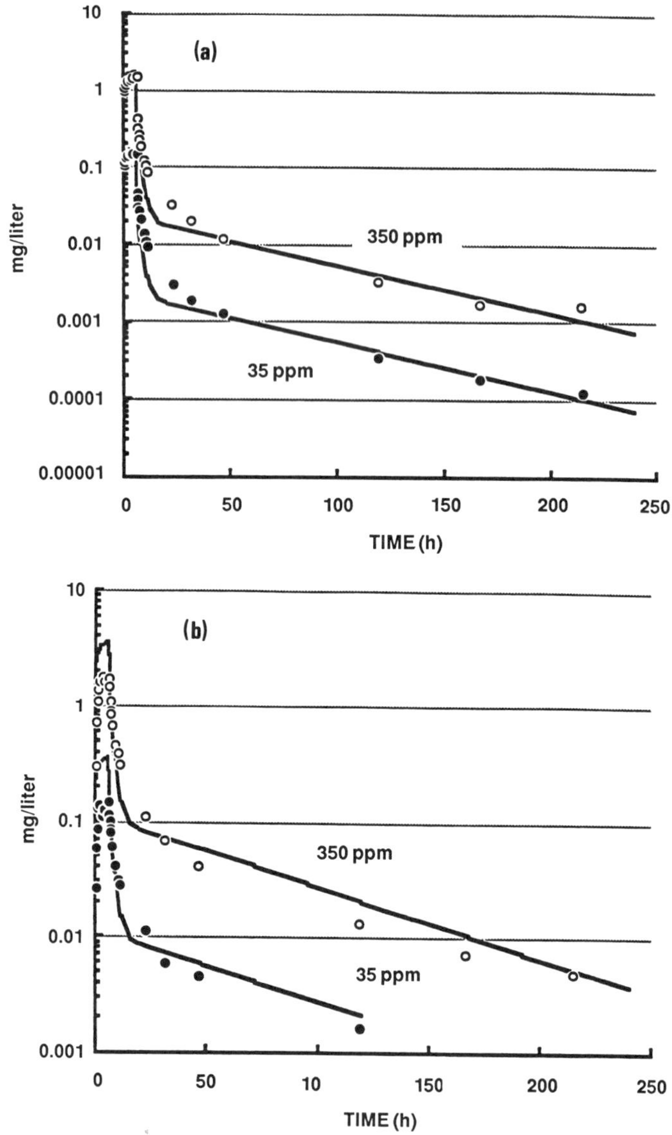

FIGURE 4 (a) Simulated and observed values of exhaled air concentration in human volunteers during and following a 6-h inhalation exposure to 35 or 350 ppm of methyl chloroform (MC). Computer predictions are shown as solid lines, while actual data are given as open circles (350 ppm) or closed circles (35 ppm). Data are milligram equivalents of MC/liter of air. (b) Simulated and observed values of venous blood concentration in human volunteers during and following a 6-h inhalation exposure to 35 or 350 ppm of methyl chloroform (MC). Computer predictions are shown as solid lines, while actual data are given as open circles (350 ppm) or closed circles (35 ppm). Data are milligram equivalents of MC/liter of air.

be administered to animals in their drinking water are limited. Because small quantities of these materials are sometimes found in human drinking water supplies, however, it would be useful to have a procedure for predicting their toxicity from studies conducted by another route. We have employed the PB-PK model described above to simulate drinking water exposure, and have collected ^{14}C disposition data for MC to check the model predictions.

In the first experiment, animals were given a bolus gavage with water containing dissolved MC. At various intervals, samples of venous blood were withdrawn and analyzed for MC, and the results are shown in Figure 5a. The model gave a reasonable simulation of the time course of MC absorbed from water in the gastrointestinal tract.

In a second experiment, animals were presented with water containing radioactive MC (saturated solution, 4.4 mg/ml) and allowed to drink ad libitum. Radioactive drinking water was available for 8 h. Samples of urine, expired air, and selected tissues were analyzed for radioactivity at various times thereafter. The average water consumption of the rats (mean weight, 250 g) was 8.1 ± 3.8 ml, corresponding to an average dose of 143 mg/kg. Overall recovery of radioactivity in these experiments (based on water consumption) was 95.2 ± 4.33%. Most of the radioactivity was recovered on the charcoal traps within the first 12 h of the experiment, with small amounts of radioactivity also seen in the urine, CO_2, carcass, and skin fractions.

For the purpose of simulation, it was assumed that rats ingested MC in their drinking water at a constant rate and that absorption of MC occurred rapidly. Input of MC was set equal to the average rate of intake (4.47 mg of MC/h) for the entire period in which the animals had access to treated water. It was assumed that all of the MC ingested by the rats entered the liver compartment directly, and the differential equations were written to reflect these assumptions. These animals were slightly larger than the young rats used in the inhalation studies; and the body weight, cardiac output, alveolar ventilation rate, and metabolic velocity were scaled as outlined previously. Values used in the model are listed in Table 2. Other than this, the PB-PK model was not changed for the simulation of drinking water exposure.

The average rate of elimination of MC (milligrams of MC/hour) during the interval was calculated and plotted at the midpoint of the excretion interval. Results are shown in Figure 5b. The time course of [^{14}C]MC in expired air was reasonably well described for the first 24 h of exposure.

The model predicted that 3.58 μmol of MC would be metabolized, and 8.19 μmol (3.0% of the administered radioactivity) was actually recovered in urine and CO_2 (Table 4). The model also predicted that 97% of the total ingested radioactivity would be eliminated within 12 h after the treated

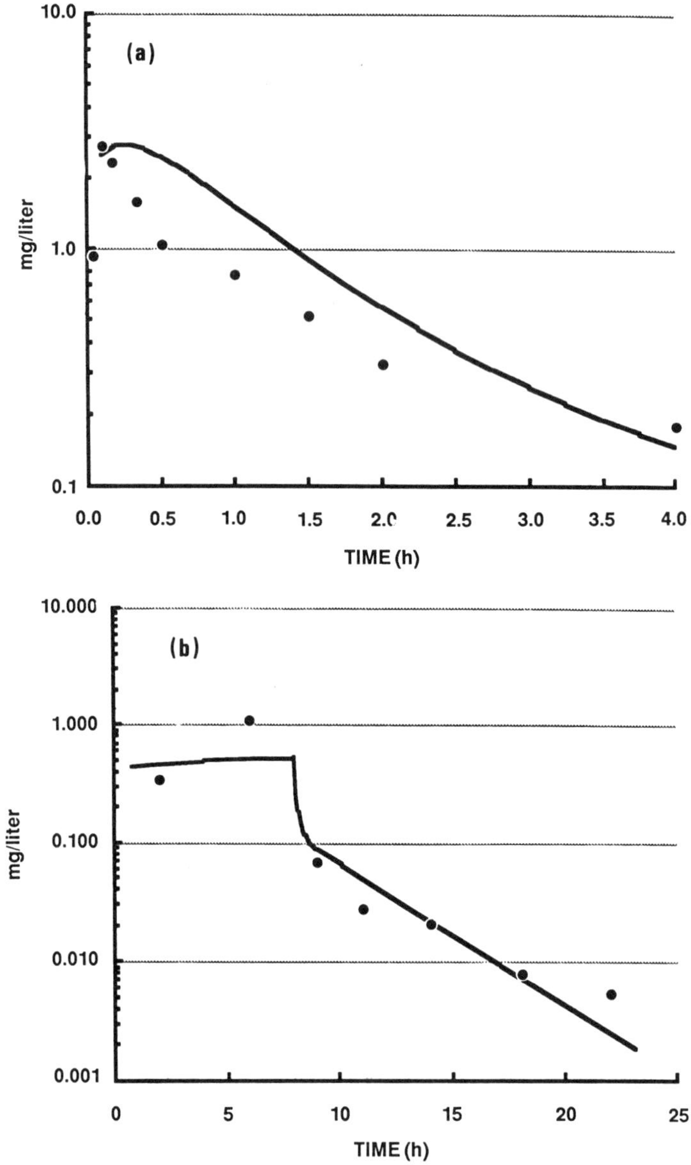

FIGURE 5 (a) Simulated and observed values of venous blood concentration in Fischer 344 male rats following gavage with a water solution of MC (14.2 mg/kg). Computer predictions are shown as a solid line, while actual data are given as closed circles. Data are milligram equivalents of MC/liter of blood. (b) Simulated and observed values of exhaled air concentration in Fischer 344 male rats during and following drinking water exposure to a saturated solution of MC (8-h exposure; dose, 143 mg/kg). Computer predictions are shown as a solid line, while actual data are given as closed circles. Data are milligram equivalents of MC/liter of air.

TABLE 4 Comparison of Values for Selected Parameters for Disposition of MC in Old (approximately 18.5 months) Rats and Old Mice as Predicted by the Models and Observed by Schumann et al. (1982b)

Parameter	Model Prediction	Actual Data	Ratio[a]
Old rat (18.5 months; 1,500-ppm inhalation exposure)			
Body burden at 6 h (μmol)	669.5	724.3	0.92
Amount metabolized (μmol)	13.8	18.8	0.73
Conc. in fat[b] (nmol/g)	4,604.0	5,685.0	0.81
Conc. in liver[b] (nmol/g)	440.4	606.0	0.73
Charcoal (μmol)			
0–3 h	142.7	185.3	0.77
3–6 h	84.4	128.2	0.66
6–12 h	125.7	138.3	0.91
12–24 h	152.0	128.8	1.18
24–36 h	76.5	63.8	1.20
Old mouse (18.5 months; 1,500-ppm inhalation exposure)			
Body burden at 6 h (μmol)	68.6	141.0	0.49
Amount metabolized (μmol)	2.21	8.04	0.27
Conc. in fat (nmol/g)	10,217.0	10,603.0	0.96
Conc. in liver (nmol/g)	491.3	1,794.0	0.27
Charcoal (μmol)			
0–3 h	29.2	93.48	0.31
3–6 h	14.9	26.5	0.56
6–12 h	14.3	11.0	1.30
12–24 h	6.99	1.80	3.88
24–36 h	0.91	0.15	6.07
Young rat (drinking water exposure, 143 mg/kg)			
Amount metabolized (μmol)	3.58	8.20	0.44
Percentage of dose eliminated			
0–12 h	97.0	91.2	1.06

[a]The ratio is equal to the model prediction column divided by the actual data column.
[b]Concentrations in fat and liver were those observed at the end of the 6-h exposure.

water was withdrawn. Actual data show that 91.2 ± 3.6% of the ingested radioactivity was recovered during this period. These data are in fair agreement with the model prediction, considering the difficulties associated with administration of MC through this route.

Frantz and Watanabe (1983) reported a detailed comparison of the fate of [^{14}C]PERC administered in drinking water and via inhalation. Although their data were not analyzed with a PB-PK model, they reported that the elimination kinetics of PERC were consistent with results generated by both gavage and inhalation exposures, and that the fate of PERC was not substantially different from the disposition resulting from other routes of administration.

These data offer encouragement that pharmacokinetic models may play an important role in understanding and predicting the toxicity of materials found in drinking water based on studies conducted by another route.

MODELING INTRAVENOUS INJECTION EXPOSURES (MC)

Another common route of exposure is intravenous injection. To see if the PB-PK model for MC was capable of describing this different route, groups of six rats were injected with 8.8 or 47.0 mg/kg of MC. Samples of venous blood were drawn at various times and analyzed for MC.

For modeling purposes, injection of MC was described as a rapid infusion into pooled venous blood. No other changes were made to model parameters. The disposition of MC administered by this route was well described by the standard PB-PK model (Figure 6).

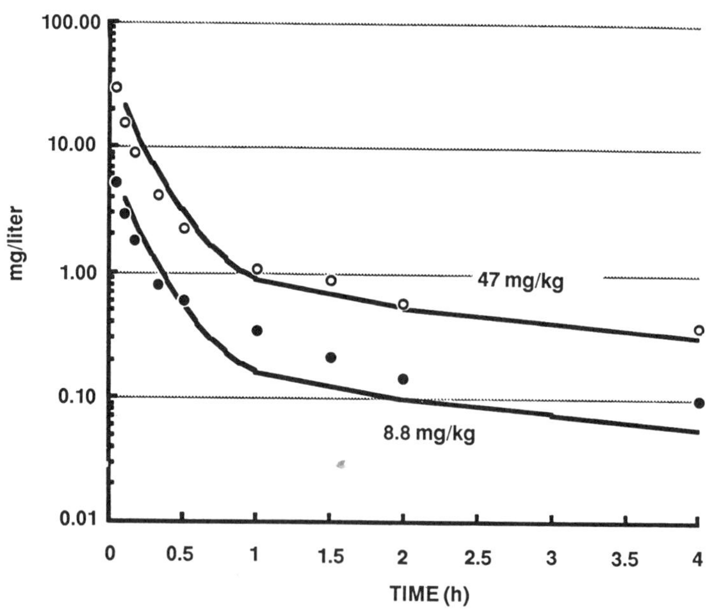

FIGURE 6 Simulated and observed values of venous blood concentrations in Fischer 344 male rats following intravenous injection of MC (dose rates, 8.8 and 47 mg/kg). Computer predictions are shown as a solid line, while actual data are given as closed circles. Data are milligram equivalents of MC/liter of venous blood.

PHARMACOKINETIC STUDIES IN OLD ANIMALS

Most pharmacokinetic studies are conducted with young adult animals, while chronic toxicology studies are conducted with animals ranging in age from very young at the beginning of an experiment to very old. We have used the PB-PK model developed for MC to determine whether we could develop reasonable hypotheses to predict the effect of aging upon disposition of this material.

Old Rats

Schumann et al. (1982b) studied the pharmacokinetics of inhaled MC in old male rats from a chronic inhalation toxicology study. The age of these animals was approximately 18.5 months, and the average body weight was 468 g. They noted that the older animals appeared to have more fat tissue than young rats, and also observed several differences between the disposition of MC in young and old animals. Data from the old rats exposed to 1,500 ppm of MC were not particularly well simulated by the standard pharmacokinetic model for young rats (simulation not shown).

Other investigators have noted this phenomenon, and have found it necessary to increase the relative percentage of fat when constructing pharmacokinetic models in older animals (Lutz et al., 1977). Consequently, the fat compartment was increased from 7.08% to 28.3% of total body weight, and the percentages of body weight assigned to the nonperfused and slowly perfused tissue compartments were reduced to compensate for this (Table 2). Otherwise, the model was constructed exactly as outlined for young rats. When the simulation was rerun under these conditions, the model described the data reasonably well.

Nine parameters were chosen for comparison of model simulations and actual data. These included the end-exposure body burden, the amount of MC metabolized, the concentration of MC in fat and liver tissue at the end of the exposure, and the amount of radioactivity (presumed to be MC) recovered on an activated charcoal trap at 0–3, 3–6, 6–12, 12–24, and 24–36 h postexposure. The results are summarized in Table 4. The ratio of predicted values to actual data varied from a low of 0.66 to a high of 1.20, and the mean ratio was 0.88 ± 0.18.

Old Mice

Schumann et al. (1982b) also collected data in geriatric male mice of similar ages (approximately 18.5 months). The model for young mice was modified by increasing the relative fat content from 4% of body weight to 16% of body weight, and the disposition of MC in old animals was simulated after this change. The data were not particularly well simulated by the model. The ratio of predicted values to actual data varied from a low of 0.27 to a high of 6.07, and the mean ration was 1.57 ± 1.92 (Table 4). It appears that age-related changes in the disposition of MC in geriatric mice may involve additional factors beyond the increase in body fat. Further research will be needed to elucidate these factors.

Summary

In summary, unified PB-PK models encompassing data from a variety of sources were developed for several of the organohalides. These models can be used to strengthen the scientific basis of risk assessment, improve experimental design of chronic studies, and give insight into mechanisms of toxicity and metabolism. Many uncertainties certainly remain in the risk assessment process, but risk assessments that properly consider the role of physiologically based pharmacokinetics should be significantly more reliable than those that do not.

REFERENCES

Ahmed, A. E., and M. W. Anders. 1978. Metabolism of dihalomethanes to formaldehyde and inorganic halide. I. In vitro studies. Drug Metab. Dispos. 4:357–361.

Andersen, M. E., H. J. Clewell, M. L. Gargas, F. A. Smith, and R. H. Reitz. 1987. Physiologically-based pharmacokinetics and the risk assessment process for methylene chloride. Toxicol. Appl. Pharmacol. 87:185–205.

Buben, J. A., and E. J. O'Flaherty. 1985. Delineation of the role of metabolism in the hepatotoxicity of trichloroethylene and perchloroethylene: A dose-effect study. Toxicol. Appl. Pharm. 78:105–122.

Burek, J. E., K. D. Nitschke, T. J. Bell, D. L. Waskerle, R. C. Childs, J. D. Beyer, D. A. Dittenber, L. W. Rampy, and M. J. McKenna. 1984. Methylene chloride: A two-year inhalation toxicity and oncogenicity study in rats and hamsters. Fund Appl. Toxicol. 4:30–47.

Caster, W. O., J. Poncelet, A. B. Simon, and W. D. Armstrong. 1956. Tissue weights of the rat. I. Normal values determined by dissection and chemical methods. Proc. Soc. Exp. Biol. Med. 91:122–126.

Davis, N. R., and W. W. Mapleson. 1981. Structure and quantification of a physiological model of the distribution of injected agents and inhaled anaesthetics. Br. J. Anaesth. 53:399–404.

Frantz, S. W., and P. G. Watanabe. 1983. Tetrachloroethylene: Balance and tissue distribution in male Sprague-Dawley rats by drinking water administration. Toxicol. Appl. Pharmacol. 69:66–72.

Gargas, M. L., H. J. Clewell, and M. E. Andersen. 1986. Metabolism of inhaled dihalomethanes in vivo: Differentiation of kinetic constants for two independent pathways. Toxicol. Appl. Pharmacol. 82:211–223.

Green, T. 1983. The metabolic activation of dichloromethane and chloro-fluoromethane in a bacterial mutation assay using Salmonella typhimurium. Mutat. Res. 118:277–288.

ICRP (International Commission on Radiation Protection). 1975. Report of the Task Group on Reference Man, ICRP Publication 23, W. S. Snyder, M. J. Cook, E. S. Nasset, L. R. Rauhausen, G. P. Howells, and I. H. Tiptom, eds. New York: Pergamon Press.

Kubic, V. L., M. W. Anders, R. R. Engel, C. H. Barlow, and W. S. Caughey. 1974. Metabolism of dihalomethanes to carbon monoxide. I. In vivo studies. Drug. Metab. Dispos. 2:53–57.

Lutz, R. J., R. L. Dedrick, H. B. Matthews, T. E. Eling, and M. W. Anderson. 1977. A preliminary pharmacokinetic model for several chlorinated biphenyls in the rat. Drug Metab. Dispos. 5:386–396.

McKenna, M. J., J. A. Zempel, and W. H. Braun. 1982. The pharmacokinetics of inhaled methylene chloride in rats. Toxicol. Appl. Pharmacol. 65:1–10.

Miller, E. C., and J. A. Miller. 1966. Mechanisms of chemical carcinogenesis-genesis: Nature of proximate carcinogens and interactions with macro-molecules. Pharmacol. Rev. 18–805.

Monster, A. C., G. Boersma, and H. Steenweg. 1979. Kinetics of 1,1,1,-trichloroethane in volunteers: Influence of exposure concentration and work load. Int. Arch. Occup. Environ. Health 42:293–302.

NTP (National Toxicology Program). 1985. NTP Technical Report on the Toxicology-Toxicology and Carcinogenesis Studies of Dichloromethane in F-344/N Rats and B6C3F1 Mice (Inhalation Studies). NTP-TR-306 (Board Draft).

Nolan, R. J., N. L. Freshour, D. L. Rick, L. P. McCarty, and J. H. Saunders. 1984. Kinetics and metabolism of inhaled methylchloroform (1,1,1-trichloroethane) in male volunteers. Fund. Appl. Toxicol. 4:654–662.

Ramsey, J. R., and M. E. Andersen. 1984. A physiologically based de-description of the inhalation pharmacokinetics of styrene in rats and humans. Toxicol. Appl. Pharmacol. 73:159–175.

Reitz, R. H., F. A. Smith, and M. E. Andersen. 1986. In vivo metabolism of ^{14}C-methylene chloride (MEC). The Toxicologist 6:260.

Sato, A., and T. Nakajima. 1979. Partition coefficients of some aromatic hydrocarbon and ketones in water, blood, and oil. Br. J. Ind. Med. 36:231–234.

Schumann, A. M. 1984. Inhalation kinetics. Food Solvents Workshop I: Methylene Chloride, Bethesda, Md. Washington, D.C.: Nutrition Foundation.

Schumann, A. M., J. F. Quast, and P. G. Watanabe. 1980. The pharmacokinetics and macromolecular interaction of perchloroethylene in mice and rats as related to oncogenicity. Toxicol. Appl. Pharm. 55:207–219.

Schumann, A. M., T. R. Fox, and P. G. Watanabe. 1982a. [^{14}C]methylchloroform (1,1,1-trichloroethane): Pharmacokinetics in rats and mice following inhalation exposure. Toxicol. Appl. Pharmacol. 62:390–401.

Schumann, A. M., T. R. Fox, and P. G. Watanabe. 1982b. A comparison of the fate of inhaled methylchloroform (1,1,1-trichloroethane) following single or repeated exposure in rats and mice. Fund. Appl. Toxicol. 2:27–32.

Serota, D., B. Ulland, and F. Carlborg. 1984a. Hazelton chronic oral study in mice. Food Solvents Workshop I: Methylene Chloride, Bethesda, Md. Washington, D.C.: Nutrition Foundation.

Serota, D., B. Ulland, and F. Carlborg. 1984b. Hazelton chronic oral study in rats, Food Solvents Workshop I: Methylene Chloride, Bethesda, Md. Washington, D.C.: Nutrition Foundation.

Methotrexate: Pharmacokinetics and Assessment of Toxicity

Paul F. Morrison, Robert L. Dedrick, and *Robert J. Lutz*

INTRODUCTION

Many of the physical and biological processes encountered in physiological pharmacokinetics play a role in determining the pharmacokinetic behavior of the antifolate compound methotrexate (MTX). Thus, results of studies with this compound, with its long and extensive history as a subject of both experimental and theoretical research, provide good examples of the quantitation of such phenomena. In addition, the biochemical origins of toxicity of this compound are reasonably well understood so that connections can be made between pharmacokinetic variables and toxic endpoints.

In this paper, we briefly summarize MTX pharmacokinetics and then focus on the dose scheduling aspects of these pharmacokinetics that affect target tissue concentrations and expected toxicity. Because of the highly nonlinear character of the pharmacokinetics and pharmacodynamics, we will see that dose-toxicity relationships are complex.

MTX PHARMACOKINETICS

Methotrexate is a folate analog that, following administration, distributes primarily to the non-fatty tissues of the body. The principal organs containing the compound are shown in Figure 1. Transport of MTX across the capillary and cell membranes of the liver, kidney, and skin is rapid, so that equilibrium ratios of tissue to plasma concentrations (plasma con-

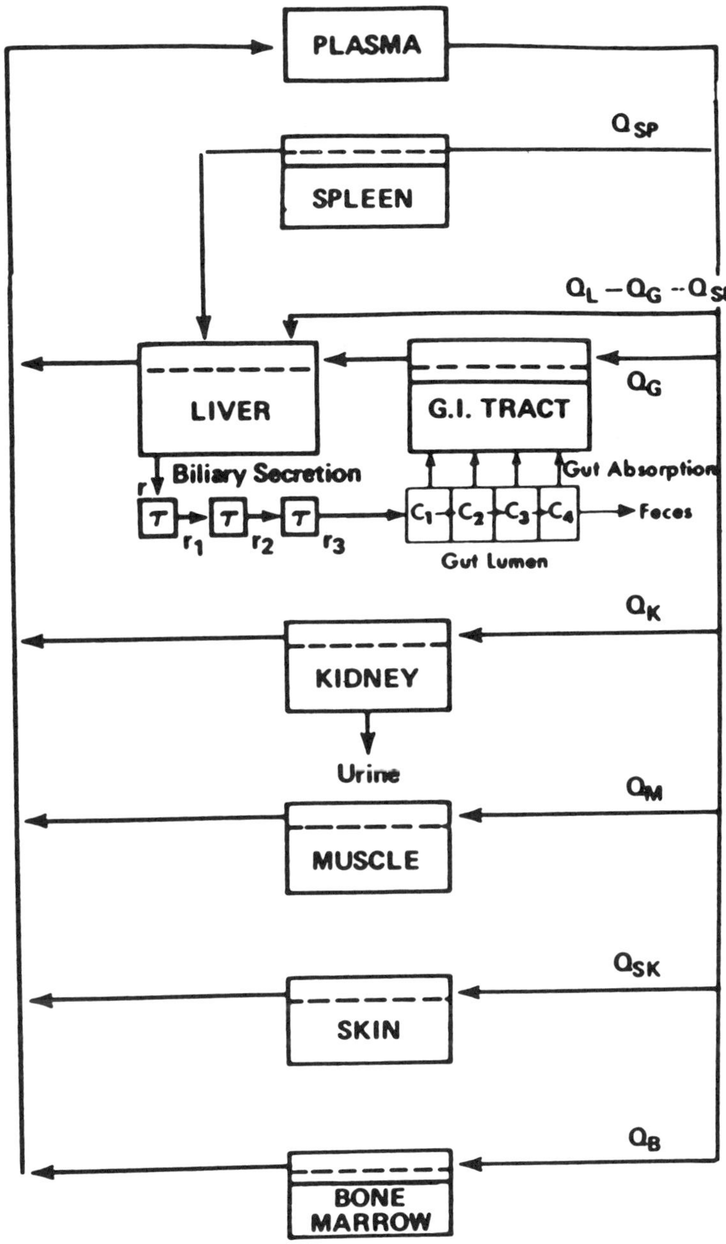

FIGURE 1 Scheme of principal organs in the MTX model.

centrations $> 1 \mu M$) are established on a time scale consistent with plasma flow limitation. This ratio is also established quickly in muscle, although transport across muscle cells is absent. The tissues of the gastrointestinal (GI) tract, spleen, and bone marrow differ from the others in that transport of MTX across cell membranes is slow. As a consequence initial delivery of MTX to these organs is membrane transport limited rather than blood flow limited.

MTX is cleared from the body through both biliary and urinary routes. Most of the drug excreted into the bile passes through the intestine and is excreted fecally, but the drug is also subject to partial intestinal reabsorption and to metabolism by enteric bacteria (Breithaupt and Kuenzlen, 1982; Zaharko and Dedrick, 1984). Excretion of MTX by the kidney is a result of both glomerular filtration and tubular secretion. The net result of plasma protein binding, filtration, and saturable secretion (and possibly reabsorption) is a kidney clearance that is of the order of, but not generally equal to, creatinine clearance. Concentration, pH, and competitive anion effects have been observed (Zaharko and Dedrick, 1984).

In some organs, most notably the liver and kidney but also in intestine, marrow, and many tumors (Baugh et al., 1973; Jacobs et al., 1977; Whitehead et al., 1975), MTX undergoes metabolism to active polyglutamate derivatives. Because these polyglutamates are capable of being retained in some tissues far longer than the parent MTX, the total exposure time to MTX and its polyglutamate derivatives may, depending on schedule, greatly exceed that to unreacted drug alone (Balinska et al., 1982; Kennedy et al., 1983; Morrison and Allegra, 1987). It has been argued that this effect is of great importance in the killing of tumor cells by high doses of MTX but of relatively lesser importance in normal intestinal mucosa and bone marrow cells (Goldman and Matherly, 1986), which do not allow large levels of MTX polyglutamyl derivatives to accumulate.

MTX also undergoes hydroxylation by liver aldehyde oxidase to form 7-hydroxymethotrexate, a metabolite with a long half-life of 24 h in humans (Breithaupt and Kuenzlen, 1982). Immediately following 6-h infusions, the concentration of this metabolite in human plasma reaches about 6% of the MTX concentration; 12 h after the infusion, the hydroxy metabolite and MTX have similar concentrations. A similar time to attainment of equal concentrations is estimated to occur following bolus injection. Hence, the antifolate activity of this compound, as well as that of MTX, must often be considered several hours after administration of the drug to humans.

MECHANISM OF TOXICITY

The overall mechanisms by which MTX induces cell toxicity are reasonably well understood, although many details of these mechanisms are still under active investigation. The primary event is the very strong intracellular binding of MTX to dihydrofolate reductase, an enzyme needed for the continued production of folate cofactors required for both thymidylate (Figure 2) and purine biosynthesis. Binding of drug prevents this enzyme from allowing continued production of DNA precursors, resulting in a cessation of DNA synthesis and, if continued long enough, in cell death. Polyglutamate derivatives of MTX, when formed, bind even more strongly to the enzyme than the parent drug and are thus potent mediators of cytotoxicity themselves (Jolivet and Chabner, 1983). Even the polyglutamates of 7-hydroxymethotrexate may have some of this activity (Goldman and Matherly, 1986).

For the drug to block reductase sufficiently for cell kill, binding of enzyme must be greater than 95% complete (Jackson and Harrap, 1973, 1979). This high percentage is derived from the presence in cells of a quantity of reductase far in excess of the amount required to maintain adequate folate cofactor production for survival. Furthermore, because of

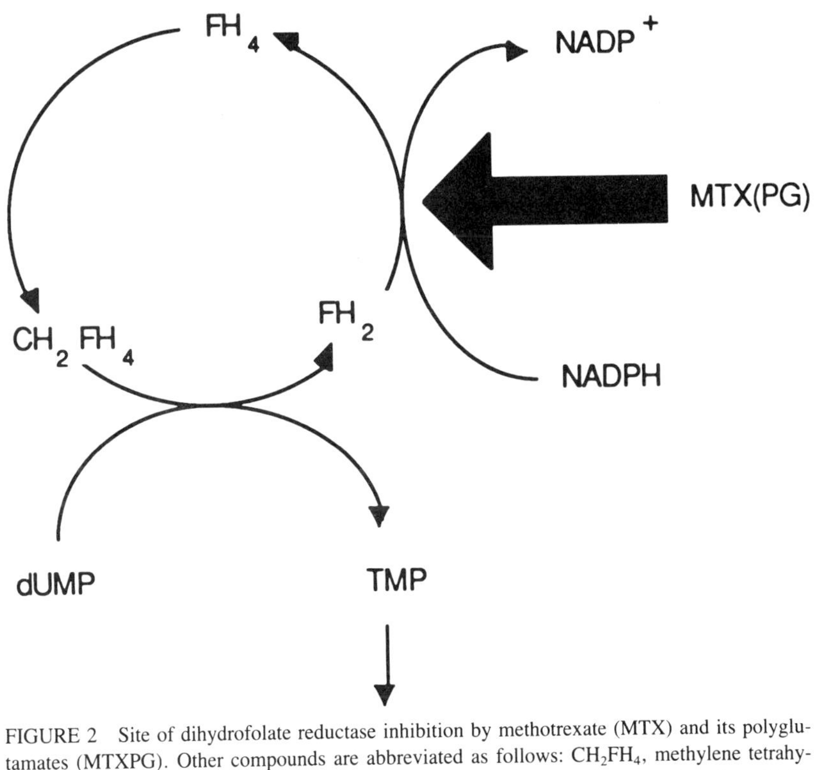

FIGURE 2 Site of dihydrofolate reductase inhibition by methotrexate (MTX) and its polyglutamates (MTXPG). Other compounds are abbreviated as follows: CH_2FH_4, methylene tetrahydrofolate; FH_2, dihydrofolate; FH_4, tetrahydrofolate; dUMP, deoxyuridine monophosphate; TMP, thymidine monophosphate; DNA, deoxyribonucleic acid; $NADP^+$ and NADPH, oxidized and reduced forms of nicotinamide adenine dinucleotide phosphate.

the very large buildup of normal dihydrofolate substrate that occurs behind the inhibited enzyme (Figure 2), such a high percentage is obtained only if the intracellular concentration of MTX (and its polyglutamates) is maintained several orders of magnitude in excess of the reductase inhibition constant (10^{-11} *M*). Hence, effective intracellular concentrations of drug are in the 10^{-7} to 10^{-8} *M* range rather than 10^{-11} *M*, and the most immediate measure of cytotoxic potential is a high free intracellular concentration of MTX.

A METHOTREXATE PHARMACOKINETIC MODEL

A mathematical model that describes the physiological pharmacokinetics of MTX in several species is summarized in Figure 1. This has been described at length by Bischoff et al. (1971) and Dedrick et al. (1973), and therefore, only a brief overview is presented here. Eight organ regions have been included in the model, although the overall pharmacokinetics is only a weak function of inclusion of the spleen. The most sensitive sites of normal tissue toxicity are the intestinal mucosa cells (GI tract) and bone marrow. The liver excretes drug into the bile which, after being delayed by transport through the biliary system, enters the intestinal lumen where some of it is reabsorbed. The model consists of the set of differential mass-balance equations constructed for each organ region.

The parameters of the model are numerous, as summarized for the rat in Table 1. These are the parameters required for the model to simulate drug behavior over 0- to 4-h periods following bolus administration. For much longer periods of time or for periods following long-term infusion of drug, more constants and differential equations accounting for hydroxylation and polyglutamation are required.

If one ultimately wishes to describe toxic drug effects in several species, these parameters must be available for each species. In general, the procedures by which these parameter values may be obtained fall into two classes. The class to which each parameter belongs is identified in Table 1. The first class consists of the extracellular volume (ECF), organ plasma flow rate (Q), organ volume (V), kidney clearance (k_K), fecal transit time (k_F), bile residence time (τ), and MTX reductase dissociation constant (ϵ). These parameters are relatively invariant in passing from one species to another (ECF and ϵ), are known for all species and do not generally depend on drug (k_F and τ), or can be scaled on the basis of body size. All remaining parameters belong to the second class and require that *in vivo* or closely allied experiments be performed on each species because no adequate a priori rules or scaling laws are available. In the case of the tissue/plasma distribution ratio R, the value for muscle is the same for all species because the drug only equilibrates between plasma and ECF in

TABLE 1 Parameters of MTX Physiologic Pharmacokinetic Model

Class 1	
Organ volume	V_i
Extracellular volume	ECF_i
Blood flow rate	Q_i
Kidney clearance	k_K
Fecal transit time	k_F
Bile residence time	τ
MTX reductase dissociation constant	ϵ
Class 2	
Tissue/plasma distribution ratio	R_i
Dihydrofolate reductase concentration	a_i
Biliary clearance rate	k_L
Intestinal reabsorption rate (saturable)	k_G, K_G
Membrane transport rate (saturable)	k_i, K_i
(+ Metabolism constants)	

this organ, but the value is quite different across species for the kidney and liver. For example, the mouse concentrates fivefold more drug in liver than does the dog (Bischoff et al., 1971). No general rule exists across species for the ability of hepatocytes to transport and bind MTX, and thus separate experiments are required for each species to determine R. The tissue-specific dihydrofolate reductase concentration (in MTX equivalents, a) has also been placed in this class of parameters. After the fact, we know that these concentrations do not vary much across species, but when MTX was first under investigation, there was no a priori reason for assuming this to be the case. Arguments based on the observation that reductase, because of its role in deoxynucleotide synthesis, is most necessary for proliferating cells, and that tissue concentrations are therefore determined by the fraction of proliferating cells, have some appeal as a guide to interspecies extrapolation. At the outset of such an extrapolation, however, the question would still have remained as to the species-specific concentration in a single proliferating cell.

Presently, there is also no straightforward way to scale biliary clearance (k_L) and intestinal reabsorption parameters from one species to the next. As an example, there appear to be significant differences between species with respect to saturation of liver clearance. In the rat, Kates and Tozer (1976) reported a Michaelis constant of 70 μM for excretion of MTX into bile; however, saturation is not observed in the mouse when plasma concentrations are in substantial excess of this value (D. S. Zaharko and R. L. Dedrick, unpublished data). Intestinal reabsorption parameters also seem to require a species-specific investigation. Intestinal absorption appears to be saturable, as reviewed by Zaharko and Dedrick (1984). For compounds that are not actively absorbed at the intestinal wall, general models may eventually become available for interspecies extrapolation based upon molecular and solution features such as molecular weight, charge, ionic strength, and species-specific mucous composition and thickness; however, as yet, these models are only in their early stages of development (Peppas et al., 1984).

Remaining parameters such as membrane transport constants (k, K) and metabolism rate constants for hydroxylation and polyglutamation are still other quantities that require species-specific work. *In vitro* studies on cell lines derived from different animals indicated that the Michaelis transport constant K was only weakly species dependent. The other transport constant k, reflecting membrane carrier density and maximum rate of transmembrane transport, is more variable across cell lines and, furthermore, is not easily obtained from *in vitro* studies of normal transport-limited tissues. *In vitro* experiments designed to measure MTX uptake rates in perfused tissue specimens might be attempted, but *in vivo* testing raises fewer questions about the representativeness of the experimental model. Metabolism rate constants are particularly difficult to extrapolate across species. Metabolic enzymes are subject to major interspecies variability, such as that due to differences in regulation or to gross structural differences that affect substrate binding. It is nearly impossible to know a priori if such differences exist between species without direct testing. In the case of MTX, in which polyglutamation and hydroxylation reactions occur in the liver, *in vitro* assessment of enzyme parameters from cultured hepatocytes derived from various species may provide an initial look at whether a single set of tissue parameters has interspecies applicability. It is well known, however, that hepatocytes in culture rapidly diverge from their behavior in intact liver (Balinska et al., 1982), and standardization of assays across species could prove difficult.

DOSE SCALING

In strict form, dose scaling refers to the ability to estimate a tissue concentration at an arbitrary dose level by scaling the known concentration at some other level by the dose ratio. As long as the pharmacokinetics of the tissue region is governed by linear differential equations describing drug distribution and metabolism, this is an allowable procedure.

MTX plasma concentrations are, in fact, scalable over a wide dose range after bolus administration (Dedrick et al., 1970). Figure 3 shows plasma concentrations in the rat over a dose range of 0.05 to 25 mg/kg administered intravenously (i.v.) (Dedrick et al., 1973), in which the last three doses (b through d) differ by factors of 10. It is apparent in these log-linear plots that the plasma concentration curves are virtually identical

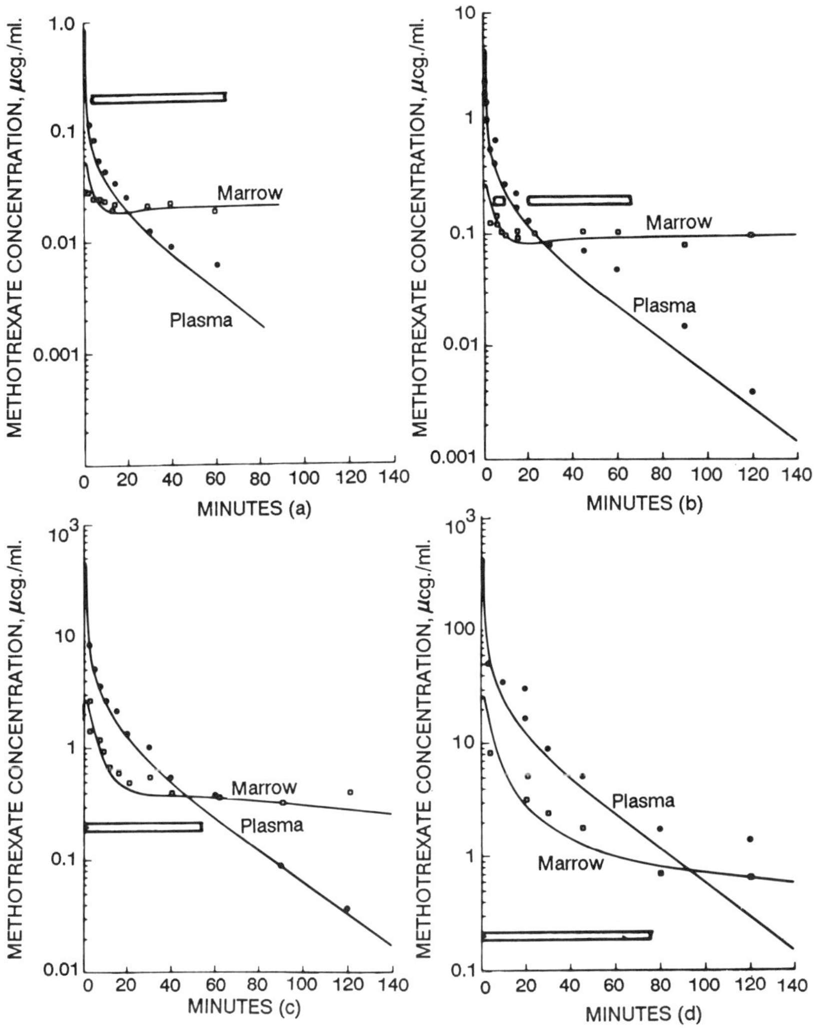

FIGURE 3 Plasma and bone marrow concentrations of MTX in rat. The bone marrow concentrations are total tissue concentrations and are thus an average over the extracellular and intracellular regions. These concentrations are nonzero and dose scalable at short times because equilibration between plasma and the extracellular space is essentially instantaneous, and saturability only occurs during transport from the extracellular to intracellular space. Solid lines represent model simulations. Data points were obtained from one rat at each time. Key: (a) 0.05 mg/kg i.v.; (b) 0.25 mg/kg i.v.; (c) 2.5 mg/kg i.v.; (d) 25 mg/kg i.v. SOURCE: Dedrick et al. (1973).

except for a decade difference in ordinate scale. The first curve (a) also scales with dose by the appropriate factor of 5. This scalability arises, in spite of the presence of nonlinearities in the complete scheme, because over 95% of the volume of MTX distribution is composed of non-transport-saturating, rapidly equilibrating compartments, principally the kidney, liver, plasma, skin, and extracellular volume of the muscle. In addition, the small compartments that do exhibit nonlinear distribution, the intracellular spaces of the gut, spleen, and bone marrow, do not rapidly transport drug into their cells. Hence, several plasma half-lives in the rat ($t_{1/2}$ = 0.3 h) can pass after administration of drug in the therapeutic dose range before plasma levels fall to the point where return of the small amount of drug in these deep nonlinear compartments could influence plasma kinetics.

On the other hand, if one is interested in assessing the drug delivery to the gut, spleen, and bone marrow, then nonlinear pharmacokinetics prevents dose scaling from applying. These organs exhibit two strong nonlinearities in the first few hours after drug administration: saturable uptake of MTX from plasma, and strong binding of MTX (and its polyglutamates) to the target enzyme dihydrofolate reductase. This is exhibited in Figure 3, in which total MTX concentrations in rat bone marrow (extracellular MTX + intracellular MTX) are plotted as a function of time for four doses (Dedrick et al., 1973). The bar denotes the concentration of reductase in this tissue. Note that plasma-to-marrow concentration ratios are not constant over the 140 min shown here, and thus that marrow concentration does not scale with dose. The total marrow concentration shown in Figure 3 does scale with dose at short times, but this only reflects the rapid equilibrium attained between plasma and extracellular space (the dominant transport before cell uptake becomes significant). Note also that the marrow curves of the two lower dose levels are flat at long times and lie below the reductase content bar, thereby reflecting strong nonlinear enzyme binding of the drug that enters the cells in the first few minutes of exposure. Because the tight binding prevents drug efflux from occurring, the mass of enzyme-bound drug reflects the cumulative result of transport into the cell. At 0.05 and 0.25 mg/kg, this mass scales with dose (0.02 versus 0.10 μg/ml) and thus transport is linear. However, attempts at using linear transport to extrapolate from 0.25 to 2.5 mg/kg fail, an observation that is consistent with plasma levels at the higher dose exceeding a Michaelis transport constant of about 1 μ*M*, a value characteristic of the range found in a variety of cell lines (Goldman, 1969, 1971; Schilsky et al., 1981). Figure 4 shows the magnitude of this saturable transport effect by dose level as the difference between the dashed and solid lines. The solid line shows rat marrow concentrations when saturation

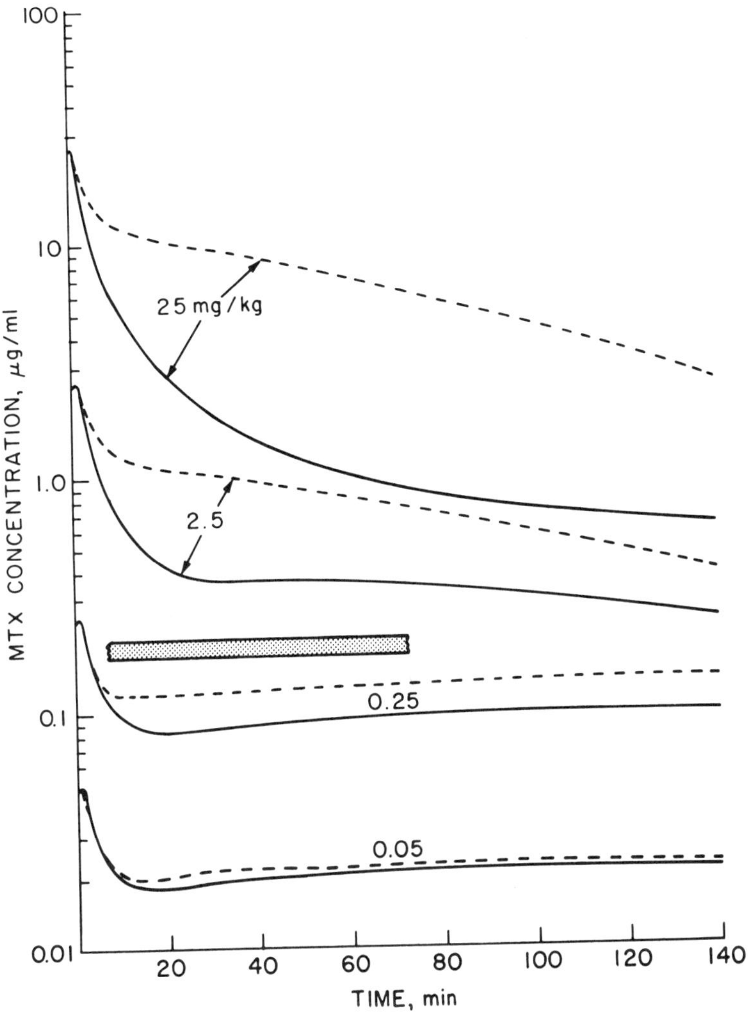

FIGURE 4 Comparison of model simulation of linear and saturable transport in bone marrow at several doses. The dashed lines represent the model simulations for linear transport to the intracellular compartment of bone marrow. The solid lines represent model simulations with saturable transport. SOURCE: Dedrick et al. (1973).

is operative, while the dashed line, providing a poor fit to the data (not shown), shows the result when linear transport is assumed.

Hence, it can be concluded that, for sufficiently short times (e.g., <4 h in the rat), many organ regions are dose scalable, while others, for example, the gut, marrow, and spleen, are not.

DOSE SCHEDULING

We next turn our attention to the rather dramatic effects that dose scheduling has on toxic response and to the formal connection between MTX pharmacokinetics and toxic response. Up to this point, the discussion has mainly involved distributional events that occur after bolus dosage. We will now see that the time of inhibition of dihydrofolate reductase is the primary correlate with toxicity.

Table 2 shows acute toxic response in terms of the lethal dose for 50% of mice (LD_{50}) exposed to a variety of drug schedules (Zaharko, 1975). The first entry is for a bolus dose of 350 mg/kg, while the next entries correspond to divided doses of decreasing total dose, and the final entry corresponds to a 96-h infusion of a 3-mg/kg total dose. These results immediately show that response does not directly correlate with total dose. Decreasing dose by more than a factor of 100 led to an increase, rather than a substantial decrease, in toxicity. Furthermore, the area under the plasma concentration-time curve, a frequently used metric, does not correlate with toxic response. This can be seen in Figure 5 (Zaharko, 1975). The steep curve shows the MTX plasma concentration following the 350-mg/kg bolus dose, while the flat curve shows the concentration following the 3-mg/kg 96-h infusion. The area under the bolus curve is about 2 orders of magnitude greater than that under the infusion curve, yet toxic response is greater with infusion.

The principal correlate with response is the length of time that dihydrofolate reductase and, consequently, DNA synthesis are strongly inhibited. If inhibition of DNA synthesis is a good correlate, then one would expect that the onset of toxicity should be observable after the same inhibition time, regardless of dose schedule. For at least two infusion schedules, this has been observed experimentally in mouse small intestine (Zaharko et al., 1977). Figure 6 shows the recovery of DNA synthesis in mice, as measured by deoxyuridine incorporation into DNA, following an infusion of 1 μg of MTX/h for 48 h, a schedule that just barely avoids

TABLE 2 Schedule Dependence of Methotrexate Toxicity in Mice

Individual dose (mg/kg)	Schedule	Total dose (mg/kg)	Peak plasma concentration (*M*)	Effect
350	Single dose	350	10^{-3}	LD_{50}
25	Twice daily	50	10^{-4}	LD_{50}
3	Every 3 h, 5 times, rest 8 h, and then every 3 h, 3 times	24	10^{-5}	$>LD_{50}$
0.5	Every 3 h, 20 times	10	10^{-6}	$>LD_{50}$
0.8 μg/h	Infusion 96 h	3	10^{-8}	$>LD_{50}$

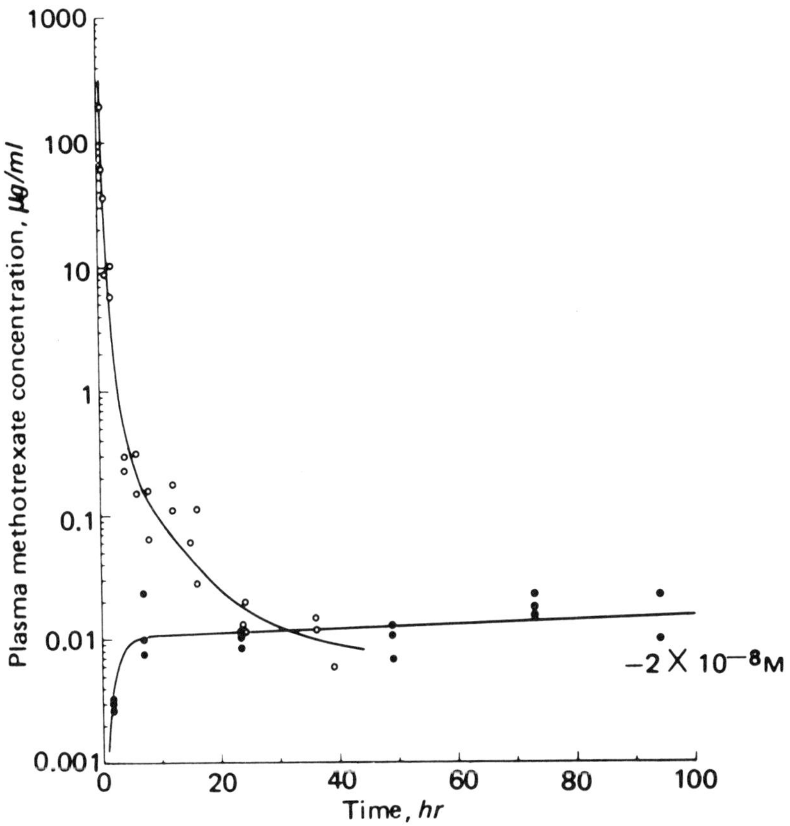

FIGURE 5 Concentration of methotrexate in mouse plasma. Open circles, single dose of 350 mg/kg i.v.; closed circles, constant infusion of 0.8 μg/h. SOURCE: Zaharko (1975).

lethal toxicity. If severe inhibition of greater than 90% is considered, DNA synthesis is inhibited relative to control for 35 h. Figure 7 shows similar data for a 10-fold higher rate of infusion (10 μg/h) but of shorter duration (17 h). Like the previous schedule, this one is designed to just barely avoid lethal toxicity at the end of the infusion period (Table 2 of Zaharko et al., 1977). An inhibition time virtually identical to that above was observed, a period of about 30 h.

Further indication of DNA synthesis inhibition time as an appropriate measure of toxicity comes from observing the trend in lethality if, at a fixed infusion rate, infusion times are lengthened. As expected, lethality increases. For 10 μg of MTX/h, lethality jumps from 20% following a 24-h infusion to 90% following a 56-h infusion (Zaharko et al., 1977).

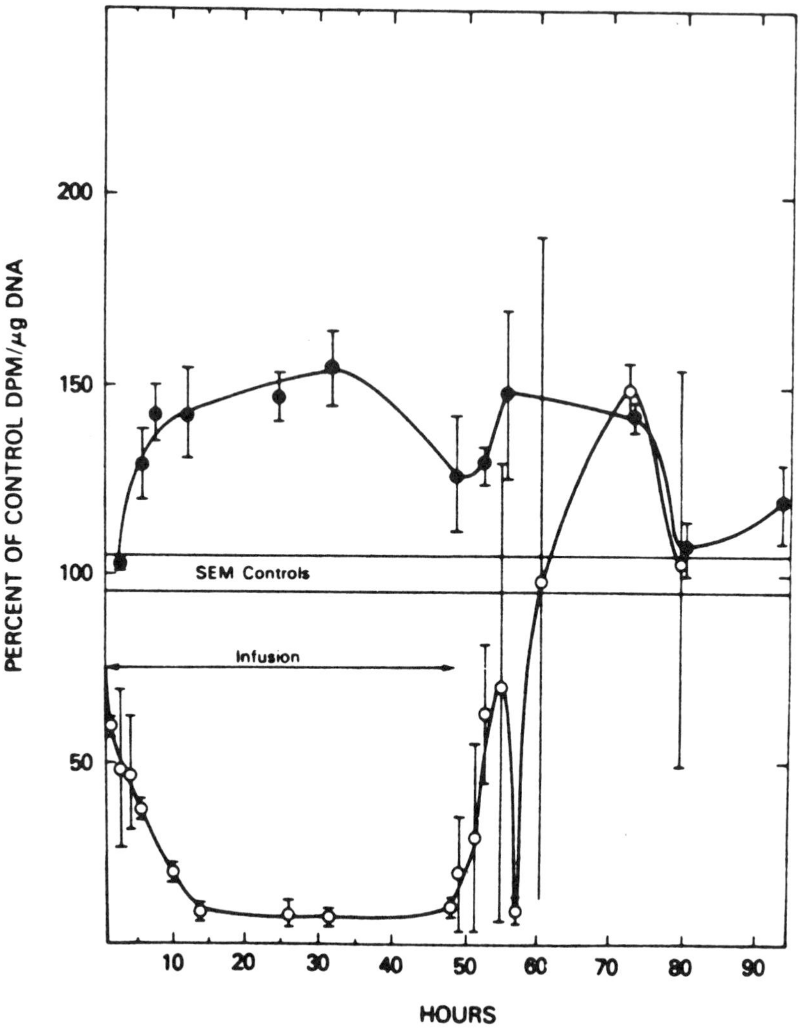

FIGURE 6 Incorporation of ^3H into DNA of small intestine following an MTX infusion of 1 μg/h for 48 h and an injection of [^3H]thymidine (dark circle) or [^3H]uridine injection (open circle). Vertical standard error bars indicate the range of the two mice used per point. At least six controls were used for each experiment. The first 10 cm of the small intestine was used from each mouse. SOURCE: Zaharko et al. (1977).

Lethality correlation with total dose exists for a fixed infusion rate, but not when extended over a range of infusion rates.

To predict toxic response from a given dose and schedule, the pharmacokinetic model of MTX must be coupled to the inhibition of DNA

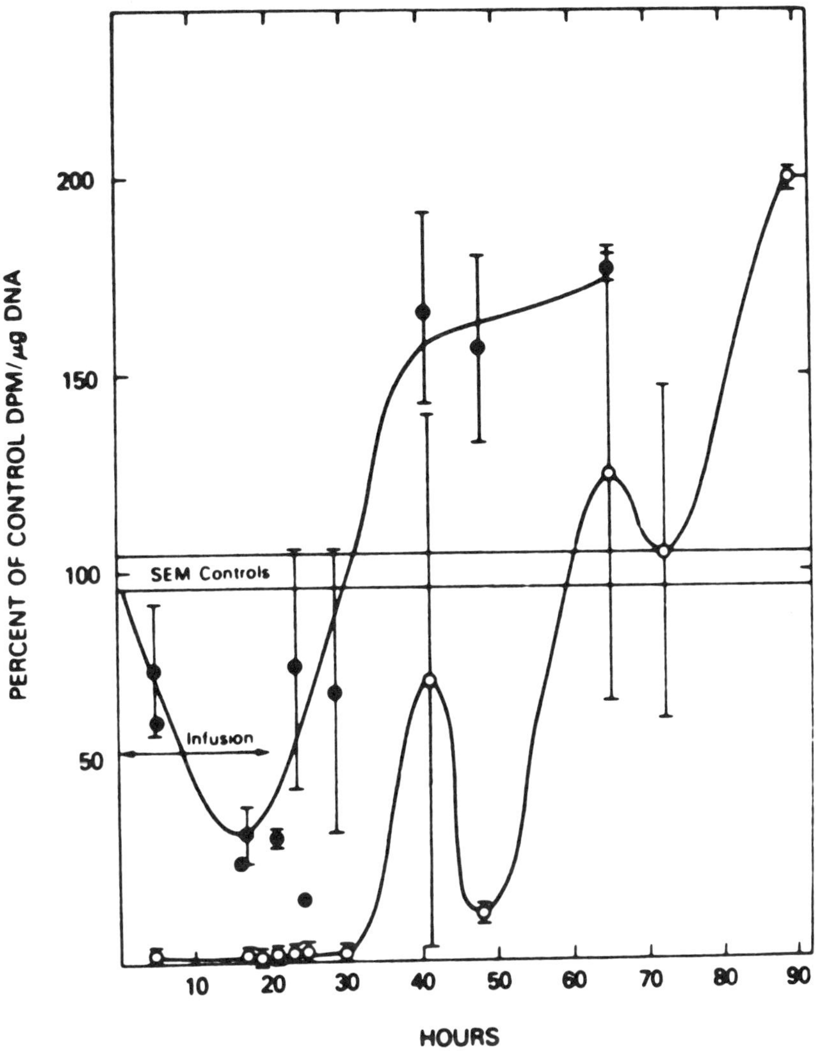

FIGURE 7 Same as described in the legend to Figure 6, except that the MTX infusion was 10 μg/h for 17 h. Solid dots without vertical error bars are an identical experiment with MTX (intraperitoneal dose of 25 mg/kg) given at the start of infusion; mean of two mice; the range was not included, but it was similar to those in other experiments. SOURCE: Zaharko et al. (1977).

synthesis. In the mid-1970s, this was accomplished very simply by observing that recovery of DNA synthesis occurred when MTX plasma concentrations, as measured by a competitive binding assay, fell below 10^{-8} M (Chabner and Young, 1973). Figure 8 shows that recovery in

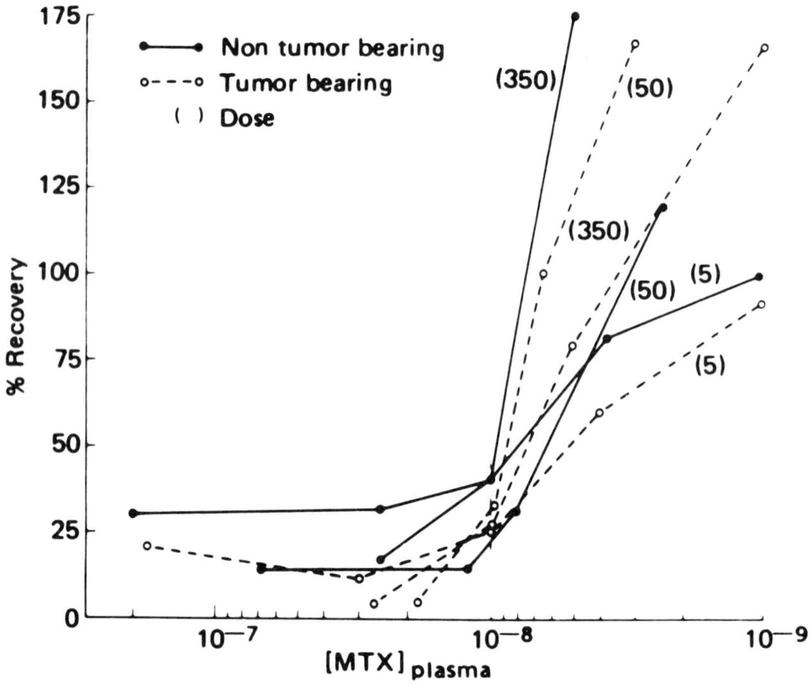

FIGURE 8 Incorporation of [^{3}H]uridine into DNA of mouse bone marrow as a percentage of the pretreatment rate. The indicated doses are in milligrams/kilogram, administered intraperitoneally. SOURCE: Chabner and Young (1973).

bone marrow was a strong function of this pseudo-threshold following bolus dosing from 5 to 350 mg/kg. Thus, the pharmacokinetic model outlined earlier only needed to be solved for the length of time that the plasma concentration remained above 10^{-8} *M* to infer toxic response. The 10^{-8} *M* value was interpreted as the free MTX concentration in equilibrium with just sufficient free reductase (about 5% of the total) to allow resumption of thymidylate synthesis.

Since the mid-1970s, the interpretation of these observations has become much more complex. The reason for this was the discovery of significant metabolism of MTX. When only low bolus doses were considered, cytotoxic concentrations of MTX did not exist for a sufficiently long period for significant metabolism to be detected. With the introduction of high-dose protocols, and their long infusion times, in the mid-1970s (Frei et al., 1975; Jaffe et al., 1978), metabolism became apparent, and new assays were developed for the detection of metabolites.

Some of these metabolic events, particularly the polyglutamation of MTX (Balinska et al., 1982; Baugh et al., 1973; Jolivet et al., 1982;

Morrison and Allegra, 1987) were shown to create very active drug forms that cleared from the intracellular milieu of cells at rates that were far slower than the clearance of parent drug from the plasma. Hence, a rationale was needed to explain just why 10^{-8} M plasma concentrations have been observed to correlate so well with recovery of DNA synthesis and, more importantly, to identify the conditions under which this correlation might break down and when another approach to bridging MTX pharmacokinetics to toxicity would be required. Thus, there is no theoretical reason for expecting that when the 10^{-8} M concentration is reached, DNA synthesis should immediately resume under all protocols.

A full rationale requires more research, but preliminary analysis indicates that, under single-bolus conditions (as employed in Figure 8), cells are incapable of producing truly inhibiting quantities of polyglutamates, leaving the parent drug form alone to account for inhibition and recovery of DNA synthesis as observed. Under long-term infusion or closely spaced multiple bolus conditions, however, cells have much more time to produce polyglutamates and attain inhibiting levels of these compounds. Under these circumstances, MTX pharmacokinetic models need to be expanded to include polyglutamation kinetics, and intracellular MTX polyglutamate concentrations, rather than plasma concentrations of parent compound, need to be correlated with levels of DNA synthesis.

SUMMARY

In summary, we have seen that (1) only about half of the parameters of the (nonmetabolizing) physiological pharmacokinetic model can be obtained from chemical invariants and interspecies scalings; (2) dose scaling applies to the large-volume organ regions (e.g., kidney, liver, plasma) for a few hours after injection, but never to the principal organs of toxicity, the sensitive tissues of the marrow and intestinal mucosa; and (3) toxic response is a strong function of dose scheduling, correlating neither with total drug dose nor area under the MTX-plasma concentration curve, but with the time that MTX (and polyglutamate) concentrations remain above inhibiting levels of dihydrofolate reductase.

REFERENCES

Balinska, M., Z. Nimec, and J. Galivan. 1982. Characteristics of methotrexate polyglutamate formation in cultured hepatic cells. Arch. Biochem. 216:466–476.

Baugh, C. M., C. L. Krumdieck, and M. G. Nair. 1973. Polygammaglutamyl metabolites of methotrexate. Biochem. Biophys. Res. Commun. 52:27–34.

Bischoff, K. B., R. L. Dedrick, D. S. Zaharko, and J. A. Longstreth. 1971. Methotrexate pharmacokinetics. J. Pharm. Sci. 60:1128–1133.

Breithaupt, H., and E. Kuenzlen. 1982. Pharmacokinetics of methotrexate and 7-hydroxy-methotrexate following infusions of high-dose methotrexate. Cancer Treatment Rep. 66:1733–1741.

Chabner, B. A., and R. C. Young. 1973. Threshold methotrexate concentration for in vivo inhibition of DNA synthesis in normal and tumorous target tissues. J. Clin. Invest. 52:1804–1811.

Dedrick, R. L., K. B. Bischoff, and D. C. Zaharko. 1970. Interspecies correlation of plasma concentration history of methotrexate (NCS-740) Cancer Treatment Rep. 54:95–101.

Dedrick, R. L., D. S. Zaharko, and R. J. Lutz. 1973. Transport and binding of methotrexate in vivo. J. Pharm. Sci. 62:882–890.

Frei, E., N. Jaffe, M. H. N. Tattersall, S. Pitman, and L. Parker, 1975. New approach to cancer chemotherapy with methotrexate. N. Engl. J. Med. 292:846–851.

Goldman, I. D. 1969. Transport energetics of the folic acid analogue, methotrexate, in L1210 leukemia cells. J. Biol. Chem. 244:3779–3785.

Goldman, I. D. 1971. The characteristics of the membrane transport of amethopterin and the naturally occurring folates. Ann. N.Y. Acad. Sci. 186:400–422.

Goldman, I. D., and L. H. Matherly. 1986. The cellular pharmacology of methotrexate. Pp. 283–308 in Membrane Transport of Antineoplastic Agents, I. D. Goldman, ed. New York: Pergamon.

Jackson, R. C., and K. R. Harrap. 1973. Studies with a mathematical model of folate metabolism. Arch. Biochem. Biophys. 158:827–841.

Jackson, R. C., and K. R. Harrap. 1979. Computer models of anticancer drug interaction. Pharmacol. Ther. 4:245–280.

Jacobs, S. A., C. J. Derr, and D. G. Johns. 1977. Accumulation of methotrexate diglutamate in human liver during methotrexate therapy. Biochem. Pharmacol. 26:2310–2313.

Jaffe, N., E. Frei, H. Watts, and D. Traggis. 1978. High-dose methotrexate in osteogenic sarcoma; a 5-year experience. Cancer Treatment Rep. 62:259–264.

Jolivet, J., and B. A. Chabner. 1983. Intracellular pharmacokinetics of methotrexate polyglutamates in human breast cancer cells. J. Clin. Invest. 72:773–778.

Jolivet, J., R. L. Schilsky, B. D. Bailey, J. C. Drake, and B. A. Chabner. 1982. Synthesis, retention, and biological activity of methotrexate polyglutamates in cultured human breast cancer cell. J. Clin. Invest. 70:351–360.

Kates, R. E., and T. N. Tozer. 1976. Biliary secretion of methotrexate in rats and its inhibition by probenecid. J. Pharm. Sci. 65:1348–1352.

Kennedy, D. G., R. Clarke, H. W. van den Berg, and R. F. Murphy. 1983. The kinetics of methotrexate polyglutamate formation and efflux in a human breast cancer cell line: The effect of insulin. Biochem. Pharmacol. 32:41–46.

Morrison, P. F., and C. J. Allegra. 1987. The kinetics of methotrexate polyglutamation in human breast cancer cells. Arch. Biochem. Biophys. 254:597–610.

Peppas, N. A., P. J. Hansen, and P. A. Buri. 1984. A theory of molecular diffusion in the intestinal mucus. Int. J. Pharm. 20:107–118.

Schilsky, R. L., B. D. Bailey, and B. A. Chabner. 1981. Characteristics of membrane transport of methotrexate by cultured human breast cancer cells. Biochem. Pharmacol. 30:1537–1542.

Whitehead, V. M., M. M. Perrault, and S. Stelcner. 1975. Tissue-specific synthesis of methotrexate polyglutamates in the rat. Cancer Res. 35:2985–2990.

Zaharko, D. S. 1975. The kinetics of drug action. Pp. 69–83 in Pharmacological Basis of Cancer Chemotherapy, 27th Annual Symposium on Fundamental Cancer Research 1974. Baltimore, The Williams & Wilkins Co.

Zaharko, D. S., and R. L. Dedrick. 1984. Pharmacokinetics of methotrexate in animals and man. Pp. 97–131 in Folate Antagonists as Therapeutic Agents, Vol. 2. Pharmacology, Experimental and Clinical Therapeutics, F. M. Sirotnak, J. J. Burchall, W. D. Ensminger, and J. A. Montgomery, eds. Orlando, Fla.: Academic Press.

Zaharko, D. S., W. P. Fung, and K.-H. Yang. 1977. Relative biochemical aspects of low and high doses of methotrexate in mice. Cancer Res. 37:1602–1607.

PART VII

Summary: Prospectives and Future Directions

Prospective Predictions and Validations in Anticancer Therapy

Jerry M. Collins

INTRODUCTION

At first glance, the processes of anticancer drug development and environmental risk assessment may not seem to have much conceptual overlap. However, there is a strong common thread based upon the need to make decisions regarding allowable human exposure limits. In both cases, heavy reliance is placed upon interspecies toxicological comparisons.

Risk assessment is based upon both mathematical models and experimental data. For example, the data might be the incidence of tumor formation in rodents following controlled laboratory exposures to a toxin. The role of the model is to predict the incidence of carcinogenesis in humans under a variety of occupational and/or environmental exposure conditions. The weakest link in this process is model validation. Due to ethical considerations, it is not usually possible to administer precise amounts of toxic chemicals to humans. If a human population develops an unusual form of cancer, epidemiologic detectives might be able to trace the source to a particular chemical. The human exposure data are estimated retrospectively in whatever fashion possible, but the uncertainty in these calculations is a major hurdle in quantitative analyses. Once a specific chemical becomes suspect, it would be possible to do a set of quantitative experiments in animals.

Even if we accept these examples with their imprecise estimates of human exposure, the total data base for model validation is very small. Yet a variety of needs forces us to accept these models as the basis for major prospective decisions that have an impact upon the health of citizens and the economic well-being of corporations and communities.

The preclinical toxicology phase of drug development shares some of the same facets as the safety testing of industrial pollutants or other potential environmental contaminants. For example, animals are used to determine a lethal dose, such as the lethal dose for 10% of the animals tested (LD_{10}). That estimate is then used to determine safety in humans. Perhaps the single largest difference between the development of anticancer drugs and the assessment of risks from environmental contaminants is that direct experimental evidence is obtained in humans that can be (rapidly) compared with data from animals.

The treatment of a life-threatening disease requires a rather different set of risk-to-benefit decisions than considerations of maximally allowed pollutants. The drugs used for the treatment of cancer have narrower safety margins than those used for most other diseases. In general, the ratio of a therapeutic dose to a toxic dose approaches unity.

DRUG DEVELOPMENT

Drug development consists of a progression of steps (Table 1) that starts with the discovery of a new compound and ends with a clinical determination of therapeutic utility. To begin human testing, a safe starting dose is needed. Establishment of a safe starting dose is one of the chief functions of preclinical toxicology studies. As reviewed by Grieshaber and Marsoni (1986), the current preclinical toxicology protocol for anticancer drugs provides the basis for a safe starting dose, tailored to potency in rodents. The human starting dose is 1/10 of the mouse LD_{10}, expressed on a milligrams/square meter basis. Prior to human testing, this dose is confirmed in a second species.

After the starting dose has been evaluated in patients, subsequent doses are escalated. Although there is always therapeutic intent when an anticancer drug is given to patients, the major scientific goal of initial clinical trials is to determine the acute, reversible toxicity. The endpoint of these phase I trials is called the maximum tolerated dose, or MTD. The MTD

TABLE 1 Stages of Drug Development

	Name	Function
Preclinical	Discovery	Random or planned
	Screening	Bioactivity
	Toxicology	LD_{10}; organ sites
Clinical	Phase I	Safety
	Phase II	Activity
	Phase III	Efficacy

is used to establish the dose for more detailed efficacy studies in phase II testing. The procedure used for dose escalation must achieve a balance between the desire to escalate slowly enough to be safe and the desire to escalate fast enough to be efficient. The most commonly used procedure is known as the modified Fibonacci scheme (Goldsmith et al., 1975). The initial escalation is rapid (100%, or doubling of the dose); subsequent escalations narrow down until the 30–35% range is reached (Figure 1).

In summary, there are two areas of risk assessment that are encountered in these early clinical trials: (1) selection of a safe starting dose, and (2) choosing the rate of dose escalation.

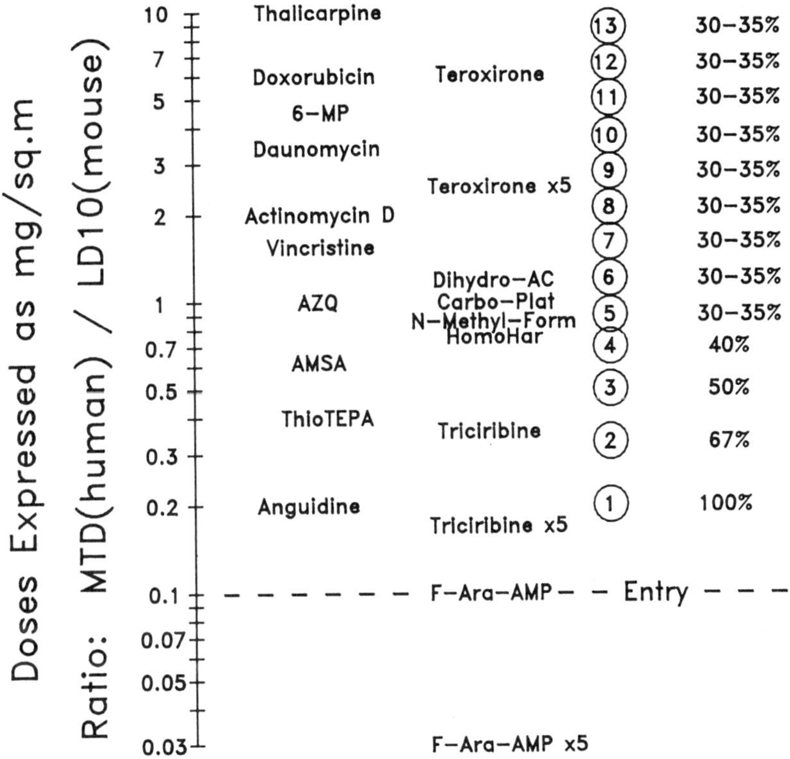

FIGURE 1 Interspecies toxicity comparison. For these 17 anticancer drugs, the median MTD in humans was equal to the mouse LD_{10}, when both doses were expressed on a milligram/square meter basis. To compare toxicity on a milligram/kilogram basis, the ordinate was multiplied by 0.083. All drugs were given as single doses ($\times 1$) or as five daily doses ($\times 5$). Data in the second column are from Grieshaber and Marsoni (1986). Data in the first column were collected from information in the literature or on file in the Toxicology and Investigational Drug Branches, Division of Cancer Treatment, National Cancer Institute. As a reference, the modified Fibonacci escalation steps are shown in the third column. Adapted from Collins et al. (1986).

COMPARISON OF HUMAN AND MURINE TOXICITY

How well does this strategy work? In Figure 1, the ratio of (human MTD)/(mouse LD_{10}) is presented for a series of anticancer drugs (Collins et al., 1986). First, it is worth noting that this particular collection of interspecies toxicology data, although not exhaustive, probably exceeds any comparable compilation for environmental contaminants or other human toxins. Rather than focusing on specific drugs at this stage, it is helpful to get some appreciation of the range of variation. It could be argued that there is considerable variation, even though the average is quite reasonable. It might also be reasonably argued, however, that this level of agreement is adequate for comparative purposes. A number of factors provide motivation to probe further. For example, patient safety might be improved by a better understanding of the sources of this variation. Also, the efficiency of early clinical testing might be raised if the toxicologic variation could be related to measurable determinants, such as plasma levels.

What are possible explanations for this variation in toxicity between mouse and man? Table 2 lists three possibilities: (1) differences in drug metabolism, elimination, and binding; (2) exposure time differences; and (3) target cell sensitivity differences.

Elimination rates determine the drug exposure, or $C \times T$, the area under the concentration versus time curve. The concept of $C \times T$ with regard to drug toxicity originated during World War I (Prentiss, 1937). German pharmacologists observed that mustard agents are equally toxic whether a high concentration is inhaled for a short time or a low concentration is inhaled for a long time. The essential feature is that the $C \times T$ is the determinant of effect rather than the absolute concentration itself.

Most toxicologists have presented their dosing information in terms of milligrams/kilogram. Based upon the relationships between body surface area and body weight (Freireich et al., 1966), it is possible to interconvert dosing data between milligrams/square meter and milligrams/kilogram. For mice, the relationship is 1 m^2 = 3 kg. For humans, it is 1 m^2 = 37 kg. Empirically, either set of units can be used to present raw data. The use of body surface area has a distinct advantage, however, when toxicity

TABLE 2 Potential Explanations for Variation in Toxicity Between Mouse and Man

1. Species differences in drug metabolism, elimination, and binding
2. Schedule dependency due to exposure time differences
3. Species differences in target cell sensitivity

comparisons are made across species. The physiological determinants of elimination rates (such as glomerular filtration rate or organ blood flow) tend to be highly correlated with body surface area. Thus, if the dose is expressed in milligrams/square meter and elimination rates (milliliter/minute/square meter) are identical in mice and men, then the drug exposure ($C \times T$) will be the same in both species at the same dose. On the other hand, if the dose is expressed in milligrams/kilogram, the dose in humans that produces equal $C \times T$ will be 1/12th the mouse dose, because a correction must be made for body surface area differences. The factor of 12 is simply the ratio of body surface area constants, 37/3.

Freireich et al. (1966), Skipper et al. (1971), and Schabel et al. (1983) have reported that with many anticancer drugs, toxicity observations carry across species on a milligram/square meter basis, as long as schedules are similar. They also were aware, however, that there are exceptions and that more complete exposure parameters such as $C \times T$ or plasma pharmacokinetics allow more useful comparisons of toxic or therapeutic responses from experimental and clinical studies.

Doxorubicin appears to be an example of metabolism/elimination differences. The MTD in man is fivefold greater than the LD_{10} in mice, on a milligram/square meter basis. Yet, as shown in Figure 2, there is considerable agreement between blood levels measured at equitoxic doses. It appears that humans are more tolerant than mice due to a higher clearance (milliliters/minute/square meter) for doxorubicin.

FACTORS OTHER THAN $C \times T$

The second factor of possible importance is a difference in exposure times. For some drugs, there are threshold concentrations or time dependencies that are related to toxicity and/or mechanisms of action. For drugs with equal clearance values in mice and humans (milliliters/minute/square meter), Skipper and colleagues (1971) have made the point that a bolus dose of equal milligrams/square meter generally produces rather different time courses in mice and man (Figure 3). If there is a threshold for action and a critical exposure time, where the threshold lies can give major differences in species response. For example, if the threshold in Figure 3 is set at 10^{-6} M, there is no effect in man. If the threshold is set at 10^{-7} M, however, the duration of effect is much longer in man than in mice. Note that the time course for doxorubicin (Figure 2) is an exception to the generalized pattern.

In a classic study by Quinn et al. (1958) 29 years ago, the threshold effect was first demonstrated for the barbiturate hexobarbital. After a standard dose of 50 to 100 mg/kg was given to mice, rats, and rabbits,

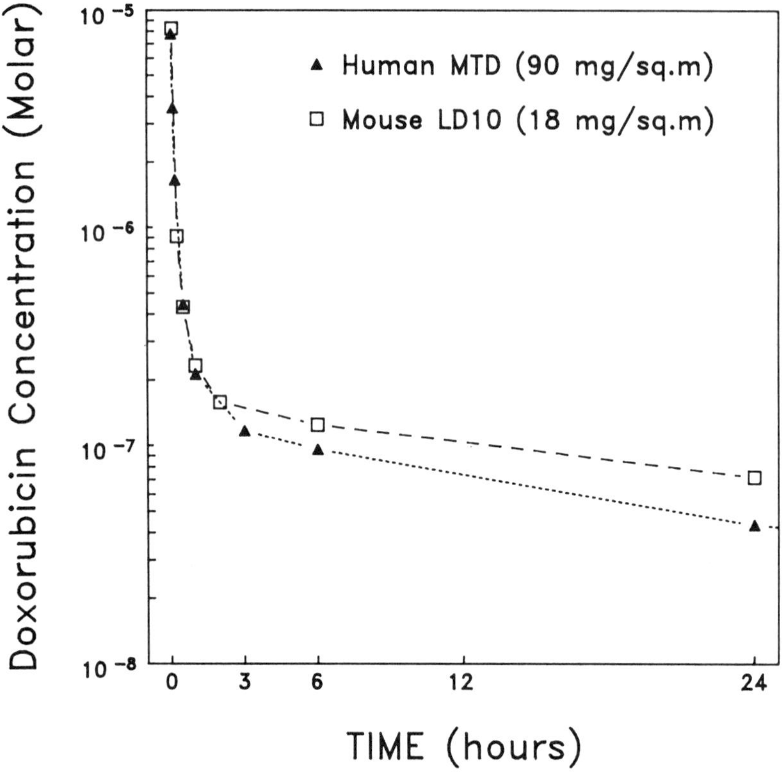

FIGURE 2 Doxorubicin plasma concentrations in mice and humans at equitoxic doses. Human data were scaled from a 75-mg/m^2 intravenous (i.v.) dose. Mouse data were scaled from a 75-mg/m^2 dose given i.v. to CDF1 mice. Reprinted with permission from Collins et al. (1986).

there was substantial variation in the drug effectiveness. The times to awakening were 12, 90, and 49 min in mice, rats, and rabbits, respectively. When the plasma pharmacokinetics of hexobarbital were investigated, it was found that the three species exhibited rather different elimination rates, or plasma half-times. At the time of awakening, however, the plasma drug concentration was similar in all three species. Thus, this is an example of a species difference in drug effect that is determined by pharmacokinetic changes in exposure patterns. Studies in dogs did not give as clear a pattern as for the other three species.

The first two reasons for the species variation in toxicology have been oriented toward plasma pharmacokinetics. The third factor essentially covers all explanations that are not related to the delivery of the drug to the site of action. There can be differences at the cellular level that determine species sensitivity. For example, if the drug needs to be activated inside the cell, there may be a species difference in activation capabilities.

Fludarabine phosphate (F-Ara-AMP) has been found to have the largest difference in dose (10- to 30-fold) between the mouse LD_{10} and the human MTD. This discrepancy is apparently an example of species differences in target cell sensitivity. The phosphate group on F-Ara-AMP makes the drug readily soluble, but it is rapidly cleaved to F-Ara-A *in vivo*. Within 5 min following administration, only the F-Ara-A form can be detected in plasma. As shown in the plasma profiles following single doses of F-Ara-AMP (Figure 4), there is no obvious plasma pharmacokinetic explanation for the species difference in toxicity. In contrast to the situation for doxorubicin (in which equitoxic doses produced similar plasma drug concentrations), it can be seen from these data that the plasma concentrations of the circulating species F-Ara-A are considerably different. Studies with bone marrow cultures *in vitro* indicate that human bone

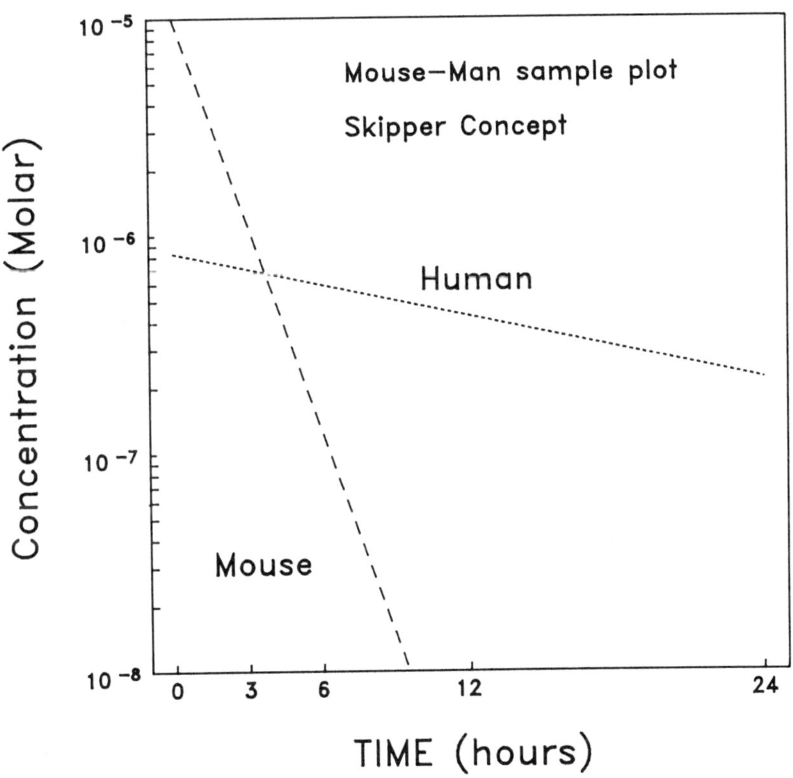

FIGURE 3 Idealized plasma concentrations in mice and humans following equal bolus doses (milligrams/square meter). Assume that volume of distribution (liters/kilogram) and clearance (milliliters/minute/square meter) are similar in both species. Adapted from Skipper et al. (1971).

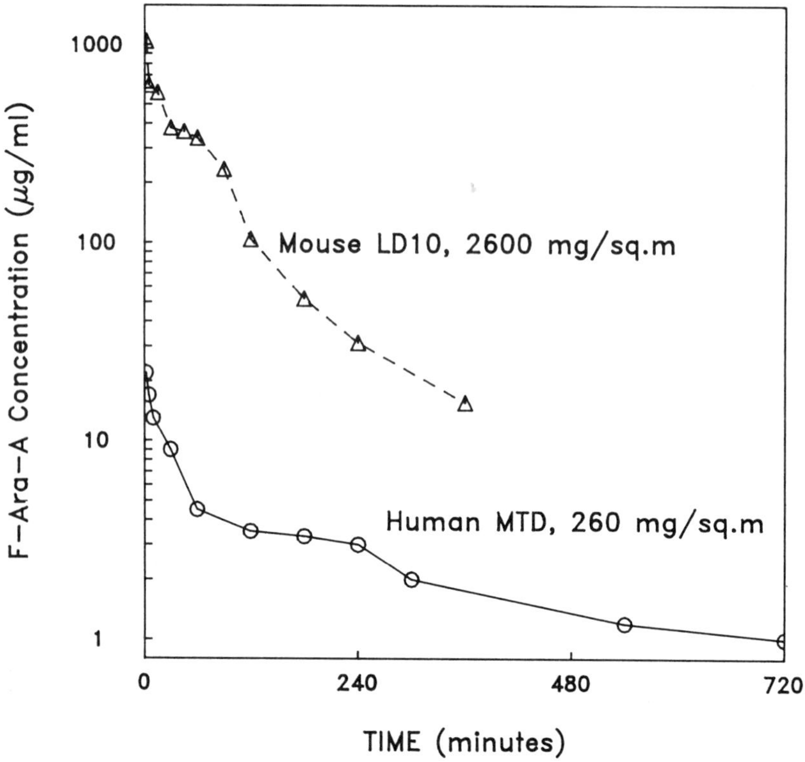

FIGURE 4 F-Ara-A plasma concentrations in mice and humans at equitoxic doses. Mouse data are from Noker et al. (1983). Human data are from Malspeis (Minutes of the Phase I Working Group, Bethesda, Md., June 13–14, 1983). Reprinted with permission from Collins et al. (1986).

marrow cells are intrinsically more sensitive to this particular compound than are mouse marrow cells (C. Poston et al., unpublished data). Because bone marrow suppression is the principal acute toxicity *in vivo,* it appears that the species differences are due to target cell differences for this drug.

SUMMARY OF DATA

$C \times T$ information is listed for 12 anticancer drugs in Table 3. For the first nine of these drugs, the $C \times T$ ratio is a useful predictor of the relative toxicity in mice and humans. For 5-azacytidine, doxorubicin (as already discussed), and teroxirone, the $C \times T$ ratio is far better than the dose ratio. For the next five drugs on this list, the $C \times T$ ratio was also a reasonable predictor, although it was about the same as the dose ratio. For thio-TEPA, the $C \times T$ ratio was also an improvement over the dose ratio.

TABLE 3 The Mouse as a Quantitative Predictor of Human Toxicity: Comparison of Dose Ratio (milligram/square meter basis) and $C \times T$ Ratio for Human MTD to Mouse LD_{10}

Drug	Dose ratio	$C \times T$ ratio
1. 5-Azacytidine	6.5	(1.1)
2. Doxorubicin	5.0	0.8
3. Teroxirone	4.3	0.8
4. Diaziquone (AZQ)	1.0	(0.7)
5. Indicine-*N*-oxide	0.9	0.6
6. Amsacrine (AMSA)	0.8	1.3
7. Deoxycoformycin	0.7	1.1
8. Tiazofurin	0.7	0.9
9. Thio-TEPA	0.4	1.0
10. PALA	2.8	3.3
11. F-Ara-AMP	0.1	0.1
12. Dihydroazacytidine	1.2	0.3

There are also some drugs for which the $C \times T$ ratio was not an effective predictor of toxicity. For PALA, both the dose ratio and the $C \times T$ ratio were overly conservative; i.e., humans were threefold more tolerant than mice. A more serious case arises when man is less tolerant than mice. Two such cases were found in our survey. F-Ara-AMP was discussed above. Dihydroazacytidine was also a case in which man was less tolerant than mice, due to severe chest pain. As pointed out by Grieshaber and Marsoni (1986), it is not easy to pick up some toxicities in a mouse toxicology study.

Thus, the use of $C \times T$ seems to be a useful starting point for understanding differences in toxicity between mouse and man, but it is not completely accurate. Further work is ongoing at the National Cancer Institute that is exploring the use of $C \times T$ data to adjust escalation rates in phase I trials (Collins et al., 1986).

CONCLUSIONS

The development of new anticancer drugs generates a unique quantitative data base that can be used for interspecies comparisons of toxicity. In addition to serving as a collection of empirical toxicity data, the data base can be used for the testing of hypotheses regarding the fundamental determinants of toxicity. For example, the role of pharmacokinetics can be probed. Finally, the insights gained from analyses of past experience can be put to practical use in the form of improvements to the drug development process.

ACKNOWLEDGMENTS

Substantial contributions have been made to the concepts and data presented in this paper by Dr. Robert Dedrick of the Division of Research Services, National Institutes of Health, and by the staff and contractors of the Division of Cancer Treatment, National Cancer Institute. Portions of this manuscript have been excerpted from Collins et al. (1986).

REFERENCES

Collins, J. M., D. S. Zaharko, R. L. Dedrick, and B. A. Chabner. 1986. Potential roles for preclinical pharmacology in phase I clinical trials. Cancer Treatment Rep. 70:73–80.

Freireich, E. J., E. A. Gehan, D. P. Rall, L. H. Schmidt, and H. E. Skipper. 1966. Quantitative comparison of toxicity of anticancer agents in mouse, rat, hamster, dog, monkey, and man. Cancer Chemother. Rep. 50:219–244.

Goldsmith, M. A., M. Slavik, and S. K. Carter. 1975. Quantitative prediction of drug toxicity in humans from toxicology in small and large animals. Cancer Res. 35:1354–1364.

Grieshaber, C. K., and S. Marsoni. 1986. The relation of preclinical toxicology to findings in early clinical trials. Cancer Treatment Rep. 70:65–72.

Noker, P. E., G. F. Duncan, S. M. L. El Dareer, and D. L. Hill. 1983. Disposition of 9-β-D-arabinofuranosyl-2-fluoroadenine 5′-phosphate in mice and dogs. Cancer Treatment Rep. 67:445–456.

Quinn, G. P., J. Axelrod, and B. B. Brodie. 1958. Species, strain, and sex differences in metabolism of hexobarbitone, amidopyrine, antipyrine and aniline. Biochem. Pharmacol. 1:152–159.

Prentiss, A. M. 1937. Chemicals in War: A Treatise on Chemical Warfare. New York: McGraw-Hill.

Schabel, F. M., Jr., D. P. Griswold, T. H. Corbett, and W. R. Laster, Jr. 1983. Increasing therapeutic response rates to anticancer drugs by applying the basic principles of pharmacology. Pharmacol. Ther. 20:283–305.

Skipper, H. E., F. M. Schabel, Jr., L. B. Mellet, J. A. Montgomery, L. J. Wilkoff, H. H. Lloyd, and R. W. Brockman. 1971. Implications of biochemical, cytokinetic, pharmacologic and toxicologic relationships in the design of optimal therapeutic schedules. Cancer Chemother. Rep. 54:431–450.

The Application of Pharmacokinetic Data in Carcinogenic Risk Assessment

Daniel Krewski, Duncan J. Murdoch, and *Jim R. Withey*

INTRODUCTION

The area of carcinogenic risk assessment has received considerable attention in recent years (Krewski and Brown, 1981), and continues to be a subject of much current debate (OSTP, 1985). Information on carcinogenic risks can be obtained by using epidemiological studies of human populations (Day, 1985) and toxicological experiments conducted in the laboratory (Bickis and Krewski, 1985). While epidemiological investigations can provide information on adverse health effects directly in man, experimental studies with animal models have the advantage of being able to predict potential health hazards in advance of actual human exposure, and thus continue to be widely used for identifying substances with carcinogenic potential.

The main disadvantage of toxicological testing is the ultimate need to translate test results obtained in animals to the human situation. Because laboratory tests are conducted at relatively high-dose levels to induce measurable rates of response in a small sample of experimental animals, it is often necessary to extrapolate these results to lower doses corresponding more closely to anticipated human exposure levels. Because the test species do not resemble the target species in all respects, it is also necessary to take into account interspecies differences when extrapolating between the animal model used and man. It may also be necessary to infer what would occur if the test compound were administered by a different route of exposure than that actually employed.

Traditional approaches to interspecies extrapolation require the selection of an appropriate scale for dose for which potency remains the same in animals and humans. For example, it may be assumed that potency is invariant across species when dose is expressed in amount per body weight per day or in amount per unit surface area. Empirical evidence to support the use of any particular scale, however, is limited (Mantel and Schneiderman, 1975).

Pharmacokinetic models can be used to describe the fate of chemical substances after they enter the body (Gehring, 1978; Withey, 1984), thereby providing useful information on the formation of the reactive metabolites responsible for the induction of toxic effects in individual tissues. These models can be used to describe the relationship between the dose administered in toxicological tests and the dose delivered to the target tissue (Gehring and Blau, 1977).

One of the first applications of pharmacokinetics to the problem of carcinogenic risk assessment was given by Cornfield (1977), who proposed a simple pharmacokinetic model with an irreversible detoxification step that predicts a threshold in the dose-response curve at steady state. Brown et al. (1978) subsequently noted that the threshold effect apparently associated with this construct may not occur in practice because of the formation of some reactive metabolite during the approach to steady state and the possibility that detoxification may be reversible. Subsequent applications of pharmacokinetics in risk assessment have been reviewed by Hoel (1985).

The pharmacokinetic models employed in these applications generally envisage biological systems that consist of a small number of compartments, with interactions among these compartments described by a corresponding set of differential equations. In contrast to such mathematical pharmacokinetic models, physiological pharmacokinetic models in which the body is considered to consist of a larger number of relevant physiological compartments have also been employed in recent years. Although they require detailed information on physiological and biochemical parameters, such physiologically based pharmacokinetic models have also been used to predict the risk associated with exposure to specific substances such as styrene and methylene chloride (Ramsey and Andersen, 1984; Andersen et al., 1987)

In this paper, we discuss the use of pharmacokinetic data in extrapolating the results of laboratory tests on carcinogenicity between doses, species, and routes of exposure. We begin with a brief review of the current status of carcinogenic risk assessment, and continue with an overview of both mathematical and physiologically based pharmacokinetic models.

The use of mathematical pharmacokinetic models to describe the conversion of the administered dose of a xenobiotic to the dose delivered to the target tissue is then considered, and the implications of saturation effects in the metabolic activation process for extrapolation to low doses are explored. It is noted that the concentration of the active form of the xenobiotic at the target tissue may vary over time, depending on the rates of absorption, distribution, and elimination, as well as the manner in which the exogenous level of exposure may change with time. The implications of these effects on estimates of carcinogenic risks are also considered.

We conclude with a series of examples for which both bioassay and pharmacokinetic data are available that illustrate the concepts discussed in this paper. In particular, data on formaldehyde, vinyl chloride, and methylene chloride are used to compare risk estimates derived on the basis of administered and delivered doses. The use of physiologically based pharmacokinetic models to extrapolate between species and routes is also illustrated by using data on methylene chloride and perchloroethylene. Our conclusions concerning the potential application of pharmacokinetic data in carcinogenic risk assessment are then summarized.

CARCINOGENIC RISK ASSESSMENT

Models of Carcinogenesis

Quantitative estimates of risk are generally derived by using a biologically based mathematical model of the process of carcinogenesis. The Armitage-Doll multistage model is based on the assumption that a tumor will be induced following completion of a k-stage process, in which each stage corresponds to the occurrence of some fundamental biological event such as a mutation at a specific gene locus (Armitage, 1985). This model provides an adequate description of the age-specific tumor incidence rates for a variety of lesions that occur in the human population, with values of k generally in the range of 2 to 7.

Other mathematical models can also be derived under different biological assumptions. Moolgavkar and Venzon (1979) and Moolgavkar and Knudson (1981), for example, postulated that two mutations each occurring at the time of cell division are necessary for a normal cell to become malignant. Unlike the multistage model, this model provides for effects on cell proliferation, which can increase the pool of initiated cells available for malignant transformation, as well as for normal tissue growth. Incorporation of these additional factors yields a two-stage model that is sufficiently flexible to describe the age-specific tumor incidence rates for many forms of cancer. Because information on cell kinetics and tissue growth is not inherent in bioassay results, however, this must be obtained from separate studies.

Risks at Low Doses

To estimate the level of risk associated with exposure to low doses of carcinogenic substances, it is necessary to extrapolate downward from the high doses used in laboratory studies by using an assumed functional relationship between dose and response. Such extrapolations can depend strongly on the mathematical model of carcinogenesis that is employed (Krewski and Van Ryzin, 1981) because of differences in the shape of the dose-response curve implied at low doses. If attention is restricted to models for which the dose-response curve is linear at low doses, however, such differences become slight (Krewski et al., 1984).

A number of arguments can be advanced in support of the hypothesis of low-dose linearity (Krewski et al., 1986; Murdoch et al., in press). For example, both the Armitage-Doll multistage model and the two-stage birth-death-mutation model of Moolgavkar and Knudson (1981) and Moolgavkar and Venzon (1979) lead to low-dose linearity when the transition intensity functions between stages are linear functions of dose. In the absence of specific model assumptions, Crump et al. (1976) demonstrated low-dose linearity in a general class of additive background models in which the test agent is considered to combine in an additive fashion with an effective background dose of that same substance. Low-dose linearity also occurs with pharmacokinetic models in those cases in which all kinetic processes are linear at low doses, provided that the probability of tumor occurrence is proportional to the dose delivered to the target tissue in the low-dose region. The assumption of low-dose linearity in carcinogenic risk assessment is supported by the U.S. Office of Science and Technology Policy, which stated that "when data and information are limited, models or procedures which incorporate low dose linearity are preferred" (OSTP, 1985, p. 10378).

Given the assumption of low-dose linearity, the estimation of carcinogenic risks at low levels of exposure essentially involves estimating the slope of the dose-response curve at the origin. This may be done by fitting the multistage model to the available data and computing upper confidence limits on risk in the low-dose region (Crump and Howe, 1984). Because these confidence limits are linear in the low-dose region, this is referred to as the linearized multistage model. This model is currently used by the Carcinogen Assessment Group of the Environmental Protection Agency for purposes of low-dose extrapolation (Anderson and the Carcinogen Assessment Group of the U.S. Environmental Protection Agency, 1983).

Krewski et al. (1986) have proposed a different approach to linear extrapolation that avoids the need for the specific parametric assumptions underlying the multistage model. This procedure involves linear interpolation between a lower confidence limit on the response rate in the control group and an upper confidence limit on the response rate in any of the exposed groups in which the observed response rate is not significantly different from that in the control group. The minimum (positive) slope then provides an upper confidence limit on the slope at the origin. In most cases, this model-free approach to linear extrapolation provides results similar to those based on the linearized multistage model. This method of robust linear extrapolation can be advantageous, however, in cases in which the multistage model provides a poor fit to the experimental data, as can occur when the dose-response curve tends to rise sharply at low doses and then plateau at higher doses.

PHARMACOKINETIC MODELS

Mathematical Pharmacokinetic Models

The description of complex phenomena that occur *in vivo* following exposure to a toxicant is accomplished with maximum simplification by means of mathematical equations derived from the conception of biological systems that consist of a small number of compartments. In this framework, the temporal relationship of the concentration of the xenobiotic in blood, plasma, urine, or expired air is usually modeled with multiexponential equations consisting of two or three terms (Gibaldi and Perrier, 1975). This provides for the subsequent evaluation of the kinetic coefficients associated with the pharmacokinetic compartmental model.

One of the first studies in which multiexponential equations were used to describe the temporal relationship of blood levels following a single dose involved acetone administered intravenously, per os, per rectum, or following intraperitoneal or subcutaneous injection (Widmark, 1919). Subsequently, mathematical equations were derived to describe the kinetics of accumulation after repeated dosing (Widmark and Tandberg, 1924).

Later developments in mathematical modeling addressed important issues involved in the assessment of pharmacological or toxic response, such as a quantitative evaluation of the nature of the dose-response relationship, the effects of different routes of administration, factors affecting uptake and elimination, and the role of metabolism with respect to the elicited response (Wagner, 1971). Information on the kinetics of the processes of absorption, distribution, metabolism, and excretion of a xenobiotic has also permitted the calculation of such parameters as the interval required between doses to achieve rapid steady-state blood levels or to predict whether a particular exposure regimen results in an accumulation of the test substance in blood or body tissues (Dittert, 1977).

The value of mathematical modeling has also been demonstrated in the interpretation of toxicity tests (Gehring, 1978; Hammer and Bozler,

1977) and in the interspecies extrapolation of data (Gillette, 1976; Reitz et al., 1978). The interpretation and quantitation of precise mechanisms that actually occur within physiological compartments in the body, however, are generally not possible by using mathematical pharmacokinetic models.

Physiologically Based Pharmacokinetic Models

One of the methods of examining the kinetics of absorption, distribution, metabolism, and excretion of a xenobiotic involves a description of the body as a series of relevant physiological compartments arranged as a system of parallel shunts between the venous and arterial blood supplies (Fiserova-Bergerova, 1983; Himmelstein and Lutz, 1979). Such models use basic physiological and biochemical information to determine the temporal relationships of disposition and distribution of an administered dose. Physiological information such as blood flow rates to each compartment, the partition coefficients between the blood and organ tissues, and the volume of the compartment allows differential mass-balance equations to be derived for each compartment. These equations describe the influx, outflow, accumulation, and disappearance of the test substance within the body.

Physiologically based pharmacokinetic (PB-PK) models require information based on the anatomy and physiology of the test animal, the solubility of the test chemical in various organs, and biochemical constants for tissue binding and metabolism in specific organs (Ramsey and Andersen, 1984). Three types of information are essential. First, partition coefficients are required to express the relative solubility of the compound in blood and various tissues in the model. Second, physiological constants are needed for tissue and organ volumes and for blood flow through these. Third, biochemical constants are used to define the rate coefficients for important biotransformation pathways.

Partition coefficients can be obtained by direct measurement in the laboratory (Sato and Nakajima, 1979) or estimated from the partition coefficient between octanol and water. Tissue volumes and blood flows to various tissues or tissue groups can often be found in the existing literature (Sato and Nakajima, 1979). Allometric relationships have also been useful in this application (Adolph, 1949). Biotransformation data are usually obtained from *in vivo* and *in vitro* kinetic studies after the principal metabolic pathways have been identified qualitatively.

Two limiting cases can be invoked to accommodate the pharmacokinetic behavior of most compounds. In the flow-limited model, the cell membrane transport rates are so rapid that the rate of blood flow is the limiting rate that determines the rate of uptake. In the membrane-limited model, on the other hand, cell membrane permeability is low compared with the on the other hand, cell membrane permeability is low compared with the blood perfusion rate. Thus, the rate of uptake to a specific tissue is limited by the rate at which it crosses cellular membranes within that tissue.

The pharmacokinetics of volatile hydrocarbons and halogenated hydrocarbons such as styrene, methylene chloride, bromochloromethane, and dibromomethane are best described by a flow-limited model (Andersen et al., 1984; McDougal et al., 1986; Ramsey and Andersen, 1984). The drugs thiopental, cytarabine, mercaptopurine, sulfobromophthalein, salicylate, lidocaine, adriamycin, digoxin, and methotrexate all show pharmacokinetic behavior that is best described by a membrane-limited model (Himmelstein and Lutz, 1979). Both flow-limited and membrane-limited compartments can exist in the same model.

The grouping of individual organs with similar blood flow, diffusion, and permeability properties into single compartments is flexible. For example, the adrenals, kidney, thyroid, brain, heart, and hepato-portal system are usually pooled into one compartment because their perfusion/volume ratios are relatively high, facilitating their classification as a vessel-rich group (VRG). The muscles and skin are usually placed in the muscle group (MG), while adipose tissue and bone marrow form the fat group (FG). Compartments with a poor vascularity, such as bones, teeth, ligaments, hair, and cartilage, are placed in a vessel-poor group (VPG), (Fiserova-Bergerova, 1983).

PB-PK models not only allow a prediction of the temporal relationships of the tissue concentrations of a xenobiotic postdosing but also accommodate physiological changes such as a change in blood flow or renal clearance for an individual subject. Interspecies and intersubject differences, nonlinear metabolic kinetics, and differences in mechanisms of uptake (for example, as a consequence of altering the route of administration) can also be accommodated by the model.

Effects in different species can be predicted by using PB-PK models developed in one species and scaling those constants such as metabolic rates that are often unknown for the target species. When available, however, information on metabolism in the species of interest should be used.

ADMINISTERED AND DELIVERED DOSES

One of the potential applications of pharmacokinetics to risk assessment involves the use of pharmacokinetic models to determine the effective dose of compound of interest that reaches the target tissue. The dose delivered to the target tissue can then be used in place of the administered dose in an attempt to obtain more accurate estimates of risk. In this section, we explore the effects of using the delivered dose as a surrogate for the administered dose in predicting carcinogenic risks.

Curvilinear Dose Response

Many xenobiotic compounds require some form of metabolic activation to exert their carcinogenic potential (Gehring and Blau, 1977). The amount of the reactive metabolite reaching the target tissue is influenced by the pharmacokinetic laws governing the overall activation process. When one or more steps in this process are saturable, the relationship between the dose d^* delivered to the target tissue and the administered dose d can be nonlinear (Hoel et al., 1983).

To explore the effects of metabolic activation in estimates of low-dose risks, consider the simple pharmacokinetic model for metabolic activation of a particular toxicant shown in Figure 1 (Krewski et al., 1986). Here, exposure occurs at a constant rate d. Once absorbed into the body, the toxin X can be either eliminated or activated to its reactive form Y. The reactive metabolite can then be eliminated or detoxified.

It is assumed that the elimination of both X and Y follows first-order linear kinetics, whereas activation and detoxification are assumed to be enzymatically mediated processes following saturable Michaelis-Menten kinetics. At steady state, the dose d^* of the reactive metabolite reaching the target tissue can be expressed as a function $d^* = f(d)$ of the administered dose.

This simple model can be used to illustrate the impact of saturation effects on the relationship between the dose d^* delivered to the target tissue and the administered dose d (Figure 2). When both activation and detoxification follow linear kinetics, d^* is proportional to d. If the detoxification process is allowed to be saturable, d^* is still proportional to d at low doses. Once the process saturates, however, the delivered dose d^* increases rapidly as a function of d. Conversely, if only the activation step is considered to be saturable, d^* levels off in relation to d as saturation occurs. A combination of both these effects can occur when both activation and detoxification are saturable.

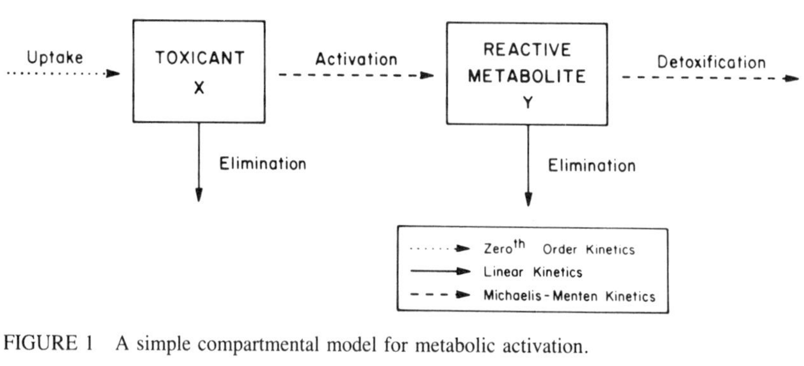

FIGURE 1 A simple compartmental model for metabolic activation.

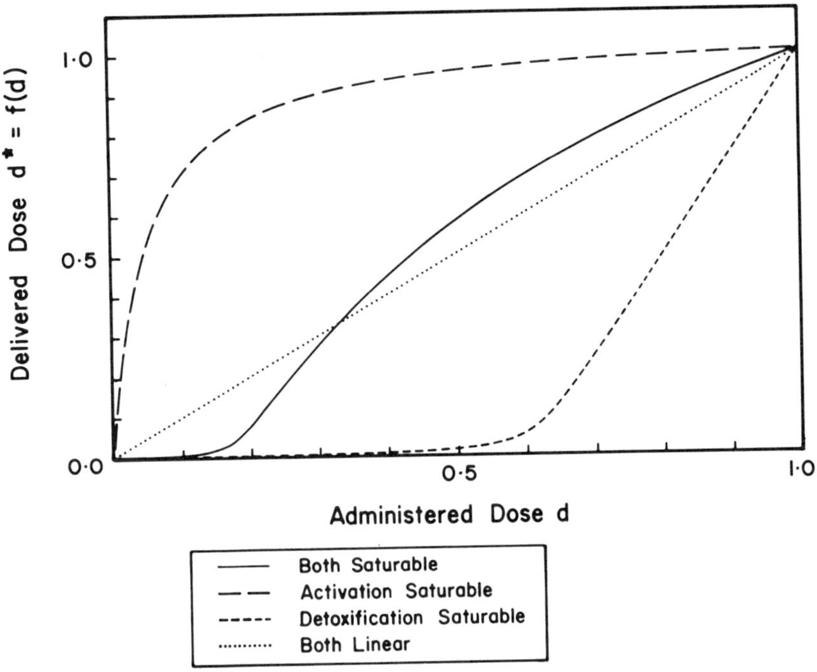

FIGURE 2 Administered and delivered doses with saturable kinetics.

To assess the implications of saturable kinetics on linear extrapolation from high to low doses using the administered dose d, we conducted a computer simulation involving the four cases illustrated in Figure 2. In this investigation, the probability of tumor induction was assumed to satisfy the simple one-stage model $P(d) = 1 - \exp(-\lambda d^*)$, where the delivered dose is in turn a function $d^* = f(d)$ of the administered dose and $\lambda = 1$. Because $P(d) \approx \lambda d^*$ for even moderately large values of d^*, the dose-response curve is essentially linear on the delivered dose scale. Under linear kinetics, d^* is proportional to d, in which case $P(d)$ is also a linear function of d. Because d^* is a nonlinear function of d in the presence of saturation effects, however, the dose-response curve $P(d)$ is a nonlinear function of d when activation or detoxification is saturable.

Using this model, experimental outcomes were generated with 50 observations at each of the four dose levels $d = 0$, 1/4, 1/2, and 1 corresponding to current National Toxicology Program (NTP) recommendations for experimental design (NTP, 1984). (Without the loss of generality, the doses used here represent fractions of the maximum tolerated dose [MTD], the highest dose normally used in a carcinogen bioassay.) For each of the

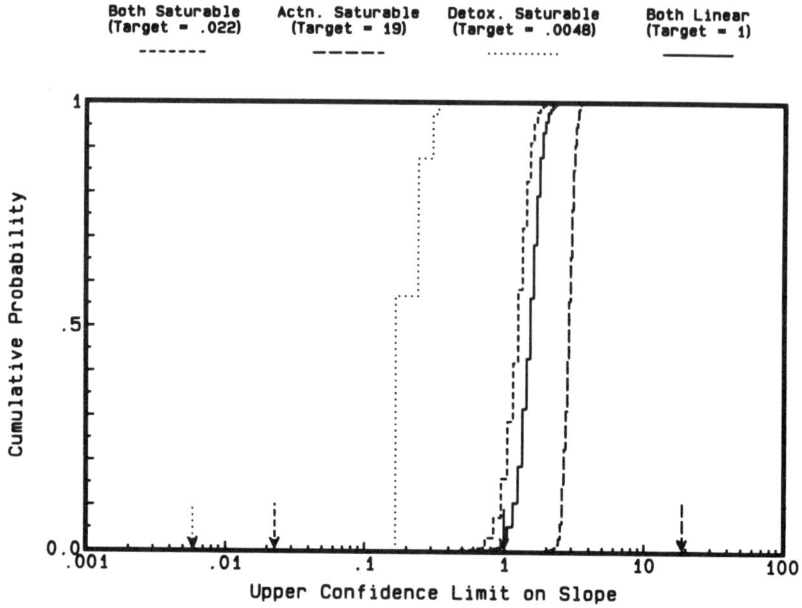

FIGURE 3 Cumulative distribution of upper confidence limits on low-dose slope based on robust linear extrapolation. Arrows indicate target slopes.

four cases depicted in Figure 2, 1,000 sets of experimental data were generated, and a 95% upper confidence limit on the slope of the dose-response curve at the origin was calculated by using robust linear extrapolation. The cumulative distributions of these upper confidence limits on the low-dose slope are shown in Figure 3.

Under linear kinetics, the dose-response curve is essentially linear, and the values are tightly clustered near the true value of unity. In the two cases in which the detoxification process is saturable, however, the upper confidence limits are well in excess of the target slopes. When only the activation process is saturable, on the other hand, all of the values lie below the true slope of 19. This latter result is due to the steepness of the dose-response curve in the low-dose region, and is less pronounced if lower doses are included in the experimental design. For example, inclusion of a group of 50 animals at a dose of 1/32 of the MTD gave simulated confidence limits ranging from 8.5 to 23, as compared with the range of 2.1 to 3.6 in Figure 3. Whittemore et al. (1986) showed that the biases in the confidence limits can be reduced if information about the pharmacokinetics is available.

Time-Dependent Exposure

In the preceding section, the dose delivered to the target tissue was constant over time because exposure was by continuous infusion, with the system assumed to be in steady state. If steady state has not been achieved, or if the level of exposure $d(t)$ varies with time t, the delivered dose $d^*(t)$ is time dependent.

To illustrate this, consider the simple compartmental model shown in Figure 4. Here, exposure can occur at a constant rate d or intermittently at different points in time. Once absorbed, the toxin X can be either eliminated or detoxified. As in the previous model, elimination is presumed to follow linear kinetics, whereas detoxification can be a saturable process. Because metabolic activation of the toxin is not required, our interest is in the concentration $d^*(t)$ of the toxin in the target tissue over time t.

The temporal profile of the dose delivered to the target tissue is shown in Figure 5a for four different exposure scenarios under the assumption that the detoxification process is nonsaturable (Withey and Murdoch, 1987). Curve A presents the case of continuous inhalation exposure to a constant dose d = 60 μg/h over an 8-h working day. In this case, a plateau or steady-state level is achieved following 30 min of exposure, with clearance being essentially complete within 1.5 h following the termination of exposure. The systemic availability, or area under the concentration-time curve, is 19.3 μgh/ml.

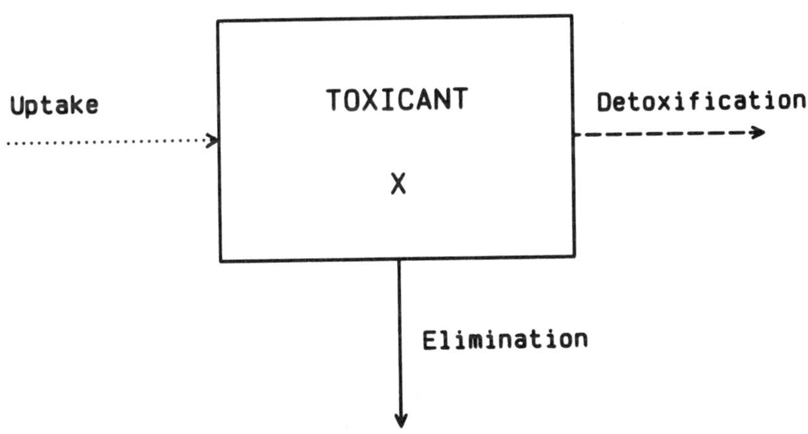

FIGURE 4 A simple compartmental model for a direct-acting toxicant.

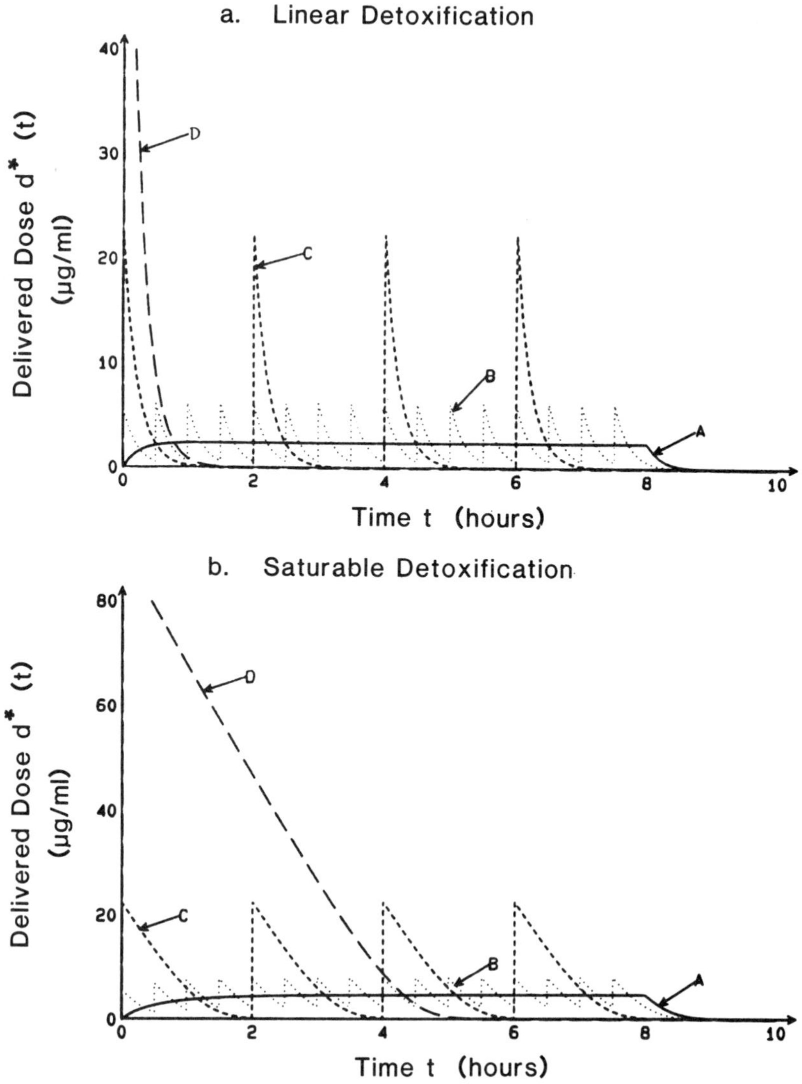

FIGURE 5 Time-dependent exposure patterns with saturable kinetics.

Curve B is derived so as to have the same total administered dose of 480 μg following 16 equal oral doses of 30 μg given every 30 min, assuming that there is instantaneous absorption. Curves C and D are similar, corresponding to exposures every 2 h and once at the beginning of the day, respectively. Because the kinetics are nonsaturable, the sys-

TABLE 1 Systemic Availability (μg h/ml) Under Linear and Saturable Detoxification

Number of Doses	Detoxification (μg h/ml)	
	Linear	Saturable
Continuous	19.3	36.4
16	19.3	38.9
4	19.3	63.6
1	19.3	193.7

temic availability under these three exposure regimens is the same as under the regimen shown in curve A.

Even though the systemic availability is identical under all four exposure scenarios, the peak tissue concentrations vary appreciably. The higher peak exposures associated with the regimens shown in curves C and D may be expected to lead to more severe acute toxic effects than from the lower peak exposures in the regimens shown in curves A and B, as occurs, for example, with hydrogen sulfide (Evans, 1967). As demonstrated in the next section, however, this may not be the case with carcinogenic effects.

Saturation of detoxification or other elimination processes can substantially increase systemic availability (Figure 5b). With saturation effects chosen to approximate the linear detoxification at low doses assumed previously, systemic availability is nearly doubled in the case of continuous infusion with a Michaelis constant of 5 μg/ml (Table 1). With a single oral dose, saturation increases the area under the concentration-time curve by 10-fold.

Risks with Time-Dependent Exposure Patterns

The risk associated with exposure to a carcinogenic substance may depend not only on the total dose experienced over a period of time but also on the pattern of exposure during that period. For example, under a multistage model in which only the first stage is dose dependent, exposure early in life is more effective than exposure later in life (Day and Brown, 1980). Conversely, if only the last stage is dose related, exposure later in life will be of greater concern.

To explore the effects of time-dependent exposure patterns, consider a multistage model in which only one stage is dose dependent. As shown in the Appendix, the carcinogenic risk associated with a delivered dose $d^*(t)$ over $0 < t < T$ will be the same as that for an equivalent constant dose of

$$d_T^* = \int_0^T d^*(t)\, w(t)\, dt, \tag{1}$$

where w (t) is a weighting function depending on which stage of the model is affected. For example, consider a six-stage model, in which only the first stage is dose dependent, and a constant dose d^* over the first quarter of an individual's expected lifetime followed by no exposure thereafter (Figure 6). Because dosing late in life is relatively ineffective, the equivalent constant dose d_T^* is close to the actual dose $d^*(t)$ that reaches the target tissue during the period of exposure. Note that a simple time-weighted average dose

$$\bar{d}_T^* = \int_0^T d^*(t) dt / T \tag{2}$$

is notably lower than the equivalent constant dose, and hence would underestimate the risk corresponding to the actual exposure pattern. Although the degree of such underestimation is in general no more than k-fold, where k denotes the number of stages in a multistage model with a single dose-dependent stage, this result cannot be guaranteed with other

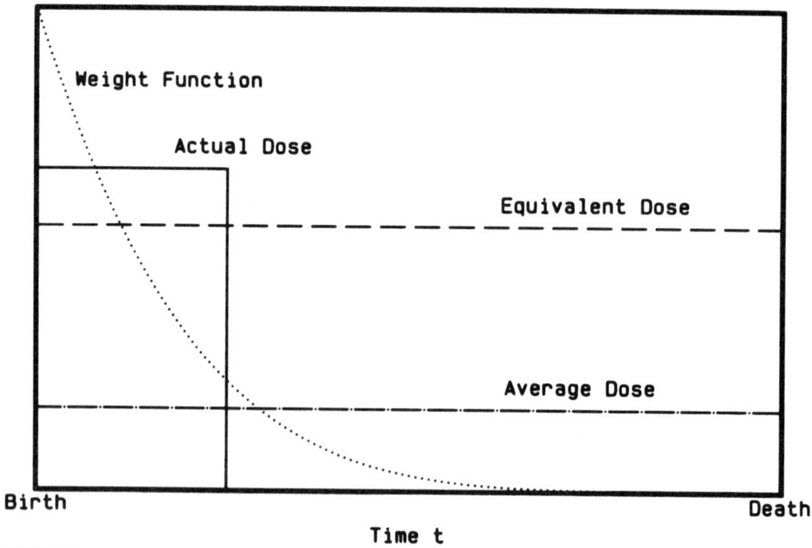

FIGURE 6 Equivalent constant dose with time-dependent exposure (six-stage model, with the first-stage being dose dependent).

models such as the birth-death-mutation model discussed earlier (Murdoch and Krewski, 1987).

These results can be used to evaluate the risks associated with the time-dependent exposure patterns discussed above. Because clearance from the target tissue following cessation of exposure is sufficiently rapid so as to preclude bioaccumulation, systemic availability following repeated daily exposures is obtained simply by multiplying the values in Table 1 by the number of days during which exposure takes place. Because the weight function $w(\cdot)$ can be treated as a constant over a short period of time, such as a single day, it follows from Equation 1 that the equivalent constant dose is nearly proportional to the area under the concentration-time curve for a single day. In the absence of bioaccumulation, systemic availability thus provides a good measure of delivered dose on which estimates of carcinogenic risk can be based.

APPLICATIONS

In the previous sections of this paper, we have explored the application of the delivered dose concept in carcinogenic risk assessment in general terms. In this section, we consider several compounds that have demonstrated carcinogenic potential in long-term bioassays and for which some pharmacokinetic information on delivered dose is also available. These latter data permit a comparison estimate of low-dose cancer risks based on administered and delivered doses, and may also facilitate extrapolation between different species and routes of exposure.

Formaldehyde

Formaldehyde has been shown to induce cancerous lesions in the nasal cavity of both rats and mice, predominantly squamous cell carcinomas (Albert et al., 1982; Swenberg et al., 1980). In addition to these bioassay results, information on delivered dose is available in the form of covalent binding to respiratory mucosal DNA in rats (Casanova-Schmitz et al., 1984). Assuming that covalently bound formaldehyde is associated with neoplastic change, these data can be used to evaluate the impact of the use of the delivered rather than the administered dose in predicting cancer risks at low doses.

For purposes of illustration, consider the data on tumor incidence following inhalation exposure of 6 h/day, 5 days/week for 24 months shown in Table 2, as previously considered in this context by Starr and Buck (1984). Experimental data on covalently bound formaldehyde as reported by Starr and Buck (1984) are also shown. These results indicate that the relationship between the delivered and administered doses is nonlinear,

TABLE 2 Squamous Cell Carcinomas of the Nasal Cavity Induced in Male and Female Rats Exposed to Formaldehyde

Administered Dose (ppm)	Covalently Bound Formaldehyde (nmol/mg)	Tumor Incidence
0.0	0.0	0/160 (0%)
0.3	0.002	
2.0	0.022	0/160 (0%)
5.6	0.212[a]	2/160 (1.3%)
6.0	0.233	
10.0	0.406	
14.3	0.600[a]	87/160 (52.5%)
15.0	0.631	

[a]Based on linear interpolation between adjacent points.

with a marked increase in the level of covalent binding above administered doses of 2 ppm (Figure 7). The dose-response curve for squamous cell carcinomas is also highly nonlinear on either the administered or delivered dose scale, with a large number of tumors observed only in the high-dose group.

Although the relationship between the delivered and administered doses is unknown below 2 ppm, the data are not inconsistent with a linear relationship in this region with a slope of approximately 0.01 nmol/mg/ppm. Under this assumption, robust linear extrapolation on the delivered dose scale leads to an upper confidence limit on the slope of the dose-response curve at the origin of 0.23 mg/nmol, equivalent to a slope of $0.0023\ \text{ppm}^{-1} = 0.01\ \text{nmol/mg/ppm} \times 0.23\ \text{mg/nmol}$ on the administered dose scale. The corresponding value based on linear extrapolation directly on the administered dose scale is $0.0089\ \text{ppm}^{-1}$. Given the assumption that covalent binding provides a relevant measure of delivered dose, the upper limit based on delivered dose is roughly fourfold lower than that based on administered dose.

Vinyl Chloride

Vinyl chloride monomer (VCM) is known to induce a variety of tumors in rats, mice, and hamsters at diverse sites (Maltoni et al., 1981). In one study, male and female Sprague-Dawley rats were exposed to doses up to 10,000 ppm of VCM by inhalation 4 h/day, 5 days/week over a period of 1 year. The occurrence of angiosarcomas of the liver is summarized in Table 3. Note that these results indicate that the incidence of these lesions does not continue to increase at doses above 2,500 ppm.

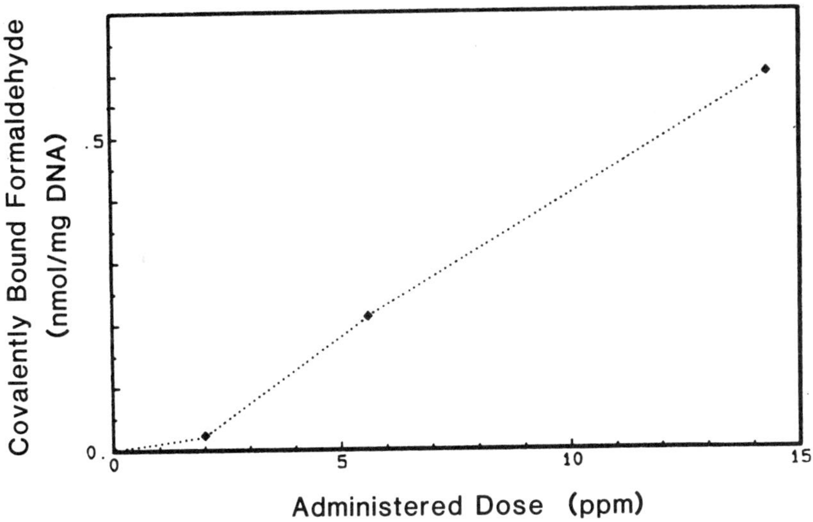

FIGURE 7 Covalently bound formaldehyde in respiratory mucosal DNA in rats.

Although VCM is known to form a number of different metabolites, Gehring et al. (1978) noted that the total amount d^* of VCM metabolized in the liver appeared to depend on the level of exposure d via the Michaelis-Menten equation:

$$d^* = \frac{Vd}{K + d}, \tag{3}$$

TABLE 3 Angiosarcomas Induced in Male and Female Sprague-Dawley Rats Exposed to Vinyl Chloride Monomer

Administered Dose (ppm)	Metabolized VCM (μg/4 h)	Tumor Incidence
0	0	0/461 (0%)
1	17	0/118 (0%)
5	84	0/119 (0%)
10	164	1/119 (0.8%)
25	395	5/120 (4.2%)
50	739	15/354 (4.2%)
100	1,309	1/120 (0.8%)
150	1,761	6/119 (5.0%)
200	2,129	15/120 (10.0%)
250	2,434	3/59 (5.1%)
500	3,413	6/60 (10.0%)
2,500	5,029	13/60 (21.7%)
6,000	5,402	13/59 (22.0%)
10,000	5,520	7/70 (11.7%)

where $V = 8{,}558$ μg/6 h is the estimated maximum velocity, and $K = 336$ ppm is the estimated Michaelis constant at which half the maximum velocity is achieved. Under this model, saturation effects are apparent only at administered doses on the order of several hundred ppm or higher (Figure 8). At lower doses, the amount metabolized per unit of exposure to VCM is approximately $V/K = 25.5$ μg/6 h/ppm.

In this example, the dose-response curve is apparently linear at doses below 200–500 ppm on either the administered or delivered dose scale. Because of this, robust linear extrapolation based on delivered dose, which focuses primarily on data in this dose range, leads to an upper confidence limit on the low-dose slope of 0.00034 (μg/4 h)$^{-1}$, corresponding to 0.0058 ppm^{-1} = 0.00034 (μg/4 h)$^{-1}$ × 25.5 μ/6 h/ppm × (4 h/6 h) on the administered dose scale. This value is virtually identical to the value of 0.0056 ppm^{-1} obtained by using the administered dose.

Methylene Chloride and Perchloroethylene

Andersen et al. (1987) used a PB-PK model to assess the risk to humans of exposure to methylene chloride, also know as dichloromethane or DCM. Mice exposed to 2,000 and 4,000 ppm of DCM by inhalation 6 h/day, 5 days/week, demonstrated an excess of hepatocellular adenomas/carcinomas as compared with unexposed controls (NTP, 1986). An earlier study conducted by the National Coffee Association (EPA, 1985), in which mice were exposed to DCM in drinking water at concentrations leading

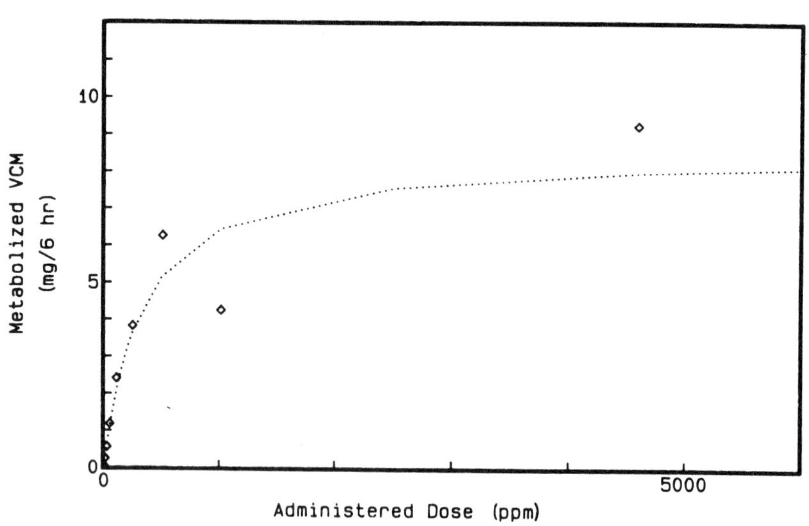

FIGURE 8 Metabolized vinyl chloride monomer in rats.

to exposures of up to 250 mg/kg/day, showed no increase in the incidence of liver tumors (Table 4).

The flow-limited PB-PK model shown in Figure 9 has been employed by Andersen et al. (1987) to explore this apparent inconsistency between routes of exposure and to allow extrapolation to humans. The paths on the right-hand side of the figure represent flows of arterial blood, while those on the left-hand side represent flows of venous blood. The lung has been divided into a first compartment in which gas exchange takes place, and a second compartment in which metabolism occurs. DCM administered in drinking water is assumed to pass through the gastrointestinal (GI) tract directly into the liver.

There are two paths involved in the metabolism of DCM: the saturable mixed-function oxidase (MFO) path, and the essentially linear glutathione-*S*-transferase (GST) path. Because the mechanism of carcinogenesis of DCM is not known, three possible dose metameters were considered: the parent compound itself, the rate of metabolism of DCM in the MFO path, and the rate of metabolism of DCM in the GST path. The latter two were used as surrogates for metabolite formation.

This model can be used to derive estimates of concentration of DCM and the rates of each of the two paths in the liver (Table 4). It can be seen that the rate of the MFO path is comparable in the two studies, whereas the other two measures of delivered dose are much higher in the inhalation experiment than in the drinking water experiment, as are the corresponding tumor incident rates. This can be taken as evidence that the MFO path is not responsible for tumor induction. Because the rate of delivery of the parent compound is proportional to the rate of the GST path, effects of the parent compound cannot be distinguished from those

TABLE 4 Hepatocellular Adenomas/Carcinomas Induced in Female Mice Exposed to Methylene Chloride

Route of Administration (dose units)	Administered Dose	Delivered Dose MFO Path (g/liter/day)	GST Path (g/liter/day)	Parent Compound (mg h/liter)	Tumor Incidence
Drinking water (mg/kg/day)	0	0.0	0.000	0.0	6/100 (6%)
	60	1.3	0.003	1.3	4/100 (4%)
	125	2.8	0.007	2.9	2/50 (4%)
	185	4.0	0.011	4.6	5/50 (10%)
	250	5.4	0.016	6.4	3/50 (6%)
Inhalation (ppm)	0	0.0	0.00	0	3/50 (6%)
	2,000	3.6	0.85	360	16/48 (33%)
	4,000	3.7	1.80	770	40/48 (83%)

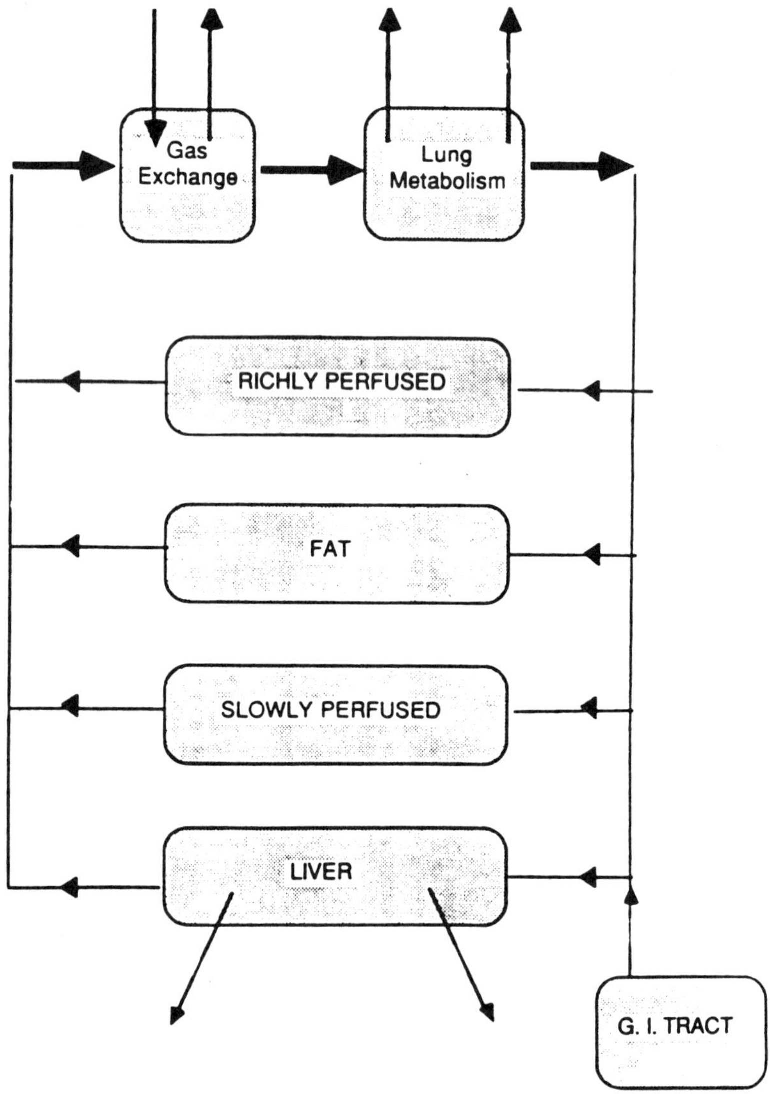

FIGURE 9 A physiologic pharmacokinetic model for methylene chloride (Andersen et al., 1987).

of the GST path. Nonetheless, on the basis of other biochemical considerations, Andersen et al. (1987) concluded that the GST surrogate is the most appropriate predictor of tumor incidence.

Based on the GST path, the delivered dose in the high-dose groups is approximately 100 times higher in the inhalation study than in the drinking water study, as compared with the tenfold difference indicated by using the administered doses. This is due to the fact that the doses in the inhalation study were all well above the level necessary to saturate the MFO path, leading to a high rate of activity in the GST path (Figure 10), whereas those in the drinking water study were not.

An upper confidence limit on the risk at low doses for female mice can be calculated by using robust linear extrapolation based on the delivered dose data from the inhalation study in Table 4 to be 0.56 (g/liter/day)$^{-1}$. At low doses, the rate of the GST path is approximately 0.036 mg/liter/day/ppm, so that the low-dose slope on the administered dose scale is 2.0×10^{-5} ppm^{-1}. Working directly on the administered dose scale, the low-dose slope value would be 2.4×10^{-4} ppm^{-1}, which is 12 times higher.

To extrapolate to humans, we follow the results of Andersen et al. (1987) and assume that the GST surrogate has the same potency across species on a body weight scale. Calculation of the amount of GST surrogate formed in humans requires determination of the human values of the model parameters by allometric scaling or, preferably, by experimental measurement. In the present example, Andersen et al. (1987) calculated many of the physiological constants and the rates of the GST path allometrically, but determined the metabolic constants involved in the saturable MFO pathway experimentally. The calculation also requires specification of the dosing regimen. At low doses, inhalation of methylene chloride for 6 h

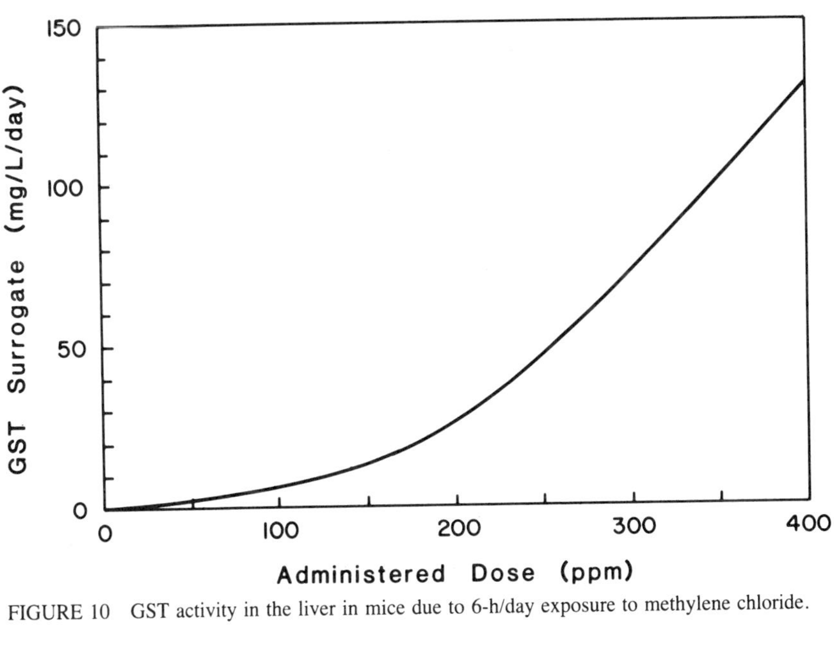

FIGURE 10 GST activity in the liver in mice due to 6-h/day exposure to methylene chloride.

daily results in GST rate per part per million of about 0.012 mg/liter/day/ppm, one-third of the rate for mice. Thus, the 95% upper confidence limit on the low-dose slope for humans is 6.7×10^{-6} ppm^{-1}, which is 36 times less than the corresponding value for mice calculated on the basis of the administered dose.

It is worth noting that this ratio does not apply to all dosing routes. For example, when drinking water exposure is modeled as continuous infusion to the liver, human GST activity per unit dose is 0.16 (mg/liter/day)/(mg/kg/day). This is about 3 times higher than that of mice, rather than 3 times lower as calculated above for inhalation exposure.

C. Chen and J. N. Blancato (this volume) used a similar physiological pharmacokinetic model to assess the risks of exposure to perchloroethylene (PCE). Their model had no lung metabolism compartment, because one path in the liver was assumed to account for all metabolism and to produce the active metabolite. Its rate was used as the dose surrogate. Species equivalence on surface area, body weight, liver volume, and air concentration scales were all considered. In each case, the use of the metabolized dose produced estimates of human risks 5–10 times lower than calculations based on the administered dose.

SUMMARY AND CONCLUSIONS

The process of carcinogenic risk assessment based on the results of toxicological experiments conducted in the laboratory involves certain assumptions, such as that of low-dose linearity when extrapolating from high to low doses. By using a simple mathematical pharmacokinetic model for metabolic activation in which the probability of tumor induction is proportional to the delivered dose, it was shown with a computer simulation that saturation effects in metabolic activation resulting in a curvilinear dose response can have an impact on estimates of low-dose risk obtained by linear extrapolation. In particular, saturation of detoxification processes can result in an appreciable overestimation of risk, whereas saturation of activation processes can lead to some underestimation of risk. The most accurate estimates of risk are obtained with a linear dose response.

These effects were further evaluated by using existing data on formaldehyde and vinyl chloride monomer, using covalent binding to DNA in the nasal mucosa and the level of metabolism in the liver as possible measures of delivered dose, respectively. With formaldehyde, low-dose risks predicted on the basis of delivered dose were a factor of 4 lower than those obtained by using the administered dose level, because of the nonlinear relationship between the delivered and administered doses within the experimental dose range. Because VCM metabolism is nearly proportional to the level of exposure at low to moderate doses, estimates of low-dose risk based on either the delivered or administered dose scale were virtually identical.

The dose delivered to the target tissue may vary with time, depending on both the exposure regimen and metabolic activation. By using a simple mathematical compartmental model based on linear kinetics, it was demonstrated that exposure regimens with different dosing intervals having the same systemic availability lead to different peak concentrations in the target tissue. In the absence of bioaccumulation, however, it was noted that under the multistage model of carcinogenesis, the area under the concentration-time curve is a better predictor of carcinogenic risk than are peak concentrations.

Carcinogenic risk assessment can also require extrapolations between different routes of exposure and from the animal model used to humans. This can be done with PB-PK models. With methylene chloride and perchloroethylene, it was noted that estimates of low-dose risks in humans using predictions of dose delivered to the target tissue based on such models can be substantially lower than traditional estimates based on the use of the administered dose level.

In conclusion, pharmacokinetic models can be used to obtain more accurate estimates of risk at low doses through the use of the dose delivered to the target tissue as a surrogate for the administered dose in toxicological studies, particularly when the response of interest is roughly proportional to the delivered dose. Predictions of the internal tissue dose for other routes of exposure can be obtained by using PB-PK models, provided that measurements of the physiological and biochemical constants associated with the different routes are available. Extrapolation between species can also be facilitated by using physiological models to predict the delivered dose in the species of interest. In those cases in which all of the relevant model parameters cannot be measured directly in humans, however, these must be obtained by scaling the corresponding values in animals. This approach to interspecies extrapolation is of most use when the dose-response curves for animals and humans are comparable when expressed in terms of delivered dose.

ACKNOWLEDGMENTS

We thank Dr. Richard Reitz for kindly providing the data shown in Figure 10, and for helpful comments on the original draft of this article.

REFERENCES

Adolph, E. F. 1949. Quantitative relations in the physiological constitutions of mammals. Science 109:579–585.

Albert, R., A. Sellakumar, S. Laskin, M. Kuschner, N. Nelson, and C. Snyder. 1982. Gaseous formaldehyde and hydrogen chloride induction of nasal cancer in the rat. J. Nat. Cancer Inst. 68:597–603.

Andersen, M. E., R. L. Archer, H. J. Clewell, and M. G. MacNaughton. 1984. A physiological model of the intravenous and inhalation pharmacokinetics of three dihalomethanes—CH_2Cl_2, CH_2BrCl, CH_2Br_2—in the rat. Toxicologist 4:111.

Andersen, M. E., H. J. Clewell III, M. L. Gargas, F. A. Smith, and R. H. Reitz. 1987. Physiologically based pharmacokinetics and the risk assessment process for methylene chloride. Toxicol. App. Pharmacol. 87:185–205.

Anderson, E. L., and the Carcinogen Assessment Group of the U.S. Environmental Protection Agency. 1983. Quantitative approaches in use to assess cancer risk. Risk Analysis 3:277–295.

Armitage, P. 1985. Multistage models of carcinogenesis. Environ. Health Perspect. 63:195–201.

Bickis, M., and D. Krewski. 1985. Statistical design and analysis of the long-term carcinogenicity bioassay. Pp. 125–147 in Toxicological Risk Assessment, Vol. I, Biological and Statistical Criteria, D. B. Clayson, D. Krewski, and I. C. Munro, eds. Boca Raton, Fla.: CRC Press.

Brown, C. C., T. R. Fears, M. H. Gail, M. Schneiderman, R. E. Tarone, and N. Mantel. 1978. Letter to the editor and reply by J. Cornfield. Science 202:1105–1108.

Casanova-Schmitz, M., T. B. Starr, and H. d'A. Heck. 1984. Differentiation between metabolic incorporation and covalent binding in the labeling of macromolecules in the rat nasal mucosa and bone marrow by inhaled ^{14}C- and ^{3}H-formaldehyde. Toxicol. Appl. Pharmacol. 76:26–44.

Cornfield, J. 1977. Carcinogenic risk assessment. Science 198:693–699.

Crump K. S., and R. B. Howe. 1984. The multi-stage model with a time-dependent dose pattern: Applications to carcinogenic risk assessment. Risk Analysis 4:163–176.

Crump, K. S., D. G. Hoel, C. H. Langley, and R. Peto. 1976. Fundamental carcinogenic processes and their implications for low dose risk assessment. Cancer Res. 36:2973–2979.

Davis, N. R., and W. W. Mapleson. 1981. Structure and quantification of a physiological model of the distribution of injected agents and inhaled anesthetics. Br. J. Anesthesiol. 53:399–405.

Day, N. E. 1985. Epidemiological methods for the assessment of human cancer risk. Pp. 3–15 in Toxicological Risk Assessment, Vol. II, General Criteria and Case Studies, D. B. Clayson, D. Krewski, and I. C. Munro, eds. Boca Raton, Fla.: CRC Press.

Day, N. E., and C. C. Brown. 1980. Multistage models and the primary prevention of cancer. J. Nat. Cancer Inst. 64:977–989.

Dittert, L. W. 1977. Pharmacokinetic prediction of tissue residues. J. Toxicol. Environ. Health 2:735–756.

EPA (U.S. Environmental Protection Agency). 1985. Health assessment document for dichloromethane (methylene chloride). Final Report. EPA/600/8-82/004F. Washington, D.C.: U.S. Environmental Protection Agency.

Evans, C. L. 1967. The toxicity of hydrogen sulphide and other sulphides. Q. J. Exp. Physiol. 52:231–248.

Fiserova-Bergerova, V. 1983. Physiological models for pulmonary administration of inert vapors and gases. In Modeling of Inhalation Exposure to Vapors: Uptake, Distribution and Elimination, Vol. 1, V. Fiserova-Bergerova, ed. Boca Raton, Fla.: CRC Press.

Gehring, P. J. 1978. Chemobiokinetics and metabolism. In Principles and Methods for Evaluating the Toxicity of Chemicals, Part I. Geneva: World Health Organization.

Gehring, P. J., and G. E. Blau. 1977. Mechanisms of carcinogenesis: Dose response. J Environ. Pathol. Toxicol. 1:163–179

Gehring, P. J., P. G. Watanabe, and C. N. Park. 1978. Resolution of dose-response toxicity data for chemicals requiring metabolic activation: Example—vinyl chloride. Toxicol. Appl. Pharmacol. 44:581–591.

Gibaldi, M., and D. Perrier. 1975. Pharmacokinetics. New York: Marcel Dekker.

Gillette, J. R. 1976. Application of pharmacokinetic principles in the extrapolation of animal data to human. Clin. Toxicol. 9:709–721.

Hammer, R., and G. Bozler. 1977. Pharmacokinetics as an aid in the interpretation of toxicity tests. Arzneim. Forsch. 27:555–557.

Himmelstein, K. J., and R. J. Lutz. 1979. A review of the applications of physiologically based pharmacokinetic modeling. J. Pharmacokinet. Biopharm. 7:127–145.

Hoel, D. G. 1985. Incorporation of pharmacokinetics in low-dose risk estimation. Pp. 205–214 in Toxicological Risk Assessment, Vol. I, Biological and Statistical Criteria, D. B. Clayson, D. Krewski, and I. C. Munro, eds. Boca Raton, Fla.: CRC Press.

Hoel, D. G., N. L. Kaplan, and M. W. Anderson. 1983. Implication of nonlinear kinetics on risk estimation in carcinogenesis. Science 291:1032–1037.

Krewski, D., and C. Brown. 1981. Carcinogenic risk assessment: A guide to the literature. Biometrics 37:353–366.

Krewski, D., and J. Van Ryzin. 1981. Dose response models for quantal response toxicity data. Pp. 201–231 in Statistics and Related Topics, M. Csorgo, D. Dawson, J. N. K. Rao, and E. Saleh, eds. Amsterdam: North-Holland.

Krewski, D., C. Brown, and D. Murdoch. 1984. Determining "safe" levels of exposure: Safety factors or mathematical models? Fund. Appl. Toxicol. 4:S383–S394.

Krewski, D., D. Murdoch, and A. Dewanji. 1986. Statistical modelling and extrapolation of carcinogenesis data. Pp. 259–282 in Modern Statistical Methods in Chronic Disease Epidemiology, S. H. Moolgavkar and R. L. Prentice, eds. New York: Wiley-Interscience.

Maltoni, C., G. Lefemine, A. Ciliberti, G. Cotti, and D. Carretti. 1981. Carcinogenicity bioassays of vinyl chloride monomer: A model of risk assessment on an experimental basis. Environ. Health Perspect. 41:3–29

Mantel, N., and M. A. Schneiderman. 1975. Estimating safe levels, a hazardous undertaking. Cancer Res. 35:1379–1386.

McDougal, J. N., G. W. Jepson, H. J. Clewell, M. G. MacNaughton, and M. E. Andersen. 1986. A physiological pharmacokinetic model for dermal absorption of vapours in the rat. Toxicol. Appl. Pharmacol. 85:286–294.

Moolgavkar, S. H., and A. G. Knudson, Jr. 1981. Mutation and cancer: A model for human carcinogenesis. J. Natl. Cancer Inst. 66:1037–1052.

Moolgavkar, S. H., and D. J. Venzon. 1979. Two-event models for carcinogenesis. Incidence curves for childhood and adult tumors. Math. Biosci. 47:55–77.

Murdoch, D., and D. Krewski. 1987. Carcinogenic Risk Assessment with Time-Dependent Exposure. Technical Report No. 106. Laboratory for Research on Statistics and Probability. Ottawa: Carleton University.

Murdoch, D. J., D. Krewski, and K. S. Crump. In press. Mathematical models of carcinogenesis. Pp. 61–89 in Cancer Modelling, J. R. Thompson and B. W. Brown, eds. New York: Marcel Dekker.

NTP (National Toxicology Program). 1984. Report of the NTP Ad Hoc Panel on Chemical Carcinogenesis Testing and Evaluation. Washington, D.C.: U.S. Department of Health and Human Services.

NTP (National Toxicology Program). 1986. Toxicology and Carcinogenesis Studies of Dichloromethane (Methylene Chloride) in F344/N Rats and $B6C3F_1$ Mice. CAS No. 75-09-2. Bethesda, Md.: National Institutes of Health.

OSTP (Office of Science and Technology Policy). 1985. Chemical carcinogens: A review of the science and its associated principles. Fed. Regist. 50:10372–10442.

Rall, D. P. 1969. Difficulties in extrapolating the results of toxicity studies in laboratory animals to man. Environ. Res. 2:360–367.

Ramsey, J. C., and M. E. Andersen. 1984. A physiologically based description of inhalation pharmacokinetics of styrene in rats and humans. Toxicol. Appl. Pharmacol. 73:159–175.

Reitz, R. H., P. J. Gehring, and C. N. Park. 1978. Carcinogenic risk estimation for chloroform: An alternative to EPA's procedure. Food Cosmet. Toxicol. 16:511–514.

Sato, A., and T. Nakajima. 1979. Partition coefficients of some aromatic hydrocarbons and ketones in water, blood and oil. Br. J. Ind. Med. 36:231–234.

Starr, T. B., and R. B. Buck. 1984. The importance of delivered dose in estimating low-dose cancer risk from inhalation exposure to formaldehyde. Fund. Appl. Toxicol. 4:740–753.

Swenberg, J. A., W. D. Kerns, R. E. Mitchell, E. J. Gralla, and K. L. Pavkov. 1980. Induction of squamous cell carcinomas of the rat nasal cavity by inhalation exposure to formaldehyde vapor. Cancer Res. 40:3398–3402.

Wagner, J. G. 1971. Pp. 239–241 in Biopharmaceutics and Relevant Pharmacokinetics, 1st ed. Hamilton, Ill.: Drug Intelligence Publications.

Whittemore, A. S., S. C. Grosser, and A. Silvers. 1986. Pharmacokinetics in low dose extrapolation using animal cancer data. Fund. Appl. Toxicol. 7:183–190.

Widmark, E. M. P. 1919. Studies in the concentration of indifferent narcotics in blood and tissues. Acta. Med. Scand. 52:87–164.

Widmark, E. M. P., and J. Tandberg. 1924. The limitations for the accumulation of indifferent narcotics. Theoretical calculation. Biochem. J. 147:358–369.

Withey, J. R. 1984. Pharmacokinetics and metabolism. Chapt. IV., pp. 36–45 in Current Issues in Toxicology: Interpretation and Extrapolation of Chemical and Biological Carcinogenicity Data to Establish Human Safety Standards. New York: Springer-Verlag.

Withey, J. R., and D. Murdoch. 1987. Application of pharmacokinetics in risk assessment for pesticides. Pp. 569–572 in R. Greenhalgh and T. R. Roberts, eds. Pesticide Science and Biotechnology. Philadelphia: Blackwell Scientific.

APPENDIX: THE MULTISTAGE MODEL AND TIME-DEPENDENT DOSING

Under the multistage model, the probability of a tumor occurring by time t is given by:

$$P(t) = 1 - \exp[-H(t)], \tag{A-1}$$

where

$$H(t) = \int_0^t \int_0^{u_k} \ldots \int_0^{u_2} \prod_{i=1}^{k} \lambda_i(t)du_1 \ldots du_k \tag{A-2}$$

denotes the cumulative hazard function and $\lambda_i(t)$ is the transition intensity function for stage $i = 1, \ldots k$ (Crump and Howe, 1984). To accommodate exposure to a particular xenobiotic, the intensity functions are often assumed to be linearly related to the dose $d(t)$ at time t, with

$$\lambda_i(t) = a_i + b_i d(t), \tag{A-3}$$

where $(a_i, b_i > 0)$. For constant exposure $d(t) = d$, this reduces to the usual form of the multistage model with:

$$H(t) = \frac{t^k}{k!} \prod_{i=1}^{k} (a_i + b_i d). \tag{A-4}$$

To explore the effects of time-dependent dosing under the multistage model, consider the simplest case in which only the rth stage is dose dependent. Because $b_i = 0$ for $i \neq r$, we have:

$$H(t) = a(t) + \mathrm{b}(t)\, d_t, \tag{A-5}$$

where

$$a(t) = t^k(\Pi_{i=1}^k a_i)/k!,$$

$$b(t) = (b_r/a_r)a(t), \text{ and}$$

$$d_t = E[d(U_{r:k})] = \int_0^t d(u)w(u; r, k - r + 1, t)du. \tag{A-6}$$

Here, $w(u; r, k - \mathrm{r} + 1, t)$ represents the marginal density of $U_{r:k}$, the rth order statistic in a sample of size k from the uniform distribution on $[0, t]$. This density is given by:

$$w(u; r, k - r + 1, t) = \frac{u^{r-1}(t - u)^{k-r}}{B(r, k - r + 1)t^k}, \tag{A-7}$$

$(0 < u < t)$, with $B(\cdot,\cdot)$ denoting the beta function (see Murdoch and Krewski, 1987, for further details). It follows from Equation A-5 that d_t represents that fixed dose which, if given continuously from time 0 to time t, would produce the same cumulative hazard at time t as the variable dosing regimen $d(t)$. It follows from Equation A-6 that the use of a constant time-weighted average dose

$$\bar{d}_t = \int_0^t d(u)du/t \qquad \text{(A-8)}$$

will be valid when $k = 1$, because $w(u;\ 1,\ 1,\ t) = 1/t$ for $0 < u < t$. When $k > 1$, however, the use of $\bar{d}_t$ in place of d_t does not in general lead to the same cumulative hazard at time t as that which accrues under the variable dosing regimen $d(t)$.

When only one stage is dose dependent, it follows from Equation A-5 that the excess risk is proportional to d_t. Because this function is a weighted average of the values of $d(u)$ over $0 < u < t$, the ratio R of the excess risk under variable dosing as compared with that based on a constant average dose is at most the maximum value of the density $w(u;\ r,\ k - r + 1,\ t)$ divided by its average value $1/t$. This ratio can be written as:

$$R = \frac{k!\ (r - 1)^{r-1}\ (k - r)^{k-r}}{(r - 1)!(k - r)!(k - 1)^{k-1}} \leqslant k\cdot \qquad \text{(A-9)}$$

Thus, the excess risk under variable dosing is at most k times the risk under constant dosing.

PART VIII
Perspectives

Perspectives

The objective of risk assessments is to provide society with estimates of the likelihood of toxic injury in human populations as a consequence of exposure to naturally occurring or man-made chemicals. Fundamentally, the assessment procedure involves the development of a hypothetical dose-response curve for an untested species (humans) on the basis of exposure studies in laboratory animals followed by projection of the "curve" to estimate the levels of exposure that may be considered safe.

In practice, the risk assessor is faced with the problem of predicting the likelihood of injury to humans exposed to low concentrations of materials from data gathered in studies of animals receiving high doses, often by a different route of administration. The complexity of this three-way extrapolation—dose, species, and route—presents a formidable challenge. Not the least of the risk assessor's concerns is the problem of relating an applied "dose" to the "dose" of the active material in target tissues. Traditionally, "dose" has been understood to mean the amount of test material administered to a test animal per unit time. As discussed extensively by various speakers in this workshop, the toxic response of an animal is related to exposure through two sets of factors: *pharmacokinetic factors*, which govern the concentration in target organs and the interaction of the material with the site of action; and *pharmacodynamic factors*, which govern the sequence of events that result from the interaction and lead to the manifestation of the toxic response. Clearly, if it were accepted that the severity of a toxic response depends on the concentration or the area under the curve (AUC) of the toxicant at its site of action, "dose" could be defined in terms of these pharmacokinetic parameters. Extrapolation across dose, species, and route could then be performed based on the actual concentration or AUC of the toxicant at the site of action, rather than on the applied dose. The task of the assessor would then be greatly eased, and the prediction would be much improved.

The rapid development of computer hardware and software during the last decade has markedly increased the diversity of the mathematical tools applicable to the pharmacokinetics of environmental chemicals. In particular, computers have permitted the application of physiologically based pharmacokinetic (PB-PK) models to define "dose" in pharmacokinetic terms. As was amply demonstrated in the workshop, relevant PB-PK data can be used to reduce uncertainty in extrapolation and risk assessment. It is now theoretically possible to simulate concentrations of virtually any xenobiotic substance or any of its metabolites in any organ, regardless of the complexity of the systems. PB-PK analysis has particular value when absorption of inhaled or ingested material might be incomplete and when capacity-limited enzymatic processes can cause nonlinearity between ingested dose and AUC of the parent material in blood and various organs. An improved definition of dose is especially useful when toxicity results directly from the action of a parent material. Moreover, allometric methods for extrapolation between species are most likely to be valid when such toxic materials are eliminated predominantly by exhalation or urinary excretion.

It was also made clear in the workshop that the potential power of the approach is matched by potential pitfalls and hence that simplification of extrapolation of dose, species, and route must be based on an understanding of both general principles of toxicology and specific information on the mechanism of the toxicity under consideration. The pharmacokinetics of toxic metabolites can be much more complex than those of the parent substance. Various speakers made it clear that in such cases, supplementary information on the mechanisms of action must be provided before pharmacokinetics can be usefully applied to risk assessment. Unfortunately, in current practice such critical information is often missing from the results of pharmacokinetic studies presented to risk assessors, and the usefulness of pharmacokinetic studies is correspondingly limited to estimation of the bioavailability of the parent material. Problems in applying pharmacokinetics to risk assessment tend to fall into three broad categories: inadequacies of the present data base, incomplete definition or understanding of underlying processes, and inherent uncertainties in the modeling process.

Inadequacies of the data base are largely correctable with present technology. As emerged in many of the discussion sessions of the workshop, refinement and adaptation of the protocols for both pharmacokinetic and toxicologic procedures will be needed to allow fruitful comparisons, especially when chronic administration of the test compound might induce or impair the enzymes responsible for its elimination. Because such changes are not always predictable, it would be advantageous to incorporate pharmacokinetic studies into the toxicologic protocols and, in the case of chronic exposure, to repeat the pharmacokinetic studies at intervals throughout the exposure period. If it is not feasible to integrate the pharmacokinetic and toxicologic studies directly, it will be important to ensure that the pharmacokinetic studies are done under toxicologically relevant conditions. It will also be important to develop protocols for toxicity studies that would show whether toxicity is caused solely by the parent substance or by its metabolites. Much work remains to be done to collect and improve estimates of kinetic parameters and to determine how they vary within and among human and animal populations. Answers to many of the questions in this category can be expected to accumulate as experience with PB-PK risk assessment grows and as the integration between pharmacokinetic and toxicologic studies improves.

Appreciation of problems in understanding the underlying processes can also be expected to grow as experience in the use of PB-PK modeling in risk assessment accumulates. Some of the problems can be predicted from present knowledge; others will appear as our sophistication increases. Knowledge of the kinetics of the interaction of biologically active substances with putative sites of action and of how physiologic processes modify the interactions is sparse. As it increases, we will probably be able to improve pharmacokinetic and risk-assessment equations. In addition, the role of pharmacodynamic factors needs to be defined. Examples of problems in this category include differences in longevity between test animals and human species, differences in DNA repair rates, differences in transformation frequencies, and differences in mechanisms of "promotion." Those problems cannot be addressed by the use of present procedures for risk assessment and have necessarily been largely ignored. Research stimulated by the opportunities and challenges posed by PB-PK assessment, however, could reduce uncertainties associated with the concept of "dose" and thereby allow us to focus on the roles of pharmacodynamic factors, which then may be incorporated into more detailed models. A lack of correlation between the concentration in target tissues and response would be of special interest to toxicologists, because it would disclose major interspecies or intraspecies differences in the pharmacodynamics of the toxicity, which may profoundly affect the validity of risk assessment.

As to inherent uncertainties in the modeling process, some factors are unlikely to be completely validated soon. Biologic systems are highly complex, and numerous factors influence the effective rate constants that describe the metabolic processes. Modeling consists of simplifying the system by selecting steps whose estimated rate constants will allow the best fit of a mathematical model to the biologic reality. Clearly, the more detailed the model, the greater will be the theoretical utility, but also the more numerous will be the sources of uncertainty and the data necessary to estimate values of the parameters of the models. In practice, a compromise between complexity and simplicity must be reached. It is crucial to develop procedures for identifying the critical parameters in these complex models (i.e., by performing sensitivity analysis) and to develop appropriate statistical tools for estimating the parameters and the confidence limits of these estimates in complex models. The latter task is complicated. Estimates of some of the parameters can be obtained from specific experimental data (whose experimental error can be directly estimated), but others are based on collective experience from a variety of sources (such as Bayesian input as to the prior "believability" of the values estimated by the computer during optimization). Several participants dealt with these topics and identified them as important for the development of pharmacokinetic modeling in risk assessment.

Perhaps the most important idea that emerged in the workshop was the need to integrate mechanistic information with pharmacokinetic studies and with the risk assessment process. For example, it is essential to know whether toxicity results from the action of a parent material or from its metabolites. If the ultimate toxicant is a metabolite, it is important to obtain information on the biologic disposition and stability of the metabolite and on the pharmacokinetic interrelationships among organs of formation, organs of elimination, and target sites. Availability of appropriate mechanistic information markedly increases the power of PB-PK analysis and the scientific credibility of extrapolation across doses, species, and routes of administration. As illustrated by many participants, the incorporation of toxic active/reactive metabolites into PB-PK models complicates the analysis. But the insight that can be gained from the integration of knowledge of the pharmacokinetics and the mode of action of toxic metabolites will focus attention on the dominant factors that govern the manifestation of different toxicities, and that attention will in turn lead to the identification of other important risk factors. Appreciation of the roles of such risk factors is likely to lead to recognition of highly sensitive subpopulations that are not discernible by present testing processes.

The workshop brought together a large group of people from diverse backgrounds who share an interest in the application of PB-PK in risk assessment. The size of the audience and the extent of discussions, both formal and informal, indicate a high degree of interest in the critical evaluation of present and future roles of PB-PK analysis in risk assessment. Clearly, pharmacokinetic studies will have a large role in the elucidation of mechanisms of toxicity and of the dominant factors that account for intraspecies and interspecies differences in the incidence of toxic effects. Even in the absence of knowledge of the mechanisms of toxicity, pharmacokinetic studies will be essential for assessing the bioavailability of materials subjected to toxicity tests.

As was well illustrated by this workshop, PB-PK analysis has genuine promise for improving risk assessment. The basic tools are available. It remains for us to use and improve them. In doing so, we shall explore the strengths and weaknesses of PB-PK analysis, evaluate how such analysis may improve the process of risk assessment, and identify the areas of research that will lead to further improvement of our procedures and concepts.

Index

A

B

C

E

F

G

H

I

J

K

L

M

Q

R

S